LAPLACE TRANSFORM

$f(t)$	$F(s) \equiv \mathcal{L}[f(t)] \equiv \int_0^\infty f(t)e^{-st}\,dt$	
1	$\dfrac{1}{s}$	Elementary functions (Exercises 12.2.1 and 12.2.2)
$t^n \; (n > -1)$	$n!\,s^{-(n+1)}$	
e^{at}	$\dfrac{1}{s-a}$	
$\sin \omega t$	$\dfrac{\omega}{s^2 + \omega^2}$	
$\cos \omega t$	$\dfrac{s}{s^2 + \omega^2}$	
$\sinh at = \dfrac{1}{2}(e^{at} - e^{-at})$	$\dfrac{1}{2}\left(\dfrac{1}{s-a} - \dfrac{1}{s+a}\right) = \dfrac{a}{s^2 - a^2}$	
$\cosh at = \dfrac{1}{2}(e^{at} + e^{-at})$	$\dfrac{1}{2}\left(\dfrac{1}{s-a} + \dfrac{1}{s+a}\right) = \dfrac{s}{s^2 - a^2}$	
$\dfrac{df}{dt}$	$sF(s) - f(0)$	Fundamental properties (Sec. 12.2.3 and Exercise 12.2.3)
$\dfrac{d^2 f}{dt^2}$	$s^2 F(s) - sf(0) - \dfrac{df}{dt}(0)$	
$-tf(t)$	$\dfrac{dF}{ds}$	
$e^{at}f(t)$	$F(s-a)$	
$H(t-b)f(t-b)$	$e^{-bs}F(s) \quad (b > 0)$	
$\displaystyle\int_0^t f(t-\bar{t})g(\bar{t})\,d\bar{t}$	$F(s)\,G(s)$	Convolution (Sec. 12.2.4)
$\delta(t-b)$	$e^{-bs} \quad (b \geq 0)$	Dirac delta function (Sec. 12.2.4)
$\dfrac{1}{2\pi i}\displaystyle\int_{\gamma - i\infty}^{\gamma + i\infty} F(s)e^{st}\,ds$	$F(s)$	Inverse transform (Sec. 12.7)
$t^{-1/2}e^{-a^2/4t}$	$\sqrt{\dfrac{\pi}{s}}\,e^{-a\sqrt{s}} \quad (a \geq 0)$	Miscellaneous (Exercise 12.2.9)
$t^{-3/2}e^{-a^2/4t}$	$\dfrac{2\sqrt{\pi}}{a}\,e^{-a\sqrt{s}} \quad (a > 0)$	

Second Edition

Elementary Applied Partial Differential Equations

with Fourier Series and Boundary Value Problems

RICHARD HABERMAN

Department of Mathematics
Southern Methodist University

PRENTICE-HALL, INC., Englewood Cliffs, New Jersey 07632

Library of Congress Cataloging-in-Publication Data

Haberman, Richard, (date)
 Elementary applied partial differential equations.

 Bibliography: p.
 Includes index.
 1. Differential equations, Partial.   2. Fourier
series.   3. Boundary value problems.   I. Title.
QA377.H27   1987   515.3′53   86-17051
ISBN 0-13-252875-4   ·

Editorial/production supervision and
 interior design: Maria McColligan
Cover design: Karen A. Stephens
Manufacturing buyer: John B. Hall

Printed in the United States of America

10 9 8 7 6

ISBN 0-13-252875-4   01

Prentice-Hall International, Inc., *London*
Prentice-Hall of Australia Pty. Limited, *Sydney*
Editora Prentice-Hall do Brasil, Ltda., *Rio de Janeiro*
Prentice-Hall Canada Inc., *Toronto*
Prentice-Hall Hispanoamericana, S.A., *Mexico*
Prentice-Hall of India Private Limited, *New Delhi*
Prentice-Hall of Japan, Inc., *Tokyo*
Prentice-Hall of Southeast Asia Pte. Ltd., *Singapore*

To Liz, Ken, and Vicki

Contents

Contents

Preface

This text discusses elementary partial differential equations in the engineering and physical sciences. It is suited for courses whose titles include Fourier series, orthogonal functions, or boundary value problems. It may also be used in courses on Green's functions or transform methods.

Simple models (heat flow, vibrating strings and membranes) are emphasized. Equations are formulated carefully from physical principles, motivating most mathematical topics. Solution techniques are developed patiently. Mathematical results frequently are given physical interpretations. Proofs of theorems (if given at all) are presented after explanations based on illustrative examples. Numerous exercises of varying difficulty form an essential part of this text. Answers are provided for those exercises marked with a star (*).

Standard topics such as the method of separation of variables, Fourier series, and orthogonal functions are developed with considerable detail. In addition, there is a variety of clearly presented topics, such as differentiation and integration of Fourier series, zeros of Sturm-Liouville eigenfunctions, Rayleigh quotient, multidimensional eigenvalue problems, Bessel functions for a vibrating circular membrane, eigenfunction expansions for nonhomogeneous problems, Green's functions, Fourier and Laplace transform solutions, method of characteristics, and numerical methods. Some optional advanced material of interest is also included (for example, asymptotic expansion of large eigenvalues, calculation of perturbed frequencies using the Fredholm alternative, and the dynamics of shock waves).

The text has evolved from the author's experiences teaching this material to different types of students at various institutions (M.I.T., U.C.S.D., Rutgers,

Ohio State, and S.M.U.). Prerequisites for the reader are calculus and elementary ordinary differential equations. (These are occasionally reviewed in the text, where necessary.) For the beginning student, the core material for a typical course consists of most of Chapters 1–6. This will usually be supplemented by a few other topics. The text is somewhat flexible for an instructor, since most sections in Chapters 7–13 only depend on Chapters 1–6. Chapter 10 is an exception, since it requires Chapters 8 and 9.

Most of the first edition remains. Revised derivations, which are intended to be clearer, are presented for heat flow and the vibrations of strings and membranes. Some of the more advanced mathematical theory on the convergence of a Fourier series has been deleted, so that there is room for a later section that formulates the partial differential equations of traffic flow. Although only a few new exercises have been added, answers are now provided for a substantially increased number of exercises.

My object has been to explain clearly many elementary aspects of partial differential equations as an introduction to this vast and important field. The student, after achieving a certain degree of competence and understanding, can use this text as a reference, but should be prepared to refer to other books cited in the bibliography for additional material.

Finally, it is hoped that this text enables the reader to find enjoyment in the study of the relationships between mathematics and the physical sciences.

Richard Haberman

Heat Equation

1.1 INTRODUCTION

We wish to discuss the solution of elementary problems involving partial differential equations, the kinds of problems that arise in various fields of science and engineering. A **partial differential equation** (PDE) is a mathematical equation containing partial derivatives, for example,

$$\frac{\partial u}{\partial t} + 3\frac{\partial u}{\partial x} = 0. \tag{1.1.1}$$

We could begin our study by determining what functions $u(x, t)$ satisfy (1.1.1). However, we prefer to start by investigating a physical problem. We do this for two reasons. First, our mathematical techniques probably will be of greater interest to you when it becomes clear that these methods analyze physical problems. Second, we will actually find that physical considerations will motivate many of our mathematical developments.

Many diverse subject areas in engineering and the physical sciences are dominated by the study of partial differential equations. No list could be all-inclusive. However, the following examples should give you a feeling for the type of areas that are highly dependent on the study of partial differential equations: acoustics, aerodynamics, elasticity, electrodynamics, fluid dynamics, geophysics (seismic wave propagation), heat transfer, meteorology, oceanography, optics, petroleum engineering, plasma physics (ionized liquids and gases), quantum mechanics.

We will follow a certain philosophy of applied mathematics in which the analysis of a problem will have three stages:

1. Formulation
2. Solution
3. Interpretation

We begin by formulating the equations of heat flow describing the transfer of thermal energy. Heat energy is caused by the agitation of molecular matter. Two basic processes take place in order for thermal energy to move: conduction and convection. **Conduction** results from the collisions of neighboring molecules in which the kinetic energy of vibration of one molecule is transferred to its nearest neighbor. Thermal energy is thus spread by conduction even if the molecules themselves do not move their location appreciably. In addition, if a vibrating molecule moves from one region to another, it takes its thermal energy with it. This type of movement of thermal energy is called **convection**. In order to begin our study with relatively simple problems, we will study heat flow only in cases in which the conduction of heat energy is much more significant than its convection. We will thus think of heat flow primarily in the case of solids, although heat transfer in fluids (liquids and gases) is also primarily by conduction if the fluid velocity is sufficiently small.

1.2 DERIVATION OF THE CONDUCTION OF HEAT IN A ONE-DIMENSIONAL ROD

Thermal energy density. We begin by considering a rod of constant cross-sectional area A oriented in the x-direction (from $x = 0$ to $x = L$) as illustrated in Fig. 1.2.1. We temporarily introduce the amount of thermal energy per unit volume as an unknown variable and call it the **thermal energy density**:

$$\boxed{e(x, t) \equiv \text{thermal energy density.}}$$

We assume that all thermal quantities are constant across a section; the rod is one-dimensional. The simplest way this may be accomplished is to insulate perfectly the lateral surface area of the rod. Then no thermal energy can pass through the lateral surface. The dependence on x and t corresponds to a situation in which the rod is not uniformly heated; the thermal energy density varies from one cross section to another.

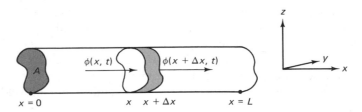

Figure 1.2.1 One-dimensional rod with heat energy flowing into and out of a thin slice.

Heat Equation   Chap. 1

Heat energy.   We consider a thin slice of the rod contained between x and $x + \Delta x$ as illustrated in Fig. 1.2.1. If the thermal energy density is constant throughout the volume, then the total energy in the slice is the product of the thermal energy density and the volume. In general, the energy density is not constant. However, if Δx is exceedingly small, then $e(x, t)$ may be approximated as a constant throughout the volume so that

$$\text{heat energy} = e(x, t)A\, \Delta x,$$

since the volume of a slice is $A\, \Delta x$.

Conservation of heat energy.   The heat energy between x and $x + \Delta x$ changes in time due only to heat energy flowing across the edges (x and $x + \Delta x$) and heat energy generated inside (due to positive or negative sources of heat energy). No heat energy changes are due to flow across the lateral surface, since we have assumed that the lateral surface is insulated. The fundamental heat flow process is described by the word equation

$$\begin{array}{l} \text{rate of change} \\ \text{of heat energy} \\ \text{in time} \end{array} = \begin{array}{l} \text{heat energy flowing} \\ \text{across boundaries} \\ \text{per unit time} \end{array} + \begin{array}{l} \text{heat energy generated} \\ \text{inside per unit time.} \end{array}$$

This is called **conservation of heat energy.** For the small slice, the rate of change of heat energy is

$$\frac{\partial}{\partial t}[e(x, t)A\, \Delta x],$$

where the partial derivative $\partial / \partial t$ is used because x is being held fixed.

Heat flux.   Thermal energy flows to the right or left in a one-dimensional rod. We introduce the **heat flux**

$$\phi(x, t) = \begin{array}{l} \text{heat flux (the amount of thermal energy } \textit{per unit} \\ \textit{time} \text{ flowing to the right } \textit{per unit surface area}). \end{array}$$

If $\phi(x, t) < 0$, it means that heat energy is flowing to the left. Heat energy flowing per unit time across the boundaries of the slice is $\phi(x, t)A - \phi(x + \Delta x, t)A$, since the heat flux is the flow per unit surface area and it must be multiplied by the surface area. If $\phi(x, t) > 0$ and $\phi(x + \Delta x, t) > 0$, as illustrated in Fig. 1.2.1, then the heat energy flowing per unit time at x contributes to an increase of the heat energy in the slice, whereas the heat flow at $x + \Delta x$ decreases the heat energy.

Heat sources.   We also allow for internal sources of thermal energy:

$$Q(x, t) = \text{heat energy per unit volume generated per unit time,}$$

perhaps due to chemical reactions or electrical heating. $Q(x, t)$ is approximately constant in space for a thin slice, and thus the total thermal energy generated per unit time in the thin slice is approximately $Q(x, t)A\,\Delta x$.

Conservation of heat energy (thin slice). The rate of change of heat energy is due to thermal energy flowing across the boundaries and internal sources:

$$\frac{\partial}{\partial t}[e(x, t)A\,\Delta x] \approx \phi(x, t)A - \phi(x + \Delta x, t)A + Q(x, t)A\,\Delta x. \quad (1.2.1)$$

Equation (1.2.1) is not precise because various quantities were assumed approximately constant for the small cross-sectional slice. We claim that (1.2.1) becomes increasingly accurate as $\Delta x \to 0$. Before giving a careful (and mathematically rigorous) derivation, we will just attempt to explain the basic ideas of the limit process, $\Delta x \to 0$. In the limit as $\Delta x \to 0$, (1.2.1) gives no interesting information, namely, $0 = 0$. However, if we first divide by Δx and then take the limit as $\Delta x \to 0$, we obtain

$$\frac{\partial e}{\partial t} = \lim_{\Delta x \to 0} \frac{\phi(x, t) - \phi(x + \Delta x, t)}{\Delta x} + Q(x, t), \quad (1.2.2)$$

where the constant cross-sectional area has been canceled. We claim that this result is exact (with no small errors), and hence we replace the $\approx$ in (1.2.1) by $=$ in (1.2.2). In this limiting process, $\Delta x \to 0$, t is being held fixed. Consequently, from the definition of a partial derivative,

$$\boxed{\frac{\partial e}{\partial t} = -\frac{\partial \phi}{\partial x} + Q.} \quad (1.2.3)$$

Conservation of heat energy (exact). An alternative derivation of conservation of heat energy has the advantage of our not being restricted to small slices. The resulting approximate calculation of the limiting process ($\Delta x \to 0$) is avoided. We consider any *finite* segment (from $x = a$ to $x = b$) of the original one-dimensional rod (see Fig. 1.2.2). We will investigate the conservation of heat energy in this region. The total heat energy is $\int_a^b e(x, t)\,dx$, the sum of the contributions of the infinitesimal slices. Again it changes only due to heat energy flowing through the side edges ($x = a$ and $x = b$) and heat energy generated inside the region, and thus (after canceling the constant A)

$$\boxed{\frac{d}{dt}\int_a^b e\,dx = \phi(a, t) - \phi(b, t) + \int_a^b Q\,dx.} \quad (1.2.4)$$

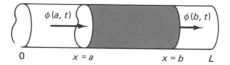

Figure 1.2.2 Heat energy flowing into and out of a finite segment of a rod.

Heat Equation   Chap. 1

Technically, an ordinary derivative d/dt appears in (1.2.4) since $\int_a^b e\, dx$ depends only on t, not also on x. However,

$$\frac{d}{dt}\int_a^b e\, dx = \int_a^b \frac{\partial e}{\partial t}\, dx,$$

if a and b are constants (and if e is continuous). This holds since inside the integral the ordinary derivative now is taken keeping x fixed, and hence it must be replaced by a partial derivative. Every term in (1.2.4) is now an ordinary integral if we notice that

$$\phi(a, t) - \phi(b, t) = -\int_a^b \frac{\partial \phi}{\partial x}\, dx,$$

(this* being valid if ϕ is continuously differentiable). Consequently,

$$\int_a^b \left(\frac{\partial e}{\partial t} + \frac{\partial \phi}{\partial x} - Q\right) dx = 0.$$

This integral must be zero *for arbitrary a and b*; the area under the curve must be zero for arbitrary limits. This is possible only if the integrand itself is identically zero.† Thus, we rederive (1.2.3) as

$$\boxed{\frac{\partial e}{\partial t} = -\frac{\partial \phi}{\partial x} + Q.} \qquad (1.2.5)$$

Equation (1.2.4), the **integral conservation law**, is more fundamental than the differential form (1.2.5). Equation (1.2.5) is valid in the usual case in which the physical variables are continuous.

A further explanation of the minus sign preceding $\partial \phi/\partial x$ is in order. For example, if $\partial \phi/\partial x > 0$ for $a \leq x \leq b$, then the heat flux ϕ is an increasing function of x. The heat is flowing greater to the right at $x = b$ than at $x = a$ (assuming that $b > a$). Thus (neglecting any effects of sources Q), the heat energy must decrease between $x = a$ and $x = b$, resulting in the minus sign in (1.2.5).

Temperature and specific heat. We usually describe materials by their **temperature**,

$$\boxed{u(x, t) = \text{temperature,}}$$

not their thermal density. Distinguishing between the concepts of temperature

---

* This is one of the fundamental theorems of calculus.

† Most proofs of this result are inelegant. Suppose that $f(x)$ is continuous and $\int_a^b f(x)\, dx = 0$ for arbitrary a and b. We wish to prove $f(x) = 0$ for all x. We can prove this by assuming that there exists a point x_0 such that $f(x_0) \neq 0$ and demonstrating a contradiction. If $f(x_0) \neq 0$ and $f(x)$ is continuous, then there exists some region near x_0 in which $f(x)$ is of one sign. Pick a and b to be in this region, and hence $\int_a^b f(x)\, dx \neq 0$ since $f(x)$ is of one sign throughout. This contradicts the statement that $\int_a^b f(x)\, dx = 0$, and hence it is impossible for $f(x_0) \neq 0$. Equation (1.2.5) follows.

and thermal energy is not necessarily a trivial task. Only in the mid-1700s did the existence of accurate experimental apparatus enable physicists to recognize that it may take different amounts of thermal energy to raise two different materials from one temperature to another larger temperature. This necessitates the introduction of the **specific heat** (or heat capacity):

$$c = \quad \text{specific heat (the heat energy that must be supplied to a unit mass of a substance to raise its temperature one unit).}$$

In general, from experiments (and our definition) the specific heat c of a material depends on the temperature u. For example, the thermal energy necessary to raise a unit mass from 0°C to 1°C could be different from that needed to raise the mass from 85°C to 86°C for the same substance. Heat flow problems with the specific heat depending on the temperature are mathematically quite complicated. (Exercise 1.2.1 briefly discusses this situation.) Often for restricted temperature intervals, the specific heat is approximately independent of the temperature. However, experiments suggest that different materials require different amounts of thermal energy to heat up. Since we would like to formulate the correct equation in situations in which the composition of our one-dimensional rod might vary from position to position, the specific heat will depend on x, $c = c(x)$. In many problems the rod is made of one material (a uniform rod), in which case we will let the specific heat c be a constant. In fact, most of the solved problems in this text (as well as other books) correspond to this approximation, c constant.

Thermal energy.   The thermal energy in a thin slice is $e(x, t)A\, \Delta x$. However, it is also defined as the energy it takes to raise the temperature from a reference temperature 0° to its actual temperature $u(x, t)$. Since the specific heat is independent of temperature, the heat energy per unit mass is just $c(x)u(x, t)$. We thus need to introduce the **mass density** $\rho(x)$:

$$\rho(x) = \text{mass density (mass per unit volume)},$$

allowing it to vary with x, possibly due to the rod being composed of nonuniform material. The total mass of the thin slice is $\rho A\, \Delta x$. The total thermal energy in any thin slice is thus $c(x)u(x, t) \cdot \rho A\, \Delta x$, so that

$$e(x, t)A\, \Delta x = c(x)u(x, t)\rho A\, \Delta x.$$

In this way we have explained the basic relationship between thermal energy and temperature:

$$e(x, t) = c(x)\rho(x)u(x, t). \tag{1.2.6}$$

This states that the thermal energy per unit volume equals the thermal energy per unit mass per unit degree times the temperature times the mass density (mass per unit volume). When the thermal energy density is eliminated using (1.2.6), conservation of thermal energy, (1.2.3) or (1.2.5), becomes

$$c(x)\rho(x)\frac{\partial u}{\partial t} = -\frac{\partial \phi}{\partial x} + Q. \qquad (1.2.7)$$

Fourier's law.   Usually, (1.2.7) is regarded as one equation in two unknowns, the temperature $u(x, t)$ and the heat flux (flow per unit surface area per unit time) $\phi(x, t)$. How and why does heat energy flow? In other words, we need an expression for the dependence of the flow of heat energy on the temperature field. First we summarize certain qualitative properties of heat flow with which we are all familiar:

1. If the temperature is constant in a region, no heat energy flows.
2. If there are temperature differences, the heat energy flows from the hotter region to the colder region.
3. The greater the temperature differences (for the same material), the greater is the flow of heat energy.
4. The flow of heat energy will vary for different materials, even with the same temperature differences.

Fourier (1768–1830) recognized properties 1 through 4 and summarized them (as well as numerous experiments) by the formula

$$\phi = -K_0 \frac{\partial u}{\partial x}, \qquad (1.2.8)$$

known as **Fourier's law of heat conduction**. Here $\partial u/\partial x$ is the derivative of the temperature; it is the slope of the temperature (as a function of x for fixed t); it represents temperature differences (per unit length). Equation (1.2.8) states that the heat flux is proportional to the temperature difference (per unit length). If the temperature u increases as x increases (i.e., the temperature is hotter to the right, $\partial u/\partial x > 0$), then we know (property 2) that heat energy flows to the left. This explains the minus sign in (1.2.8).

We designate the coefficient of proportionality K_0. It measures the ability of the material to conduct heat and is called the **thermal conductivity**. Experiments indicate that different materials conduct heat differently; K_0 depends on the particular material. The larger K_0 is, the greater the flow of heat energy with the same temperature differences. A material with a low value of K_0 would be a poor conductor of heat energy (and ideally suited for home insulation). For a rod composed of different materials, K_0 will be a function of x. Furthermore, experiments show that the ability to conduct heat for most materials is different at different temperatures, $K_0(x, u)$. However, just as with the specific heat c, the dependence on the temperature is often not important in particular problems.

Thus, throughout this text we will assume that the thermal conductivity K_0 only depends on x, $K_0(x)$. Usually, in fact, we will discuss uniform rods in which K_0 is a constant.

Heat equation. If Fourier's law, (1.2.8), is substituted into the conservation of heat energy equation, (1.2.7), a partial differential equation results:

$$c\rho \frac{\partial u}{\partial t} = \frac{\partial}{\partial x}\left(K_0 \frac{\partial u}{\partial x}\right) + Q. \qquad (1.2.9)$$

We usually think of the sources of heat energy Q as being given, and the only unknown being the temperature $u(x, t)$. The thermal coefficients c, ρ, K_0 all depend on the material and hence may be functions of x. In the special case of a uniform rod, in which c, ρ, K_0 are all constants, the partial differential equation (1.2.9) becomes

$$c\rho \frac{\partial u}{\partial t} = K_0 \frac{\partial^2 u}{\partial x^2} + Q.$$

If, in addition, there are no sources, $Q = 0$, then after dividing by the constant $c\rho$, the partial differential equation becomes

$$\frac{\partial u}{\partial t} = k \frac{\partial^2 u}{\partial x^2}, \qquad (1.2.10)$$

where the constant k,

$$k = \frac{K_0}{c\rho},$$

is called the **thermal diffusivity**, the thermal conductivity divided by the product of the specific heat and mass density. Equation (1.2.10) is often called the **heat equation**; it corresponds to no sources and constant thermal properties. If heat energy is initially concentrated in one place, (1.2.10) will describe how the heat energy spreads out, a physical process known as **diffusion**. Other physical quantities besides temperature smooth out in much the same manner, satisfying the same partial differential equation (1.2.10). For this reason (1.2.10) is also known as the **diffusion equation**. For example, the concentration $u(x, t)$ of chemicals (such as perfumes and pollutants) satisfies the diffusion equation (1.2.8) in certain one-dimensional situations.

Initial conditions. The partial differential equations describing the flow of heat energy, (1.2.9) or (1.2.10), have one time derivative. When an ordinary differential equation has one derivative, the initial value problem consists of solving the differential equation with one initial condition. Newton's law of motion for the position x of a particle yields a second-order ordinary differential equation, $md^2x/dt^2 = $ forces. It involves second derivatives. The initial value problem consists of solving the differential equation with two initial conditions, the initial position x and the initial velocity dx/dt. From these pieces of information (including the knowledge of the forces), by solving the differential equation with the initial conditions, we can predict the future motion of a particle in the x-direction. We wish to do the same process for our partial differential equation, that is, predict the future temperature. Since the heat equations have one time derivative, we must be given one **initial condition** (IC) (usually at $t = 0$), the initial temperature. It is possible that the initial temperature is not constant, but depends on x. Thus, we must be given the initial temperature distribution,

$$u(x, 0) = f(x).$$

Is this enough information to predict the future temperature? We know the initial temperature distribution and that the temperature changes according to the partial differential equation (1.2.9) or (1.2.10). However, we need to know what happens at the two boundaries, $x = 0$ and $x = L$. Without knowing this information, we cannot predict the future. Two conditions are needed corresponding to the second spatial derivatives present in (1.2.9) or (1.2.10), usually one condition at each end. We discuss these boundary conditions in the next section.

EXERCISES 1.2

1.2.1. Suppose that the specific heat is a function of position and temperature, $c(x, u)$.

    (a) Show that the heat energy per unit mass necessary to raise the temperature of a thin slice of thickness Δx from $0°$ to $u(x, t)$ is not $c(x)u(x, t)$, but instead $\int_0^u c(x, \bar{u}) \, d\bar{u}$.

    (b) Rederive the heat equation in this case. Show that (1.2.7) remains unchanged.

1.2.2. Consider conservation of thermal energy (1.2.4) for any segment of a one-dimensional rod $a \leq x \leq b$. By using the fundamental theorem of calculus

$$\left[\frac{\partial}{\partial b} \int_a^b f(x) \, dx = f(b) \right],$$

derive the heat equation (1.2.9).

*1.2.3. If $u(x, t)$ is known, give an expression for the total thermal energy contained in a rod $(0 < x < L)$.

1.2.4. Consider a thin one-dimensional rod without sources of thermal energy whose lateral surface area is not insulated.

    (a) Assume that the heat energy flowing out of the lateral sides per unit surface area per unit time is $w(x, t)$. Derive the partial differential equation for the temperature $u(x, t)$.

    (b) Assume that $w(x, t)$ is proportional to the temperature difference between the rod $u(x, t)$ and a known outside temperature $\gamma(x, t)$. Derive that

$$c\rho \frac{\partial u}{\partial t} = \frac{\partial}{\partial x}\left(K_0 \frac{\partial u}{\partial x} \right) - \frac{P}{A}[u(x, t) - \gamma(x, t)]h(x), \qquad (1.2.11)$$

    where $h(x)$ is a positive x-dependent proportionality, P is the lateral perimeter, and A is the cross-sectional area.

    (c) Compare (1.2.11) to the equation for a one-dimensional rod whose lateral surfaces are insulated, but with heat sources.

(d) Specialize (1.2.11) to a rod of circular cross section with constant thermal properties and 0° outside temperature.

*(e) Consider the assumptions in part (d). Suppose that the temperature in the rod is uniform [i.e., $u(x, t) = u(t)$]. Determine $u(t)$ if initially $u(0) = u_0$.

1.3 BOUNDARY CONDITIONS

In solving the heat equation, either (1.2.9) or (1.2.10), one **boundary condition** (BC) is needed at each end of the rod. The appropriate condition depends on the physical mechanism in effect at each end. Often the condition at the boundary depends on both the material inside and outside the rod. To avoid a more difficult mathematical problem, we will assume that the outside environment is known, not significantly altered by the rod.

Prescribed temperature.   In certain situations, the temperature of the end of the rod, for example $x = 0$, may be approximated by a **prescribed temperature**,

$$u(0, t) = u_B(t), \tag{1.3.1}$$

where $u_B(t)$ is the temperature of a fluid bath (or reservoir) with which the rod is in contact.

Insulated boundary.   In other situations it is possible to prescribe the heat flow rather than the temperature,

$$-K_0(0)\frac{\partial u}{\partial x}(0, t) = \phi(t), \tag{1.3.2}$$

where $\phi(t)$ is given. This is equivalent to giving one condition for the first derivative, $\partial u/\partial x$, at $x = 0$. The slope is given at $x = 0$. Equation (1.3.2) *cannot* be integrated in x because the slope is known only at one value of x. The simplest example of the prescribed heat flow boundary condition is when an end is **perfectly insulated** (sometimes we omit the "perfectly"). In this case there is no heat flow at the boundary. If $x = 0$ is insulated, then

$$\frac{\partial u}{\partial x}(0, t) = 0. \tag{1.3.3}$$

Newton's law of cooling.   When a one-dimensional rod is in contact at the boundary with a moving fluid (e.g., air), then neither the prescribed temperature nor the prescribed heat flow may be appropriate. For example, let us imagine a very warm rod in contact with cooler moving air. Heat will leave the rod, heating up the air. The air will then carry the heat away. This process of heat transfer is called **convection**. However, the air will be hotter near the rod. Again, this is a complicated problem; the air temperature will actually vary with distance from the rod (ranging between the bath and rod temperatures). Experiments show that, as a good approximation, the heat flow leaving the rod is proportional to the temperature difference between the bar and the prescribed external temperature. This boundary condition is called **Newton's law of cooling**. If it is valid at $x = 0$, then

$$-K_0(0)\frac{\partial u}{\partial x}(0, t) = -H[u(0, t) - u_B(t)], \qquad (1.3.4)$$

where the proportionality constant H is called the **heat transfer coefficient** (or the convection coefficient). This boundary condition* involves a linear combination of u and $\partial u/\partial x$. We must be careful with the sign of proportionality. If the rod is hotter than the bath $[u(0, t) > u_B(t)]$, then usually heat flows out of the rod at $x = 0$. Thus, heat is flowing to the left, and in this case the heat flow would be negative. That is why we introduced a minus sign in (1.3.4) (with $H > 0$). The same conclusion would have been reached had we assumed that $u(0, t) < u_B(t)$. Another way to understand the signs in (1.3.4) is to again assume that $u(0, t) > u_B(t)$. The temperature is hotter to the right at $x = 0$ and we should expect the temperature to continue to increase to the right. Thus, $\partial u/\partial x$ should be positive at $x = 0$. Equation (1.3.4) is consistent with this argument. In Exercise 1.3.1 you are asked to derive, in the same manner, that the equation for Newton's law of cooling at a right end point $x = L$ is

$$-K_0(L)\frac{\partial u}{\partial x}(L, t) = H[u(L, t) - u_B(t)], \qquad (1.3.5)$$

where $u_B(t)$ is the external temperature at $x = L$. We immediately note the significant sign difference between the left boundary (1.3.4) and the right boundary (1.3.5).

The coefficient H in Newton's law of cooling is experimentally determined. It depends on properties of the rod as well as fluid properties (including the fluid velocity). If the coefficient is very small, then very little heat energy flows across the boundary. In the limit as $H \to 0$, Newton's law of cooling approaches the insulated boundary condition. We can think of Newton's law of cooling for $H \neq 0$ as representing an imperfectly insulated boundary. If $H \to \infty$, the boundary condition approaches the one for prescribed temperature, $u(0, t) = u_B(t)$. This is most easily seen by dividing (1.3.4), for example, by H:

$$-\frac{K_0(0)}{H}\frac{\partial u}{\partial x}(0, t) = -[u(0, t) - u_B(t)].$$

Thus, $H \to \infty$ corresponds to no insulation at all.

Summary. We have described three different kinds of boundary conditions. For example, at $x = 0$:

$$u(0, t) = u_B(t) \qquad \text{prescribed temperature}$$

$$-K_0(0)\frac{\partial u}{\partial x}(0, t) = \phi(t) \qquad \text{prescribed heat flux}$$

$$-K_0(0)\frac{\partial u}{\partial x}(0, t) = -H[u(0, t) - u_B(t)] \qquad \text{Newton's law of cooling}$$

* For another situation in which (1.3.4) is valid, see Berg and McGregor [1966].

These same conditions could hold at $x = L$, noting that the change of sign $(-H$ becoming $H)$ is necessary for Newton's law of cooling. One boundary condition occurs at each boundary. It is not necessary that both boundaries satisfy the same kind of boundary condition. For example, it is possible for $x = 0$ to have a prescribed oscillating temperature

$$u(0, t) = 100 - 25 \cos t,$$

and for the right end, $x = L$, to be insulated,

$$\frac{\partial u}{\partial x}(L, t) = 0.$$

EXERCISES 1.3

1.3.1. Consider a one-dimensional rod, $0 \leqslant x \leqslant L$. Assume that the heat energy flowing out of the rod at $x = L$ is proportional to the temperature difference between the end temperature of the bar and the known external temperature. Derive (1.3.5) (briefly, physically explain why $H > 0$).

*1.3.2. Two one-dimensional rods of different materials joined at $x = x_0$ are said to be in **perfect thermal contact** if the temperature is continuous at $x = x_0$:

$$u(x_0 - , t) = u(x_0 + , t)$$

and no heat energy is lost at $x = x_0$ (i.e., the heat energy flowing out of one flows into the other). What mathematical equation represents the latter condition at $x = x_0$? Under what special condition is $\partial u / \partial x$ continuous at $x = x_0$?

*1.3.3. Consider a bath containing a fluid of specific heat c_f and mass density ρ_f which surrounds the end $x = L$ of a one-dimensional rod. Suppose that the bath is rapidly stirred in a manner such that the bath temperature is approximately uniform throughout, equaling the temperature at $x = L$, $u(L, t)$. Assume that the bath is thermally insulated except at its perfect thermal contact with the rod, where the bath may be heated or cooled by the rod. Determine an equation for the temperature in the bath. (This will be a boundary condition at the end $x = L$.) (*Hint:* See Exercise 1.3.2.)

1.4 EQUILIBRIUM TEMPERATURE DISTRIBUTION

1.4.1 Prescribed Temperature

Let us now formulate a simple, but typical, problem of heat flow. If the thermal coefficients are constant and there are no sources of thermal energy, then the temperature $u(x, t)$ in a one-dimensional rod $0 \leqslant x \leqslant L$ satisfies

$$\frac{\partial u}{\partial t} = k \frac{\partial^2 u}{\partial x^2}. \tag{1.4.1}$$

The solution of this partial differential equation must satisfy the initial condition

$$u(x, 0) = f(x) \tag{1.4.2}$$

and one boundary condition at each end. For example, each end might be in contact with different large baths, such that the temperature at each end is prescribed

$$u(0, t) = T_1(t)$$

$$u(L, t) = T_2(t).$$

(1.4.3)

One aim of this text is to enable the reader to solve the problem specified by (1.4.1)–(1.4.3).

Equilibrium temperature distribution. Before we begin to attack such an initial and boundary value problem for partial differential equations, we discuss a physically related question for ordinary differential equations. Suppose that the boundary conditions at $x = 0$ and $x = L$ were **steady** (i.e., independent of time),

$$u(0, t) = T_1 \quad \text{and} \quad u(L, t) = T_2,$$

where T_1 and T_2 are given constants. We define an **equilibrium** or **steady-state** solution to be a temperature distribution that does not depend on time, that is, $u(x, t) = u(x)$. Since $\partial/\partial t \, u(x) = 0$, the partial differential equation becomes $k(\partial^2 u/\partial x^2) = 0$, but partial derivatives are not necessary, and thus

$$\frac{d^2u}{dx^2} = 0.$$

(1.4.4)

The boundary conditions are

$$u(0) = T_1$$

(1.4.5a)

$$u(L) = T_2.$$

(1.4.5b)

In doing steady-state calculations, the initial conditions are usually ignored. Equation (1.4.4) is a rather trivial second-order ordinary differential equation. Its general solution may be obtained by integrating twice. Integrating (1.4.4) yields $du/dx = C_1$, and integrating a second time shows that

$$u(x) = C_1 x + C_2.$$

(1.4.6)

We recognize (1.4.6) as the general equation of a straight line. Thus, from the boundary conditions (1.4.5) the equilibrium temperature distribution is the straight line that equals T_1 at $x = 0$ and T_2 at $x = L$, as sketched in Fig. 1.4.1. Geometrically, there is a unique equilibrium solution for this problem. Algebraically, we can

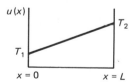

Figure 1.4.1 Equilibrium temperature distribution.

determine the two arbitrary constants, C_1 and C_2, by applying the boundary conditions, $u(0) = T_1$ and $u(L) = T_2$:

$$u(0) = T_1 \quad \text{implies} \quad T_1 = C_2 \tag{1.4.7}$$

$$u(L) = T_2 \quad \text{implies} \quad T_2 = C_1 L + C_2.$$

It is easy to solve (1.4.7) for the constants $C_2 = T_1$ and $C_1 = (T_2 - T_1)/L$. Thus, the unique equilibrium solution for the steady-state heat equation with these fixed boundary conditions is

$$\boxed{u(x) = T_1 + \frac{T_2 - T_1}{L}x.} \tag{1.4.8}$$

Approach to equilibrium. For the time-dependent problem, (1.4.1) and (1.4.2), with steady boundary conditions (1.4.5), we expect the temperature distribution $u(x, t)$ to change in time; it will not remain equal to its initial distribution $f(x)$. If we wait a very, very long time, we would imagine that the influence of the two ends should dominate. The initial conditions are usually forgotten. Eventually, the temperature is physically expected to approach the equilibrium temperature distribution, since the boundary conditions are independent of time:

$$\lim_{t \to \infty} u(x, t) = u(x) = T_1 + \frac{T_2 - T_1}{L}x. \tag{1.4.9}$$

In Sec. 7.2 we will solve the time-dependent problem and show that (1.4.9) is satisfied. However, if a steady state is approached, it is more easily obtained by directly solving the equilibrium problem.

1.4.2 Insulated Boundaries

As a second example of a steady-state calculation, we consider a one-dimensional rod again with no sources and with constant thermal properties, but this time with insulated boundaries at $x = 0$ and $x = L$. The formulation of the time-dependent problem is

$$\text{PDE:} \qquad \frac{\partial u}{\partial t} = k \frac{\partial^2 u}{\partial x^2} \tag{1.4.10a}$$

$$\text{IC:} \qquad u(x, 0) = f(x) \tag{1.4.10b}$$

$$\text{BC:} \qquad \frac{\partial u}{\partial x}(0, t) = 0 \tag{1.4.10c}$$

$$\frac{\partial u}{\partial x}(L, t) = 0. \tag{1.4.10d}$$

The equilibrium problem is derived by setting $\partial / \partial t = 0$. The equilibrium temperature distribution satisfies

$$\text{ODE:} \qquad \boxed{\frac{d^2 u}{dx^2} = 0} \tag{1.4.11a}$$

$$\text{BC:} \quad \boxed{\begin{aligned} \frac{du}{dx}(0) &= 0 \\[2mm] \frac{du}{dx}(L) &= 0, \end{aligned}} \qquad \begin{aligned} &(1.4.11b) \\[4mm] &(1.4.11c) \end{aligned}$$

where the initial condition is neglected (for the moment). The general solution of $d^2u/dx^2 = 0$ is again an arbitrary straight line,

$$u = C_1 x + C_2. \tag{1.4.12}$$

The boundary conditions imply that the slope must be zero at both ends. Geometrically, any straight line that is flat (zero slope) will satisfy (1.4.11) as illustrated in Fig. 1.4.2. The solution is any constant temperature. Algebraically, from (1.4.12), $du/dx = C_1$ and both boundary conditions imply $C_1 = 0$. Thus,

$$u(x) = C_2 \tag{1.4.13}$$

for any constant C_2. Unlike the first example (with fixed temperatures at both ends), here there is not a unique equilibrium temperature. Any constant temperature is an equilibrium temperature distribution for insulated boundary conditions.

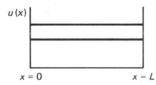

Figure 1.4.2 Various constant equilibrium temperature distributions (with insulated ends).

Thus, for the time-dependent initial value problem, we expect

$$\lim_{t \to \infty} u(x, t) = C_2;$$

if we wait long enough a rod with insulated ends should approach a constant temperature. This seems physically quite reasonable. However, it does not make sense that the solution should approach an arbitrary constant; we ought to know what constant it approaches. In this case, the lack of uniqueness was caused by the complete neglect of the initial condition. In general, the equilibrium solution will not satisfy the initial condition. However, the particular constant equilibrium solution is determined by considering the initial condition for the time-dependent problem (1.4.10). Since both ends are insulated, the total thermal energy is constant. This follows from the integral conservation of thermal energy of the entire rod [see (1.2.4)]:

$$\frac{d}{dt} \int_0^L c\rho u \, dx = -K_0 \frac{\partial u}{\partial x}(0, t) + K_0 \frac{\partial u}{\partial x}(L, t). \tag{1.4.14}$$

Since both ends are insulated,

$$\int_0^L c\rho u \, dx = \text{constant}. \tag{1.4.15}$$

One implication of (1.4.15) is that the initial thermal energy must equal the final ($\lim_{t \to \infty}$) thermal energy. The initial thermal energy is $c\rho \int_0^L f(x) \, dx$ since

$u(x, 0) = f(x)$, while the equilibrium thermal energy is $c\rho \int_0^L C_2 \, dx = c\rho C_2 L$ since the equilibrium temperature distribution is a constant $u(x, t) = C_2$. The constant C_2 is determined by equating these two expressions for the constant total thermal energy, $c\rho \int_0^L f(x) \, dx = c\rho C_2 L$. Solving for C_2 shows that the desired unique steady-state solution should be

$$u(x) = C_2 = \frac{1}{L} \int_0^L f(x) \, dx, \qquad (1.4.16)$$

the **average** of the initial temperature distribution. It is as though the initial condition is not entirely forgotten. Later we will find a $u(x, t)$ that satisfies (1.4.10) and show that $\lim_{t \to \infty} u(x, t)$ is given by (1.4.16).

EXERCISES 1.4

1.4.1. Determine the equilibrium temperature distribution for a one-dimensional rod with constant thermal properties with the following sources and boundary conditions:

*(a) $Q = 0$, $u(0) = 0$, $u(L) = T$

(b) $Q = 0$, $u(0) = T$, $u(L) = 0$

(c) $Q = 0$, $\dfrac{\partial u}{\partial x}(0) = 0$, $u(L) = T$

*(d) $Q = 0$, $u(0) = T$, $\dfrac{\partial u}{\partial x}(L) = \alpha$

(e) $\dfrac{Q}{K_0} = 1$, $u(0) = T_1$, $u(L) = T_2$

*(f) $\dfrac{Q}{K_0} = x^2$, $u(0) = T$, $\dfrac{\partial u}{\partial x}(L) = 0$

(g) $Q = 0$, $u(0) = T$, $\dfrac{\partial u}{\partial x}(L) + u(L) = 0$

*(h) $Q = 0$, $\dfrac{\partial u}{\partial x}(0) - [u(0) - T] = 0$, $\dfrac{\partial u}{\partial x}(L) = \alpha$

In these you may assume that $u(x, 0) = f(x)$.

1.4.2. Consider the equilibrium temperature distribution for a uniform one-dimensional rod with sources $Q/K_0 = x$ of thermal energy, subject to the boundary conditions $u(0) = 0$ and $u(L) = 0$.

*(a) Determine the heat energy generated per unit time inside the entire rod.

(b) Determine the heat energy flowing out of the rod per unit time at $x = 0$ and at $x = L$.

(c) What relationships should exist between the answers in parts (a) and (b)?

1.4.3. Determine the equilibrium temperature distribution for a one-dimensional rod composed of two different materials in perfect thermal contact at $x = 1$. For $0 < x < 1$, there is one material ($c\rho = 1$, $K_0 = 1$) with a constant source ($Q = 1$), whereas for the other $1 < x < 2$ there are no sources ($Q = 0$, $c\rho = 2$, $K_0 = 2$) (see Exercise 1.3.2), with $u(0) = 0$ and $u(2) = 0$.

1.4.4. If both ends of a rod are insulated, derive *from the partial differential equation* that the total thermal energy in the rod is constant.

1.4.5. Consider a one-dimensional rod $0 \leq x \leq L$ of known length and known constant thermal properties without sources. Suppose that the temperature is an *unknown* constant T at $x = L$. Determine T if we know (in the steady state) both the temperature and the heat flow at $x = 0$.

1.4.6. The two ends of a uniform rod of length L are insulated. There is a constant source of thermal energy $Q_0 \neq 0$ and the temperature is initially $u(x, 0) = f(x)$.
  (a) Show mathematically that there does not exist any equilibrium temperature distribution. Briefly explain physically.
  (b) Calculate the total thermal energy in the entire rod.

1.4.7. For the following problems, determine an equilibrium temperature distribution (if one exists). For what values of β are there solutions? Explain physically.

*(a) $\dfrac{\partial u}{\partial t} = \dfrac{\partial^2 u}{\partial x^2} + 1$ $u(x, 0) = f(x)$ $\dfrac{\partial u}{\partial x}(0, t) = 1$ $\dfrac{\partial u}{\partial x}(L, t) = \beta$

(b) $\dfrac{\partial u}{\partial t} = \dfrac{\partial^2 u}{\partial x^2}$ $u(x, 0) = f(x)$ $\dfrac{\partial u}{\partial x}(0, t) = 1$ $\dfrac{\partial u}{\partial x}(L, t) = \beta$

(c) $\dfrac{\partial u}{\partial t} = \dfrac{\partial^2 u}{\partial x^2} + x - \beta$ $u(x, 0) = f(x)$ $\dfrac{\partial u}{\partial x}(0, t) = 0$ $\dfrac{\partial u}{\partial x}(L, t) = 0$

1.5 DERIVATION OF THE HEAT EQUATION IN TWO OR THREE DIMENSIONS

Introduction. In Sec. 1.2 we showed that for the conduction of heat in a one-dimensional rod the temperature $u(x, t)$ satisfies

$$c\rho \frac{\partial u}{\partial t} = \frac{\partial}{\partial x}\left(K_0 \frac{\partial u}{\partial x}\right) + Q.$$

In cases in which there are no sources ($Q = 0$) and the thermal properties are constant, the partial differential equation becomes

$$\frac{\partial u}{\partial t} = k\frac{\partial^2 u}{\partial x^2},$$

where $k = K_0/c\rho$. Before we solve problems involving these partial differential equations, we will formulate partial differential equations corresponding to heat flow problems in two or three spatial dimensions. We will find the derivation to be similar to the one used for one-dimensional problems, although important differences will emerge. We propose to derive new and more complex equations (before solving the simpler ones) so that, when we do discuss *techniques* for the solutions of PDEs, we will have more than one example to work with.

Heat energy. We begin our derivation by considering any *arbitrary subregion R* as illustrated in Fig. 1.5.1. As in the one-dimensional case, conservation of heat energy is summarized by the following word equation:

rate of change of heat energy	=	heat energy flowing across the boundaries per unit time	+	heat energy generated inside per unit time,

Figure 1.5.1 Three-dimensional subregion R.

where the heat energy within an arbitrary subregion R is

$$\text{heat energy} = \iiint\limits_{R} c\rho u \, dV,$$

instead of the one-dimensional integral used in Sec. 1.2.

Heat flux vector and normal vectors. We need an expression for the flow of heat energy. In a one-dimensional problem the heat flux ϕ is defined to the right ($\phi < 0$ means flowing to the left). In a three-dimensional problem the heat flows in some direction, and hence the **heat flux is a vector ϕ**. The magnitude of ϕ is the amount of heat energy flowing per unit time per unit surface area. However, in considering conservation of heat energy it is only the heat flowing *across the boundaries* per unit time that is important. If, as at point A in Fig. 1.5.2, the heat flow is parallel to the boundary, then there is no heat energy *crossing* the boundary at that point. In fact, it is only the normal component of the heat flow that contributes (as illustrated by point B in Fig. 1.5.2). At any point there are two normal vectors, an inward and an outward normal $\mathbf{n}$. We will use the convention of only utilizing the **unit outward normal vector $\hat{\mathbf{n}}$** (where the ^ stands for a unit vector).

Figure 1.5.2 Outward normal component of heat flux vector.

Conservation of heat energy. At each point the amount of heat energy flowing *out* of the region R per unit time per unit surface area is the outward normal component of the heat flux vector. From Fig. 1.5.2 at point B, the outward normal component of the heat flux vector is $|\phi| \cos \theta = \phi \cdot \mathbf{n}/|\mathbf{n}| = \phi \cdot \hat{\mathbf{n}}$. If the heat flux vector ϕ is directed inward, then $\phi \cdot \hat{\mathbf{n}} < 0$ and the outward flow of heat energy is negative. To calculate the total heat energy flowing out of R per unit time, we must multiply $\phi \cdot \hat{\mathbf{n}}$ by the differential surface area dS and ''sum'' over the entire surface that encloses the region R. This*

---

* Sometimes the notation ϕ_n is used instead of $\phi \cdot \hat{\mathbf{n}}$, meaning the outward normal component of ϕ.

is indicated by the closed surface integral $\oiint \phi \cdot \hat{\mathbf{n}} \, dS$. This is the amount of heat energy (per unit time) leaving the region R and (if positive) results in a decreasing of the total heat energy within R. If Q is the rate of heat energy generated per unit volume, then the total heat energy generated per unit time is $\iiint_R Q \, dV$. Consequently, conservation of heat energy for an arbitrary three-dimensional region R becomes

$$\frac{d}{dt} \iiint_R c\rho u \, dV = -\oiint \phi \cdot \hat{\mathbf{n}} \, dS + \iiint_R Q \, dV. \qquad (1.5.1)$$

Divergence theorem. In one dimension, a way in which we derived a partial differential relationship from the integral conservation law was to notice (via the fundamental theorem of calculus) that

$$\phi(a) - \phi(b) = -\int_a^b \frac{\partial \phi}{\partial x} \, dx;$$

that is, the flow through the boundaries can be expressed as an integral over the entire region for one-dimensional problems. We claim that the divergence theorem is an analogous procedure for functions of three variables. The divergence theorem deals with a vector $\mathbf{A}$ (with components A_x, A_y, and A_z; i.e., $\mathbf{A} = A_x \hat{\mathbf{i}} + A_y \hat{\mathbf{j}} + A_z \hat{\mathbf{k}}$) and its divergence defined as follows:

$$\nabla \cdot \mathbf{A} \equiv \frac{\partial}{\partial x} A_x + \frac{\partial}{\partial y} A_y + \frac{\partial}{\partial z} A_z. \qquad (1.5.2)$$

Note that the divergence of a vector is a scalar. **The divergence theorem states that the volume integral of the divergence of any continuously differentiable vector A is the closed surface integral of the outward normal component of A:**

$$\iiint_R \nabla \cdot \mathbf{A} \, dV = \oiint \mathbf{A} \cdot \hat{\mathbf{n}} \, dS. \qquad (1.5.3)$$

This is also known as Gauss's theorem. It can be used to relate certain surface integrals to volume integrals, and vice versa. It is very important and very useful (both immediately and later in this text). We omit a derivation, which may be based on repeating the one-dimensional fundamental theorem in all three dimensions.

Application of the divergence theorem to heat flow. In particular, the closed surface integral that arises in the conservation of heat energy (1.5.1), corresponding to the heat energy flowing across the boundary per unit time, can be written as a volume integral according to the divergence theorem, (1.5.3). Thus, (1.5.1) becomes

$$\frac{d}{dt} \iiint_R c\rho u \, dV = -\iiint_R \nabla \cdot \phi \, dV + \iiint_R Q \, dV. \qquad (1.5.4)$$

We note that the time derivative in (1.5.4) can be put inside the integral (since R is fixed in space) if the time derivative is changed to a partial derivative. Thus, all the expressions in (1.5.4) are volume integrals over the same volume, and they can be combined into one integral:

$$\iiint_R \left[c\rho \frac{\partial u}{\partial t} + \nabla \cdot \boldsymbol{\phi} - Q \right] dV = 0. \tag{1.5.5}$$

Since this integral is zero for all regions R, it follows (as it did for one-dimensional integrals) that the integrand itself must be zero:

$$c\rho \frac{\partial u}{\partial t} + \nabla \cdot \boldsymbol{\phi} - Q = 0$$

or, equivalently,

$$c\rho \frac{\partial u}{\partial t} = -\nabla \cdot \boldsymbol{\phi} + Q. \tag{1.5.6}$$

Equation (1.5.6) reduces to (1.2.3) in the one-dimensional case.

Fourier's law of heat conduction. In one-dimensional problems, from experiments according to Fourier's law, the heat flux ϕ is proportional to the derivative of the temperature, $\phi = -K_0 \, \partial u/\partial x$. The minus sign is related to the fact that thermal energy flows from hot to cold. $\partial u/\partial x$ is the change in temperature per unit length. These same ideas are valid in three dimensions. In an appendix, we derive that the heat flux vector $\boldsymbol{\phi}$ is proportional to the temperature gradient $\left(\nabla u \equiv \frac{\partial u}{\partial x} \hat{\mathbf{i}} + \frac{\partial u}{\partial y} \hat{\mathbf{j}} + \frac{\partial u}{\partial z} \hat{\mathbf{k}} \right)$:

$$\boldsymbol{\phi} = -K_0 \nabla u, \tag{1.5.7}$$

known as **Fourier's law of heat conduction**, where again K_0 is called the thermal conductivity. Thus, in three dimensions the gradient ∇u replaces $\partial u/\partial x$.

Heat equation. When the heat flux vector, (1.5.7), is substituted into the conservation of heat energy equation, (1.5.6), a partial differential equation for the temperature results:

$$c\rho \frac{\partial u}{\partial t} = \nabla \cdot (K_0 \nabla u) + Q. \tag{1.5.8}$$

In the cases in which there are no sources of heat energy ($Q = 0$) and the thermal coefficients are constant, (1.5.8) becomes

$$\frac{\partial u}{\partial t} = k \nabla \cdot (\nabla u), \tag{1.5.9}$$

where $k = K_0/c\rho$ is again called the thermal diffusivity. From their definitions, we calculate the divergence of the gradient of u:

$$\nabla \cdot (\nabla u) = \frac{\partial}{\partial x}\left(\frac{\partial u}{\partial x}\right) + \frac{\partial}{\partial y}\left(\frac{\partial u}{\partial y}\right) + \frac{\partial}{\partial z}\left(\frac{\partial u}{\partial z}\right)$$

(1.5.10)

$$= \boxed{\frac{\partial^2 u}{\partial x^2} + \frac{\partial^2 u}{\partial y^2} + \frac{\partial^2 u}{\partial z^2} \equiv \nabla^2 u.}$$

This expression $\nabla^2 u$ is defined to be the **Laplacian** of u. Thus, in this case

$$\boxed{\frac{\partial u}{\partial t} = k \nabla^2 u.}$$

(1.5.11)

Equation (1.5.11) is often known as the **heat** or **diffusion equation** in three spatial dimensions. The notation $\nabla^2 u$ is often used to emphasize the role of the del operator ∇:

$$\nabla \equiv \frac{\partial}{\partial x}\hat{\mathbf{i}} + \frac{\partial}{\partial y}\hat{\mathbf{j}} + \frac{\partial}{\partial z}\hat{\mathbf{k}}.$$

Note that ∇u is ∇ operating on u, while $\nabla \cdot \mathbf{A}$ is the vector dot product of del with $\mathbf{A}$. Furthermore, $\nabla^2 u$ is the dot product of the del operator with itself or

$$\nabla \cdot \nabla = \frac{\partial}{\partial x}\left(\frac{\partial}{\partial x}\right) + \frac{\partial}{\partial y}\left(\frac{\partial}{\partial y}\right) + \frac{\partial}{\partial z}\left(\frac{\partial}{\partial z}\right)$$

operating on u, hence the notation del squared, ∇^2.

Initial boundary value problem.   In addition to (1.5.8) or (1.5.11), the temperature satisfies a given initial distribution,

$$u(x, y, z, 0) = f(x, y, z).$$

The temperature also satisfies a boundary condition at every point on the surface that encloses the region of interest. The boundary condition can be of various types (as in the one-dimensional problem). The temperature could be prescribed,

$$u(x, y, z, t) = T(x, y, z, t),$$

everywhere *on the boundary*, where T is a known function of t at each point of the boundary. It is also possible that the flow across the boundary is prescribed. Frequently, we might have the boundary (or part of the boundary) **insulated**. This means that there is no heat flow across that portion of the boundary. Since the heat flux vector is $-K_0 \nabla u$, the heat flowing out will be the unit outward normal component of the heat flow vector, $-K_0 \nabla u \cdot \hat{\mathbf{n}}$, where $\hat{\mathbf{n}}$ is a unit outward normal to the boundary surface. Thus, at an insulated surface,

$$\nabla u \cdot \hat{\mathbf{n}} = 0.$$

Recall that $\nabla u \cdot \hat{\mathbf{n}}$ is the directional derivative of u in the outward normal direction; it is also called the normal derivative.*

* Sometimes (in other books and references) the notation $\partial u / \partial n$ is used. However, to calculate $\partial u / \partial n$ one usually calculates the dot product of the two vectors, ∇u and $\hat{\mathbf{n}}$, $\nabla u \cdot \hat{\mathbf{n}}$, so we will not use the notation $\partial u / \partial n$ in this text.

Often Newton's law of cooling is a more realistic condition at the boundary. It states that the heat energy flowing out per unit time per unit surface area is proportional to the difference between the temperature at the surface u and the temperature outside the surface u_b. Thus, if Newton's law of cooling is valid, then at the boundary

$$-K_0 \nabla u \cdot \hat{\mathbf{n}} = H(u - u_b). \qquad (1.5.12)$$

Note that usually the proportionality constant $H > 0$, since if $u > u_b$, then we expect that heat energy will flow out and $-K_0 \nabla u \cdot \hat{\mathbf{n}}$ will be greater than zero. Equation (1.5.12) verifies the two forms of Newton's law of cooling for one-dimensional problems. In particular, at $x = 0$, $\hat{\mathbf{n}} = -\hat{\mathbf{i}}$ and the left-hand side (l.h.s) of (1.5.12) becomes $K_0 \, \partial u/\partial x$, while at $x = L$, $\hat{\mathbf{n}} = \hat{\mathbf{i}}$ and the l.h.s. of (1.5.12) becomes $-K_0 \, \partial u/\partial x$ [see (1.3.4) and (1.3.5)].

Steady state. If the boundary conditions and any sources of thermal energy are independent of time, it is possible that there exist steady-state solutions to the heat equation satisfying the given steady boundary condition:

$$0 = \nabla \cdot (K_0 \nabla u) + Q.$$

Note that an equilibrium temperature distribution $u(x, y, z)$ satisfies a partial differential equation when more than one spatial dimension is involved. In the case with constant thermal properties, the equilibrium temperature distribution will satisfy

$$\nabla^2 u = -\frac{Q}{K_0}, \qquad (1.5.13)$$

known as **Poisson's equation**.

If, in addition, there are no sources ($Q = 0$), then

$$\boxed{\nabla^2 u = 0;} \qquad (1.5.14)$$

the Laplacian of the temperature distribution is zero. Equation (1.5.14) is known as **Laplace's equation**. It is also known as the **potential equation**, since the gravitational and electrostatic potentials satisfy (1.5.14) if there are no sources. We will solve a number of problems involving Laplace's equation in later sections.

Two-dimensional problems. All the previous remarks about three-dimensional problems are valid if the geometry is such that the temperature only depends on x, y and t. For example, Laplace's equation in two dimensions, x and y, corresponding to equilibrium heat flow with no sources (and constant thermal properties) is

$$\nabla^2 u = \frac{\partial^2 u}{\partial x^2} + \frac{\partial^2 u}{\partial y^2} = 0,$$

since $\partial^2 u/\partial z^2 = 0$. Two-dimensional results can be derived directly (without taking a limit of three-dimensional problems), by using fundamental principles in two dimensions. We will not repeat the derivation. However, we can easily outline the results. Every time a volume integral ($\iiint_R \cdots dV$) appears, it must be replaced by a surface integral over the entire two-dimensional plane region ($\iint_R \cdots dS$). Similarly, the boundary contribution for three-dimensional problems, which is the closed surface integral $\oiint \cdots dS$, must be replaced by the closed line integral $\oint \cdots d\tau$, an integration over the boundary of the two-dimensional plane surface. These results are not difficult to derive since the divergence theorem in three dimensions,

$$\iiint_R \nabla \cdot \mathbf{A} \, dV = \oiint \mathbf{A} \cdot \hat{\mathbf{n}} \, dS, \tag{1.5.15}$$

is valid in two dimensions, taking the form

$$\iint_R \nabla \cdot \mathbf{A} \, dS = \oint \mathbf{A} \cdot \hat{\mathbf{n}} \, d\tau. \tag{1.5.16}$$

Sometimes (1.5.16) is called *Green's theorem*, but we prefer to refer to it as the two-dimensional divergence theorem. In this way only one equation need be familiar to the reader, namely (1.5.15); *the conversion to two-dimensional form involves only changing the number of integral signs.*

Polar and cylindrical coordinates. The Laplacian,

$$\nabla^2 u = \frac{\partial^2 u}{\partial x^2} + \frac{\partial^2 u}{\partial y^2} + \frac{\partial^2 u}{\partial z^2}, \tag{1.5.17}$$

is important for the heat equation (1.5.11) and its steady-state version (1.5.14), as well as for other significant problems in science and engineering. Equation (1.5.17) written as above in Cartesian coordinates is most useful when the geometrical region under investigation is a rectangle or a rectangular box. Other coordinate systems are frequently useful. In practical applications, one may need the formula that expresses the Laplacian in the appropriate coordinate system. In circular cylindrical coordinates, with r the radial distance from the z-axis and θ the angle

$$\boxed{\begin{aligned} x &= r \cos \theta \\ y &= r \sin \theta \\ z &= z, \end{aligned}} \tag{1.5.18}$$

the Laplacian can be shown to equal the following formula:

$$\boxed{\nabla^2 u = \frac{1}{r} \frac{\partial}{\partial r} \left(r \frac{\partial u}{\partial r} \right) + \frac{1}{r^2} \frac{\partial^2 u}{\partial \theta^2} + \frac{\partial^2 u}{\partial z^2}.} \tag{1.5.19}$$

There may be no need to memorize this formula, as it can often be looked up in a reference book. As an aid in minimizing errors, it should be noted that every term in the Laplacian has the dimension of u divided by two spatial dimensions [just as in Cartesian coordinates, (1.5.17)]. Since θ is measured in radians, which have no dimensions, this remark aids in remembering to divide $\partial^2 u/\partial\theta^2$ by r^2. In polar coordinates (by which we mean a two-dimensional coordinate system with z fixed, usually $z = 0$), the Laplacian is the same as (1.5.19) with $\partial^2 u/\partial z^2 = 0$ since there is no dependence on z. Equation (1.5.19) can be derived using the chain rule for partial derivatives, applicable for changes of variables. The algebra is involved, and little is learned about PDEs; we leave it as an exercise.

In some physical situations it is known that the temperature does *not* depend on the polar angle θ; it is said to be **circularly** or **axially symmetric**. In that case

$$\nabla^2 u = \frac{1}{r}\frac{\partial}{\partial r}\left(r\frac{\partial u}{\partial r}\right) + \frac{\partial^2 u}{\partial z^2}. \tag{1.5.20}$$

EXERCISES 1.5

1.5.1. Let $c(x, y, z, t)$ denote the concentration of a pollutant (the amount per unit volume).
  (a) What is an expression for the total amount of pollutant in the region R?
  (b) Suppose that the flow **J** of the pollutant is proportional to the gradient of the concentration. (Is this reasonable?) Express conservation of the pollutant.
  (c) Derive the partial differential equation governing the diffusion of the pollutant.

*1.5.2. For conduction of thermal energy, the heat flux vector is $\boldsymbol{\phi} = -K_0\,\nabla u$. If in addition the molecules move at an average velocity **V**, a process called **convection**, then briefly explain why $\boldsymbol{\phi} = -K_0\,\nabla u + c\rho u\mathbf{V}$. Derive the corresponding equation for heat flow, including both conduction and convection of thermal energy (assuming constant thermal properties with no sources).

1.5.3. Consider the polar coordinates

$$x = r\cos\theta$$
$$y = r\sin\theta.$$

  (a) Since $r^2 = x^2 + y^2$, show that

$$\frac{\partial r}{\partial x} = \cos\theta, \quad \frac{\partial r}{\partial y} = \sin\theta, \quad \frac{\partial\theta}{\partial y} = \frac{\cos\theta}{r}, \quad \text{and} \quad \frac{\partial\theta}{\partial x} = \frac{-\sin\theta}{r}.$$

  (b) Show that

$$\hat{\mathbf{r}} = \cos\theta\,\hat{\mathbf{i}} + \sin\theta\,\hat{\mathbf{j}} \quad \text{and} \quad \hat{\boldsymbol{\theta}} = -\sin\theta\,\hat{\mathbf{i}} + \cos\theta\,\hat{\mathbf{j}}.$$

  (c) Using the chain rule, show that

$$\nabla u = \frac{\partial u}{\partial r}\hat{\mathbf{r}} + \frac{1}{r}\frac{\partial u}{\partial\theta}\hat{\boldsymbol{\theta}}.$$

  (d) If $\mathbf{A} = A_r\hat{\mathbf{r}} + A_\theta\hat{\boldsymbol{\theta}}$, show that

$$\nabla \cdot \mathbf{A} = \frac{1}{r}\frac{\partial}{\partial r}(rA_r) + \frac{1}{r}\frac{\partial}{\partial\theta}(A_\theta),$$

  since $\partial\hat{\mathbf{r}}/\partial\theta = \hat{\boldsymbol{\theta}}$ and $\partial\hat{\boldsymbol{\theta}}/\partial\theta = -\hat{\mathbf{r}}$ follows from part (b).
  (e) Show that

$$\nabla^2 u = \frac{1}{r}\frac{\partial}{\partial r}\left(r\frac{\partial u}{\partial r}\right) + \frac{1}{r^2}\frac{\partial^2 u}{\partial\theta^2}.$$

1.5.4. Using Exercise 1.5.3(a) and the chain rule for partial derivatives, derive the special case of Exercise 1.5.3(e) if $u(r)$ only.

1.5.5. Assume that the temperature is circularly symmetric, $u = u(r, t)$, where $r^2 = x^2 + y^2$. We will derive the heat equation for this problem. Consider any circular annulus $a \leqslant r \leqslant b$.
   (a) Show that the total heat energy is $2\pi \int_a^b c\rho u r \, dr$.
   (b) Show that the flow of heat energy per unit time out of the annulus at $r = b$ is $-2\pi b K_0 \, \partial u / \partial r|_{r=b}$. A similar result holds at $r = a$.
   (c) Use parts (a) and (b) to derive the circularly symmetric heat equation without sources:
$$\frac{\partial u}{\partial t} = \frac{k}{r} \frac{\partial}{\partial r} \left(r \frac{\partial u}{\partial r} \right).$$

1.5.6. Modify Exercise 1.5.5 if the thermal properties depend on r.

1.5.7. Derive the heat equation in two dimensions by using Green's theorem, (1.5.16), the two-dimensional form of the divergence theorem.

1.5.8. If Laplace's equation is satisfied in three dimensions, show that
$$\oiint \nabla u \cdot \hat{n} \, dS = 0$$
for any closed surface. (*Hint:* Use the divergence theorem.) Give a physical interpretation of this result (in the context of heat flow).

1.5.9. Determine the equilibrium temperature distribution inside a circular annulus $(r_1 \leqslant r \leqslant r_2)$:
   *(a)** if the outer radius is at temperature T_2 and the inner at T_1.
   (b) if the outer radius is insulated and the inner radius is at temperature T_1.

1.5.10. Determine the equilibrium temperature distribution inside a circle $(r \leqslant r_0)$ if the boundary is fixed at temperature T_0.

*1.5.11.** Consider
$$\frac{\partial u}{\partial t} = \frac{k}{r} \frac{\partial}{\partial r} \left(r \frac{\partial u}{\partial r} \right) \qquad a < r < b$$
subject to
$$u(r, 0) = f(r), \qquad \frac{\partial u}{\partial r}(a, t) = \beta, \quad \text{and} \quad \frac{\partial u}{\partial r}(b, t) = 1.$$
Using physical reasoning, for what value(s) of β does an equilibrium temperature distribution exist?

1.5.12. Assume that the temperature is spherically symmetric, $u = u(r, t)$, where r is the distance from a fixed point $(r^2 = x^2 + y^2 + z^2)$. Consider the heat flow (without sources) between any two concentric spheres of radii a and b.
   (a) Show that the total heat energy is $4\pi \int_a^b c\rho u r^2 \, dr$.
   (b) Show that the flow of heat energy per unit time out of the spherical shell at $r = b$ is $4\pi b^2 K_0 \, \partial u / \partial r|_{r=b}$. A similar result holds at $r = a$.
   (c) Use parts (a) and (b) to derive the spherically symmetric heat equation
$$\frac{\partial u}{\partial t} = \frac{k}{r^2} \frac{\partial}{\partial r} \left(r^2 \frac{\partial u}{\partial r} \right).$$

*1.5.13.** Determine the *steady-state* temperature distribution between two concentric *spheres* with radii 1 and 4, respectively, if the temperature of the outer sphere is maintained at 80° and the inner sphere at 0° (see Exercise 1.5.12).

1.5.14. Isobars are lines of constant temperature. Show that isobars are perpendicular to any part of the boundary that is insulated.

1.5 APPENDIX: REVIEW OF GRADIENT AND A DERIVATION OF FOURIER'S LAW OF HEAT CONDUCTION

Experimentally, for isotropic* materials (i.e., without preferential directions) **heat flows from hot to cold in the direction in which temperature differences are greatest.** The heat flow is proportional (with proportionality constant K_0, the thermal conductivity) to the rate of change of temperature in this direction.

The change in the temperature Δu is

$$\Delta u = u(\mathbf{x} + \Delta \mathbf{x}, t) - u(\mathbf{x}, t) \approx \frac{\partial u}{\partial x} \Delta x + \frac{\partial u}{\partial y} \Delta y + \frac{\partial u}{\partial z} \Delta z.$$

In the direction $\hat{\boldsymbol{\alpha}} = \alpha_1 \hat{\mathbf{i}} + \alpha_2 \hat{\mathbf{j}} + \alpha_3 \hat{\mathbf{k}}$, $\Delta \mathbf{x} = \Delta s\, \hat{\boldsymbol{\alpha}}$, where Δs is the distance between $\mathbf{x}$ and $\mathbf{x} + \Delta \mathbf{x}$. Thus, the rate of change of the temperature in the direction $\hat{\boldsymbol{\alpha}}$ is the **directional derivative**:

$$\lim_{\Delta s \to 0} \frac{\Delta u}{\Delta s} = \alpha_1 \frac{\partial u}{\partial x} + \alpha_2 \frac{\partial u}{\partial y} + \alpha_3 \frac{\partial u}{\partial z} = \hat{\boldsymbol{\alpha}} \cdot \nabla u,$$

where it has been convenient to define the following *vector*:

$$\nabla u \equiv \frac{\partial u}{\partial x} \hat{\mathbf{i}} + \frac{\partial u}{\partial y} \hat{\mathbf{j}} + \frac{\partial u}{\partial z} \hat{\mathbf{k}}, \tag{1.5.A1}$$

called the **gradient** of the temperature. From the property of dot products, if θ is the angle between $\hat{\boldsymbol{\alpha}}$ and ∇u, then the directional derivative is $|\nabla u| \cos \theta$ since $|\hat{\boldsymbol{\alpha}}| = 1$. The largest rate of change of u (the largest directional derivative) is $|\nabla u| > 0$ and it occurs if $\theta = 0$ (i.e., in the direction of the gradient). Since this derivative is positive, the temperature increase is greatest in the direction of the gradient. Since heat energy flows in the direction of decreasing temperatures, **the heat flow vector is in the opposite direction to the heat gradient.** It follows that

$$\boldsymbol{\phi} = -K_0 \nabla u, \tag{1.5.A2}$$

since $|\nabla u|$ equals the magnitude of the rate of change of u (in the direction of the gradient). This again is called Fourier's law of heat conduction. Thus, in three dimensions, the gradient ∇u replaces $\partial u / \partial x$.

Another fundamental property of the gradient is that it is normal (perpendicular) to the level surfaces. It is easier to illustrate this in a two-dimensional problem (see Fig. 1.5.3) in which the temperature is constant along level curves (rather than level surfaces). To show that the gradient is perpendicular, consider

---

* Examples of nonisotropic materials are certain crystals and grainy woods.

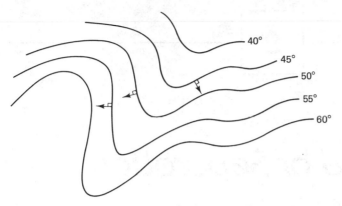

Figure 1.5.3 The gradient is perpendicular to level surfaces of the temperature.

the surface on which the temperature is the constant T_0, $u(x, y, z, t) = T_0$. We calculate the differential of both sides (at a fixed time) along the surface. Since T_0 is constant, $dT_0 = 0$. Therefore, using the chain rule of partial derivatives,

$$du = \frac{\partial u}{\partial x}dx + \frac{\partial u}{\partial y}dy + \frac{\partial u}{\partial z}dz = 0. \qquad (1.5.A3)$$

Equation (1.5.A3) can be written as

$$\left(\frac{\partial u}{\partial x}\hat{\mathbf{i}} + \frac{\partial u}{\partial y}\hat{\mathbf{j}} + \frac{\partial u}{\partial z}\hat{\mathbf{k}}\right) \cdot (dx\,\mathbf{i} + dy\,\hat{\mathbf{j}} + dz\,\hat{\mathbf{k}}) = 0$$

or

$$\nabla u \cdot (dx\,\hat{\mathbf{i}} + dy\,\hat{\mathbf{j}} + dz\,\hat{\mathbf{k}}) = 0. \qquad (1.5.A4)$$

$dx\,\hat{\mathbf{i}} + dy\,\hat{\mathbf{j}} + dz\,\hat{\mathbf{k}}$ represents any vector in the tangent plane of the level surface. From (1.5.A4), its dot product with ∇u is zero; that is, ∇u is perpendicular to the tangent plane. Thus, ∇u is perpendicular to the surface $u = $ constant.

We have thus learned two properties of the gradient, ∇u:

1. Direction: ∇u is perpendicular to the surface $u = $ constant. (∇u is also in the direction of the largest directional derivative. u increases in the direction of the gradient.)
2. Magnitude: $|\nabla u|$ is the largest value of the directional derivative.

2

Method of Separation of Variables

2.1 INTRODUCTION

In Chapter 1 we developed from physical principles an understanding of the heat equation and its corresponding initial and boundary conditions. We are ready to pursue the mathematical solution of some typical problems involving partial differential equations. We will use a technique called the method of separation of variables. You will have to become an expert in this method, and so we will discuss quite a few examples. We will emphasize problem solving techniques, but we must also understand how not to misuse the technique.

A relatively simple, but typical, problem for the equation of heat conduction occurs for a one-dimensional rod ($0 \le x \le L$) when all the thermal coefficients are constant. Then the PDE,

$$\frac{\partial u}{\partial t} = k \frac{\partial^2 u}{\partial x^2} + \frac{Q(x, t)}{c\rho}, \qquad \begin{matrix} t > 0 \\ 0 < x < L, \end{matrix} \qquad (2.1.1)$$

must be solved subject to the initial condition,

$$u(x, 0) = f(x), \qquad 0 < x < L, \qquad (2.1.2)$$

and two boundary conditions. For example, if both ends of the rod have prescribed temperature, then

$$\begin{matrix} u(0, t) = T_1(t) \\ u(L, t) = T_2(t). \end{matrix} \qquad t > 0 \qquad (2.1.3)$$

The method of separation of variables is used when the partial differential equation and the boundary conditions are linear and homogeneous, concepts we now explain.

2.2 LINEARITY

As in the study of ordinary differential equations, the concept of linearity will be very important for us. A **linear operator** L by definition satisfies

$$\boxed{L(c_1u_1 + c_2u_2) = c_1L(u_1) + c_2L(u_2)} \tag{2.2.1}$$

for any two functions u_1 and u_2, where c_1 and c_2 are arbitrary constants. $\partial/\partial t$ and $\partial^2/\partial x^2$ are examples of linear operators since they satisfy (2.2.1):

$$\frac{\partial}{\partial t}(c_1u_1 + c_2u_2) = c_1\frac{\partial u_1}{\partial t} + c_2\frac{\partial u_2}{\partial t}$$

$$\frac{\partial^2}{\partial x^2}(c_1u_1 + c_2u_2) = c_1\frac{\partial^2 u_1}{\partial x^2} + c_2\frac{\partial^2 u_2}{\partial x^2}.$$

It can be shown (see Exercise 2.2.1) that any linear combination of linear operators is a linear operator. Thus, the **heat operator**

$$\frac{\partial}{\partial t} - k\frac{\partial^2}{\partial x^2}$$

is also a linear operator.

A **linear equation** for u is of the form

$$L(u) = f, \tag{2.2.2}$$

where L is a linear operator and f is known. Examples of *linear partial differential equations* are

$$\frac{\partial u}{\partial t} = k\frac{\partial^2 u}{\partial x^2} + f(x, t) \tag{2.2.3a}$$

$$\frac{\partial u}{\partial t} = k\frac{\partial^2 u}{\partial x^2} + \alpha(x, t)u + f(x, t) \tag{2.2.3b}$$

$$\frac{\partial^2 u}{\partial x^2} + \frac{\partial^2 u}{\partial y^2} = 0 \tag{2.2.3c}$$

$$\frac{\partial u}{\partial t} = \frac{\partial^3 u}{\partial x^3} + \alpha(x, t)u. \tag{2.2.3d}$$

Examples of *nonlinear* partial differential equations are

$$\frac{\partial u}{\partial t} = k\frac{\partial^2 u}{\partial x^2} + \alpha(x, t)u^4 \tag{2.2.3e}$$

$$\frac{\partial u}{\partial t} + u\frac{\partial u}{\partial x} = \frac{\partial^3 u}{\partial x^3}. \tag{2.2.3f}$$

That u^4 and $u\partial u/\partial x$ terms are nonlinear; they do not satisfy (2.2.1).

If $f = 0$, then (2.2.2) becomes $L(u) = 0$, called a **linear homogeneous** equation. Examples of linear homogeneous partial differential equations include the heat equation,

$$\frac{\partial u}{\partial t} - k\frac{\partial^2 u}{\partial x^2} = 0, \tag{2.2.4}$$

as well as (2.2.3c) and (2.2.3d). From (2.2.1) it follows that $L(0) = 0$ (let $c_1 = c_2 = 0$). Therefore, $u = 0$ is always a solution of a linear homogeneous equation. For example, $u = 0$ satisfies the heat equation (2.2.4). We call $u = 0$ the *trivial* solution of a linear homogeneous equation. The simplest way to test whether an equation is homogeneous is to substitute the function u identically equal to zero. If $u \equiv 0$ satisfies a linear equation, then it must be that $f = 0$ and hence the linear equation is homogeneous. Otherwise, the equation is said to be **non-homogeneous** [e.g., (2.2.3a) and (2.2.3b)].

The fundamental property of linear operators (2.2.1) allows solutions of linear equations to be added together in the following sense:

Principle of Superposition
If u_1 and u_2 satisfy a linear *homogeneous* equation, then an arbitrary linear combination of them, $c_1 u_1 + c_2 u_2$, also satisfies the same linear homogeneous equation.

The proof of this depends on the definition of a linear operator. Suppose that u_1 and u_2 are two solutions of a linear homogeneous equation. That means that $L(u_1) = 0$ and $L(u_2) = 0$. Let us calculate $L(c_1 u_1 + c_2 u_2)$. From the definition of a linear operator,

$$L(c_1 u_1 + c_2 u_2) = c_1 L(u_1) + c_2 L(u_2).$$

Since u_1 and u_2 are homogeneous solutions, it follows that $L(c_1 u_1 + c_2 u_2) = 0$. This means that $c_1 u_1 + c_2 u_2$ satisfies the linear homogeneous equation $L(u) = 0$ if u_1 and u_2 satisfy the *same* linear homogeneous equation.

The concepts of linearity and homogeneity also apply to boundary conditions, in which case the variables are evaluated at specific points. Examples of *linear* boundary conditions are the conditions we have discussed:

$$u(0, t) = f(t) \tag{2.2.5a}$$

$$\frac{\partial u}{\partial x}(L, t) = g(t) \tag{2.2.5b}$$

$$\frac{\partial u}{\partial x}(0, t) = 0 \tag{2.2.5c}$$

$$-K_0 \frac{\partial u}{\partial x}(L, t) = h[u(L, t) - g(t)]. \tag{2.2.5d}$$

A *nonlinear* boundary condition, for example, would be

$$\frac{\partial u}{\partial x}(L, t) = u^2(L, t). \tag{2.2.5e}$$

Only (2.2.5c) is satisfied by $u \equiv 0$ (of the linear conditions) and hence is homogeneous. It is not necessary that a boundary condition be $u(0, t) = 0$ for $u \equiv 0$ to satisfy it.

EXERCISES 2.2

2.2.1. Show that any linear combination of linear operators is a linear operator.

2.2.2. (a) Show that $L(u) = \dfrac{\partial}{\partial x}\left[K_0(x)\dfrac{\partial u}{\partial x}\right]$ is a linear operator.

(b) Show that usually $L(u) = \dfrac{\partial}{\partial x}\left[K_0(x, u)\dfrac{\partial u}{\partial x}\right]$ is not a linear operator.

2.2.3. Show that $\dfrac{\partial u}{\partial t} = k\dfrac{\partial^2 u}{\partial x^2} + Q(u, x, t)$ is linear if $Q = \alpha(x, t)u + \beta(x, t)$ and in addition homogeneous if $\beta(x, t) = 0$.

2.2.4. In this exercise we derive superposition principles for nonhomogeneous problems.
(a) Consider $L(u) = f$. If u_p is a particular solution, $L(u_p) = f$, and if u_1 and u_2 are homogeneous solutions, $L(u_i) = 0$, show that $u = u_p + c_1u_1 + c_2u_2$ is another particular solution.
(b) If $L(u) = f_1 + f_2$, where u_{p_i} is a particular solution corresponding to f_i, what is a particular solution for u?

2.2.5 If L is a linear operator, show that $L(\sum_{n=1}^{M} c_nu_n) = \sum_{n=1}^{M} c_nL(u_n)$. Use this result to show that the principle of superposition may be extended to any finite number of homogeneous solutions.

2.3 HEAT EQUATION WITH ZERO TEMPERATURES AT FINITE ENDS

2.3.1 Introduction

Partial differential equation (2.1.1) is linear, but it is homogeneous only if there are no sources, $Q(x, t) = 0$. The boundary conditions (2.1.3) are also linear, and they too are homogeneous only if $T_1(t) = 0$ and $T_2(t) = 0$. We thus first propose to study

PDE:	$\dfrac{\partial u}{\partial t} = k\dfrac{\partial^2 u}{\partial x^2}$ $\begin{array}{l}0 < x < L \\ t > 0\end{array}$	(2.3.1)
BC:	$u(0, t) = 0$	(2.3.2a)
	$u(L, t) = 0$	(2.3.2b)
IC:	$u(x, 0) = f(x).$	(2.3.3)

The problem consists of a linear homogeneous partial differential equation with linear homogeneous boundary conditions. There are two reasons for our investigating this type of problem, (2.3.1)–(2.3.3), besides the fact that we claim it can be solved by the method of separation of variables. First, this problem is a relevant physical problem corresponding to a one-dimensional rod ($0 < x$

$< L$) with no sources and both ends immersed in a 0° temperature bath. We are very interested in predicting how the initial thermal energy (represented by the initial condition) changes in this relatively simple physical situation. Second, it will turn out that in order to solve the nonhomogeneous problem (2.1.1)– (2.1.3), we will need to know how to solve the homogeneous problem, (2.3.1)– (2.3.3).

2.3.2 Separation of Variables

In the **method of separation of variables**, we attempt to determine solutions in the product form

$$u(x, t) = \phi(x)G(t),$$

(2.3.4)

where $\phi(x)$ is only a function of x and $G(t)$ only a function of t. Equation (2.3.4) must satisfy the *linear homogeneous* partial differential equation (2.3.1) and boundary conditions (2.3.2), but for the moment we set aside (ignore) the initial condition. The **product solution**, (2.3.4), does not satisfy the initial conditions. Later we will explain how to satisfy the initial conditions.

Let us be clear from the beginning—we do not give any reasons why we choose the form (2.3.4). (Daniel Bernoulli invented this technique in the 1700s. IT WORKS, as we shall see.) We substitute the assumed product form, (2.3.4), into the partial differential equation (2.3.1):

$$\frac{\partial u}{\partial t} = \phi(x)\frac{dG}{dt}$$

$$\frac{\partial^2 u}{\partial x^2} = \frac{d^2\phi}{dx^2}G(t),$$

and consequently the heat equation (2.3.1) implies that

$$\phi(x)\frac{dG}{dt} = k\frac{d^2\phi}{dx^2}G(t).$$

(2.3.5)

We note that we can "separate variables" by dividing both sides of (2.3.5) by $\phi(x)G(t)$:

$$\frac{1}{G}\frac{dG}{dt} = k\frac{1}{\phi}\frac{d^2\phi}{dx^2}.$$

Now the variables have been "separated" in the sense that the left-hand side is only a function of t and the right-hand side only a function of x. We can continue in this way, but it is *convenient* (i.e., not necessary) also to divide by the constant k, and thus

$$\underbrace{\frac{1}{kG}\frac{dG}{dt}}_{\substack{\text{function} \\ \text{of } t \text{ only}}} = \underbrace{\frac{1}{\phi}\frac{d^2\phi}{dx^2}}_{\substack{\text{function} \\ \text{of } x \text{ only}}}.$$

(2.3.6)

This could be obtained directly from (2.3.5) by dividing by $k\phi(x)G(t)$. How is it possible for a function of time to equal a function of space? If x and t are both to be arbitrary *independent* variables, then x cannot be a function of t (or t a function of x) as seems to be specified by (2.3.6). The important idea is that we claim it is necessary that both sides of (2.3.6) must equal the same constant:

$$\frac{1}{kG}\frac{dG}{dt} = \frac{1}{\phi}\frac{d^2\phi}{dx^2} = -\lambda, \tag{2.3.7}$$

where λ is an arbitrary *constant* known as the **separation constant**.* We will explain momentarily the mysterious minus sign, which was introduced only for convenience.

Equation (2.3.7) yields two ordinary differential equations, one for $G(t)$ and one for $\phi(x)$:

$$\frac{d^2\phi}{dx^2} = -\lambda\phi \tag{2.3.8}$$

$$\frac{dG}{dt} = -\lambda kG. \tag{2.3.9}$$

We reiterate that λ is a constant and it is the same constant that appears in both (2.3.8) and (2.3.9). The product solutions, $u(x, t) - \phi(x)G(t)$, must also satisfy the two homogeneous boundary conditions. For example, $u(0, t) = 0$ implies that $\phi(0)G(t) = 0$. There are two possibilities. Either $G(t) \equiv 0$ (the meaning of $\equiv$ is identically zero, for all t) or $\phi(0) = 0$. If $G(t) = 0$, then from (2.3.4), the assumed product solution is identically zero, $u(x, t) \equiv 0$. This is not very interesting. [$u(x, t) \equiv 0$ is called the **trivial solution** since $u(x, t) = 0$ automatically satisfies any homogeneous PDE and any homogeneous BC.] Instead, we look for nontrivial solutions. For nontrivial solutions, we must have

$$\phi(0) - 0. \tag{2.3.10}$$

By applying the other boundary condition, $u(L, t) = 0$, we obtain in a similar way that

$$\phi(L) = 0. \tag{2.3.11}$$

Product solutions, in addition to satisfying two ordinary differential equations, (2.3.8) and (2.3.9), must also satisfy boundary conditions (2.3.10) and (2.3.11).

---

* As further explanation for the constant in (2.3.7), let us say the following. Suppose that the left-hand side of (2.3.7) is some function of t, $(1/kG)\,dG/dt = w(t)$. If we differentiate with respect to x, we get zero: $0 = d/dx\,(1/\phi\,d^2\phi/dx^2)$. Since $1/\phi\,d^2\phi/dx^2$ is only a function of x, this implies that $1/\phi\,d^2\phi/dx^2$ must be a constant, its derivative equaling zero. In this way (2.3.7) follows.

2.3.3 Time-Dependent Equation

The advantage of the product method is that it transforms a partial differential equation, which we do not know how to solve, into two ordinary differential equations. The boundary conditions impose two conditions on the x-dependent ordinary differential equation (ODE). The time-dependent equation has no additional conditions, just

$$\frac{dG}{dt} = -\lambda k G. \tag{2.3.12}$$

Let us solve (2.3.12) first before we discuss solving the x-dependent ODE with its two homogeneous boundary conditions. Equation (2.3.12) is a first-order linear homogeneous differential equation *with constant coefficients*. We can obtain its general solution quite easily. Nearly all constant-coefficient (linear and homogeneous) ODEs can be solved by seeking exponential solutions, $G = e^{rt}$, where in this case by substitution the characteristic polynomial is $r = -\lambda k$. Therefore, the general solution of (2.3.12) is

$$G(t) = ce^{-\lambda k t}. \tag{2.3.13}$$

We have remembered that for linear homogeneous equations, if $e^{-\lambda k t}$ is a solution, then $ce^{-\lambda k t}$ is a solution (for any arbitrary multiplicative constant c). The time-dependent solution is a simple exponential. Recall that λ is the separation constant, which for the moment is arbitrary. However, eventually we will discover that only certain values of λ are allowable. If $\lambda > 0$, the solution exponentially decays as t increases (since $k > 0$). If $\lambda < 0$, the solution exponentially increases, and if $\lambda = 0$, the solution remains constant in time. Since this is a heat conduction problem and the temperature $u(x, t)$ is proportional to $G(t)$, we do not *expect* the solution to grow exponentially in time. Thus, we expect $\lambda \geq 0$; we have not proved that statement, but will do so later. When we see any parameter we often automatically assume that it is positive, *even though we shouldn't*. Thus, it is rather convenient that we have discovered that we expect $\lambda \geq 0$. In fact, that is why we introduced the expression $-\lambda$ when we separated variables [see (2.3.7)]. If we had introduced μ (instead of $-\lambda$), then our previous arguments would have suggested that $\mu \leq 0$. In summary, when separating variables in (2.3.7), we mentally solve the time-dependent equation and see that $G(t)$ does not exponentially grow only if the separation constant was ≤ 0. We then introduce $-\lambda$ for convenience, since we would now *expect* $\lambda \geq 0$. We next show how we actually determine all allowable separation constants. We will verify mathematically that $\lambda \geq 0$, as we expect by the physical arguments presented above.

2.3.4 Boundary Value Problem

The x-dependent part of the assumed product solution, $\phi(x)$, satisfies a second-order ODE with two homogeneous boundary conditions:

$$\frac{d^2\phi}{dx^2} = -\lambda\phi$$
$$\phi(0) = 0$$
$$\phi(L) = 0.$$

(2.3.14)

We call (2.3.14) a **boundary value problem** for ordinary differential equations. In the usual first course in ordinary differential equations, only initial value problems are specified. For example (think of Newton's law of motion for a particle), we solve second-order differential equations ($m\, d^2y/dt^2 = F$) subject to two initial conditions [$y(0)$ and $dy/dt(0)$ given] *both at the same time*. Initial value problems are quite nice, as *usually* there *exist unique* solutions to initial value problems. However, (2.3.14) is quite different. It is a boundary value problem, since the two conditions are not given at the same place (e.g., $x = 0$) but at two different places, $x = 0$ and $x = L$. There is no simple theory which guarantees that the solution exists or is unique to this type of problem. In particular, we note that $\phi(x) \equiv 0$ satisfies the ODE and both homogeneous boundary conditions, no matter what the separation constant λ is, even if $\lambda < 0$; it is referred to as the **trivial solution** of the boundary value problem. It corresponds to $u(x, t) \equiv 0$, since $u(x, t) = \phi(x)G(t)$. If solutions of (2.3.14) had been unique, then $\phi(x) \equiv 0$ would be the only solution; we would not be able to obtain nontrivial solutions of a linear homogeneous PDE by the product (separation of variables) method. Fortunately, there are other solutions of (2.3.14). However, there do not exist nontrivial solutions of (2.3.14) for all values of λ. Instead, we will show that there are certain special values of λ, called **eigenvalues*** of the boundary value problem (2.3.14), for which there are nontrivial solutions, $\phi(x)$. A nontrivial $\phi(x)$, which exists only for certain values of λ, is called an **eigenfunction** corresponding to the eigenvalue λ.

Let us try to determine the eigenvalues λ. In other words, for what values of λ are there nontrivial solutions of (2.3.14). We solve (2.3.14) directly. The second-order ODE is linear and homogeneous with constant coefficients; two independent solutions are usually obtained in the form of exponentials, $\phi = e^{rx}$. Substituting this exponential into the differential equation yields the characteristic polynomial $r^2 = -\lambda$. The solutions corresponding to the two roots have significantly different properties depending on the value of λ. There are four cases:

1. $\lambda > 0$, in which the two roots are purely imaginary and are complex conjugates of each other, $r = \pm i\sqrt{\lambda}$.
2. $\lambda = 0$, in which the two roots coalesce and are equal, $r = 0, 0$.
3. $\lambda < 0$, in which the two roots are real and unequal, $r = \pm\sqrt{-\lambda}$, one positive and one negative. (Note that in this case $-\lambda$ is positive, so that the square root operation is well defined.)
4. λ itself complex.

* The word *eigenvalue* comes from the German word *eigenwert*, meaning characteristic value.

We will ignore the last case (as most of you would have done anyway) since we will later (Chapter 5) prove that λ is real in order for a nontrivial solution of the boundary value problem (2.3.14) to exist. From the time-dependent solution, using physical reasoning, we expect that $\lambda \geq 0$; perhaps then it will be unnecessary to analyze case 3. However, we will demonstrate a mathematical reason for the omission of this case.

Eigenvalues and eigenfunctions ($\lambda > 0$). Let us first consider the case in which $\lambda > 0$. The boundary value problem is

$$\frac{d^2\phi}{dx^2} = -\lambda\phi \qquad (2.3.15)$$

$$\phi(0) = 0 \qquad (2.3.16a)$$

$$\phi(L) = 0. \qquad (2.3.16b)$$

If $\lambda > 0$, exponential solutions have imaginary exponents, $e^{\pm i\sqrt{\lambda}x}$. In this case, the solutions oscillate. If we desire real independent solutions, the choices $\cos \sqrt{\lambda}\, x$ and $\sin \sqrt{\lambda}\, x$ are usually made ($\cos \sqrt{\lambda}\, x$ and $\sin \sqrt{\lambda}\, x$ are each linear combinations of $e^{\pm i\sqrt{\lambda}x}$). Thus, the general solution of (2.3.15) is

$$\phi = c_1 \cos \sqrt{\lambda}\, x + c_2 \sin \sqrt{\lambda}\, x, \qquad (2.3.17)$$

an arbitrary linear combination of two independent solutions. (The linear combination may be chosen from any two independent solutions.) $\cos \sqrt{\lambda}\, x$ and $\sin \sqrt{\lambda}\, x$ are usually the most convenient, but $e^{i\sqrt{\lambda}x}$ and $e^{-i\sqrt{\lambda}x}$ can be used. In some examples, other independent solutions are chosen. For example, Exercise 2.3.2(f) illustrates the advantage of sometimes choosing $\cos \sqrt{\lambda}\, (x - a)$ and $\sin \sqrt{\lambda}\, (x - a)$ as independent solutions.

We now apply the boundary conditions. $\phi(0) = 0$ implies that

$$0 = c_1.$$

The cosine term vanishes, since the solution must be zero at $x = 0$. Thus, $\phi(x) = c_2 \sin \sqrt{\lambda}\, x$. Only the boundary condition at $x = L$ has not been satisfied. $\phi(L) = 0$ implies that

$$0 = c_2 \sin \sqrt{\lambda}\, L.$$

Either $c_2 = 0$ or $\sin \sqrt{\lambda}\, L = 0$. If $c_2 = 0$, then $\phi(x) \equiv 0$ since we already determined that $c_1 = 0$. This is the trivial solution, and we are searching for those values of λ that have nontrivial solutions. The eigenvalues λ must satisfy

$$\sin \sqrt{\lambda}\, L = 0. \qquad (2.3.18)$$

$\sqrt{\lambda}\, L$ must be a zero of the sine function. A sketch of $\sin z$ (see Fig. 2.3.1) or our knowledge of the sine function shows that $\sqrt{\lambda}\, L = n\pi$. $\sqrt{\lambda}\, L$ must equal an integral multiple of π, where n is a positive integer since $\sqrt{\lambda} > 0$ ($n = 0$ is not appropriate since we assumed that $\lambda > 0$ in this derivation). The **eigenvalues** λ are

$$\lambda = \left(\frac{n\pi}{L}\right)^2, \qquad n = 1, 2, 3, \ldots \qquad (2.3.19)$$

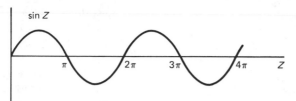

Figure 2.3.1 Zeros of sin z.

The eigenfunction corresponding to the eigenvalue $\lambda = (n\pi/L)^2$ is

$$\phi(x) = c_2 \sin \sqrt{\lambda}\, x = c_2 \sin \frac{n\pi x}{L}, \qquad (2.3.20)$$

where c_2 is an arbitrary multiplicative constant. Often we pick a convenient value for c_2, for example $c_2 = 1$. We should remember, though, that any specific eigenfunction can always be multiplied by an arbitrary constant, since the PDE and BCs are linear and homogeneous.

Eigenvalue ($\lambda = 0$). Now we will determine if $\lambda = 0$ is an eigenvalue for (2.3.15) subject to the boundary conditions (2.3.16). $\lambda = 0$ is a special case. If $\lambda = 0$, (2.3.15) implies that

$$\phi = c_1 + c_2 x,$$

corresponding to the double-zero roots, $r = 0, 0$, of the characteristic polynomial.* To determine whether $\lambda = 0$ is an eigenvalue, the homogeneous boundary conditions must be applied. $\phi(0) = 0$ implies that $0 = c_1$, and thus $\phi = c_2 x$. In addition, $\phi(L) = 0$ implies that $0 = c_2 L$. Since the length L of the rod is positive ($\neq 0$), $c_2 = 0$ and thus $\phi(x) \equiv 0$. This is the trivial solution, so we say that $\lambda = 0$ is *not* an eigenvalue, for *this* problem [(2.3.15) and (2.3.16)]. Be wary, though; $\lambda = 0$ is an eigenvalue for other problems and should be looked at individually for any new problem you may encounter.

Eigenvalues ($\lambda < 0$). Are there any negative eigenvalues? If $\lambda < 0$, the solution of

$$\frac{d^2\phi}{dx^2} = -\lambda\phi \qquad (2.3.21)$$

is not difficult, but you may have to be careful. The roots of the characteristic polynomial are $r = \pm\sqrt{-\lambda}$, so solutions are $e^{\sqrt{-\lambda}x}$ and $e^{-\sqrt{-\lambda}x}$. If you do not like the notation $\sqrt{-\lambda}$, you may prefer what is equivalent (if $\lambda < 0$), namely $\sqrt{|\lambda|}$. However, $\sqrt{|\lambda|} \neq \sqrt{\lambda}$, since $\lambda < 0$. It is convenient to let

$$\lambda = -s,$$

---

* Please do *not* say that $\phi = c_1 \cos \sqrt{\lambda}\, x + c_2 \sin \sqrt{\lambda}\, x$ is the general solution for $\lambda = 0$. If you do that, you find for $\lambda = 0$ that the general solution is an arbitrary constant. Although an arbitrary constant solves (2.3.15) when $\lambda = 0$, (2.3.15) is still a linear second-order differential equation; its general solution must be a linear combination of *two* independent solutions. It is possible to choose $\sin \sqrt{\lambda}\, x/\sqrt{\lambda}$ as a second independent solution so that as $\lambda \to 0$ it agrees with the solution x. However, this involves too much work. It is better just to consider $\lambda = 0$ as a separate case.

in the case in which $\lambda < 0$. Then $s > 0$, and the differential equation (2.3.21) becomes

$$\frac{d^2\phi}{dx^2} = s\phi. \tag{2.3.22}$$

Two independent solutions are $e^{+\sqrt{s}\,x}$ and $e^{-\sqrt{s}\,x}$, since $s > 0$. The general solution is

$$\phi = c_1 e^{\sqrt{s}\,x} + c_2 e^{-\sqrt{s}\,x}. \tag{2.3.23}$$

Frequently, we instead use the hyperbolic functions. As a review, the definitions of the hyperbolic functions are

$$\cosh z \equiv \frac{e^z + e^{-z}}{2} \quad \text{and} \quad \sinh z \equiv \frac{e^z - e^{-z}}{2},$$

simple linear combinations of exponentials. These are sketched in Fig. 2.3.2. Note that $\sinh 0 = 0$ and $\cosh 0 = 1$ (the results analogous to those for trigonometric functions). Also note that $d/dz \cosh z = \sinh z$ and $d/dz \sinh z = \cosh z$, quite similar to trigonometric functions, but easier to remember because of the lack of the annoying appearance of any minus signs in the differentiation formulas. If hyperbolic functions are used instead of exponentials, the general solution of (2.3.22) can be written as

$$\phi = c_3 \cosh \sqrt{s}\, x + c_4 \sinh \sqrt{s}\, x, \tag{2.3.24}$$

a form equivalent to (2.3.23). To determine if there are any negative eigenvalues ($\lambda < 0$, but $s > 0$ since $\lambda = -s$), we again apply the boundary conditions. Either form (2.3.23) or (2.3.24) can be used; the same answer is obtained either way. From (2.3.24), $\phi(0) = 0$ implies that $0 = c_3$, and hence $\phi = c_4 \sinh \sqrt{s}\, x$. The other boundary condition, $\phi(L) = 0$, implies that $c_4 \sinh \sqrt{s}\, L = 0$. Since $\sqrt{s}\, L > 0$ and since sinh is never zero for a positive argument (see Fig. 2.3.2), it follows that $c_4 = 0$. Thus, $\phi(x) \equiv 0$. The only solution of (2.3.22) for $\lambda < 0$ that solves the homogeneous boundary conditions is the trivial solution. Thus, there are no negative eigenvalues. For this example, the existence of negative eigenvalues would have corresponded to exponential growth in time. We did not expect such solutions on physical grounds, and here we have verified mathematically in an explicit manner that there cannot be any negative eigenvalues for this problem. In some other problems there can be negative eigenvalues. Later (Sec. 5.3) we will formulate a theory, involving the Rayleigh quotient, in which we will know before we start many problems that

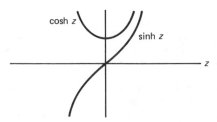

Figure 2.3.2 Hyperbolic functions.

there cannot be negative eigenvalues. This will at times eliminate calculations such as the ones just performed.

Eigenfunctions—summary.   We summarize our results for the boundary value problem resulting from separation of variables:

$$\frac{d^2\phi}{dx^2} + \lambda\phi = 0$$
$$\phi(0) = 0$$
$$\phi(L) = 0.$$

This boundary value problem will arise many times in the text. It is helpful to nearly memorize the result that the eigenvalues λ are all positive (not zero or negative),

$$\lambda = \left(\frac{n\pi}{L}\right)^2,$$

where n is any positive integer, $n = 1, 2, 3, \ldots$, and the corresponding eigenfunctions are

$$\phi(x) - \sin\frac{n\pi x}{L}.$$

If we introduce the notation λ_1 for the first (or lowest) eigenvalue, λ_2 for the next, and so on, we see that $\lambda_n = (n\pi/L)^2$, $n = 1, 2, \ldots$ The corresponding eigenfunctions are sometimes denoted $\phi_n(x)$, the first few of which are sketched in Fig. 2.3.3. All eigenfunctions are (of course) zero at both $x = 0$ and $x = L$. Notice that $\phi_1(x) = \sin \pi x/L$ has no zeros for $0 < x < L$, and $\phi_2(x) - \sin 2\pi x/L$ has one zero for $0 < x < L$. In fact, $\phi_n(x) = \sin n\pi x/L$ has $n - 1$ zeros for $0 < x < L$. We will claim later (see Sec. 5.3) that, remarkably, this is a general property of eigenfunctions.

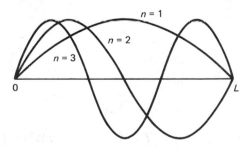

Figure 2.3.3 Eigenfunctions $\sin n\pi x/L$ and their zeros.

Spring–mass analog.   We have obtained solutions of $d^2\phi/dx^2 = -\lambda\phi$. Here we present the analog of this to a spring–mass system, which some of you may find helpful. A spring–mass system subject to Hooke's law satisfies $md^2y/dt^2 = -ky$, where $k > 0$ is the spring constant. Thus, if $\lambda > 0$, the ODE (2.3.15) may be thought of as a spring–mass system with a restoring force. Thus, if $\lambda > 0$, the solution should oscillate. It should not be surprising that the BCs (2.3.16) can be satisfied for $\lambda > 0$; a nontrivial solution of the ODE, which is zero at $x = 0$, has a chance of being zero again at $x = L$ since there is a restoring force and the solution of the ODE oscillates. We have shown that this can happen for specific values of $\lambda > 0$. However, if $\lambda < 0$, then the force is not restoring. It would seem less likely that a nontrivial solution which is zero at $x = 0$ could possibly be zero again at $x = L$. We must not always trust our intuition entirely, so we have verified these facts mathematically.

2.3.5 Product Solutions and the Principle of Superposition

In summary, we obtained product solutions of the heat equation, $\partial u/\partial t = k\,\partial^2 u/\partial x^2$, satisfying the specific homogeneous boundary conditions $u(0, t) = 0$ and $u(L, t) = 0$ only corresponding to $\lambda > 0$. These solutions, $u(x, t) = \phi(x)G(t)$, have $G(t) = ce^{-\lambda kt}$ and $\phi(x) = c_2 \sin \sqrt{\lambda}\, x$, where we determined from the boundary conditions $[\phi(0) = 0$ and $\phi(L) = 0]$ the allowable values of the separation constant λ, $\lambda = (n\pi/L)^2$. Here n is a positive integer. Thus, product solutions of the heat equation are

$$\boxed{u(x, t) = B \sin \frac{n\pi x}{L} e^{-k(n\pi/L)^2 t},} \qquad n = 1, 2, \ldots, \qquad (2.3.25)$$

where B is an arbitrary constant ($B = cc_2$). This is a different solution for each n. Note that as t increases, these special solutions exponentially decay. In particular, for these solutions, $\lim_{t\to\infty} u(x, t) = 0$. In addition, $u(x, t)$ satisfies a special initial condition, $u(x, 0) = B \sin n\pi x/L$.

Initial value problems.   We can use the simple product solutions, (2.3.25), to satisfy an initial value problem if the initial condition happens to be just right. For example, suppose that we wish to solve the following initial value problem:

PDE: $\qquad \dfrac{\partial u}{\partial t} = k\dfrac{\partial^2 u}{\partial x^2}$

BC: $\qquad \begin{aligned} u(0, t) &= 0 \\ u(L, t) &= 0 \end{aligned}$

IC: $\qquad u(x, 0) = 4 \sin \dfrac{3\pi x}{L}$

Our product solution $u(x, t) = B \sin n\pi x/L\; e^{-k(n\pi/L)^2 t}$ satisfies the initial condition $u(x, 0) = B \sin n\pi x/L$. Thus, by picking $n = 3$ and $B = 4$, we will have satisfied the initial condition. Our solution of this example is thus

$$u(x, t) = 4 \sin \frac{3\pi x}{L} e^{-k(3\pi/L)^2 t}.$$

It can be proved that this physical problem (as well as most we consider) has a unique solution. Thus, it does not matter what procedure we used to obtain the solution.

Principle of superposition. The product solutions appear to be very special, since they may be used directly only if the initial condition happens to be of the appropriate form. However, we wish to show that these solutions are useful in many other situations; in fact, in all situations. Consider the same PDE and BCs, but instead subject to the initial condition

$$u(x, 0) = 4 \sin \frac{3\pi x}{L} + 7 \sin \frac{8\pi x}{L}.$$

The solution of this problem can be obtained by adding together two simpler solutions obtained by the product method:

$$u(x, t) = 4 \sin \frac{3\pi x}{L} e^{-k(3\pi/L)^2 t} + 7 \sin \frac{8\pi x}{L} e^{-k(8\pi/L)^2 t}.$$

We immediately see that this solves the initial condition (substitute $t = 0$) as well as the boundary conditions (substitute $x = 0$ and $x = L$). Only slightly more work shows that the partial differential equation has been satisfied. This is an illustration of the principle of superposition.

Superposition (extended). The principle of superposition can be extended to show that if $u_1, u_2, u_3, \ldots, u_M$ are solutions of a linear homogeneous problem, then any linear combination of these is also a solution, $c_1 u_1 + c_2 u_2 + c_3 u_3 + \cdots + c_M u_M = \sum_{n=1}^{M} c_n u_n$, where c_n are arbitrary constants. Since we know from the method of separation of variables that $\sin n\pi x/L e^{-k(n\pi/L)^2 t}$ is a solution of the heat equation (solving zero boundary conditions) for all positive n, it follows that any linear combination of these solutions is also a solution of the linear homogeneous heat equation. Thus,

$$u(x, t) = \sum_{n=1}^{M} B_n \sin \frac{n\pi x}{L} e^{-k(n\pi/L)^2 t} \tag{2.3.26}$$

solves the heat equation (with zero boundary conditions) for any finite M. We have added solutions to the heat equation, keeping in mind that the "amplitude" B could be different for each solution, yielding the subscript B_n. Equation (2.3.26) shows that we can solve the heat equation if initially

$$u(x, 0) = f(x) = \sum_{n=1}^{M} B_n \sin \frac{n\pi x}{L}, \tag{2.3.27}$$

that is, if the initial condition equals a finite sum of the appropriate sine functions. What should we do in the usual situation in which $f(x)$ is *not* a finite linear combination of the appropriate sine functions? We claim that the theory of Fourier series (to be described with considerable detail in Chapter 3) states that:

1. Any function $f(x)$ (with certain very reasonable restrictions, to be discussed later) can be approximated (in some sense) by a finite linear combination of $\sin n\pi x/L$.

2. The approximation may not be very good for small M, but gets to be a better and better approximation as M is increased (see Sec. 5.10).

3. Furthermore, if we consider the limit as $M \to \infty$, then not only is (2.3.27) the best approximation to $f(x)$ using combinations of the eigenfunctions, but (again in some sense) the resulting infinite series will converge to $f(x)$ [with some restrictions on $f(x)$, to be discussed].

We thus claim (and clarify and make precise in Chapter 3) that "any" initial condition $f(x)$ can be written as an infinite linear combination of $\sin n\pi x/L$, known as a type of **Fourier series**:

$$f(x) = \sum_{n=1}^{\infty} B_n \sin \frac{n\pi x}{L}. \tag{2.3.28}$$

What is more important is that we also claim that the corresponding infinite series is the solution of our heat conduction problem:

$$u(x, t) = \sum_{n=1}^{\infty} B_n \sin \frac{n\pi x}{L} e^{-k(n\pi/L)^2 t}. \tag{2.3.29}$$

Analyzing infinite series such as (2.3.28) and (2.3.29) is not easy. We must discuss the convergence of these series as well as briefly discuss the validity of an infinite series solution of our entire problem. For the moment, let us ignore these somewhat theoretical issues and concentrate on the construction of these infinite series solutions.

2.3.6 Orthogonality of Sines

One very important practical point has been neglected. Equation (2.3.29) is our solution with the coefficients B_n satisfying (2.3.28) (from the initial conditions), but how do we determine the coefficients B_n? We assume it is possible that

$$f(x) = \sum_{n=1}^{\infty} B_n \sin \frac{n\pi x}{L}, \tag{2.3.30}$$

where this is to hold over the region of the one-dimensional rod, $0 \le x \le L$. We will assume that standard mathematical operations are also valid for infinite series. Equation (2.3.30) represents one equation in an infinite number of unknowns, but it should be valid at every value of x. If we substitute a thousand different values of x into (2.3.30), each of the thousand equations would hold, but there would still be an infinite number of unknowns. This is not an efficient way to determine the B_n. Instead, we frequently will employ an extremely important technique based on noticing (perhaps from a table of integrals) that the eigenfunctions $\sin n\pi x/L$ satisfy the following integral property:

$$\int_0^L \sin \frac{n\pi x}{L} \sin \frac{m\pi x}{L} \, dx = \begin{cases} 0 & m \neq n \\ L/2 & m = n, \end{cases}$$

(2.3.31a)
(2.3.31b)

where m and n are positive integers.

To use these conditions, (2.3.31), to determine B_n, we multiply both sides of (2.3.30) by $\sin m\pi x/L$ (for any fixed integer m, independent of the "dummy" index n):

$$f(x) \sin \frac{m\pi x}{L} = \sum_{n=1}^{\infty} B_n \sin \frac{n\pi x}{L} \sin \frac{m\pi x}{L}.$$

(2.3.32)

Next we integrate (2.2.32) from $x = 0$ to $x = L$:

$$\int_0^L f(x) \sin \frac{m\pi x}{L} \, dx = \sum_{n=1}^{\infty} B_n \int_0^L \sin \frac{n\pi x}{L} \sin \frac{m\pi x}{L} \, dx.$$

(2.3.33)

For finite series, the integral of a sum of terms equals the sum of the integrals. We assume that this is valid for this infinite series. Now we evaluate the infinite sum. From the integral property (2.3.31), we see that each term of the sum is zero whenever $n \neq m$. In summing over n, eventually n equals m. It is only for that one value of n, i.e. $n = m$, that there is a contribution to the infinite sum. The only term that appears on the right-hand side of (2.3.33) occurs when n is replaced by m:

$$\int_0^L f(x) \sin \frac{m\pi x}{L} \, dx = B_m \int_0^L \sin^2 \frac{m\pi x}{L} \, dx.$$

Since the integral on the right equals $L/2$, we can solve for B_m:

$$B_m = \frac{\int_0^L f(x) \sin (m\pi x/L) \, dx}{\int_0^L \sin^2 (m\pi x/L) \, dx} = \frac{2}{L} \int_0^L f(x) \sin \frac{m\pi x}{L} \, dx.$$

(2.3.34)

This result is very important and so is the method by which it was obtained. Try to learn both. The integral in (2.3.34) is considered to be known since $f(x)$ is the given initial condition. The integral cannot usually be evaluated, in which case numerical integrations (on a computer) may need to be performed to get explicit numbers for B_m, $m = 1, 2, 3, \ldots$.

You will find that the formula (2.3.31b), $\int_0^L \sin^2 n\pi x/L \, dx = L/2$, is quite useful in many different circumstances, including applications having nothing to do with the material of this text. One reason for its applicability is that there are many periodic phenomena in nature ($\sin \omega t$), and usually energy or power is proportional to the square ($\sin^2 \omega t$). The average energy is then proportional to $\int_0^{2\pi/\omega} \sin^2 \omega t \, dt$ divided by the period $2\pi/\omega$. It is worthwhile to memorize that the average over a full period of sine or cosine squared is $\frac{1}{2}$. Thus, *the integral over any number of complete periods of the square of a sine or cosine is one-half the length of the interval.* In this way $\int_0^L \sin^2 n\pi x/L \, dx = L/2$, since the interval 0 to L is either a complete or a half period of $\sin n\pi x/L$.

Orthogonality. Whenever $\int_0^L A(x)B(x) \, dx = 0$ we say that the functions $A(x)$ and $B(x)$ are **orthogonal** over the interval $0 \leq x \leq L$. We borrow the terminology "orthogonal" from perpendicular vectors because $\int_0^L A(x) B(x) \, dx = 0$ is analogous to a zero dot product, as is explained further in the appendix to this section. A set of functions each member of which is orthogonal to every *other* member is called an **orthogonal set of functions.** An example is that of the functions $\sin n\pi x/L$, the eigenfunctions of the boundary value problem

$$\frac{d^2\phi}{dx^2} + \lambda\phi = 0 \quad \text{with} \quad \phi(0) = 0 \quad \text{and} \quad \phi(L) = 0.$$

They are mutually orthogonal because of (2.3.31a). Therefore, we call (2.3.31) an orthogonality condition.

In fact, we will discover that for most other boundary value problems, the eigenfunctions will form an orthogonal set of functions (with certain modifications discussed in Chapter 5 with respect to Sturm-Liouville eigenvalue problems).

2.3.7 Formulation, Solution, and Interpretation of an Example

As an example, let us analyze our solution in the case in which the initial temperature is constant, 100°C. This corresponds to a physical problem that is easy to reproduce in the laboratory. Take a one-dimensional rod and place the entire rod in a large tub of boiling water (100°C). Let it sit there for a long time. After a while (we expect) the rod will be at 100°C throughout. Now insulate the lateral sides (if that had not been done earlier) and suddenly (at $t = 0$) immerse the two ends in large well-stirred baths of ice water, 0°C. The mathematical problem is

PDE: $\qquad \dfrac{\partial u}{\partial t} = k \dfrac{\partial^2 u}{\partial x^2} \qquad\qquad t > 0, \qquad 0 < x < L \qquad$ (2.3.35a)

BC: $\qquad \begin{aligned} u(0, t) &= 0 \\ u(L, t) &= 0 \end{aligned} \qquad\qquad\qquad t > 0 \qquad\qquad\qquad$ (2.3.35b)

IC: $\qquad u(x, 0) = 100 \qquad\qquad 0 < x < L.$ $\qquad\qquad$ (2.3.35c)

According to (2.3.29) and (2.3.34), the solution is

$$u(x, t) = \sum_{n=1}^{\infty} B_n \sin \frac{n\pi x}{L} e^{-k(n\pi/L)^2 t}, \qquad (2.3.36)$$

where

$$B_n = \frac{2}{L} \int_0^L f(x) \sin \frac{n\pi x}{L} \, dx \qquad (2.3.37)$$

and $f(x) = 100$. Recall that the coefficient B_n was determined by having (2.3.36) satisfy the initial condition,

$$f(x) = \sum_{n=1}^{\infty} B_n \sin \frac{n\pi x}{L}. \tag{2.3.38}$$

We calculate the coefficients B_n from (2.3.37):

$$B_n = \frac{2}{L} \int_0^L 100 \sin \frac{n\pi x}{L} \, dx = \frac{200}{L} \left(-\frac{L}{n\pi} \cos \frac{n\pi x}{L} \right) \Big|_0^L$$

$$= \frac{200}{n\pi} (1 - \cos n\pi) = \begin{cases} 0 & n \text{ even} \\ \dfrac{400}{n\pi} & n \text{ odd} \end{cases} \tag{2.3.39}$$

since $\cos n\pi = (-1)^n$ which equals 1 for n even and -1 for n odd. The series (2.3.38) will be studied further in Chapter 3. In particular, we must explain the intriguing situation that the initial temperature equals 100 everywhere, but the series (2.3.38) equals 0 at $x = 0$ and $x = L$ (due to the boundary conditions).

Approximations to the initial value problem. We have now obtained the solution to the initial value problem (2.3.35) for the heat equation with zero boundary conditions ($x = 0$ and $x = L$) and initial temperature distribution equaling 100. The solution is (2.3.36), with B_n given by (2.3.39). The solution is quite complicated, involving an infinite series. What can we say about it? First, we notice that $\lim_{t \to \infty} u(x, t) = 0$. The temperature distribution approaches a steady state, $u(x, t) = 0$. This is not surprising physically since both ends are at $0°$; we expect all the initial heat energy contained in the rod to flow out the ends. The equilibrium problem, $d^2u/dx^2 = 0$ with $u(0) = 0$ and $u(L) = 0$, has a unique solution, $u \equiv 0$, agreeing with the limit as t tends to infinity of the time-dependent problem.

One question of importance that we can answer is the manner in which the solution approaches steady state. If t is large, what is the approximate temperature distribution, and how does it differ from the steady state $0°$? We note that each term in (2.3.36) decays at a different rate. The more oscillations in space, the faster the decay. If t is such that $kt(\pi/L)^2$ is large, then each succeeding term is much smaller than the first. We can then approximate the infinite series by only the first term:

$$u(x, t) \approx \frac{400}{\pi} \sin \frac{\pi x}{L} e^{-k(\pi/L)^2 t}. \tag{2.3.40}$$

The larger t is, the better this is as an approximation. Even if $kt(\pi/L)^2 = \frac{1}{2}$, this is not a bad approximation since

$$\frac{e^{-k(3\pi/L)^2 t}}{e^{-k(\pi/L)^2 t}} = e^{-8(\pi/L)^2 kt} = e^{-4} = 0.018 \ldots .$$

Thus, if $kt(\pi/L)^2 \geq \frac{1}{2}$, we can use the simple approximation. We see that for these times the spatial dependence of the temperature is just the simple rise and fall of $\sin \pi x/L$, as illustrated in Fig. 2.3.4. The peak amplitude, occurring in

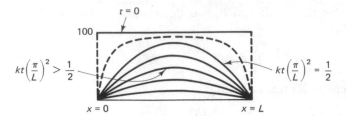

Figure 2.3.4 Decay of the temperature distribution.

the middle $x = L/2$, decays exponentially in time. For $kt(\pi/L)^2$ less than $\frac{1}{2}$, the spatial dependence cannot be approximated by one simple sinusoidal function; more terms are necessary in the series. The solution can be easily computed, using a finite number of terms. In some cases many terms may be necessary, and there would be better ways to calculate $u(x, t)$.

2.3.8 Summary

Let us summarize the method of separation of variables as it appears for the one example:

$$\text{PDE:} \qquad \frac{\partial u}{\partial t} = k \frac{\partial^2 u}{\partial x^2}$$

$$\text{BC:} \qquad \begin{aligned} u(0, t) &= 0 \\ u(L, t) &= 0 \end{aligned}$$

$$\text{IC:} \qquad u(x, 0) = f(x).$$

1. Make sure that you have a linear and homogeneous PDE with linear and homogeneous BC.
2. Temporarily ignore the nonzero IC.
3. Separate variables (determine differential equations implied by the assumption of product solutions) and introduce a separation constant.
4. Determine separation constants as the eigenvalues of a boundary value problem.
5. Solve other differential equations. Record all product solutions of the PDE obtainable by this method.
6. Apply the principle of superposition (form a linear combination of all product solutions).
7. Attempt to satisfy the initial condition.
8. Determine coefficients using the orthogonality of the eigenfunctions.

These steps should be *understood*, not memorized. It is important to note that:

1. The principle of superposition applies to solutions of the PDE (do not add up solutions of various different ordinary differential equations).
2. Do not apply the initial condition $u(x, 0) = f(x)$ until *after* the principle of superposition.

EXERCISES 2.3

2.3.1. For the following partial differential equations, what ordinary differential equations are implied by the method of separation of variables?

*(a) $\dfrac{\partial u}{\partial t} = \dfrac{k}{r}\dfrac{\partial}{\partial r}\left(r\dfrac{\partial u}{\partial r}\right)$

(b) $\dfrac{\partial u}{\partial t} = k\dfrac{\partial^2 u}{\partial x^2} - v_0\dfrac{\partial u}{\partial x}$

*(c) $\dfrac{\partial^2 u}{\partial x^2} + \dfrac{\partial^2 u}{\partial y^2} = 0$

(d) $\dfrac{\partial u}{\partial t} = \dfrac{k}{r^2}\dfrac{\partial}{\partial r}\left(r^2\dfrac{\partial u}{\partial r}\right)$

*(e) $\dfrac{\partial u}{\partial t} = k\dfrac{\partial^4 u}{\partial x^4}$

*(f) $\dfrac{\partial^2 u}{\partial t^2} = c^2\dfrac{\partial^2 u}{\partial x^2}$

2.3.2. Consider the differential equation

$$\frac{d^2\phi}{dx^2} + \lambda\phi = 0.$$

Determine the eigenvalues λ (and corresponding eigenfunctions) if ϕ satisfies the following boundary conditions. Analyze three cases ($\lambda > 0, \lambda = 0, \lambda < 0$). You may assume that the eigenvalues are real.

(a) $\phi(0) = 0$ and $\phi(\pi) = 0$

*(b) $\phi(0) = 0$ and $\phi(1) = 0$

(c) $\dfrac{d\phi}{dx}(0) = 0$ and $\dfrac{d\phi}{dx}(L) = 0$ (If necessary, see Sec. 2.4.1.)

*(d) $\phi(0) = 0$ and $\dfrac{d\phi}{dx}(L) = 0$

(e) $\dfrac{d\phi}{dx}(0) = 0$ and $\phi(L) - 0$

*(f) $\phi(a) = 0$ and $\phi(b) = 0$ (You may assume that $\lambda > 0$.)

(g) $\phi(0) = 0$ and $\dfrac{d\phi}{dx}(L) + \phi(L) = 0$ (If necessary, see Sec. 5.8.)

2.3.3. Consider the heat equation

$$\frac{\partial u}{\partial t} = k\frac{\partial^2 u}{\partial x^2},$$

subject to the boundary conditions

$$u(0, t) = 0$$
$$u(L, t) = 0.$$

Solve the initial value problem if the temperature is initially

(a) $u(x, 0) = 6\sin\dfrac{9\pi x}{L}$

(b) $u(x, 0) = 3\sin\dfrac{\pi x}{L} - \sin\dfrac{3\pi x}{L}$

*(c) $u(x, 0) = 2\cos\dfrac{3\pi x}{L}$

(d) $u(x, 0) = \begin{cases} 1 & 0 < x \leqslant L/2 \\ 2 & L/2 < x < L \end{cases}$

(Your answer in part (c) may involve certain integrals that do not need to be evaluated.)

2.3.4. Consider

$$\frac{\partial u}{\partial t} = k\frac{\partial^2 u}{\partial x^2}$$

subject to $u(0, t) = 0$, $u(L, t) = 0$, and $u(x, 0) = f(x)$.

*(a) What is the total heat energy in the rod as a function of time?

(b) What is the flow of heat energy out of the rod at $x = 0$? at $x = L$?

*(c) What relationship should exist between parts (a) and (b)?

2.3.5. Evaluate (be careful if $n = m$)

$$\int_0^L \sin \frac{n\pi x}{L} \sin \frac{m\pi x}{L} \, dx \qquad \text{for } n > 0, m > 0.$$

Use the trigonometric identity

$$\sin a \sin b = \tfrac{1}{2}[\cos (a - b) - \cos (a + b)].$$

*2.3.6. Evaluate

$$\int_0^L \cos \frac{n\pi x}{L} \cos \frac{m\pi x}{L} \, dx \qquad \text{for } n \geq 0, m \geq 0.$$

Use the trigonometric identity

$$\cos a \cos b = \tfrac{1}{2}[\cos (a + b) + \cos (a - b)].$$

(Be careful if $a - b = 0$ or $a + b = 0$.)

2.3.7. Consider the following boundary value problem (if necessary, see Sec. 2.4.1):

$$\frac{\partial u}{\partial t} = k \frac{\partial^2 u}{\partial x^2} \quad \text{with} \quad \frac{\partial u}{\partial x}(0, t) = 0, \quad \frac{\partial u}{\partial x}(L, t) = 0, \quad \text{and } u(x, 0) = f(x)$$

(a) Give a one-sentence physical interpretation of this problem.

(b) Solve by the method of separation of variables. First show that there are no separated solutions which exponentially grow in time. [*Hint:* The answer is

$$u(x, t) = A_0 + \sum_{n=1}^{\infty} A_n e^{-\lambda_n kt} \cos \frac{n\pi x}{L}.$$

What is λ_n?

(c) Show that the initial condition, $u(x, 0) = f(x)$, is satisfied if

$$f(x) = A_0 + \sum_{n=1}^{\infty} A_n \cos \frac{n\pi x}{L}.$$

(d) Using Exercise 2.3.6, solve for A_0 and A_n ($n \geq 1$).

(e) What happens to the temperature distribution as $t \to \infty$? Show that it approaches the steady-state temperature distribution (see Sec. 1.4).

*2.3.8. Consider

$$\frac{\partial u}{\partial t} = k \frac{\partial^2 u}{\partial x^2} - \alpha u.$$

This corresponds to a one-dimensional rod either with heat loss through the lateral sides with outside temperature $0°$ ($\alpha > 0$, see Exercise 1.2.4) or with insulated lateral sides with a heat source proportional to the temperature. Suppose that the boundary conditions are

$$u(0, t) = 0 \quad \text{and} \quad u(L, t) = 0.$$

(a) What are the possible equilibrium temperature distributions if $\alpha > 0$?

(b) Solve the time-dependent problem [$u(x, 0) = f(x)$] if $\alpha > 0$. Analyze the temperature for large time ($t \to \infty$) and compare to part (a).

*2.3.9. Redo Exercise 2.3.8 if $\alpha < 0$. [Be especially careful if $-\alpha/k = (n\pi/L)^2$.]

2.3.10. For two- and three-dimensional vectors, a fundamental property of dot products, $\mathbf{A} \cdot \mathbf{B} = |\mathbf{A}||\mathbf{B}| \cos \theta$, implies that

$$|\mathbf{A} \cdot \mathbf{B}| \leq |\mathbf{A}||\mathbf{B}|. \tag{2.3.41}$$

In this exercise we generalize this to n-dimensional vectors and functions, in which case (2.3.41) is known as **Schwarz's inequality**. [The names of Cauchy and Buniakovsky are also associated with (2.3.41).]

(a) Show that $|\mathbf{A} - \gamma\mathbf{B}|^2 > 0$ implies (2.3.41), where $\gamma = \mathbf{A} \cdot \mathbf{B}/\mathbf{B} \cdot \mathbf{B}$.

(b) Express the inequality using both

$$\mathbf{A} \cdot \mathbf{B} = \sum_{n=1}^{\infty} a_n b_n = \sum_{n=1}^{\infty} a_n c_n \frac{b_n}{c_n}.$$

*(c) Generalize (2.3.41) to functions. [*Hint:* Let $\mathbf{A} \cdot \mathbf{B}$ mean $\int_0^L A(x)B(x)\, dx$.]

2.3.11. Solve Laplace's equation inside a rectangle:

$$\nabla^2 u = \frac{\partial^2 u}{\partial x^2} + \frac{\partial^2 u}{\partial y^2} = 0$$

subject to the boundary conditions

$$\begin{aligned} u(0, y) &= g(y) & u(x, 0) &= 0 \\ u(L, y) &= 0 & u(x, H) &= 0. \end{aligned}$$

(*Hint:* If necessary, see Sec. 2.5.1.)

2.3 APPENDIX: ORTHOGONALITY OF FUNCTIONS

Two vectors $\mathbf{A}$ and $\mathbf{B}$ are orthogonal if $\mathbf{A} \cdot \mathbf{B} = 0$. In component form: $\mathbf{A} = a_1\hat{\mathbf{i}} + a_2\hat{\mathbf{j}} + a_3\hat{\mathbf{k}}$ and $\mathbf{B} = b_1\hat{\mathbf{i}} + b_2\hat{\mathbf{j}} + b_3\hat{\mathbf{k}}$; $\mathbf{A}$ and $\mathbf{B}$ are orthogonal if $\sum_i a_i b_i = 0$. A function $A(x)$ can be thought of as a vector. If only three values of x are important, x_1, x_2, and x_3, then the components of the function $A(x)$ (thought of as a vector) are $A(x_1) \equiv a_1$, $A(x_2) \equiv a_2$, and $A(x_3) \equiv a_3$. The function $A(x)$ is orthogonal to the function $B(x)$ (by definition) if $\sum_i a_i b_i = 0$. However, in our problems, all values of x between 0 and L are important. *The function $A(x)$ can be thought of as an infinite-dimensional vector*, whose components are $A(x_i)$ for all x_i on some interval. In this manner the function $A(x)$ would be said to be orthogonal to $B(x)$ if $\sum_i A(x_i)B(x_i) = 0$, where the summation was to include all points between 0 and L. It is thus natural to *define* the function $A(x)$ to be orthogonal to $B(x)$ if $\int_0^L A(x)B(x)\, dx = 0$. The integral replaces the vector dot product; both are examples of "inner products."

In vectors, we have the three mutually perpendicular (orthogonal) unit vectors $\hat{\mathbf{i}}$, $\hat{\mathbf{j}}$, and $\hat{\mathbf{k}}$ known as the standard basis vectors. In component form

$$\mathbf{A} = a_1\hat{\mathbf{i}} + a_2\hat{\mathbf{j}} + a_3\hat{\mathbf{k}}.$$

a_1 is the projection of $\mathbf{A}$ in the $\hat{\mathbf{i}}$ direction, and so on. Sometimes we wish to represent $\mathbf{A}$ in terms of other mutually orthogonal vectors (which may not be unit vectors) $\mathbf{u}$, $\mathbf{v}$, and $\mathbf{w}$, called an *orthogonal set of vectors*. Then

$$\mathbf{A} = \alpha_u\mathbf{u} + \alpha_v\mathbf{v} + \alpha_w\mathbf{w}. \tag{2.3.A1}$$

To determine the coordinates α_u, α_v, α_w with respect to this orthogonal set, $\mathbf{u}$, $\mathbf{v}$, $\mathbf{w}$, we can form certain dot products. For example,

$$\mathbf{A} \cdot \mathbf{u} = \alpha_u\mathbf{u} \cdot \mathbf{u} + \alpha_v\mathbf{v} \cdot \mathbf{u} + \alpha_w\mathbf{w} \cdot \mathbf{u}.$$

Note that $\mathbf{v} \cdot \mathbf{u} = 0$ and $\mathbf{w} \cdot \mathbf{u} = 0$, since we assumed that this new set was mutually orthogonal. Thus, we can easily solve for the coordinate α_u of $\mathbf{A}$ in the $\mathbf{u}$ direction,

$$\alpha_u = \frac{\mathbf{A} \cdot \mathbf{u}}{\mathbf{u} \cdot \mathbf{u}}. \tag{2.3.A2}$$

($\alpha_u \mathbf{u}$ is the vector projection of $\mathbf{A}$ in the $\mathbf{u}$ direction.)

For functions, we can do a similar thing. If $f(x)$ can be represented by a linear combination of the orthogonal set, $\sin n\pi x/L$, then

$$f(x) = \sum_{n=1}^{\infty} B_n \sin \frac{n\pi x}{L},$$

where the B_n may be interpreted as the coordinates of $f(x)$ with respect to the "direction" (or basis vector) $\sin n\pi x/L$. To determine these coordinates we take the inner product with an arbitrary basis function (vector) $\sin m\pi x/L$, where the inner product of two functions is the integral of their product. Thus, as before,

$$\int_0^L f(x) \sin \frac{m\pi x}{L} \, dx = \sum_{n=1}^{\infty} B_n \int_0^L \sin \frac{n\pi x}{L} \sin \frac{m\pi x}{L} \, dx.$$

Since $\sin m\pi x/L$ is an orthogonal set of functions, $\int_0^L \sin n\pi x/L \sin m\pi x/L \, dx = 0$ for $n \neq m$. Hence, we solve for the coordinate (coefficient) B_n:

$$B_n = \frac{\displaystyle\int_0^L f(x) \sin n\pi x/L \, dx}{\displaystyle\int_0^L \sin^2 n\pi x/L \, dx}. \tag{2.3.A3}$$

This is seen to be the same idea as the projection formula (2.3.A2). Our standard formula (2.3.31b), $\int_0^L \sin^2 n\pi x/L \, dx = L/2$, returns (2.3.A3) to the more familiar form,

$$B_n = \frac{2}{L} \int_0^L f(x) \sin \frac{n\pi x}{L} \, dx. \tag{2.3.A4}$$

Both formulas (2.3.A2) and (2.3.A3) are divided by something. In (2.3.A2) it is $\mathbf{u} \cdot \mathbf{u}$, or the length of the vector $\mathbf{u}$ squared. Thus, $\int_0^L \sin^2 n\pi x/L \, dx$ may be thought of as the length squared of $\sin n\pi x/L$ (although here length means nothing other than the square root of the integral). In this manner the length squared of the function $\sin n\pi x/L$ is $L/2$, which is an explanation of the appearance of the term $2/L$ in (2.3.A4).

2.4 WORKED EXAMPLES WITH THE HEAT EQUATION (OTHER BOUNDARY VALUE PROBLEMS)

2.4.1 Heat Conduction in a Rod with Insulated Ends

Let us work out in detail the solution (and its interpretation) of the following problem defined for $0 \leq x \leq L$ and $t \geq 0$:

PDE:	$$\frac{\partial u}{\partial t} = k\frac{\partial^2 u}{\partial x^2}$$	(2.4.1a)

BC:	$$\frac{\partial u}{\partial x}(0, t) = 0$$	(2.4.1b)
	$$\frac{\partial u}{\partial x}(L, t) = 0$$	(2.4.1c)

IC:	$u(x, 0) = f(x).$	(2.4.1d)

As a review, this is a heat conduction problem in a one-dimensional rod with constant thermal properties and no sources. This problem is quite similar to the problem treated in Sec. 2.3, the only difference being the boundary conditions. Here the ends are insulated, whereas in Sec. 2.3 the ends were fixed at 0°. Both the partial differential equation and the boundary conditions are linear and homogeneous. Consequently, we apply the method of separation of variables. We may follow the general procedure described in Sec. 2.3.8. The assumption of product solutions,

$$u(x, t) = \phi(x)G(t), \tag{2.4.2}$$

implies from the PDE as before that

$$\frac{dG}{dt} = \lambda k G \tag{2.4.3}$$

$$\frac{d^2\phi}{dx^2} = -\lambda\phi, \tag{2.4.4}$$

where λ is the separation constant. Again,

$$G(t) = ce^{-\lambda kt}. \tag{2.4.5}$$

The insulated boundary conditions, (2.4.1b) and (2.4.1c), imply that the separated solutions must satisfy $d\phi/dx(0) = 0$ and $d\phi/dx(L) = 0$. The separation constant λ is then determined by finding those λ for which nontrivial solutions exist for the following boundary value problem:

	$$\frac{d^2\phi}{dx^2} = -\lambda\phi$$	(2.4.6a)
	$$\frac{d\phi}{dx}(0) = 0$$	(2.4.6b)
	$$\frac{d\phi}{dx}(L) = 0.$$	(2.4.6c)

Although the ordinary differential equation for the boundary value problem is the same one as previously analyzed, the boundary conditions are different. We must repeat some of the analysis. Once again three cases should be discussed: $\lambda > 0$, $\lambda = 0$, $\lambda < 0$ (since we will assume the eigenvalues are real).

For $\lambda > 0$, the general solution of (2.4.6a) is again

$$\phi = c_1 \cos \sqrt{\lambda} \, x + c_2 \sin \sqrt{\lambda} \, x. \tag{2.4.7}$$

We need to calculate $d\phi/dx$ to satisfy the boundary conditions:

$$\frac{d\phi}{dx} = \sqrt{\lambda} \, (-c_1 \sin \sqrt{\lambda} \, x + c_2 \cos \sqrt{\lambda} \, x). \tag{2.4.8}$$

The boundary condition $d\phi/dx(0) = 0$ implies that $0 = c_2\sqrt{\lambda}$, and hence $c_2 = 0$, since $\lambda > 0$. Thus, $\phi = c_1 \cos \sqrt{\lambda} \, x$ and $d\phi/dx = -c_1\sqrt{\lambda} \sin \sqrt{\lambda} \, x$. The eigenvalues λ and their corresponding eigenfunctions are determined from the remaining boundary condition, $d\phi/dx(L) = 0$:

$$0 = -c_1\sqrt{\lambda} \sin \sqrt{\lambda} \, L.$$

As before, for nontrivial solutions, $c_1 \neq 0$, and hence $\sin \sqrt{\lambda} \, L = 0$. The eigenvalues for $\lambda > 0$ are the same as the previous problem, $\sqrt{\lambda} \, L = n\pi$ or

$$\lambda = \left(\frac{n\pi}{L}\right)^2, \qquad n = 1, 2, 3, \ldots, \tag{2.4.9}$$

but the corresponding eigenfunctions are cosines (not sines),

$$\phi(x) = c_1 \cos \frac{n\pi x}{L}, \qquad n = 1, 2, 3, \ldots. \tag{2.4.10}$$

The resulting product solutions of the PDE are

$$u(x, t) = A \cos \frac{n\pi x}{L} e^{-(n\pi/L)^2 kt}, \qquad n = 1, 2, 3, \ldots, \tag{2.4.11}$$

where A is an arbitrary multiplicative constant.

Before applying the principle of superposition, we must see if there are any other eigenvalues. If $\lambda = 0$, then

$$\phi = c_1 + c_2 x, \tag{2.4.12}$$

where c_1 and c_2 are arbitrary constants. The derivative of ϕ is

$$\frac{d\phi}{dx} = c_2.$$

Both boundary conditions, $d\phi/dx(0) = 0$ and $d\phi/dx(L) = 0$, give the same condition, $c_2 = 0$. Thus, there are nontrivial solutions of the boundary value problem for $\lambda = 0$, namely $\phi(x)$ equaling any constant,

$$\phi(x) = c_1. \tag{2.4.13}$$

The time-dependent part is also a constant, since $e^{-\lambda kt}$ for $\lambda = 0$ equals 1. Thus, another product solution of both the linear homogeneous PDE and BCs is $u(x, t) = A$, where A is any constant.

We do not expect there to be any eigenvalues for $\lambda < 0$, since in this case the time-dependent part grows exponentially. In addition, it seems unlikely that we would find a nontrivial linear combination of exponentials which would have a zero slope at both $x = 0$ and $x = L$. In Exercise 2.4.4 you are asked to show that there are no eigenvalues for $\lambda < 0$.

In order to satisfy the initial condition, we use the principle of superposition. We should take a linear combination of *all* product solutions of the PDE (not just those corresponding to $\lambda > 0$). Thus,

$$u(x, t) = A_0 + \sum_{n=1}^{\infty} A_n \cos \frac{n\pi x}{L} e^{-(n\pi/L)^2 kt}. \qquad (2.4.14)$$

It is interesting to note that this is equivalent to

$$u(x, t) = \sum_{n=0}^{\infty} A_n \cos \frac{n\pi x}{L} e^{-(n\pi/L)^2 kt}, \qquad (2.4.15)$$

since $\cos 0 = 1$ and $e^0 = 1$. In fact, (2.4.15) is often easier to use in practice. We prefer the form (2.4.14) *in the beginning stages of the learning process*, since it more clearly shows that the solution consists of terms arising from the analysis of two somewhat distinct cases, $\lambda = 0$ and $\lambda > 0$.

The initial condition $u(x, 0) = f(x)$ is satisfied if

$$f(x) = A_0 + \sum_{n=1}^{\infty} A_n \cos \frac{n\pi x}{L}, \qquad (2.4.16)$$

for $0 \le x \le L$. The validity of (2.4.16) will also follow from the theory of Fourier series. Let us note that in the previous problem $f(x)$ was represented by a series of sines. Here $f(x)$ consists of a series of cosines and the constant term. The two cases are different due to the different boundary conditions. To complete the solution we need to determine the arbitrary coefficients A_0 and A_n ($n \ge 1$). Fortunately, from integral tables it is known that $\cos n\pi x/L$ satisfies the following **orthogonality relation**:

$$\int_0^L \cos \frac{n\pi x}{L} \cos \frac{m\pi x}{L} \, dx = \begin{cases} 0 & n \ne m \\ \dfrac{L}{2} & n = m \ne 0 \\ L & n = m = 0 \end{cases} \qquad (2.4.17)$$

for n and m nonnegative integers. Note that $n = 0$ or $m = 0$ corresponds to a constant 1 contained in the integrand. The constant $L/2$ is another application of the statement that the average of the square of a sine or cosine function is $\frac{1}{2}$. The constant L in (2.4.17) is quite simple since for $n = m = 0$, (2.4.17) becomes $\int_0^L dx = L$. Equation (2.4.17) states that the **cosine functions** (including the constant function) **form an orthogonal set of functions**. We can use that idea, in the same way as before, to determine the coefficients. Multiplying (2.4.16) by cos $m\pi x/L$ and integrating from 0 to L yields

$$\int_0^L f(x) \cos \frac{m\pi x}{L}\, dx = \sum_{n=0}^{\infty} A_n \int_0^L \cos \frac{n\pi x}{L} \cos \frac{m\pi x}{L}\, dx.$$

This holds for all m, $m = 0, 1, 2, \ldots$. The case in which $m = 0$ corresponds just to integrating (2.4.16) directly. Using the orthogonality results, it follows that only the mth term in the infinite sum contributes,

$$\int_0^L f(x) \cos \frac{m\pi x}{L}\, dx = A_m \int_0^L \cos^2 \frac{m\pi x}{L}\, dx.$$

The factor $\int_0^L \cos^2 m\pi x/L\, dx$ has two different cases, $m = 0$ and $m \neq 0$. Solving for A_m yields

$$A_0 = \frac{1}{L} \int_0^L f(x)\, dx \tag{2.4.18a}$$

$(m \geq 1)$

$$A_m = \frac{2}{L} \int_0^L f(x) \cos \frac{m\pi x}{L}\, dx. \tag{2.4.18b}$$

The two different formulas is a somewhat annoying feature of this series of cosines. It is simply caused by the factors $L/2$ and L in (2.4.17).

There is a significant difference between the solutions of the PDE for $\lambda > 0$ and the solution for $\lambda = 0$. All the solutions for $\lambda > 0$ decay exponentially in time, whereas the solution for $\lambda = 0$ remains constant in time. Thus, as $t \to \infty$ the complicated infinite series solution (2.4.14) approaches steady state,

$$\lim_{t \to \infty} u(x, t) = A_0 = \frac{1}{L} \int_0^L f(x)\, dx.$$

Not only is the steady-state temperature constant, A_0, but we recognize the constant A_0 as the average of the initial temperature distribution. This agrees with information obtained previously. Recall from Sec. 1.4 that the equilibrium temperature distribution for the problem with insulated boundaries is not unique. Any constant temperature is an equilibrium solution, but using the ideas of conservation of total thermal energy, we know that the constant must be the average of the initial temperature.

2.4.2 Heat Conduction in a Thin Circular Ring

We have investigated a heat flow problem whose eigenfunctions are sines and one whose eigenfunctions are cosines. In this subsection we illustrate a heat flow problem whose eigenfunctions are both sines *and* cosines.

Let us formulate the appropriate initial boundary value problem if a thin wire (with lateral sides insulated) is bent into the shape of a circle, as illustrated in Fig. 2.4.1. For reasons that will not be apparent for a while, we let the wire have length $2L$ (rather than L as for the two previous heat conduction problems).

Figure 2.4.1 Thin circular ring.

Since the circumference of a circle is $2\pi r$, the radius is $r = 2L/2\pi = L/\pi$. If the wire is thin enough, it is reasonable to assume that the temperature in the wire is constant along cross sections of the bent wire. In this situation the wire should satisfy a one-dimensional heat equation, where the distance is actually the arc length x along the wire:

$$\frac{\partial u}{\partial t} = k\frac{\partial^2 u}{\partial x^2}.$$

(2.4.19)

We have assumed that the wire has constant thermal properties and no sources. It is convenient in this problem to measure the arc length x, such that x ranges from $-L$ to $+L$ (instead of the more usual 0 to $2L$).

Let us assume that the wire is very tightly connected to itself at the ends ($x = -L$ to $x = +L$). The conditions of perfect thermal contact should hold there (see Exercise 1.3.2). The temperature $u(x, t)$ is continuous there,

$$u(-L, t) = u(L, t).$$

(2.4.20a)

Also, since the heat flux must be continuous there (and the thermal conductivity is constant everywhere), the derivative of the temperature is also continuous:

$$\frac{\partial u}{\partial x}(-L, t) = \frac{\partial u}{\partial x}(L, t).$$

(2.4.20b)

The two boundary conditions for the partial differential equation are (2.4.20a) and (2.4.20b). The initial condition is that the initial temperature is a given function of the position along the wire,

$$u(x, 0) = f(x).$$

(2.4.21)

The mathematical problem consists of the linear homogeneous PDE (2.4.19) subject to linear homogeneous BCs (2.4.20). As such, we will proceed in the usual way to apply the method of separation of variables. Product solutions $u(x, t) = \phi(x)G(t)$ for the heat equation have been obtained previously, where $G(t) = ce^{-\lambda kt}$. The corresponding boundary value problem is

$$\frac{d^2\phi}{dx^2} = -\lambda\phi \qquad (2.4.22a)$$

$$\phi(-L) = \phi(L) \qquad (2.4.22b)$$

$$\frac{d\phi}{dx}(-L) = \frac{d\phi}{dx}(L) \qquad (2.4.22c)$$

The boundary conditions (2.4.22b) and (2.4.22c) each involve both boundaries (sometimes called the **mixed** type). The specific boundary conditions (2.4.22b) and (2.4.22c) are referred to as **periodic boundary conditions** since although the problem can be thought of physically as being defined only for $-L < x < L$, it is often thought of as being defined periodically for all x; the temperature will be periodic ($x = x_0$ is the same physical point as $x = x_0 + 2L$, and hence must have the same temperature).

If $\lambda > 0$, the general solution of (2.4.22a) is again

$$\phi = c_1 \cos \sqrt{\lambda}\, x + c_2 \sin \sqrt{\lambda}\, x.$$

The boundary condition $\phi(-L) = \phi(L)$ implies that

$$c_1 \cos \sqrt{\lambda}\,(-L) + c_2 \sin \sqrt{\lambda}\,(-L) = c_1 \cos \sqrt{\lambda}\, L + c_2 \sin \sqrt{\lambda}\, L.$$

Since cosine is an even function, $\cos \sqrt{\lambda}\,(-L) = \cos \sqrt{\lambda}\, L$, and since sine is an odd function, $\sin \sqrt{\lambda}\,(-L) = -\sin \sqrt{\lambda}\, L$, it follows that $\phi(-L) = \phi(L)$ is satisfied only if

$$c_2 \sin \sqrt{\lambda}\, L = 0. \qquad (2.4.23)$$

Before solving (2.4.23), we analyze the second boundary condition, which involves the derivative,

$$\frac{d\phi}{dx} = \sqrt{\lambda}\,(-c_1 \sin \sqrt{\lambda}\, x + c_2 \cos \sqrt{\lambda}\, x).$$

Thus, $d\phi/dx(-L) = d\phi/dx(L)$ is satisfied only if

$$c_1\sqrt{\lambda} \sin \sqrt{\lambda}\, L = 0, \qquad (2.4.24)$$

where the evenness of cosines and the oddness of sines has again been used. Conditions (2.4.23) and (2.4.24) are easily solved. If $\sin \sqrt{\lambda}\, L \neq 0$, then $c_1 = 0$ and $c_2 = 0$, which is just the trivial solution. Thus, for nontrivial solutions,

$$\sin \sqrt{\lambda}\, L = 0,$$

which determines the eigenvalues λ. We find (as before) that $\sqrt{\lambda}\, L = n\pi$ or equivalently that

$$\lambda = \left(\frac{n\pi}{L}\right)^2, \; n = 1, 2, 3, \ldots. \qquad (2.4.25)$$

We chose the wire to have length $2L$ so that the eigenvalues have the same formula as before (this will mean less to remember, as all our problems have a similar answer). However, in this problem (unlike the others) there are no additional constraints that c_1 and c_2 must satisfy. Both are arbitrary. We say that both $\sin n\pi x/L$ and $\cos n\pi x/L$ are eigenfunctions corresponding to the eigenvalue $\lambda = (n\pi/L)^2$,

$$\phi(x) = \cos \frac{n\pi x}{L}, \quad \sin \frac{n\pi x}{L}, \quad n = 1, 2, 3, \ldots . \tag{2.4.26a}$$

In fact, any linear combination of $\cos n\pi x/L$ and $\sin n\pi x/L$ is an eigenfunction,

$$\phi(x) = c_1 \cos \frac{n\pi x}{L} + c_2 \sin \frac{n\pi x}{L}, \tag{2.4.26b}$$

but this is always to be understood when the statement is made that both are eigenfunctions. There are thus two infinite families of product solutions of the partial differential equation, $n = 1, 2, 3, \ldots ,$

$$u(x, t) = \cos \frac{n\pi x}{L} e^{-(n\pi/L)^2 kt} \quad \text{and} \quad u(x, t) = \sin \frac{n\pi x}{L} e^{-(n\pi/L)^2 kt}. \tag{2.4.27a, b}$$

All of these correspond to $\lambda > 0$.

If $\lambda = 0$, the general solution of (2.4.22a) is

$$\phi = c_1 + c_2 x.$$

The boundary condition $\phi(-L) = \phi(L)$ implies that

$$c_1 - c_2 L = c_1 + c_2 L.$$

Thus, $c_2 = 0$, $\phi(x) = c_1$ and $d\phi/dx = 0$. The remaining boundary condition, (2.4.22b), is automatically satisfied. We see that

$$\phi(x) = c_1,$$

any constant, is an eigenfunction, corresponding to the eigenvalue zero. Sometimes we say that $\phi(x) = 1$ is the eigenfunction, since it is known that any multiple of an eigenfunction is always an eigenfunction. Product solutions $u(x, t)$ are also constants in this case. Note that there is only one independent eigenfunction corresponding to $\lambda = 0$, while for each positive eigenvalue in this problem, $\lambda = (n\pi/L)^2$, there are two independent eigenfunctions, $\sin n\pi x/L$ and $\cos n\pi x/L$. Not surprisingly, it can be shown that there are no eigenvalues in which $\lambda < 0$.

The principle of superposition must be used before applying the initial condition. The most general solution obtainable by the method of separation of variables consists of an arbitrary linear combination of all product solutions:

$$\begin{aligned} u(x, t) = A_0 &+ \sum_{n=1}^{\infty} A_n \cos \frac{n\pi x}{L} e^{-(n\pi/L)^2 kt} \\ &+ \sum_{n=1}^{\infty} B_n \sin \frac{n\pi x}{L} e^{-(n\pi/L)^2 kt} \end{aligned} \tag{2.4.28}$$

The constant A_0 is the product solution corresponding to $\lambda = 0$, whereas two families of arbitrary coefficients, A_n and B_n, are needed for $\lambda > 0$. The initial condition $u(x, 0) = f(x)$ is satisfied if

$$f(x) = A_0 + \sum_{n=1}^{\infty} A_n \cos \frac{n\pi x}{L} + \sum_{n=1}^{\infty} B_n \sin \frac{n\pi x}{L}. \qquad (2.4.29)$$

Here the function $f(x)$ is a linear combination of both sines and cosines (and a constant), unlike the previous problems, where either sines or cosines (including the constant term) were used. Another crucial difference is that (2.4.29) should be valid for the entire ring, which means that $-L \leq x \leq L$, whereas the series of just sines or cosines was valid for $0 \leq x \leq L$. The theory of Fourier series will show that (2.4.29) is valid, and, more important, that the previous series of just sines or cosines are but special cases of the series in (2.4.29).

For now we wish just to determine the coefficients A_0, A_n, B_n from (2.4.29). Again the eigenfunctions form an orthogonal set since integral tables verify the following orthogonality conditions:

$$\int_{-L}^{L} \cos \frac{n\pi x}{L} \cos \frac{m\pi x}{L} \, dx = \begin{cases} 0 & n \neq m \\ L & n = m \neq 0 \\ 2L & n = m = 0 \end{cases} \qquad (2.4.30a)$$

$$\int_{-L}^{L} \sin \frac{n\pi x}{L} \sin \frac{m\pi x}{L} \, dx = \begin{cases} 0 & n \neq m \\ L & n = m \neq 0 \end{cases} \qquad (2.4.30b)$$

$$\int_{-L}^{L} \sin \frac{n\pi x}{L} \cos \frac{m\pi x}{L} \, dx = 0, \qquad (2.4.30c)$$

where n and m are arbitrary (nonnegative) integers. The constant eigenfunction corresponds to $n = 0$ or $m = 0$. Integrals of the square of sines or cosines ($n = m$) are evaluated again by the "half the length of the interval" rule. The last of these formulas, (2.4.30c), is particularly simple to derive, since sine is an odd function and cosine is an even function.* Note that, for example, $\cos n\pi x/L$ is orthogonal to every *other* eigenfunction [sines from (2.4.30c), cosines and the constant eigenfunction from (2.4.30a)].

The coefficients are derived in the same manner as before. A few steps are saved by noting (2.4.29) is equivalent to

$$f(x) = \sum_{n=0}^{\infty} A_n \cos \frac{n\pi x}{L} + \sum_{n=1}^{\infty} B_n \sin \frac{n\pi x}{L}.$$

If we multiply this by both $\cos m\pi x/L$ and $\sin m\pi x/L$ and then integrate *from* $x = -L$ to $x = +L$, we obtain

---

* The product of an odd and an even function is odd. By antisymmetry the integral of an odd function over a symmetric interval is zero.

$$\int_{-L}^{L} f(x) \left\{ \begin{array}{c} \cos \dfrac{m\pi x}{L} \\ \sin \dfrac{m\pi x}{L} \end{array} \right\} dx = \sum_{n=0}^{\infty} A_n \int_{-L}^{L} \cos \dfrac{n\pi x}{L} \left\{ \begin{array}{c} \cos \dfrac{m\pi x}{L} \\ \sin \dfrac{m\pi x}{L} \end{array} \right\} dx$$

$$+ \sum_{n=1}^{\infty} B_n \int_{-L}^{L} \sin \dfrac{n\pi x}{L} \left\{ \begin{array}{c} \cos \dfrac{m\pi x}{L} \\ \sin \dfrac{m\pi x}{L} \end{array} \right\} dx.$$

If we utilize (2.4.30), we find that

$$\int_{-L}^{L} f(x) \cos \frac{m\pi x}{L} \, dx = A_m \int_{-L}^{L} \cos^2 \frac{m\pi x}{L} \, dx$$

$$\int_{-L}^{L} f(x) \sin \frac{m\pi x}{L} \, dx = B_m \int_{-L}^{L} \sin^2 \frac{m\pi x}{L} \, dx.$$

Solving for the coefficients in a manner that we are now familiar with yields

$$(m \geq 1) \qquad \boxed{\begin{array}{l} A_0 = \dfrac{1}{2L} \displaystyle\int_{-L}^{L} f(x) \, dx \\[2ex] A_m = \dfrac{1}{L} \displaystyle\int_{-L}^{L} f(x) \cos \dfrac{m\pi x}{L} \, dx \\[2ex] B_m = \dfrac{1}{L} \displaystyle\int_{-L}^{L} f(x) \sin \dfrac{m\pi x}{L} \, dx. \end{array}} \qquad (2.4.31)$$

The solution to the problem is (2.4.28), where the coefficients are given by (2.4.31).

2.4.3 Summary of Boundary Value Problems

In many problems, including the ones we have just discussed, the specific simple constant-coefficient differential equation,

$$\frac{d^2\phi}{dx^2} = -\lambda\phi,$$

forms the fundamental part of the boundary value problem. Let us collect in one place the relevant formulas for the eigenvalues and eigenfunctions for the typical boundary conditions already discussed. You will find it helpful to understand these results because of their enormous applicability throughout this text:

BOUNDARY VALUE PROBLEMS FOR $\dfrac{d^2\phi}{dx^2} = -\lambda\phi$

Boundary conditions	$\phi(0) = 0$ $\phi(L) = 0$	$\dfrac{d\phi}{dx}(0) = 0$ $\dfrac{d\phi}{dx}(L) = 0$	$\phi(-L) = \phi(L)$ $\dfrac{d\phi}{dx}(-L) = \dfrac{d\phi}{dx}(L)$
Eigenvalues λ_n	$\left(\dfrac{n\pi}{L}\right)^2$ $n = 1, 2, 3, \ldots$	$\left(\dfrac{n\pi}{L}\right)^2$ $n = 0, 1, 2, 3, \ldots$	$\left(\dfrac{n\pi}{L}\right)^2$ $n = 0, 1, 2, 3, \ldots$
Eigenfunctions	$\sin\dfrac{n\pi x}{L}$	$\cos\dfrac{n\pi x}{L}$	$\sin\dfrac{n\pi x}{L}$ and $\cos\dfrac{n\pi x}{L}$
Series	$f(x) = \displaystyle\sum_{n=1}^{\infty} B_n \sin\dfrac{n\pi x}{L}$	$f(x) = \displaystyle\sum_{n=0}^{\infty} A_n \cos\dfrac{n\pi x}{L}$	$f(x) = \displaystyle\sum_{n=0}^{\infty} a_n \cos\dfrac{n\pi x}{L}$ $+ \displaystyle\sum_{n=1}^{\infty} b_n \sin\dfrac{n\pi x}{L}$
Coefficients	$B_n = \dfrac{2}{L}\displaystyle\int_0^L f(x) \sin\dfrac{n\pi x}{L}\,dx$	$A_0 = \dfrac{1}{L}\displaystyle\int_0^L f(x)\,dx$ $A_n = \dfrac{2}{L}\displaystyle\int_0^L f(x) \cos\dfrac{n\pi x}{L}\,dx$	$a_0 = \dfrac{1}{2L}\displaystyle\int_{-L}^L f(x)\,dx$ $a_n = \dfrac{1}{L}\displaystyle\int_{-L}^L f(x) \cos\dfrac{n\pi x}{L}\,dx$ $b_n = \dfrac{1}{L}\displaystyle\int_{-L}^L f(x) \sin\dfrac{n\pi x}{L}\,dx$

It is important to note that, in these cases, whenever $\lambda = 0$ is an eigenvalue, a constant is the eigenfunction (corresponding to $n = 0$ in $\cos n\pi x/L$).

EXERCISES 2.4

*2.4.1. Solve the heat equation $\partial u/\partial t = k\,\partial^2 u/\partial x^2$, $0 < x < L$, $t > 0$, subject to

$$\frac{\partial u}{\partial x}(0, t) = 0 \qquad t > 0$$

$$\frac{\partial u}{\partial x}(L, t) = 0 \qquad t > 0.$$

(a) $u(x, 0) = \begin{cases} 0 & x < L/2 \\ 1 & x > L/2 \end{cases}$ **(b)** $u(x, 0) = 6 + 4\cos\dfrac{3\pi x}{L}$

(c) $u(x, 0) = -2\sin\dfrac{\pi x}{L}$ **(d)** $u(x, 0) = -3\cos\dfrac{8\pi x}{L}$

*2.4.2. Solve

$$\frac{\partial u}{\partial t} = k \frac{\partial^2 u}{\partial x^2} \quad \text{with} \quad \frac{\partial u}{\partial x}(0, t) = 0$$

$$u(L, t) = 0$$

$$u(x, 0) = f(x).$$

For this problem you may assume that no solutions of the heat equation exponentially grow in time. You may also guess appropriate orthogonality conditions for the eigenfunctions.

*2.4.3. Solve the eigenvalue problem

$$\frac{d^2\phi}{dx^2} = -\lambda\phi$$

subject to

$$\phi(0) = \phi(2\pi) \quad \text{and} \quad \frac{d\phi}{dx}(0) = \frac{d\phi}{dx}(2\pi).$$

2.4.4. Explicitly show there are no negative eigenvalues for

$$\frac{d^2\phi}{dx^2} = -\lambda\phi \quad \text{subject to} \quad \frac{d\phi}{dx}(0) = 0 \quad \text{and} \quad \frac{d\phi}{dx}(L) = 0.$$

2.4.5. This problem presents an alternative derivation of the heat equation for a thin wire. The equation for a circular wire of finite thickness is the two-dimensional heat equation (in polar coordinates). Show that this reduces to (2.4.19) if the temperature does not depend on r and if the wire is very thin.

2.4.6. Determine the equilibrium temperature distribution for the thin circular ring of Section 2.4.2:
(a) Directly from the equilibrium problem (see Sec. 1.4).
(b) By computing the limit as $t \to \infty$ of the time-dependent problem.

2.4.7. Solve Laplace's equation inside a circle of radius a,

$$\nabla^2 u = \frac{1}{r}\frac{\partial}{\partial r}\left(r\frac{\partial u}{\partial r}\right) + \frac{1}{r^2}\frac{\partial^2 u}{\partial \theta^2} = 0,$$

subject to the boundary condition

$$u(a, \theta) = f(\theta).$$

(*Hint:* If necessary, see Sec. 2.5.2.)

2.5 LAPLACE'S EQUATION: SOLUTIONS AND QUALITATIVE PROPERTIES

2.5.1 Laplace's Equation Inside a Rectangle

In order to obtain more practice, we will consider a different kind of problem which can be analyzed by the method of separation of variables. We consider steady-state heat conduction in a two-dimensional region. To be specific, consider the equilibrium temperature inside a rectangle ($0 \leqslant x \leqslant L$, $0 \leqslant y \leqslant H$) when the temperature is a prescribed function of position (independent of time) on the boundary. The equilibrium temperature $u(x, y)$ satisfies Laplace's equation with the following boundary conditions:

PDE:
$$\frac{\partial^2 u}{\partial x^2} + \frac{\partial^2 u}{\partial y^2} = 0 \qquad \text{(2.5.1a)}$$

BC:
$$u(0, y) = g_1(y) \qquad \text{(2.5.1b)}$$
$$u(L, y) = g_2(y) \qquad \text{(2.5.1c)}$$
$$u(x, 0) = f_1(x) \qquad \text{(2.5.1d)}$$
$$u(x, H) = f_2(x), \qquad \text{(2.5.1e)}$$

where $f_1(x)$, $f_2(x)$, $g_1(y)$, and $g_2(y)$ are given functions of x and y, respectively. Here the partial differential equation is linear and homogeneous, but the boundary conditions, although linear, are not homogeneous. We will not be able to apply the method of separation of variables to this problem in its present form, because when we separate variables the boundary value problem (determining the separation constant) must have homogeneous boundary conditions. In this example all the boundary conditions are nonhomogeneous. We can get around this difficulty by noting that the original problem is nonhomogeneous due to the four nonhomogeneous boundary conditions. The idea behind the principle of superposition can be used sometimes for nonhomogeneous problems (see Exercise 2.2.4). We break our problem up into four problems each having one nonhomogeneous condition. We let

$$u(x, y) = u_1(x, y) + u_2(x, y) + u_3(x, y) + u_4(x, y), \qquad \text{(2.5.2)}$$

where each $u_i(x, y)$ satisfies Laplace's equation with one nonhomogeneous boundary condition and the related three homogeneous boundary conditions, as diagrammed in Fig. 2.5.1. Instead of directly solving for u, we will indicate how to solve for u_1, u_2, u_3, and u_4. Why does the sum satisfy our problem? We check to see that the PDE and the four nonhomogeneous BCs will be satisfied. Since

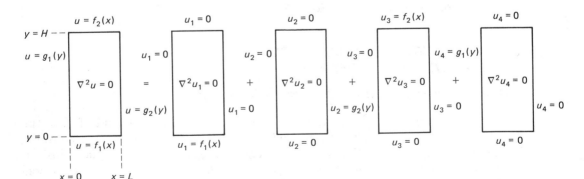

Figure 2.5.1 Laplace's equation inside a rectangle.

Method of Separation of Variables   Chap. 2

u_1, u_2, u_3, and u_4 satisfy Laplace's equation, which is linear and homogeneous, $u \equiv u_1 + u_2 + u_3 + u_4$ also satisfies the same linear and homogeneous PDE by the principle of superposition. At $x = 0$, $u_1 = 0$, $u_2 = 0$, $u_3 = 0$, and $u_4 = g_1(y)$. Therefore, at $x = 0$, $u = u_1 + u_2 + u_3 + u_4 = g_1(y)$, the desired nonhomogeneous condition. In a similar manner we can check that all four nonhomogeneous conditions have been satisfied.

The method to solve for any of the $u_i(x, y)$ is the same; only certain details differ. We will only solve for $u_4(x, y)$, and leave the rest for the exercises:

PDE:

$$\frac{\partial^2 u_4}{\partial x^2} + \frac{\partial^2 u_4}{\partial y^2} = 0 \qquad (2.5.3a)$$

BC:

$$u_4(0, y) = g_1(y) \qquad (2.5.3b)$$
$$u_4(L, y) = 0 \qquad (2.5.3c)$$
$$u_4(x, 0) = 0 \qquad (2.5.3d)$$
$$u_4(x, H) = 0. \qquad (2.5.3e)$$

We propose to solve this problem by the method of separation of variables. We begin by ignoring the nonhomogeneous condition $u_4(0, y) = g_1(y)$. Eventually, we will add together product solutions to synthesize $g_1(y)$. We look for product solutions

$$u_4(x, y) = h(x)\phi(y). \qquad (2.5.4)$$

From the three homogeneous boundary conditions, we see that

$$h(L) = 0 \qquad (2.5.5a)$$
$$\phi(0) = 0 \qquad (2.5.5b)$$
$$\phi(H) = 0. \qquad (2.5.5c)$$

Thus, the y-dependent solution $\phi(y)$ has two homogeneous boundary conditions, whereas the x-dependent solution $h(x)$ has only one. If (2.5.4) is substituted into Laplace's equation, we obtain

$$\phi(y)\frac{d^2 h}{dx^2} + h(x)\frac{d^2 \phi}{dy^2} = 0.$$

The variables can be separated by dividing by $h(x)\phi(y)$, so that

$$\frac{1}{h}\frac{d^2 h}{dx^2} = -\frac{1}{\phi}\frac{d^2 \phi}{dy^2}. \qquad (2.5.6)$$

The left-hand side is only a function of x, while the right-hand side is only a function of y. Both must equal a separation constant. Do we want to use $-\lambda$ or λ? One will be more convenient. If the separation constant is negative (as it was before), (2.5.6) implies that $h(x)$ oscillates and $\phi(y)$ is composed of exponentials. This seems doubtful, since the homogeneous boundary conditions (2.5.5) show that the y-dependent solution satisfies two homogeneous conditions;

$\phi(y)$ must be zero at $y = 0$ and at $y = H$. Exponentials in y are not expected to work. On the other hand, if the separation constant is positive, (2.5.6) implies that $h(x)$ is composed of exponentials and $\phi(y)$ oscillates. This seems more reasonable, and we thus introduce the separation constant λ (but we do *not* assume $\lambda \geqslant 0$):

$$\frac{1}{h}\frac{d^2h}{dx^2} = -\frac{1}{\phi}\frac{d^2\phi}{dy^2} = \lambda. \qquad (2.5.7)$$

This results in two ordinary differential equations:

$$\frac{d^2h}{dx^2} = \lambda h$$

$$\frac{d^2\phi}{dy^2} = -\lambda\phi.$$

The x-dependent problem is *not* a boundary value problem, since it does not have two homogeneous boundary conditions,

$$\frac{d^2h}{dx^2} = \lambda h \qquad (2.5.8a)$$

$$h(L) = 0. \qquad (2.5.8b)$$

However, the y-dependent problem is a boundary value problem and will be used to determine the eigenvalues λ (separation constants):

$$\frac{d^2\phi}{dy^2} = -\lambda\phi \qquad (2.5.9a)$$

$$\phi(0) = 0 \qquad (2.5.9b)$$
$$\phi(H) = 0. \qquad (2.5.9c)$$

This boundary value problem is one that has arisen before, but here the length of the interval is H. All the eigenvalues are positive, $\lambda > 0$. The eigenfunctions are clearly sines, since $\phi(0) = 0$. Furthermore, the condition $\phi(H) = 0$ implies that

$$\lambda = \left(\frac{n\pi}{H}\right)^2$$
$$\phi(y) = \sin\frac{n\pi y}{H} \qquad n = 1, 2, 3, \ldots. \qquad (2.5.10)$$

To obtain product solutions we now must solve (2.5.8). Since $\lambda = (n\pi/H)^2$,

$$\frac{d^2h}{dx^2} = \left(\frac{n\pi}{H}\right)^2 h. \qquad (2.5.11)$$

The general solution is a linear combination of exponentials or a linear combination of hyperbolic functions. Either can be used, but neither is particularly suited for solving the homogeneous boundary condition $h(L) = 0$. We can obtain our solution more expeditiously, if we note that both $\cosh n\pi(x - L)/H$ and $\sinh n\pi(x - L)/H$ are linearly independent solutions of (2.5.11). The general solution can be written as a linear combination of these two:

$$h(x) = a_1 \cosh \frac{n\pi}{H}(x - L) + a_2 \sinh \frac{n\pi}{H}(x - L), \qquad (2.5.12)$$

although it should now be clear that $h(L) = 0$ implies that $a_1 = 0$ (since $\cosh 0 = 1$ and $\sinh 0 = 0$). As we could have guessed originally,

$$h(x) = a_2 \sinh \frac{n\pi}{H}(x - L). \qquad (2.5.13)$$

The reason (2.5.12) is the solution (besides the fact that it solves the DE) is that it is a simple translation of the more familiar solution, $\cosh n\pi x/L$ and $\sinh n\pi x/L$. We are allowed to translate solutions of differential equations only if the differential equation does not change (said to be *invariant*) upon translation. Since (2.5.11) has constant coefficients, thinking of the origin being at $x = L$ (namely $x' = x - L$) does not affect the differential equation, since $d^2h/dx'^2 = (n\pi/H)^2 h$ according to the chain rule. For example, $\cosh n\pi x'/H = \cosh n\pi(x - L)/H$ is a solution.

Product solutions are

$$u_4(x, y) = A \sin \frac{n\pi y}{H} \sinh \frac{n\pi}{H}(x - L). \qquad (2.5.14)$$

You might now check that Laplace's equation is satisfied as well as the three required homogeneous conditions. It is interesting to note that one part (the y) oscillates and the other (the x) does not. This is a general property of Laplace's equation, not restricted to this geometry (rectangle) or to these boundary conditions.

We want to use these product solutions to satisfy the remaining condition, the nonhomogeneous boundary condition $u_4(0, y) = g_1(y)$. Product solutions do *not* satisfy nonhomogeneous conditions. Instead, we again use the principle of superposition. If (2.5.14) is a solution, so is

$$u_4(x, y) = \sum_{n=1}^{\infty} A_n \sin \frac{n\pi y}{H} \sinh \frac{n\pi}{H}(x - L). \qquad (2.5.15)$$

Evaluating at $x = 0$ will determine the coefficients A_n from the nonhomogeneous boundary condition:

$$g_1(y) = \sum_{n=1}^{\infty} A_n \sin \frac{n\pi y}{H} \sinh \frac{n\pi}{H}(-L).$$

This is the same kind of series of sine functions we have already briefly discussed, if we associate $A_n \sinh n\pi(-L)/H$ as its coefficients. Thus (by the orthogonality of $\sin n\pi y/H$ for y between 0 and H),

$$A_n \sinh \frac{n\pi}{H}(-L) = \frac{2}{H} \int_0^H g_1(y) \sin \frac{n\pi y}{H} \, dy.$$

Since $\sinh n\pi(-L)/H$ is never zero, we can divide by it and obtain finally a formula for the coefficients:

$$A_n = \frac{2}{H \sinh n\pi(-L)/H} \int_0^H g_1(y) \sin \frac{n\pi y}{H} \, dy. \qquad (2.5.16)$$

Equation (2.5.15) with coefficients determined by (2.5.16) is only the solution for $u_4(x, y)$. The original $u(x, y)$ is obtained by adding together four such solutions.

2.5.2 Laplace's Equation for a Circular Disk

Suppose that we had a thin circular disk of radius a (with constant thermal properties and no sources) with the temperature prescribed on the boundary as illustrated in Fig. 2.5.2. If the temperature on the boundary is independent of time, then it is reasonable to determine the equilibrium temperature distribution. The temperature satisfies Laplace's equation, $\nabla^2 u = 0$. The geometry of this problem suggests that we use polar coordinates, so that $u = u(r, \theta)$. In particular, on the circle $r = a$ the temperature distribution is a prescribed function of θ, $u(a, \theta) = f(\theta)$. The problem we want to solve is

PDE:
$$\nabla^2 u = \frac{1}{r}\frac{\partial}{\partial r}\left(r\frac{\partial u}{\partial r}\right) + \frac{1}{r^2}\frac{\partial^2 u}{\partial \theta^2} = 0 \qquad (2.5.17a)$$

BC:
$$u(a, \theta) = f(\theta). \qquad (2.5.17b)$$

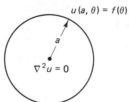

$u(a, \theta) = f(\theta)$

a

$\nabla^2 u = 0$

Figure 2.5.2 Laplace's equation inside a circular disk.

At first glance it would appear that we cannot use separation of variables because there are no homogeneous subsidiary conditions. However, the introduction of polar coordinates requires some discussion that will illuminate the use of the method of separation of variables. If we solve Laplace's equation on a rectangle (see Sec. 2.5.1), $0 \leq x \leq L$, $0 \leq y \leq H$, then conditions are necessary at the endpoints of definition of the variables, $x = 0, L$ and $y = 0, H$. Fortunately, these coincide with the physical boundaries. However, for polar coordinates, $0 \leq r \leq a$ and $-\pi \leq \theta \leq \pi$ (where there is some freedom in our definition of the angle θ). Mathematically, we need conditions at the endpoints of the coordinate system, $r = 0, a$ and $\theta = -\pi, \pi$. Here, only $r = a$ corresponds to a physical boundary. Thus, we need conditions motivated by considerations of the physical problem at $r = 0$ and at $\theta = \pm\pi$. Polar coordinates are singular at $r = 0$; for physical reasons we will prescribe that the temperature is finite or, equivalently, **bounded** there:

$$\text{boundedness} \atop \text{at origin} \qquad \boxed{|u(0, \theta)| < \infty.} \qquad (2.5.18)$$

Conditions are needed at $\theta = \pm\pi$ for mathematical reasons. It is similar to the circular wire situation. $\theta = -\pi$ corresponds to the same points as $\theta = \pi$. Although there really is not a boundary, we say that the temperature is continuous there and the heat flow in the θ-direction is continuous, which imply:

$$\text{periodicity} \qquad \boxed{\begin{array}{l} u(r, -\pi) = u(r, \pi) \\[2mm] \dfrac{\partial u}{\partial \theta}(r, -\pi) = \dfrac{\partial u}{\partial \theta}(r, \pi), \end{array}} \qquad \begin{array}{l}(2.5.19a) \\[4mm] (2.5.19b)\end{array}$$

as though the two regions were in perfect thermal contact there (see Exercise 1.3.2). Equations (2.5.19a) and (2.5.19b) are called **periodicity conditions**; they are equivalent to $u(r, \theta) = u(r, \theta + 2\pi)$. We note that subsidiary conditions (2.5.18) and (2.5.19) are all linear and homogeneous (it's easy to check that $u \equiv 0$ satisfies these three conditions). In this form the mathematical problem appears somewhat similar to Laplace's equation inside a rectangle. There are four conditions. Here, fortunately, only one is nonhomogeneous, $u(a, \theta) = f(\theta)$. This problem is thus suited for the method of separation of variables.

We look for special product solutions,

$$u(r, \theta) = \phi(\theta)G(r), \qquad (2.5.20)$$

which satisfy the PDE (2.5.17a) and the three homogeneous conditions (2.5.18) and (2.5.19). Note that (2.5.20) does *not* satisfy the nonhomogeneous boundary condition (2.5.17b). Substituting (2.5.20) into the periodicity conditions shows that

$$\phi(-\pi) = \phi(\pi)$$
$$\frac{d\phi}{d\theta}(-\pi) = \frac{d\phi}{d\theta}(\pi); \qquad (2.5.21)$$

the θ-dependent part also satisfies the **periodic boundary conditions**. The product form will satisfy Laplace's equation if

$$\frac{1}{r}\frac{d}{dr}\left(r\frac{dG}{dr}\right)\phi(\theta) + \frac{1}{r^2}G(r)\frac{d^2\phi}{d\theta^2} = 0.$$

The variables are *not* separated by dividing by $G(r)\phi(\theta)$ since $1/r^2$ remains multiplying the θ-dependent terms. Instead, divide by $(1/r^2)G(r)\phi(\theta)$, in which case,

$$\frac{r}{G}\frac{d}{dr}\left(r\frac{dG}{dr}\right) = -\frac{1}{\phi}\frac{d^2\phi}{d\theta^2} = \lambda. \qquad (2.5.22)$$

The separation constant is introduced as λ (rather than $-\lambda$) since there are two homogeneous conditions in θ, (2.5.21), and we therefore expect oscillations in θ. Equation (2.5.22) yields two ordinary differential equations. The boundary value problem to determine the separation constant is

$$\frac{d^2\phi}{d\theta^2} = -\lambda\phi$$

$$\phi(-\pi) = \phi(\pi)$$
$$\frac{d\phi}{d\theta}(-\pi) = \frac{d\phi}{d\theta}(\pi).$$

(2.5.23)

The eigenvalues λ are determined in the usual way. In fact, this is one of the three standard problems, the identical problem as for the circular wire (with $L = \pi$). Thus, the eigenvalues are

$$\lambda = \left(\frac{n\pi}{L}\right)^2 = n^2,$$

(2.5.24a)

with the corresponding eigenfunctions being both

$$\sin n\theta \quad \text{and} \quad \cos n\theta.$$

(2.5.24b)

The case $n = 0$ must be included (with only a constant being the eigenfunction).

The r-dependent problem is

$$\frac{r}{G}\frac{d}{dr}\left(r\frac{dG}{dr}\right) = \lambda = n^2,$$

(2.5.25a)

which when written in the more usual form becomes

$$r^2\frac{d^2G}{dr^2} + r\frac{dG}{dr} - n^2G = 0.$$

(2.5.25b)

Here, the condition at $r = 0$ has already been discussed. We have prescribed $|u(0, \theta)| < \infty$. For the product solutions, $u(r, \theta) = \phi(\theta)G(r)$, it follows that the condition at the origin is that $G(r)$ must be bounded there,

$$|G(0)| < \infty.$$

(2.5.26)

Equation (2.5.25) is linear and homogeneous but has nonconstant coefficients. There are exceedingly few second-order linear equations wih nonconstant coefficients that we can solve easily. Equation (2.5.25b) is one such case, an example of an equation known by a number of different names: **equidimensional** or **Cauchy** or **Euler**. The simplest way to solve (2.5.25b) is to note that for the linear differential operator in (2.5.25b) any power $G = r^p$ reproduces itself.* On substituting $G = r^p$ into (2.5.25b), we determine that $[p(p-1) + p - n^2]r^p = 0$. Thus, there usually are two distinct solutions

$$p = \pm n,$$

---

* For constant-coefficient linear differential operators, exponentials reproduce themselves.

except when $n = 0$, in which case there is only one independent solution in the form r^p. For $n \neq 0$, the general solution of (2.5.25b) is

$$G = c_1 r^n + c_2 r^{-n}. \tag{2.5.27a}$$

For $n = 0$ (and $n = 0$ is important since $\lambda = 0$ is an eigenvalue in this problem), one solution is $r^0 = 1$ or any constant. A second solution for $n = 0$ is most easily obtained from (2.5.25a). If $n = 0$, $\dfrac{d}{dr}\left(r\dfrac{dG}{dr}\right) = 0$. By integration, rdG/dr is constant, or equivalently dG/dr is proportional to $1/r$. The second independent solution is thus $\ln r$. Thus, for $n = 0$, the general solution of (2.5.25b) is

$$G = \bar{c}_1 + \bar{c}_2 \ln r. \tag{2.5.27b}$$

Equation (2.5.25b) has only one homogeneous condition to be imposed, $|G(0)| < \infty$, so it is not an eigenvalue problem. The boundedness condition would not have imposed any restrictions on the problems we have studied previously. However, here (2.5.27) shows that solutions may approach ∞ as $r \to 0$. Thus, for $|G(0)| < \infty$, $c_2 = 0$ in (2.5.27a) and $\bar{c}_2 = 0$ in (2.5.27b). The r-dependent solution (which is bounded at $r = 0$) is

$$G(r) = c_1 r^n, \qquad n \geq 0,$$

where for $n = 0$ this reduces to just an arbitrary constant.

Product solutions by the method of separation of variables, which satisfy the three homogeneous conditions, are

$$r^n \cos n\theta \quad (n \geq 0) \qquad \text{and} \qquad r^n \sin n\theta \quad (n \geq 1).$$

Note that as in rectangular coordinates for Laplace's equation, oscillations occur in one variable (here θ) and do not occur in the other variable (r). By the principle of superposition, the following solves Laplace's equation inside a circle:

$$\boxed{u(r, \theta) = \sum_{n=0}^{\infty} A_n r^n \cos n\theta + \sum_{n=1}^{\infty} B_n r^n \sin n\theta} \qquad \begin{array}{l} 0 \leq r < a \\ -\pi < \theta \leq \pi. \end{array} \tag{2.5.28a}$$

In order to solve the nonhomogeneous condition, $u(a, \theta) = f(\theta)$,

$$f(\theta) = \sum_{n=0}^{\infty} A_n a^n \cos n\theta + \sum_{n=1}^{\infty} B_n a^n \sin n\theta, \qquad -\pi < \theta \leq \pi. \tag{2.5.28b}$$

The prescribed temperature is a linear combination of all sines and cosines (including a constant term, $n = 0$). This is exactly the same question that we answered in Sec. 2.4.2 with $L = \pi$ if we let $A_n a^n$ be the coefficient of $\cos n\theta$ and $B_n a^n$ be the coefficient of $\sin n\theta$. Using the orthogonality formulas it follows that

$$A_0 = \frac{1}{2\pi} \int_{-\pi}^{\pi} f(\theta)\, d\theta$$

$$(n \geq 1) \qquad A_n a^n = \frac{1}{\pi} \int_{-\pi}^{\pi} f(\theta) \cos n\theta\, d\theta \tag{2.5.29}$$

$$B_n a^n = \frac{1}{\pi} \int_{-\pi}^{\pi} f(\theta) \sin n\theta\, d\theta.$$

Since $a^n \neq 0$, the coefficients A_n and B_n can be uniquely solved for from (2.5.29).

Equation (2.5.28a) with coefficients given by (2.5.29) determines the steady-state temperature distribution inside a circle. The solution is relatively complicated, often requiring the numerical evaluation of two infinite series. For additional interpretations of this solution, see Chapter 8 on Green's functions.

2.5.3 Qualitative Properties of Laplace's Equation

Sometimes the method of separation of variables will not be appropriate. If quantitative information is desired, numerical methods (see Chapter 13) may be necessary. In this subsection we briefly describe some qualitative properties that may be derived for Laplace's equation.

Mean value theorem. Our solution of Laplace's equation inside a circle, obtained in Sec. 2.5.2 by the method of separation of variables, yields an important result. If we evaluate the temperature at the origin, $r = 0$, we discover from (2.5.28a) that

$$u(0, \theta) = A_0 = \frac{1}{2\pi} \int_{-\pi}^{\pi} f(\theta) \, d\theta;$$

the temperature there equals the average value of the temperature at the edges of the circle. This is called the **mean value property** for Laplace's equation. It holds in general in the following specific sense. Suppose that we wish to solve Laplace's equation in any region R (see Fig. 2.5.3). Consider *any* point P inside R and a circle of any radius r_o (such that the circle is inside R). Let the temperature on the circle be $f(\theta)$, using polar coordinates centered at P. Our previous analysis still holds and thus **the temperature at any point is the average of the temperature along any circle of radius r_o** (*lying inside R*) **centered at that point.**

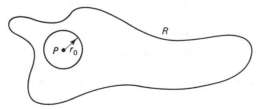

Figure 2.5.3 Circle within any region.

Maximum principles. We can use this to prove the **maximum principle** for Laplace's equation: **in steady state the temperature cannot attain its maximum in the interior** (unless the temperature is a constant everywhere) assuming no sources. The proof is by contradiction. Suppose that the maximum was at point P, as illustrated in Fig. 2.5.3. However, this should be the average of all points on any circle (consider the circle drawn). It is impossible for the temperature at P to be larger. This contradicts the original assumption, which thus cannot hold. We should not be surprised by the maximum principle. If the temperature was largest at point P, then in time the concentration of heat energy would diffuse and in steady state the maximum could not be in the interior. By letting $\psi = -u$, we can also show that the temperature cannot attain its minimum in the interior. If follows that in **steady state the maximum and minimum temperatures occur on the boundary.**

Well-posedness and uniqueness. The maximum principle is a very important tool for further analysis of partial differential equations, especially in establishing qualitative properties (see, e.g., Protter and Weinberger [1967]). We say that a problem is **well-posed** if there exists a unique solution that depends continuously on the nonhomogeneous data (i.e., the solution varies a small amount if the data are slightly changed). This is an important concept for physical problems. If the solution changed dramatically with only a small change in the data, then any physical measurement would have to be exact in order for the solution to be reliable. Fortunately, most standard problems in partial differential equations are well-posed. For example, the maximum principle can be used to prove that Laplace's equation $\nabla^2 u = 0$ with u specified as $u = f(\mathbf{x})$ on the boundary is well-posed. Suppose that we vary the boundary data a small amount such that

$$\nabla^2 v = 0 \qquad \text{with} \qquad v = g(\mathbf{x})$$

on the boundary, where $g(\mathbf{x})$ is nearly the same as $f(\mathbf{x})$ everywhere on the boundary. We consider the difference between these two solutions, $w = u - v$. Due to the linearity,

$$\nabla^2 w = 0 \qquad \text{with} \qquad w = f(\mathbf{x}) - g(\mathbf{x})$$

on the boundary. The maximum (and minimum) principles for Laplace's equation imply that the maximum and minimum occur on the boundary. Thus, at any point inside,

$$\min\,(f(\mathbf{x}) - g(\mathbf{x})) \leqslant w \leqslant \max\,(f(\mathbf{x}) - g(\mathbf{x})). \tag{2.5.30}$$

Since $g(\mathbf{x})$ is nearly the same as $f(\mathbf{x})$ everywhere, w is small, and thus the solution v is nearly the same as u; the solution of Laplace's equation slightly varies if the boundary data are slightly altered.

We can also prove that the solution of Laplace's equation is unique. We prove this by contradiction. Suppose that there are two solutions, u and v as above, which satisfy the same boundary condition [i.e., let $f(\mathbf{x}) = g(\mathbf{x})$]. If we again consider the difference ($w = u - v$), then the maximum and minimum principles imply [see (2.5.30)] that inside the region

$$0 \leqslant w \leqslant 0.$$

We conclude that $w = 0$ everywhere inside, and thus $u = v$, proving that if a solution exists, it must be unique. These properties (uniqueness and continuous dependence on the data) show that Laplace's equation with u specified on the boundary is a well-posed problem.

Solvability condition. If on the boundary the heat flow $-K_0 \nabla u \cdot \hat{\mathbf{n}}$ is specified instead of the temperature, Laplace's equation may have no solutions [for a one-dimensional example, see Exercise 1.4.7(b)]. To show this, we integrate $\nabla^2 u = 0$ over the entire two-dimensional region

$$0 = \iint \nabla^2 u \, dx \, dy = \iint \nabla \cdot (\nabla u) \, dx \, dy.$$

Using the (two-dimensional) divergence theorem, we conclude that (see Exercise 1.5.8)

$$0 = \oint \nabla u \cdot \hat{n} \, ds. \tag{2.5.31}$$

Since $\nabla u \cdot \hat{n}$ is proportional to the heat flow through the boundary, (2.5.31) implies that the *net* heat flow through the boundary must be zero in order for a steady state to exist. This is clear physically, because otherwise there would be a change (in time) of the thermal energy inside, violating the steady-state assumption. Equation (2.5.31) is called the **solvability condition** or **compatibility condition** for Laplace's equation.

EXERCISES 2.5

2.5.1. Solve Laplace's equation inside a rectangle $0 \le x \le L$, $0 \le y \le H$, with the following boundary conditions:

*(a) $\dfrac{\partial u}{\partial x}(0, y) = 0$, $\quad \dfrac{\partial u}{\partial x}(L, y) = 0$, $\quad u(x, 0) = 0$, $\quad\quad u(x, H) = f(x)$

(b) $\dfrac{\partial u}{\partial x}(0, y) = g(y)$, $\quad \dfrac{\partial u}{\partial x}(L, y) = 0$, $\quad u(x, 0) = 0$, $\quad\quad u(x, H) = 0$

*(c) $\dfrac{\partial u}{\partial x}(0, y) = 0$, $\quad u(L, y) = g(y)$, $\quad u(x, 0) = 0$, $\quad\quad u(x, H) = 0$

(d) $u(0, y) = g(y)$, $\quad u(L, y) = 0$, $\quad \dfrac{\partial u}{\partial y}(x, 0) = 0$, $\quad\quad u(x, H) = 0$

*(e) $u(0, y) = 0$, $\quad u(L, y) = 0$, $\quad u(x, 0) - \dfrac{\partial u}{\partial y}(x, 0) = 0$, $\quad u(x, H) = f(x)$

(f) $u(0, y) = f(y)$, $\quad u(L, y) = 0$, $\quad \dfrac{\partial u}{\partial y}(x, 0) = 0$, $\quad\quad \dfrac{\partial u}{\partial y}(x, H) = 0$

(g) $\dfrac{\partial u}{\partial x}(0, y) = 0$, $\quad \dfrac{\partial u}{\partial x}(L, y) = 0$, $\quad u(x, 0) = \begin{cases} 0 & x > L/2 \\ 1 & x < L/2 \end{cases}$, $\quad \dfrac{\partial u}{\partial y}(x, H) = 0$

2.5.2. Consider $u(x, y)$ satisfying Laplace's equation inside a rectangle $(0 < x < L$, $0 < y < H)$ subject to the boundary conditions

$$\frac{\partial u}{\partial x}(0, y) = 0 \qquad \frac{\partial u}{\partial y}(x, 0) = 0$$

$$\frac{\partial u}{\partial x}(L, y) = 0 \qquad \frac{\partial u}{\partial y}(x, H) = f(x).$$

*(a) *Without* solving this problem, briefly explain the physical condition under which there is a solution to this problem.

(b) Solve this problem by the method of separation of variables. Show that the solution exists only under the condition of part (a).

(c) The solution [part (b)] has an arbitrary constant. Determine it by consideration of the time-dependent heat equation (1.5.11) subject to the initial condition
$$u(x, y, 0) = g(x, y).$$

*2.5.3. Solve Laplace's equation *outside* a circular disk ($r \geq a$) subject to the boundary condition:
(a) $u(a, \theta) = \ln 2 + 4 \cos 3\theta$
(b) $u(a, \theta) = f(\theta)$
You may assume that $u(r, \theta)$ remains finite as $r \to \infty$.

*2.5.4. For Laplace's equation inside a circular disk ($r \leq a$), using (2.5.28a) and (2.5.29), show that
$$u(r, \theta) = \frac{1}{\pi} \int_{-\pi}^{\pi} f(\bar{\theta}) \left[-\frac{1}{2} + \sum_{n=0}^{\infty} \left(\frac{r}{a} \right)^n \cos n(\theta - \bar{\theta}) \right] d\bar{\theta}.$$

Using $\cos z = \text{Re} \, [e^{iz}]$, sum the resulting geometric series to obtain **Poisson's integral formula.**

2.5.5. Solve Laplace's equation inside the quarter-circle of radius 1 ($0 \leq \theta \leq \pi/2$, $0 \leq r \leq 1$) subject to the boundary conditions:

*(a) $\dfrac{\partial u}{\partial \theta}(r, 0) = 0,$ $u(r, \pi/2) = 0,$ $u(1, \theta) = f(\theta)$

(b) $\dfrac{\partial u}{\partial \theta}(r, 0) = 0,$ $\dfrac{\partial u}{\partial \theta}(r, \pi/2) = 0,$ $u(1, \theta) = f(\theta)$

*(c) $u(r, 0) = 0,$ $u(r, \pi/2) = 0,$ $\dfrac{\partial u}{\partial r}(1, \theta) = f(\theta)$

(d) $\dfrac{\partial u}{\partial \theta}(r, 0) = 0,$ $\dfrac{\partial u}{\partial \theta}(r, \pi/2) = 0,$ $\dfrac{\partial u}{\partial r}(1, \theta) = g(\theta)$

Show that the solution [part (d)] exists only if $\int_0^{\pi/2} g(\theta) \, d\theta = 0$. Explain this condition physically.

2.5.6. Solve Laplace's equation inside a semicircle of radius a ($0 < r < a, 0 < \theta < \pi$) subject to the boundary conditions:
*(a) $u = 0$ on the diameter and $u(a, \theta) = g(\theta)$.
(b) the diameter is insulated and $u(a, \theta) = g(\theta)$.

2.5.7. Solve Laplace's equation inside a 60° wedge of radius a subject to the boundary conditions:

(a) $u(r, 0) = 0,$ $u\left(r, \dfrac{\pi}{3}\right) = 0,$ $u(a, 0) = f(0)$

*(b) $\dfrac{\partial u}{\partial \theta}(r, 0) = 0,$ $\dfrac{\partial u}{\partial \theta}\left(r, \dfrac{\pi}{3}\right) = 0,$ $u(a, \theta) = f(\theta)$

2.5.8. Solve Laplace's equation inside a circular annulus ($a < r < b$) subject to the boundary conditions:

*(a) $u(a, \theta) = f(\theta),$ $u(b, \theta) = g(\theta)$

(b) $\dfrac{\partial u}{\partial r}(a, \theta) = 0,$ $u(b, \theta) = g(\theta)$

(c) $\dfrac{\partial u}{\partial r}(a, \theta) = f(\theta),$ $\dfrac{\partial u}{\partial r}(b, \theta) = g(\theta)$

If there is a solvability condition, state it and explain it physically.

*2.5.9. Solve Laplace's equation inside a 90° sector of a circular annulus ($a < r < b$, $0 < \theta < \pi/2$) subject to the boundary conditions:

(a) $u(r, 0) = 0$ $u(a, \theta) = 0$
     $u(r, \pi/2) = 0,$ $u(b, \theta) = f(\theta)$

(b) $u(r, 0) = 0$ $u(a, \theta) = 0$
     $u(r, \pi/2) = f(r),$ $u(b, \theta) = 0$

2.5.10. Using the maximum principles for Laplace's equation, prove that the solution of Poisson's equation, $\nabla^2 u = g(\mathbf{x})$, subject to $u = f(\mathbf{x})$ on the boundary, is unique.

2.5.11. Do Exercise 1.5.8.

2.5.12. (a) Using the divergence theorem, determine an alternative expression for $\iiint u \, \nabla^2 u \, dx \, dy \, dz$.

(b) Using part (a), prove that the solution of Laplace's equation $\nabla^2 u = 0$ (with u given on the boundary) is unique.

(c) Modify part (b) if $\nabla u \cdot \hat{\mathbf{n}} = 0$ on the boundary.

(d) Modify part (b) if $\nabla u \cdot \hat{\mathbf{n}} + hu = 0$ on the boundary. Show that Newton's law of cooling corresponds to $h < 0$.

2.5.13. Prove that the temperature satisfying Laplace's equation cannot attain its minimum in the interior.

2.5.14. Show that the "backwards" heat equation

$$\frac{\partial u}{\partial t} = -k\frac{\partial^2 u}{\partial x^2},$$

subject to $u(0, t) = u(L, t) = 0$ and $u(x, 0) = f(x)$, is *not* well posed. [*Hint:* Show that if the data are changed an arbitrarily small amount, for example

$$f(x) \quad \longrightarrow \quad f(x) + \frac{1}{n}\sin\frac{n\pi x}{L}$$

for large n, then the solution $u(x, t)$ changes by a large amount.]

2.5.15. Solve Laplace's equation inside a semi-infinite strip ($0 < x < \infty$, $0 < y < H$) subject to the boundary conditions:

(a) $\dfrac{\partial u}{\partial y}(x, 0) = 0,$ $\dfrac{\partial u}{\partial y}(x, H) = 0,$ $u(0, y) = f(y)$

(b) $u(x, 0) = 0,$ $u(x, H) = 0,$ $u(0, y) = f(y)$

(c) $u(x, 0) = 0,$ $u(x, H) = 0,$ $\dfrac{\partial u}{\partial x}(0, y) = f(y)$

(d) $\dfrac{\partial u}{\partial y}(x, 0) = 0,$ $\dfrac{\partial u}{\partial y}(x, H) = 0,$ $\dfrac{\partial u}{\partial x}(0, y) = f(y)$

Show that the solution [part (d)] exists only if $\int_0^H f(y) \, dy = 0$.

2.5.16. Consider Laplace's equation inside a rectangle $0 \leq x \leq L$, $0 \leq y \leq H$, with the boundary conditions

$$\frac{\partial u}{\partial x}(0, y) = 0, \quad \frac{\partial u}{\partial x}(L, y) = g(y), \quad \frac{\partial u}{\partial y}(x, 0) = 0, \quad \frac{\partial u}{\partial y}(x, H) = f(x).$$

(a) What is the solvability condition and its physical interpretation?

(b) Show that $u(x, y) = A(x^2 - y^2)$ is a solution if $f(x)$ and $g(y)$ are constants [under the conditions of part (a)].

(c) Under the conditions of part (a), solve the general case [nonconstant $f(x)$ and $g(y)$]. [*Hints:* Use part (b) and the fact that $f(x) = f_{av} + [f(x) - f_{av}]$, where
$$f_{av} = \frac{1}{L}\int_0^L f(x) \, dx.]$$

3

Fourier Series

3.1 INTRODUCTION

In solving partial differential equations by the method of separation of variables, we have discovered that important conditions [e.g., the initial condition, $u(x, 0) = f(x)$] could be satisfied only if $f(x)$ could be equated to an infinite linear combination of eigenfunctions of a given boundary value problem. Three specific cases have been investigated. One yielded a series involving sine functions, one yielded a series of cosines only (including a constant term), and the third yielded a series which included all of these previous terms.

We will begin by investigating series with both sines and cosines, because we will show that the others are just special cases of this more general series. For problems with the periodic boundary conditions on the interval $-L \leq x \leq L$, we asked whether the following infinite series (known as a **Fourier series**) makes sense:

$$f(x) = a_0 + \sum_{n=1}^{\infty} a_n \cos \frac{n\pi x}{L} + \sum_{n=1}^{\infty} b_n \sin \frac{n\pi x}{L}. \qquad (3.1.1)$$

Does the infinite series converge? Does it converge to $f(x)$? Is the resulting infinite series really a solution of the partial differential equation (and does it also satisfy all the other subsidiary conditions)? Mathematicians tell us that none of these questions have simple answers. Nonetheless, *Fourier series usually work quite well* (especially in situations where they arise naturally from physical problems). Joseph Fourier developed this type of series in his famous treatise on heat flow in the early 1800s.

The first difficulty that arises is that we claim (3.1.1) will not be valid for *all* functions $f(x)$. However, (3.1.1) will hold for some kinds of functions and will need only a small modification for other kinds of functions. In order to communicate various concepts easily, we will discuss only functions $f(x)$ that are piecewise smooth. A function $f(x)$ is **piecewise smooth** (on some interval) if the interval can be broken up into pieces (or sections) such that in each piece the function $f(x)$ is continuous* and its derivative df/dx is also continuous. The function $f(x)$ may not be continuous, but the only kind of discontinuity allowed is a finite number of jump discontinuities. A function $f(x)$ has a **jump discontinuity** at a point $x = x_0$ if the limit from the left $[f(x_0^-)]$ and the limit from the right $[f(x_0^+)]$ both exist (and are *unequal*), as illustrated in Fig. 3.1.1. An example of a piecewise smooth function is sketched in Fig. 3.1.2. Note that $f(x)$ has two jump discontinuities, at $x = x_1$ and at $x = x_3$. Also, $f(x)$ is continuous for $x_1 \leq x \leq x_3$, but df/dx is not continuous for $x_1 \leq x \leq x_3$. Instead, df/dx is continuous for $x_1 \leq x \leq x_2$ and $x_2 \leq x \leq x_3$. The interval can be broken up into pieces in which both $f(x)$ and df/dx are continuous. In this case there are four pieces, $x \leq x_1$, $x_1 \leq x \leq x_2$, $x_2 \leq x \leq x_3$, and $x \geq x_3$. Almost all functions occurring in practice (and certainly most that we discuss in this book) will be piecewise smooth. Let us briefly give an example of a function that is not piecewise smooth. Consider $f(x) = x^{1/3}$, as sketched in Fig. 3.1.3. It is *not*

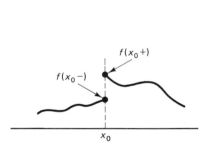

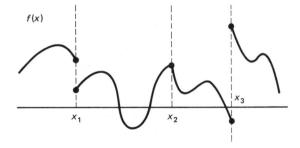

Figure 3.1.1 Jump discontinuity at $x = x_0$.

Figure 3.1.2 Example of a piecewise smooth function.

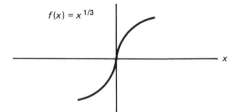

Figure 3.1.3 Example of a function that is not piecewise smooth.

---

* We do not give a definition of a continuous function here. However, one known useful fact is that if a function approaches ∞ at some point, then it is *not* continuous in any interval including that point.

piecewise smooth on any interval that includes $x = 0$, because $df/dx = 1/3x^{-2/3}$ is ∞ at $x = 0$. In other words, any region including $x = 0$ cannot be broken up into pieces such that df/dx is continuous.

Each function in the Fourier series is periodic with period $2L$. Thus, the *Fourier series of $f(x)$ on the interval* $-L \leq x \leq L$ *is periodic with period* $2L$. The function $f(x)$ does not need to be periodic. We need the **periodic extension** of $f(x)$. To sketch the periodic extension of $f(x)$, simply sketch $f(x)$ for $-L \leq x \leq L$ and then continually repeat the same pattern with period $2L$ by translating the original sketch for $-L \leq x \leq L$. For example, let us sketch in Fig. 3.1.4 the periodic extension of $f(x) = \frac{3}{2}x$ [the function $f(x) = \frac{3}{2}x$ is sketched in dotted lines for $|x| > L$]. Note the difference between $f(x)$ and its periodic extension.

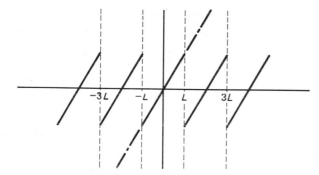

Figure 3.1.4 Periodic extension of $f(x) = \frac{3}{2}x$.

3.2 STATEMENT OF CONVERGENCE THEOREM

Definitions of Fourier coefficients and a Fourier series. We will be forced to distinguish carefully between a function $f(x)$ and its Fourier series over the interval $-L \leq x \leq L$:

$$\text{Fourier series} = a_0 + \sum_{n=1}^{\infty} a_n \cos \frac{n\pi x}{L} + \sum_{n=1}^{\infty} b_n \sin \frac{n\pi x}{L}. \qquad (3.2.1)$$

The infinite series may not even converge, and if it converges, it may not converge to $f(x)$. However, if the series converges, we learned in Chapter 2 how to determine the **Fourier coefficients** a_0, a_n, b_n using certain orthogonality integrals, (2.3.31). We will use those results as the definition of the Fourier coefficients:

$$\begin{aligned}
a_0 &= \frac{1}{2L}\int_{-L}^{L} f(x)\, dx \\[2mm]
a_n &= \frac{1}{L}\int_{-L}^{L} f(x) \cos \frac{n\pi x}{L}\, dx \\[2mm]
b_n &= \frac{1}{L}\int_{-L}^{L} f(x) \sin \frac{n\pi x}{L}\, dx.
\end{aligned} \qquad (3.2.2)$$

The Fourier series of $f(x)$ over the interval $-L \leq x \leq L$ is defined to be the infinite series (3.2.1) where the Fourier coefficients are given by (3.2.2). We immediately note that a Fourier series does not exist unless for example a_0 exists [i.e., unless $|\int_{-L}^{L} f(x)\, dx| < \infty$]. This eliminates certain functions from our consideration. For example, we do not ask what the Fourier series of $f(x) = 1/x^2$ is.

Even in situations in which $\int_{-L}^{L} f(x)\, dx$ exists, the infinite series may not converge; furthermore, if it converges, it may not converge to $f(x)$. We use the notation

$$f(x) \sim a_0 + \sum_{n=1}^{\infty} a_n \cos \frac{n\pi x}{L} + \sum_{n=1}^{\infty} b_n \sin \frac{n\pi x}{L},$$ (3.2.3)

where $\sim$ means that $f(x)$ is on the left-hand side and the Fourier series of $f(x)$ (on the interval $-L \leq x \leq L$) is on the right-hand side (even if the series diverges), but the two functions may be completely different. The symbol $\sim$ is read as "has the Fourier series (on a given interval)."

Convergence theorem for Fourier series. At first we state a theorem summarizing certain properties of Fourier series:

If $f(x)$ is *piecewise smooth* on the interval $-L \leq x \leq L$, then the Fourier series of $f(x)$ converges

1. to the *periodic extension* of $f(x)$, where the periodic extension is continuous;
2. to the average of the two limits, usually
$$\tfrac{1}{2}[f(x+) + f(x-)],$$
   where the periodic extension has a *jump discontinuity*.

We refer to this as **Fourier's theorem**. It is proved in many of the references listed in the bibliography.

Mathematically, if $f(x)$ is piecewise smooth, then for $-L < x < L$ (excluding the end points),

$$\frac{f(x+) + f(x-)}{2} = a_0 + \sum_{n=1}^{\infty} a_n \cos \frac{n\pi x}{L} + \sum_{n=1}^{\infty} b_n \sin \frac{n\pi x}{L},$$ (3.2.4)

where the Fourier coefficients are given by (3.2.2). *At points where $f(x)$ is continuous,* $f(x+) = f(x-)$ and hence (3.2.4) implies that for $-L < x < L$,

$$f(x) = a_0 + \sum_{n=1}^{\infty} a_n \cos \frac{n\pi x}{L} + \sum_{n=1}^{\infty} b_n \sin \frac{n\pi x}{L}.$$

The Fourier series actually converges to $f(x)$ at points between $-L$ and $+L$, where $f(x)$ is continuous. At the end points, $x = L$ or $x = -L$, the infinite series converges to the average of the two values of the periodic extension.

Outside the range $-L \leq x \leq L$, the Fourier series converges to a value easily determined using the known periodicity (with period $2L$) of the Fourier series.

Sketching Fourier series.   Now we are ready to apply Fourier's theorem. To sketch the Fourier series of $f(x)$ (on the interval $-L \leq x \leq L$), we:

> 1. Sketch $f(x)$ (preferably for $-L \leq x \leq L$ only).
> 2. Sketch the periodic extension of $f(x)$.

According to Fourier's theorem, the Fourier series converges (here converge means equals) to the periodic extension, where the periodic extension is continuous (which will be almost everywhere). However, at points of jump discontinuity of the periodic extension, the Fourier series converges to the average. Therefore, there is a third step:

> 3. Mark an "$\times$" at the average of the two values at any jump discontinuity of the periodic extension.

Example.   Consider

$$
f(x) = \begin{cases} 0 & x < \dfrac{L}{2} \\[2mm] 1 & x > \dfrac{L}{2} \end{cases} \tag{3.2.5}
$$

We would like to determine the Fourier series of $f(x)$ on $-L \leq x \leq L$. We begin by sketching $f(x)$ for all x in Fig. 3.2.1 (although we only need the sketch for $-L \leq x \leq L$.) Note that $f(x)$ is piecewise smooth, so we can apply Fourier's theorem. The periodic extension of $f(x)$ is sketched in Fig. 3.2.2. Often the

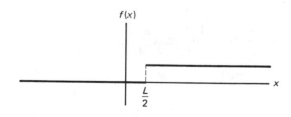

Figure 3.2.1   Sketch of $f(x)$.

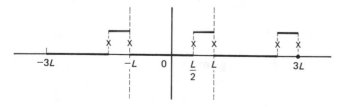

Figure 3.2.2   Fourier series of $f(x)$.

understanding of the process is made clearer by sketching at least three full periods, $-3L \leqslant x \leqslant 3L$, even though in the applications to partial differential equations only the interval $-L \leqslant x \leqslant L$ is absolutely needed. The Fourier series of $f(x)$ equals the periodic extension of $f(x)$, wherever the periodic extension is continuous (i.e., at all x except the points of jump discontinuity, which are $x = L/2, L, L/2 + 2L, -L, L/2 - 2L$, etc.). According to Fourier's theorem, at these points of jump discontinuity, the Fourier series of $f(x)$ must converge to the average. These points should be marked, perhaps with an $\times$, as in Fig. 3.2.2. At $x = L/2$ and $x = L$ (as well as $x = L/2 \pm 2nL$ and $x = L \pm 2nL$), the Fourier series converges to the average, $\frac{1}{2}$. In summary for this example,

$$\sum_{n=0}^{\infty} a_n \cos \frac{n\pi x}{L} + \sum_{n=1}^{\infty} b_n \sin \frac{n\pi x}{L} = \begin{cases} \frac{1}{2} & x = -L \\ 0 & -L < x < L/2 \\ \frac{1}{2} & x = L/2 \\ 1 & L/2 < x < L \\ \frac{1}{2} & x = L \end{cases}$$

Fourier series can converge to rather strange functions, but they are not so different from the original function.

Fourier coefficients. For a given $f(x)$, it is *not* necessary to calculate the Fourier coefficients in order to sketch the Fourier series of $f(x)$. However, it is important to know how to calculate the Fourier coefficients, given by (3.2.2). The calculation of Fourier coefficients can be an algebraically involved process. Sometimes it is an exercise in the method of integration by parts. Often, calculations can be simplified by judiciously using integral tables. In any event, we can always use a computer to approximate the coefficients numerically. As an overly simple example but one that illustrates some important points, consider $f(x)$ given by (3.2.5). From (3.2.2), the coefficients are

$$a_0 = \frac{1}{2L} \int_{-L}^{L} f(x) \, dx = \frac{1}{2L} \int_{L/2}^{L} dx = \frac{1}{4}$$

$$a_n = \frac{1}{L} \int_{-L}^{L} f(x) \cos \frac{n\pi x}{L} \, dx = \frac{1}{L} \int_{L/2}^{L} \cos \frac{n\pi x}{L} \, dx = \frac{1}{n\pi} \sin \frac{n\pi x}{L} \Big|_{L/2}^{L}$$

$$= \frac{1}{n\pi} \left(\sin n\pi - \sin \frac{n\pi}{2} \right)$$

$$b_n = \frac{1}{L} \int_{-L}^{L} f(x) \sin \frac{n\pi x}{L} \, dx = \frac{1}{L} \int_{L/2}^{L} \sin \frac{n\pi x}{L} \, dx = \frac{-1}{n\pi} \cos \frac{n\pi x}{L} \Big|_{L/2}^{L}$$

$$= \frac{1}{n\pi} \left(\cos \frac{n\pi}{2} - \cos n\pi \right)$$

We omit simplifications that arise by noting that $\sin n\pi = 0$, $\cos n\pi = (-1)^n$, and so on.

EXERCISES 3.2

3.2.1. For the following functions, sketch the Fourier series of $f(x)$ (on the interval $-L \leq x \leq L$). Compare $f(x)$ to its Fourier series:

(a) $f(x) = 1$ ***(b)** $f(x) = x^2$

(c) $f(x) = 1 + x$ ***(d)** $f(x) = e^x$

(e) $f(x) = \begin{cases} x & x < 0 \\ 2x & x > 0 \end{cases}$ ***(f)** $f(x) = \begin{cases} 0 & x < 0 \\ 1 + x & x > 0 \end{cases}$

(g) $f(x) = \begin{cases} x & x < L/2 \\ 0 & x > L/2 \end{cases}$

3.2.2. For the following functions, sketch the Fourier series of $f(x)$ (on the interval $-L \leq x \leq L$) and determine the Fourier coefficients:

*(a) $f(x) = x$ **(b)** $f(x) = e^{-x}$

*(c) $f(x) = \sin \dfrac{\pi x}{L}$ **(d)** $f(x) = \begin{cases} 0 & x < 0 \\ x & x > 0 \end{cases}$

(e) $f(x) = \begin{cases} 1 & |x| < L/2 \\ 0 & |x| > L/2 \end{cases}$ ***(f)** $f(x) = \begin{cases} 1 & x > 0 \\ 0 & x < 0 \end{cases}$

(g) $f(x) = \begin{cases} 1 & x < 0 \\ 2 & x > 0 \end{cases}$

3.2.3. Show that the Fourier series operation is linear; that is, show that the Fourier series of $c_1 f(x) + c_2 g(x)$ is the sum of c_1 times the Fourier series of $f(x)$ and c_2 times the Fourier series of $g(x)$.

3.2.4. Suppose that $f(x)$ is piecewise smooth. What value does the Fourier series of $f(x)$ converge to at the end point $x = -L$? at $x = L$?

3.3 FOURIER COSINE AND SINE SERIES

In this section we show that the series of sines only (and the series of cosines only) are special cases of a Fourier series.

3.3.1 Fourier Sine Series

Odd functions. An odd function is a function with the property $f(-x) = -f(x)$. The sketch of an odd function for $x < 0$ will be minus the mirror image of $f(x)$ for $x > 0$, as illustrated in Fig. 3.3.1. Examples of odd functions are $f(x) = x^3$ (in fact, any odd power) and $f(x) = \sin 4x$. The integral of an

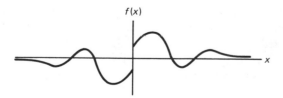

f(x)

x

Figure 3.3.1 An odd function.

odd function over a symmetric interval is zero (any contribution from $x > 0$ will be canceled by a contribution from $x < 0$).

Fourier series of odd functions. Let us calculate the Fourier coefficients of an odd function:

$$a_0 = \frac{1}{2L} \int_{-L}^{L} f(x)\,dx = 0$$

$$a_n = \frac{1}{L} \int_{-L}^{L} f(x) \cos \frac{n\pi x}{L}\,dx = 0.$$

Both are zero because the integrand, $f(x) \cos n\pi x/L$, is odd [being the product of an even function $\cos n\pi x/L$ and an odd function $f(x)$]. Since $a_n = 0$, all the cosine functions (which are even) will not appear in the Fourier series of an odd function. The Fourier series of an odd function is an infinite series of odd functions (sines):

$$f(x) \sim \sum_{n=1}^{\infty} b_n \sin \frac{n\pi x}{L}, \tag{3.3.1}$$

if $f(x)$ is odd. In this case formulas for the Fourier coefficients b_n may be simplified:

$$b_n = \frac{1}{L} \int_{-L}^{L} f(x) \sin \frac{n\pi x}{L}\,dx = \frac{2}{L} \int_{0}^{L} f(x) \sin \frac{n\pi x}{L}\,dx, \tag{3.3.2}$$

since the integral of an even function over the symmetric interval $-L$ to $+L$ is twice the integral from 0 to L. For odd functions information about $f(x)$ is needed only for $0 \leqslant x \leqslant L$.

Fourier sine series. However, only occasionally are we given an odd function and asked to compute its Fourier series. Instead, frequently series of only sines arise in the context of separation of variables. Recall that the temperature in a one-dimensional rod $0 < x < L$ with zero temperature ends [$u(0, t) = u(L, t) = 0$] satisfies

$$u(x, t) = \sum_{n=1}^{\infty} B_n \sin \frac{n\pi x}{L} e^{-(n\pi/L)^2 kt}, \tag{3.3.3}$$

where the initial condition $u(x, 0) = f(x)$ is satisfied if

$$f(x) = \sum_{n=1}^{\infty} B_n \sin \frac{n\pi x}{L}. \tag{3.3.4}$$

$f(x)$ must be represented as a series of sines; (3.3.4) appears in the same form as (3.3.1). However, there is a significant difference. In (3.3.1) $f(x)$ is given as an odd function and defined for $-L \leqslant x \leqslant L$. In (3.3.4) $f(x)$ is only defined for $0 \leqslant x \leqslant L$ (it is just the initial temperature distribution); $f(x)$ is certainly not necessarily odd. If $f(x)$ is only given for $0 \leqslant x \leqslant L$, then it can be *extended* as an odd function; see Fig. 3.3.2, called the **odd extension of $f(x)$**. The odd extension of $f(x)$ is defined for $-L \leqslant x \leqslant L$. Fourier's theorem will apply [if

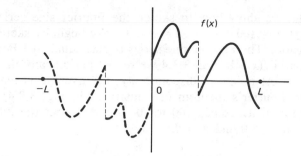

$f(x)$

$-L$

0

L

Figure 3.3.2 Odd extension of $f(x)$.

the odd extension of $f(x)$ is piecewise smooth, which just requires that $f(x)$ is piecewise smooth for $0 \leq x \leq L$]. Moreover, since the odd extension of $f(x)$ is certainly odd, its Fourier series only involves sines:

the odd extension of $f(x) \sim \sum_{n=1}^{\infty} B_n \sin n\pi x/L, \quad -L \leq x \leq L,$

where B_n are given by (3.3.2). However, we are only interested in what happens between $x = 0$ and $x = L$. In that region $f(x)$ is identical to its odd extension:

$$f(x) \sim \sum_{n=1}^{\infty} B_n \sin \frac{n\pi x}{L} \qquad 0 \leq x \leq L, \qquad (3.3.5)$$

where

$$B_n = \frac{2}{L} \int_0^L f(x) \sin \frac{n\pi x}{L} \, dx. \qquad (3.3.6)$$

We call this the **Fourier sine series of $f(x)$** (on the interval $0 \leq x \leq L$). This series (3.3.5) is nothing but an example of a Fourier series. As such, we can simply apply Fourier's theorem; just remember that $f(x)$ is only defined for $0 \leq x \leq L$. We may think of $f(x)$ as being odd (although it is not necessarily) by extending $f(x)$ as an odd function. Formula (3.3.6) is very important, but does not need to be memorized. It can be *derived* from the formulas for a Fourier series simply by assuming that $f(x)$ is odd. [It is more accurate to say that we consider the odd extension of $f(x)$]. Formula (3.3.6) is a factor of 2 larger than the Fourier series coefficients since the integrand is even. In (3.3.6) the integrals are only from $x = 0$ to $x = L$.

According to Fourier's theorem, sketching the Fourier sine series of $f(x)$ is easy:

1. Sketch $f(x)$ (for $0 < x < L$).
2. Sketch the odd extension of $f(x)$.
3. Extend as a periodic function (with period $2L$).
4. Mark an $\times$ at the average at points where the odd periodic extension of $f(x)$ has a jump discontinuity.

Example. As an example, we show how to sketch the Fourier sine series of $f(x) = 100$. We consider $f(x) = 100$ only for $0 \leqslant x \leqslant L$. We begin by sketching in Fig. 3.3.3 its odd extension. The Fourier sine series of $f(x)$ equals the Fourier series of the odd extension of $f(x)$. In Fig. 3.3.4 we repeat periodically the odd extension (with period $2L$). At points of discontinuity, the average is marked with an $\times$. According to Fourier's theorem (as illustrated in Fig. 3.3.4), the Fourier sine series of 100 actually equals 100 for $0 < x < L$, but the infinite series does not equal 100 at $x = 0$ and $x = L$:

$$100 = \sum_{n=1}^{\infty} b_n \sin \frac{n\pi x}{L}, \qquad 0 < x < L. \tag{3.3.7}$$

At $x = 0$, Fig. 3.3.4 shows that the Fourier sine series converges to 0, because at $x = 0$ the odd property of the sine series yields the average of 100 and -100, which is 0. For similar reasons, the Fourier sine series also converges to 0 at $x = L$. These observations agree with the result of substituting $x = 0$ (and $x = L$) into the infinite series of sines. The Fourier coefficients are determined from (3.3.6) as before [see (2.3.39)]

$$b_n = \frac{2}{L} \int_0^L f(x) \sin \frac{n\pi x}{L} \, dx = \frac{200}{L} \int_0^L \sin \frac{n\pi x}{L} \, dx = \begin{cases} 0 & n \text{ even} \\ \dfrac{400}{n\pi} & n \text{ odd.} \end{cases} \tag{3.3.8}$$

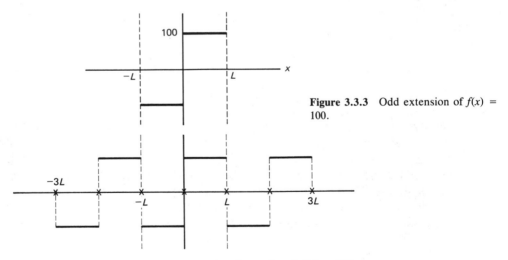

Figure 3.3.3 Odd extension of $f(x) = 100$.

Figure 3.3.4 Fourier sine series of $f(x) = 100$.

Physical example. One of the simplest examples is the Fourier sine series of a constant. This problem arose in trying to solve the one-dimensional heat equation with zero boundary conditions and constant initial temperature, $100°$:

$$\text{PDE:} \qquad \frac{\partial u}{\partial t} = k\frac{\partial^2 u}{\partial x^2}, \qquad 0 < x < L, \quad t > 0$$

BC: $u(0, t) = 0$
$u(L, t) = 0$

IC: $u(x, 0) = f(x) = 100°$

We recall from Sec. 2.3 that the method of separation of variables implied that

$$u(x, t) = \sum_{n=1}^{\infty} b_n \sin \frac{n\pi x}{L} e^{-(n\pi/L)^2 kt}. \tag{3.3.9}$$

The initial conditions are satisfied if

$$100 = f(x) = \sum_{n=1}^{\infty} b_n \sin \frac{n\pi x}{L}, \qquad 0 < x < L.$$

This may be interpreted as the Fourier sine series of $f(x) = 100$ [see (3.3.8)]. Equivalently, b_n may be determined from the orthogonality of $\sin n\pi x/L$ [see (2.3.39)].

Mathematically, the Fourier series of the initial condition has a rather bizarre behavior at $x = 0$ (and at $x = L$). In fact, for this problem, the physical situation is not very well defined at $x = 0$ (at $t = 0$). This might be illustrated in a space–time diagram, Fig. 3.3.5. We note that Fig. 3.3.5 shows that the domain of our problem is $t \geq 0$ and $0 \leq x \leq L$. However, there is a conflict that occurs at $x = 0$, $t = 0$ between the initial condition and the boundary condition. The initial condition ($t = 0$) prescribes the temperature to be 100° even as $x \to 0$, whereas the boundary condition ($x = 0$) prescribes the temperature to be 0° even as $t \to 0$. Thus, the physical problem has a discontinuity at $x = 0$, $t = 0$. In the actual physical world, the temperature cannot be discontinuous.

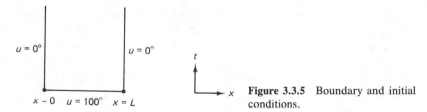

Figure 3.3.5 Boundary and initial conditions.

We introduced a discontinuity into our mathematical model by "instantaneously" transporting (at $t = 0$) the rod from a 100° bath to a 0° bath at $x = 0$. It actually takes a finite time, and the temperature would be continuous. Nevertheless, the transition from 0° to 100° would occur over an exceedingly small distance and time. We introduce the temperature discontinuity to approximate the more complicated real physical situation. Fourier's theorem thus illustrates how the physical discontinuity at $x = 0$ (initially, at $t = 0$) is reproduced mathematically. The Fourier sine series of 100° (which represents the physical solution at $t = 0$) has the nice property that it equals 100° for all x inside the rod, $0 < x < L$ (thus satisfying the initial condition there), but it equals 0° at the boundaries, $x = 0$ and $x = L$ (thus also satisfying the boundary conditions). The Fourier sine series of 100° is a strange mathematical function, but so is the physical approximation it is needed for.

Fourier series computations and the Gibbs phenomenon. Let us gain some confidence in the validity of Fourier series. The Fourier sine series of $f(x) = 100$ states that

$$100 = \frac{400}{\pi} \left(\frac{\sin \pi x/L}{1} + \frac{\sin 3\pi x/L}{3} + \frac{\sin 5\pi x/L}{5} + \cdots \right). \qquad (3.3.10)$$

Do we believe (3.3.10)? Certainly, it is not valid at $x = 0$ (as well as the other boundary $x = L$), since at $x = 0$ every term in the infinite series is zero (they cannot add to 100). However, the theory of Fourier series claims that (3.3.10) is valid everywhere except the two ends. For example, we claim it is valid at $x = L/2$. Substituting $x = L/2$ into (3.3.10) shows that

$$100 = \frac{400}{\pi} \left(1 - \frac{1}{3} + \frac{1}{5} - \frac{1}{7} + \frac{1}{9} - \frac{1}{11} + \cdots \right) \qquad \text{or}$$

$$\frac{\pi}{4} = 1 - \frac{1}{3} + \frac{1}{5} - \frac{1}{7} + \frac{1}{9} - \frac{1}{11} + \cdots.$$

At first this may seem strange. However, it is Euler's formula for π. It can be used to *compute* π (although very inefficiently); it can also be shown to be true without relying on the theory of infinite trigonometric series (see Exercise 3.3.17). The validity of (3.3.10) for other values of x, $0 < x < L$, may also surprise you. We will sketch the left- and right-hand sides of (3.3.10), hopefully convincing you of their equality. We will sketch the r.h.s. by adding up the contribution of each term of the series. Of course, we cannot add up the required infinite number of terms; we will settle for a finite number of terms. In fact, we will sketch the sum of the first few terms to see how the series approaches the constant 100 as the number of terms increases. It is helpful to know that $400/\pi = 127.32395\ldots$ (although for rough sketching 125 or 130 will do). The first term $(400/\pi) \sin \pi x/L$ by itself is the basic first rise and fall of a sine function; it is not a good approximation to the constant 100 as illustrated in Fig. 3.3.6. On the other hand, for just one term in an infinite series it is not such a bad approximation. The next term to be added is $(400/3\pi) \sin 3\pi x/L$. This is a sinusoidol oscillation, with one-third the amplitude and one-third the period of the first term. It is positive near $x = 0$ and $x = L$, where the approximation needs to be increased, and it is negative near $x = L/2$, where the approximation needs to be decreased. It is sketched in dashed lines and then added to the first term in Fig. 3.3.7. Note that the sum of the two nonzero terms already seems to be a considerable improvement over the first term. A computer plot of all the partial sums, up to and including the first six nonzero terms, is given in Fig. 3.3.8.

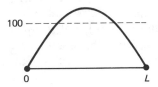

100 ---

0 L **Figure 3.3.6** First term of Fourier sine series of $f(x) = 100$.

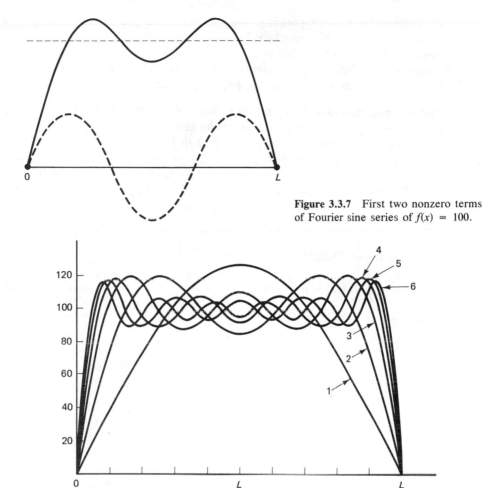

Figure 3.3.7 First two nonzero terms of Fourier sine series of $f(x) = 100$.

Figure 3.3.8 Various partial sums of Fourier sine series of $f(x) = 100$.

Actually, a lot can be learned from Fig. 3.3.8. Perhaps now it does seem reasonable that the infinite series converges to 100 for $0 < x < L$. The worst places (where the finite series differs most from 100) are getting closer and closer to $x = 0$ and $x = L$ as the number of terms increases. *For a finite number of terms in the series*, the solution starts from zero at $x = 0$ and shoots up beyond 100, what we call the primary **overshoot**. It is interesting to note that Fig. 3.3.8 illustrates the overshoot vividly. We can even extrapolate to guess what happens for 1000 terms. The series should become more and more accurate as the number of terms increases. We might expect the overshoot to vanish as $n \to \infty$, but put a straight edge on the points of maximum overshoot. It just does not seem to approach 100. Instead, it is far away from that, closer to 118. This overshoot is an example of the **Gibbs phenomenon**. In general (for large n), there is an overshoot (and corresponding undershoot) of approximately 9% of the jump discontinuity. In this case (see Fig. 3.3.4), the Fourier *sine* series of $f(x) = 100$ jumps from -100 to $+100$ at $x = 0$. Thus, the finite series will overshoot by about 9% of 200, or approximately 18. The Gibbs phenomenon occurs only when a *finite* series of eigenfunctions approximates a *discontinuous* function.

Further example of a Fourier sine series. We consider the Fourier sine series of $f(x) = x$. $f(x) = x$ is sketched on the interval $0 \leq x \leq L$ in Fig. 3.3.9a. The odd-periodic extension of $f(x)$ is sketched in Fig. 3.3.9b. The jump discontinuity of the odd-periodic extension at $x = (2n - 1)L$ shows that, for example, the Fourier sine series of $f(x) = x$ converges to zero at $x = L$, while $f(L) \neq 0$. We note that the Fourier sine series of $f(x) = x$ actually equals x for $-L < x < L$,

$$x = \sum_{n=1}^{\infty} b_n \sin \frac{n\pi x}{L}, \qquad -L < x < L. \tag{3.3.11a}$$

The Fourier coefficients are determined from (3.3.6)

$$b_n = \frac{2}{L} \int_0^L f(x) \sin \frac{n\pi x}{L}\, dx = \frac{2}{L} \int_0^L x \sin \frac{n\pi x}{L}\, dx = \frac{2L}{n\pi}(-1)^{n+1}, \tag{3.3.11b}$$

where the integral can be evaluated by integration by parts (or by a table).

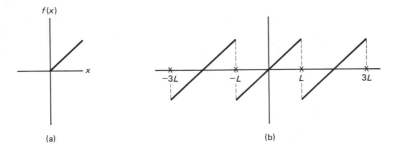

f(x)

−3L   −L   L   3L

(a)   (b)

Figure 3.3.9 (a) $f(x) = x$ and (b) its Fourier sine series.

Example. We now consider the Fourier sine series of $f(x) = \cos \pi x/L$. This may seem to ask for a sine series expansion of an even function, but in applications often the function is only given from $0 \leq x \leq L$ and *must* be expanded in a series of sines due to the boundary conditions. $\cos \pi x/L$ is sketched in Fig. 3.3.10a. It is an even function, but its odd extension is sketched in Fig. 3.3.10b. The Fourier sine series of $f(x)$ equals the Fourier series of the odd extension of $f(x)$. Thus, we repeat the sketch in Fig. 3.3.10b periodically (see Fig. 3.3.11), placing an $\times$ at the average of the two values at the jump discontinuities. The Fourier sine series representation of $\cos \pi x/L$ is

$$\cos \frac{\pi x}{L} \sim \sum_{n=1}^{\infty} b_n \sin \frac{n\pi x}{L}, \qquad 0 \leq x \leq L,$$

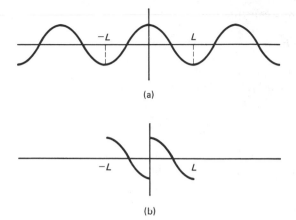

(a)

(b)

Figure 3.3.10 (a) $f(x) = \cos \pi x / L$ and (b) its odd extension.

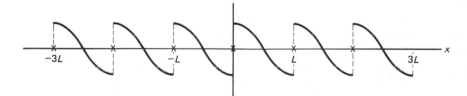

Figure 3.3.11 Fourier sine series of $f(x) = \cos \pi x / L$.

where with some effort we obtain

$$b_n = \frac{2}{L} \int_0^L \cos \frac{\pi x}{L} \sin \frac{n\pi x}{L} \, dx = \begin{cases} 0 & n \text{ odd} \\ \dfrac{4n}{\pi(n^2 - 1)} & n \text{ even} \end{cases} \qquad (3.3.12)$$

According to Fig. 3.3.11 (based on Fourier's theorem), equality holds for $0 < x < L$, but not at $x = 0$ and not at $x = L$:

$$\cos \frac{\pi x}{L} = \sum_{n=1}^{\infty} b_n \sin \frac{n\pi x}{L}, \qquad 0 < x < L.$$

At $x = 0$ and at $x = L$, the infinite series must converge to 0, since all terms in the series are zero there. Figure 3.3.11 agrees with this. You may be a bit puzzled by an aspect of this problem. You may have recalled that $\sin n\pi x / L$ is orthogonal to $\cos m\pi x / L$, and thus expected all the b_n in (3.3.12) to be zero. However, $b_n \neq 0$. The subtle point is that you should remember that $\cos m\pi x / L$ and $\sin n\pi x / L$ are orthogonal *on the interval* $-L \le x \le L$, $\int_{-L}^{L} \cos m\pi x / L \sin n\pi x / L \, dx = 0$; they are not orthogonal on $0 \le x \le L$.

3.3.2 Fourier Cosine Series

Even functions. Similar ideas are valid for even functions, in which $f(-x) = f(x)$. Let us develop the basic results. The sine coefficients of a Fourier series will be zero for an even function,

$$b_n = \frac{1}{L} \int_{-L}^{L} f(x) \sin \frac{n\pi x}{L} \, dx = 0,$$

since $f(x)$ is even. The Fourier series of an even function is a representation of $f(x)$ involving an infinite sum of only even functions (cosines):

$$f(x) \sim \sum_{n=0}^{\infty} a_n \cos \frac{n\pi x}{L}, \qquad (3.3.13a)$$

if $f(x)$ is even. The coefficients of the cosines may be evaluated using information about $f(x)$ only between $x = 0$ and $x = L$, since

$$a_0 = \frac{1}{2L} \int_{-L}^{L} f(x) \, dx = \frac{1}{L} \int_{0}^{L} f(x) \, dx \qquad (3.3.13b)$$

$$(n \geq 1) \quad a_n = \frac{1}{L} \int_{-L}^{L} f(x) \cos \frac{n\pi x}{L} \, dx = \frac{2}{L} \int_{0}^{L} f(x) \cos \frac{n\pi x}{L} \, dx, \qquad (3.3.13c)$$

using the fact that for $f(x)$ even, $f(x) \cos n\pi x/L$ is even.

Often, $f(x)$ is not given as an even function. Instead, in trying to represent an arbitrary function $f(x)$ using an infinite series of $\cos n\pi x/L$, the eigenfunctions of the boundary value problem $d^2\phi/dx^2 = -\lambda\phi$ with $d\phi/dx(0) = 0$ and $d\phi/dx(L) = 0$, we wanted

$$f(x) = \sum_{n=0}^{\infty} a_n \cos \frac{n\pi x}{L}, \qquad (3.3.14)$$

only for $0 < x < L$. We had previously determined the coefficients a_n to be the same as given by (3.3.13), but our reason was because of the orthogonality of $\cos n\pi x/L$. To relate (3.3.14) to a Fourier series, we simply introduce the **even extension of** $f(x)$, an example being illustrated in Fig. 3.3.12. If $f(x)$ is piecewise smooth for $0 \leq x \leq L$, then its even extension will also be piecewise smooth, and hence Fourier's theorem can be applied to the even extension of $f(x)$. Since the even extension of $f(x)$ is an even function, the Fourier series of the even extension of $f(x)$ will have only cosines:

$$\text{even extension of } f(x) \sim \sum_{n=0}^{\infty} a_n \cos \frac{n\pi x}{L}, \qquad -L \leq x \leq L,$$

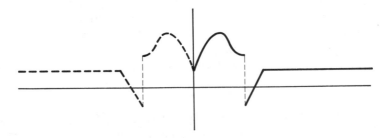

Figure 3.3.12 Even extension of $f(x)$.

where a_n is given by (3.3.13). In the region of interest, $0 \leqslant x \leqslant L$, $f(x)$ is identical to the even extension. The resulting series in that region is called the **Fourier cosine series of $f(x)$** (on the interval $0 \leqslant x \leqslant L$):

$$f(x) \sim \sum_{n=0}^{\infty} a_n \cos \frac{n \pi x}{L}, \qquad 0 \leqslant x \leqslant L \qquad (3.3.15)$$

$$a_0 = \frac{1}{L} \int_0^L f(x) \, dx \qquad (3.3.16a)$$

$$a_n = \frac{2}{L} \int_0^L f(x) \cos \frac{n \pi x}{L} \, dx. \qquad (3.3.16b)$$

The Fourier cosine series of $f(x)$ is exactly the Fourier series of the even extension of $f(x)$. Since we can apply Fourier's theorem, we have an algorithm to sketch the Fourier cosine series of $f(x)$:

1. Sketch $f(x)$ (for $0 < x < L$).
2. Sketch the even extension of $f(x)$.
3. Extend as a periodic function (with period $2L$).
4. Mark $\times$ at points of discontinuity at the average.

Example. We consider the Fourier cosine series of $f(x) = x$. $f(x)$ is sketched in Fig. 3.3.13a [note that $f(x)$ is odd!]. We consider $f(x)$ only from $x = 0$ to $x = L$, and then extend it in Fig. 3.3.13b as an even function. Next, we sketch the Fourier series of the even extension, by periodically extending the even extension (see Fig. 3.3.14). Note that between $x = 0$ and $x = L$ the Fourier cosine series has no jump discontinuities. The Fourier cosine series of $f(x) = x$ actually equals x, so that

$$x = \sum_{n=0}^{\infty} a_n \cos \frac{n \pi x}{L}, \qquad 0 \leqslant x \leqslant L. \qquad (3.3.17a)$$

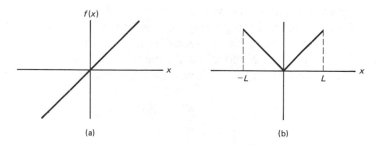

Figure 3.3.13 (a) $f(x) = x$, (b) its even extension.

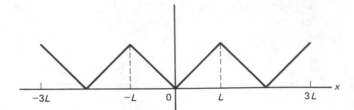

Figure 3.3.14 Fourier cosine series of the even extension of $f(x)$.

The coefficients are given by the following integrals:

$$a_0 = \frac{1}{L} \int_0^L x \, dx = \frac{1}{L} \frac{1}{2} x^2 \Big|_0^L = \frac{L}{2} \tag{3.3.17b}$$

$$a_n = \frac{2}{L} \int_0^L x \cos \frac{n\pi x}{L} \, dx = \frac{2L}{(n\pi)^2} (\cos n\pi - 1). \tag{3.3.17c}$$

The latter integral can be evaluated by tables or integration by parts. We omit the details.

3.3.3 Representing f(x) by Both a Sine and Cosine Series

It may be apparent that any function $f(x)$ (which is piecewise smooth) may be represented both as a Fourier sine series and as a Fourier cosine series. The one you would use is dictated by the boundary conditions (if the problem arose in the context of a solution to a partial differential equation using the method of separation of variables). It is also possible to use a Fourier series (including both sines and cosines). As an example, we consider the sketches of the Fourier, Fourier sine, and Fourier cosine series of

$$f(x) = \begin{cases} -\dfrac{L}{2} \sin \dfrac{\pi x}{L} & x < 0 \\ x & 0 < x < \dfrac{L}{2} \\ L - x & x > \dfrac{L}{2}. \end{cases}$$

The graph of $f(x)$ is sketched for $-L < x < L$ in Fig. 3.3.15. The Fourier series of $f(x)$ is sketched by repeating this pattern with period $2L$. On the other hand for the Fourier sine (cosine) series, first sketch the odd (even) extension of the function $f(x)$, before repeating the pattern. These three are sketched in Fig. 3.3.16. Note that for $-L \leq x \leq L$ only the Fourier series of $f(x)$ actually equals

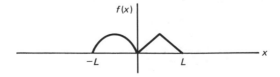

Figure 3.3.15 The graph of $f(x)$ for $-L < x < L$.

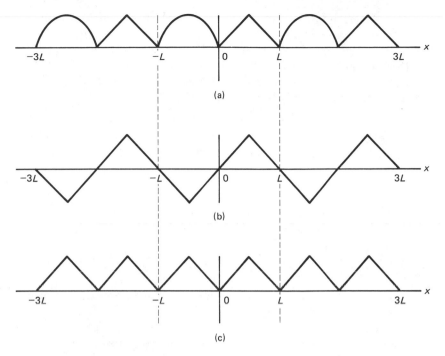

Figure 3.3.16 (a) Fourier series of the $f(x)$; (b) Fourier sine series of the $f(x)$; (c) Fourier cosine series of the $f(x)$.

$f(x)$. However, for all three cases the series equals $f(x)$ over the region $0 \leq x \leq L$.

3.3.4 Even and Odd Parts

Let us consider the Fourier series of a function $f(x)$ which is not necessarily even or odd:

$$f(x) \sim a_0 + \sum_{n=1}^{\infty} a_n \cos \frac{n\pi x}{L} + \sum_{n=1}^{\infty} b_n \sin \frac{n\pi x}{L}, \qquad (3.3.18)$$

where

$$a_0 = \frac{1}{2L} \int_{-L}^{L} f(x) \, dx$$

$$a_n = \frac{1}{L} \int_{-L}^{L} f(x) \cos \frac{n\pi x}{L} \, dx$$

$$b_n = \frac{1}{L} \int_{-L}^{L} f(x) \sin \frac{n\pi x}{L} \, dx.$$

It is interesting to see that a Fourier series is the sum of a series of cosines and a series of sines. For example, $\sum_{n=1}^{\infty} b_n \sin n\pi x/L$ is not in general the Fourier sine series of $f(x)$, because the coefficients, $b_n = 1/L \int_{-L}^{L} f(x) \sin n\pi x/L \, dx$, are *not* in general the same as the coefficients of a Fourier sine series

$[2/L \int_0^L f(x) \sin n\pi x/L \, dx]$. This series of sines by itself ought to be the Fourier sine series of some function; let us determine what function.

Equation (3.3.18) shows that $f(x)$ is represented as a sum of an even function (for the series of cosines must be an even function) and an odd function (similarly, the sine series must be odd). This is a general property of functions, since for any function it is rather obvious that

$$f(x) = \tfrac{1}{2}[f(x) + f(-x)] + \tfrac{1}{2}[f(x) - f(-x)]. \tag{3.3.19}$$

Note that the first bracketed term is an even function; we call it the **even part of** $f(x)$. The second bracketed term is an odd function, called the **odd part of** $f(x)$:

$$f_e(x) \equiv \tfrac{1}{2}[f(x) + f(-x)] \quad \text{and} \quad f_o(x) \equiv \tfrac{1}{2}[f(x) - f(-x)]. \tag{3.3.20}$$

In this way, any function is written as the sum of an odd function (the odd part) and an even function (the even part). For example, if $f(x) = 1/(1 + x)$,

$$\frac{1}{1+x} = \frac{1}{2}\left[\frac{1}{1+x} + \frac{1}{1-x}\right] + \frac{1}{2}\left[\frac{1}{1+x} - \frac{1}{1-x}\right] = \frac{1}{1-x^2} - \frac{x}{1-x^2};$$

the sum of an even function, $1/(1 - x^2)$, and an odd function, $-x/(1 - x^2)$. Consequently, the Fourier series of $f(x)$ equals the Fourier series of $f_e(x)$ [which is a cosine series since $f_e(x)$ is even] plus the Fourier series of $f_o(x)$ (which is a sine series since $f_o(x)$ is odd). This shows that the series of sines (cosines) that appears in (3.3.14) is the Fourier sine (cosine) series of $f_o(x)$ ($f_e(x)$). We summarize our result with the statement:

> *The Fourier series of $f(x)$ equals the Fourier sine series of $f_o(x)$ plus the Fourier cosine series of $f_e(x)$, where $f_e(x) = \tfrac{1}{2}[f(x) + f(-x)]$ and $f_o(x) = \tfrac{1}{2}[f(x) - f(-x)]$.*

Please do not confuse this result with even and odd *extensions*. For example, the *even part* of $f(x) = \tfrac{1}{2}[f(x) + f(-x)]$, while the

$$even \; extension \; \text{of} \; f(x) = \begin{cases} f(x), & x > 0 \\ f(-x), & x < 0. \end{cases}$$

3.3.5 Continuous Fourier Series

The convergence theorem for Fourier series shows that the Fourier series of $f(x)$ may be a different function than $f(x)$. Nevertheless, over the interval of interest, they are the same except at those few points where the periodic extension of $f(x)$ has a jump discontinuity. Sine (cosine) series are analyzed in the same way, where instead the odd (even) periodic extension must be considered. In addition to points of jump discontinuity of $f(x)$ itself, the various extensions of $f(x)$ may introduce a jump discontinuity. From the examples in the preceding section, we observe that *sometimes* the resulting series does not have any jump discontinuities. In these cases the Fourier series of $f(x)$ will actually equal $f(x)$

in the range of interest. Also, the Fourier series itself will be a continuous function.

It is worthwhile to summarize the conditions under which a Fourier series is continuous:

For piecewise smooth $f(x)$, the Fourier series of $f(x)$ is continuous for $-L \leq x \leq L$ if and only if $f(x)$ is continuous and $f(-L) = f(L)$.

It is necessary for $f(x)$ to be continuous; otherwise, there will be a jump discontinuity [and the Fourier series of $f(x)$ will converge to the average]. In Fig. 3.3.17 we illustrate the significance of the condition $f(-L) = f(L)$. We illustrate two *continuous* functions, only one of which satisfies $f(-L) = f(L)$. The condition $f(-L) = f(L)$ insists that the repeated pattern (with period $2L$) will be continuous at the end points. The boxed statement above is a fundamental result for all Fourier series. It explains the following similar theorems for Fourier sine and cosine series.

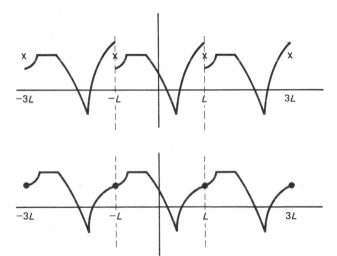

Figure 3.3.17 Fourier series of a continuous function with (a) $f(-L) \neq f(L)$ and (b) $f(-L) = f(L)$.

Consider the Fourier cosine series of $f(x)$ [$f(x)$ has been extended as an even function]. If $f(x)$ is continuous, is the Fourier cosine series continuous? An example that is continuous for $-L \leq x \leq L$ is sketched in Fig. 3.3.18. First we extend $f(x)$ evenly and then periodically. It is easily seen that:

For piecewise smooth $f(x)$, the Fourier cosine series of $f(x)$ is continuous for $0 \leq x \leq L$ if and only if $f(x)$ is continuous.

We note that no additional conditions on $f(x)$ are necessary for the cosine series to be continuous (besides $f(x)$ being continuous). One reason for this result is

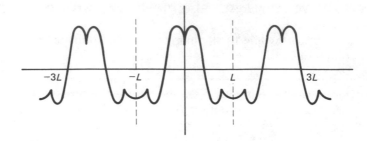

Figure 3.3.18 Fourier cosine series of a continuous function.

that if $f(x)$ is continuous for $0 \leq x \leq L$, then the even extension will be continuous for $-L \leq x \leq L$. Also note that the even extension is the same at $\pm L$. Thus, the periodic extension will automatically be continuous at the end points.

Compare this result to what happens for a Fourier sine series. Four examples are considered in Fig. 3.3.19, all continuous functions for $0 \leq x \leq L$. From the

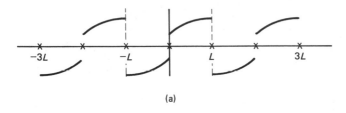

(a)

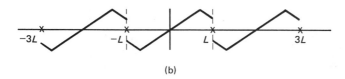

(b)

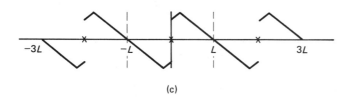

(c)

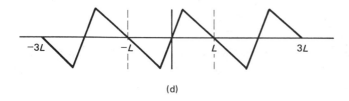

(d)

Figure 3.3.19 Fourier sine series of a continuous function with (a) $f(0) \neq 0$ and $f(L) \neq 0$; (b) $f(0) = 0$ but $f(L) \neq 0$; (c) $f(L) = 0$ but $f(0) \neq 0$; and (d) $f(0) = 0$ and $f(L) = 0$.

first three figures, we see that it is possible for the Fourier sine series of a continuous function to be discontinuous. It is seen that:

> For piecewise smooth functions $f(x)$, the Fourier sine series of $f(x)$ is continuous for $0 \leqslant x \leqslant L$ if and only if $f(x)$ is continuous and both $f(0) = 0$ *and* $f(L) = 0$.

If $f(0) \neq 0$, then the odd extension of $f(x)$ will have a jump discontinuity at $x = 0$, as illustrated in Fig. 3.3.19a and c. If $f(L) \neq 0$, then the odd extension at $x = -L$ will be of opposite sign from $f(L)$. Thus, the periodic extension will not be continuous at the end points if $f(L) \neq 0$ as in Fig. 3.3.19a and b.

EXERCISES 3.3

3.3.1. For the following functions, sketch $f(x)$, the Fourier series of $f(x)$, the Fourier sine series of $f(x)$, and the Fourier cosine series of $f(x)$.

(a) $f(x) = 1$ (b) $f(x) = 1 + x$

(c) $f(x) = \begin{cases} x & x < 0 \\ 1 + x & x > 0 \end{cases}$ *(d) $f(x) = e^x$

(e) $f(x) = \begin{cases} 2 & x < 0 \\ e^{-x} & x > 0 \end{cases}$

3.3.2. For the following functions, sketch the Fourier sine series of $f(x)$ and determine its Fourier coefficients.

(a) $f(x) = \cos \pi x/L$
(Verify the formula on p. 89)   (b) $f(x) = \begin{cases} 1 & x < L/6 \\ 3 & L/6 < x < L/2 \\ 0 & x > L/2 \end{cases}$

(c) $f(x) = \begin{cases} 0 & x < L/2 \\ x & x > L/2 \end{cases}$ *(d) $f(x) = \begin{cases} 1 & x < L/2 \\ 0 & x > L/2 \end{cases}$

3.3.3. For the following functions, sketch the Fourier sine series of $f(x)$. Also, roughly sketch the sum of a *finite* number of nonzero terms (at least the first two) of the Fourier sine series:

(a) $f(x) = \cos \pi x/L$ (Use the formula on p. 89.)
(b) $f(x) = \begin{cases} 1 & x < L/2 \\ 0 & x > L/2 \end{cases}$
(c) $f(x) = x$ (Use the formula on p. 88.)

3.3.4. Sketch the Fourier cosine series of $f(x) = \sin \pi x/L$. Briefly discuss.

3.3.5. For the following functions, sketch the Fourier cosine series of $f(x)$ and determine its Fourier coefficients:

(a) $f(x) = x^2$

(b) $f(x) = \begin{cases} 1 & x < L/6 \\ 3 & L/6 < x < L/2 \\ 0 & x > L/2 \end{cases}$

(c) $f(x) = \begin{cases} 0 & x < L/2 \\ x & x > L/2 \end{cases}$

3.3.6. For the following functions, sketch the Fourier cosine series of $f(x)$. Also roughly sketch the sum of a *finite* number of nonzero terms (at least the first two) of the Fourier cosine series:

(a) $f(x) = x$ (Use the formulas on p. 92.)

(b) $f(x) = \begin{cases} 0 & x < L/2 \\ 1 & x > L/2 \end{cases}$ (Use the formulas on p. 80.)

(c) $f(x) = \begin{cases} 1 & x < L/2 \\ 0 & x > L/2 \end{cases}$ [*Hint:* Add the functions in parts (b) and (c).]

3.3.7. Show that e^x is the sum of an even and an odd function.

3.3.8. (a) Determine formulas for the even extension of any $f(x)$. Compare to the formula for the even part of $f(x)$.

(b) Do the same for the odd extension of $f(x)$ and the odd part of $f(x)$.

(c) Calculate and sketch the four functions of parts (a) and (b) if

$$f(x) = \begin{cases} x & x > 0 \\ x^2 & x < 0 \end{cases}$$

Graphically add the even and odd parts of $f(x)$. What occurs? Similarly, add the even and odd extensions. What occurs then?

3.3.9. What is the sum of the Fourier sine series of $f(x)$ and the Fourier cosine series of $f(x)$? [What is the sum of the even and odd extensions of $f(x)$?]

*3.3.10. If $f(x) = \begin{cases} x^2 & x < 0 \\ e^{-x} & x > 0 \end{cases}$, what are the even and odd parts of $f(x)$?

3.3.11. Given a sketch of $f(x)$, describe a procedure to sketch the even and odd parts of $f(x)$.

3.3.12. (a) Graphically show that the even terms (n even) of the Fourier sine series of any function on $0 \leq x \leq L$ are odd (antisymmetric) around $x = L/2$.

(b) Consider a function $f(x)$ that is odd around $x = L/2$. Show that the odd coefficients (n odd) of the Fourier sine series of $f(x)$ on $0 \leq x \leq L$ are zero.

*3.3.13. Consider a function $f(x)$ that is even around $x = L/2$. Show that the even coefficients (n even) of the Fourier sine series of $f(x)$ on $0 \leq x \leq L$ are zero.

3.3.14. (a) Consider a function $f(x)$ which is even around $x = L/2$. Show that the odd coefficients (n odd) of the Fourier cosine series of $f(x)$ on $0 \leq x \leq L$ are zero.

(b) Explain the result of part (a) by considering a Fourier cosine series of $f(x)$ on the interval $0 \leq x \leq L/2$.

3.3.15. Consider a function $f(x)$ that is odd around $x = L/2$. Show that the even coefficients (n even) of the Fourier cosine series of $f(x)$ on $0 \leq x \leq L$ are zero.

3.3.16. Fourier series can be defined on other intervals besides $-L \leq x \leq L$. Suppose that $g(y)$ is defined for $a \leq y \leq b$. Represent $g(y)$ using periodic trigonometric functions with period $b - a$. Determine formulas for the coefficients. [*Hint:* Use the linear transformation

$$y = \frac{a + b}{2} + \frac{b - a}{2L}x.]$$

3.3.17. Consider

$$\int_0^1 \frac{dx}{1 + x^2}.$$

(a) Evaluate explicitly.

(b) Use the Taylor series of $1/(1 + x^2)$ (itself a geometric series) to obtain an infinite series for the integral.

(c) Equate part (a) to part (b) in order to derive a formula for π.

3.3.18. For continuous functions:

(a) Under what conditions does $f(x)$ equal its Fourier series for all x, $-L \leqslant x \leqslant L$?

(b) Under what conditions does $f(x)$ equal its Fourier sine series for all x, $0 \leqslant x \leqslant L$?

(c) Under what conditions does $f(x)$ equal its Fourier cosine series for all x, $0 \leqslant x \leqslant L$?

3.4 TERM-BY-TERM DIFFERENTIATION OF FOURIER SERIES

In solving partial differential equations by the method of separation of variables, the homogeneous boundary conditions sometimes suggest that the desired solution is either an infinite series of sines or cosines. For example, we consider one-dimensional heat conduction with zero boundary conditions. As before, we want to solve the initial boundary value problem

$$\frac{\partial u}{\partial t} = k \frac{\partial^2 u}{\partial x^2} \tag{3.4.1a}$$

$$u(0, t) = 0, \qquad u(L, t) = 0, \qquad u(x, 0) = f(x) \tag{3.4.1b,c,d}$$

By the method of separation of variables combined with the principle of superposition (taking a *finite* linear combination of solutions), we know that

$$u(x, t) - \sum_{n=1}^{N} b_n \sin \frac{n\pi x}{L} e^{-(n\pi/L)^2 kt}$$

solves the partial differential equation and the two homogeneous boundary conditions. To satisfy the initial conditions, in general an infinite series is needed. Does the infinite series

$$u(x, t) = \sum_{n=1}^{\infty} b_n \sin \frac{n\pi x}{L} e^{-(n\pi/L)^2 kt} \tag{3.4.2}$$

satisfy our problem? The theory of Fourier sine series shows that the Fourier coefficients b_n can be determined to satisfy any (piecewise smooth) initial condition [i.e., $b_n = 2/L \int_0^L f(x) \sin n\pi x/L \, dx$]. To see if the infinite series actually satisfies the partial differential equation, we substitute (3.4.2) into (3.4.1a). *If* the infinite Fourier series can be differentiated term by term, then

$$\frac{\partial u}{\partial t} = -\sum_{n=1}^{\infty} k \left(\frac{n\pi}{L}\right)^2 b_n \sin \frac{n\pi x}{L} e^{-(n\pi/L)^2 kt}$$

and

$$\frac{\partial^2 u}{\partial x^2} = -\sum_{n=1}^{\infty} \left(\frac{n\pi}{L}\right)^2 b_n \sin \frac{n\pi x}{L} e^{-(n\pi/L)^2 kt}.$$

Thus, the heat equation ($\partial u/\partial t = k \, \partial^2 u/\partial x^2$) is satisfied by the infinite Fourier series obtained by the method of separation of variables, if term-by-term differentiation of a Fourier series is valid.

Term-by-term differentiation of infinite series. Unfortunately, infinite series (even convergent infinite series) cannot always be differentiated term by term. It is not always true that

$$\frac{d}{dx} \sum_{n=1}^{\infty} c_n u_n = \sum_{n=1}^{\infty} c_n \frac{du_n}{dx};$$

the interchange of operations of differentiation and infinite summation is not always justified. However, we will find that in solving partial differential equations all the procedures we have performed on the infinite Fourier series are valid. We will state and prove some needed theorems concerning the validity of term-by-term differentiation of just the type of Fourier series that arise in solving partial differential equations.

Counterexample. Even for Fourier series, term-by-term differentiation is not always valid. To illustrate the difficulty in term-by-term differentiation, consider the Fourier sine series of x (on the interval $0 \leq x \leq L$) sketched in Fig. 3.4.1,

$$x = 2 \sum_{n=1}^{\infty} \frac{L}{n\pi}(-1)^{n+1} \sin \frac{n\pi x}{L}, \qquad \text{on } 0 \leq x < L,$$

as obtained earlier [see (3.3.11)]. If we differentiate the function on the left-hand side, then we have the function 1. However, if we formally differentiate term by term the function on the right, then we arrive at

$$2 \sum_{n=1}^{\infty} (-1)^{n+1} \cos \frac{n\pi x}{L}.$$

This is a cosine series, but it is not the cosine series of $f(x) = 1$ (the cosine series of 1 is just 1). Thus, Fig. 3.4.1 is an example where we cannot differentiate term by term.*

Figure 3.4.1 Fourier sine series of $f(x) = x$.

Fourier series. We claim this difficulty occurs any time the Fourier series of $f(x)$ has a jump discontinuity. Term-by-term differentiation is not justified in these situations. Instead, we claim (and prove in an exercise) that:

> *A Fourier series that is continuous can be differentiated term by term* if $f'(x)$ is piecewise smooth.

---

* In addition, the resulting infinite series does not even converge anywhere, since the nth term does not approach zero.

The result of term-by-term differentiation is the Fourier series of $f'(x)$, which may not be continuous. Similar results for sine and cosine series are of more frequent interest to the solution of our partial differential equations.

Fourier cosine series. For Fourier cosine series:

> If $f'(x)$ is piecewise smooth, then a continuous Fourier cosine series of $f(x)$ can be differentiated term by term.

The result of term-by-term differentiation is the Fourier sine series of $f'(x)$, which may not be continuous. Recall that $f(x)$ only needs to be continuous for its Fourier cosine series to be continuous. Thus, this theorem can be stated in the following alternative form:

> If $f'(x)$ is piecewise smooth, then *the Fourier cosine series of a continuous function $f(x)$ can be differentiated term by term.*

These statements apply to the Fourier cosine series of $f(x)$,

$$f(x) = \sum_{n=0}^{\infty} a_n \cos \frac{n\pi x}{L} \qquad 0 \leqslant x \leqslant L, \qquad (3.4.3)$$

where the $=$ sign means that the infinite series converges to $f(x)$ for all x $(0 \leqslant x \leqslant L)$ since $f(x)$ is continuous. Mathematically, these theorems state that term-by-term differentiation is valid,

$$f'(x) \sim -\sum_{n=1}^{\infty} \left(\frac{n\pi}{L}\right) a_n \sin \frac{n\pi x}{L}, \qquad (3.4.4)$$

where $\sim$ means equality where the Fourier sine series of $f'(x)$ is continuous and means the series converges to the average where the Fourier sine series of $f'(x)$ is discontinuous.

Example. Consider the Fourier cosine series of x [see (3.3.17)],

$$x = \frac{L}{2} - \frac{4L}{\pi^2} \sum_{\substack{n \text{ odd} \\ \text{only}}} \frac{1}{n^2} \cos \frac{n\pi x}{L}, \qquad 0 \leqslant x \leqslant L, \qquad (3.4.5)$$

as sketched in Fig. 3.4.2. Note the continuous nature of this series for $0 \leqslant x \leqslant L$, which results in the $=$ sign in (3.4.5). The derivative of this Fourier cosine series is sketched in Fig. 3.4.3; it is the Fourier sine series of $f(x) = 1$.

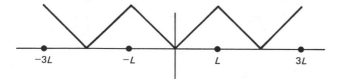

$-3L$ $-L$ L $3L$

Figure 3.4.2 Fourier cosine series of $f(x) = x$.

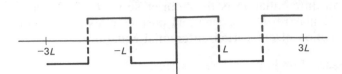

Figure 3.4.3 Fourier sine series of df/dx.

The Fourier sine series of $f(x) = 1$ can be obtained by term-by-term differentiation of the Fourier cosine series of $f(x) = x$. Assuming that term-by-term differentiation of (3.4.5) is valid as claimed, it follows that

$$1 \sim \frac{4}{\pi} \sum_{\substack{n \text{ odd} \\ \text{only}}} \frac{1}{n} \sin \frac{n\pi x}{L}, \tag{3.4.6}$$

which is in fact correct [see (3.3.8)].

Fourier sine series. A similar result is valid for Fourier sine series:

If $f'(x)$ is piecewise smooth, then a continuous Fourier sine series of $f(x)$ can be differentiated term by term.

However, if $f(x)$ is continuous, then the Fourier sine series is continuous only if $f(0) = 0$ and $f(L) = 0$. Thus, we must be careful in differentiating term by term a Fourier sine series. In particular,

If $f'(x)$ is piecewise smooth, then *the Fourier sine series of a continuous function $f(x)$ can only be differentiated term by term if $f(0) = 0$ and $f(L) = 0$.*

Proofs. The proofs of these theorems are all quite similar. We include one since it provides a way to learn more about Fourier series and their differentiability. We will prove the validity of term-by-term differentiation of the Fourier sine series of a *continuous* function $f(x)$:

$$f(x) \sim \sum_{n=1}^{\infty} b_n \sin \frac{n\pi x}{L}. \tag{3.4.7}$$

An equality holds in (3.4.7) only if $f(0) = f(L) = 0$. If $f'(x)$ is piecewise smooth, then $f'(x)$ has a Fourier cosine series

$$f'(x) \sim A_0 + \sum_{n=1}^{\infty} A_n \cos \frac{n\pi x}{L}. \tag{3.4.8}$$

This series will not converge to $f'(x)$ at points of discontinuity of $f'(x)$. We will have succeeded in showing a Fourier sine series may be term-by-term differentiated *if* we can verify that

$$f'(x) \sim \sum_{n=1}^{\infty} \left(\frac{n\pi}{L} \right) b_n \cos \frac{n\pi x}{L}$$

[i.e., if $A_0 = 0$ and $A_n = (n\pi/L)b_n$]. The Fourier cosine series coefficients are derived from (3.4.8). If we integrate by parts, we obtain

$$A_0 = \frac{1}{L}\int_0^L f'(x)\,dx = \frac{1}{L}[f(L) - f(0)] \tag{3.4.9a}$$

$$(n \neq 0) \quad A_n = \frac{2}{L}\int_0^L f'(x)\cos\frac{n\pi x}{L}\,dx$$

$$= \frac{2}{L}\left[f(x)\cos\frac{n\pi x}{L}\Big|_0^L + \frac{n\pi}{L}\int_0^L f(x)\sin\frac{n\pi x}{L}\,dx\right]. \tag{3.4.9b}$$

But from (3.4.7), b_n is the Fourier sine series coefficient of $f(x)$,

$$b_n = \frac{2}{L}\int_0^L f(x)\sin\frac{n\pi x}{L}\,dx,$$

and thus for $n \neq 0$,

$$A_n = \frac{n\pi}{L}b_n + \frac{2}{L}[(-1)^n f(L) - f(0)]. \tag{3.4.10}$$

We thus see by comparing Fourier cosine coefficients that the Fourier sine series can be term-by-term differentiated only if both $f(L) - f(0) = 0$ (so that $A_0 = 0$) and $(-1)^n f(L) - f(0) = 0$ [so that $A_n = (n\pi/L)\,b_n$]. Both of these conditions hold only if

$$f(0) = f(L) = 0,$$

exactly the conditions for a Fourier sine series of a continuous function to be continuous. Thus, we have completed the proof. However, this demonstration has given us more information. Namely, it gives the formula to differentiate the Fourier sine series of a continuous function when the series is not continuous. We have that:

If $f'(x)$ is piecewise smooth, then *the Fourier sine series of a continuous function $f(x)$,*

$$f(x) \sim \sum_{n=1}^{\infty} b_n \sin\frac{n\pi x}{L}$$

cannot, in general, be differentiated term by term. However,

$$f'(x) \sim \frac{1}{L}[f(L) - f(0)]$$

$$+ \sum_{n=1}^{\infty}\left[\frac{n\pi}{L}b_n + \frac{2}{L}((-1)^n f(L) - f(0))\right]\cos\frac{n\pi x}{L}. \tag{3.4.11}$$

Do not have to memorize.

In this proof, it may appear that we never needed $f(x)$ to be continuous. However, we applied integration by parts in order to derive (3.4.8). In the usual presentation in calculus, integration by parts is stated as being valid if both $u(x)$ and $v(x)$ *and their derivatives* are continuous. This is overly restrictive for our work. As is clarified somewhat in an exercise, we state that integration by parts is valid if only $u(x)$ and $v(x)$ are continuous. It is not necessary for their derivatives to be continuous. Thus, the result of integration by parts is valid only if $f(x)$ is continuous.

Example. Let us reconsider the Fourier sine series of $f(x) = x$,

$$x \sim 2 \sum_{n=1}^{\infty} \frac{L}{n\pi} (-1)^{n+1} \sin \frac{n\pi x}{L}. \tag{3.4.12}$$

We already know that $(d/dx)x = 1$ does not have a Fourier cosine series which results from term-by-term differentiation of (3.4.12) since $f(L) \neq 0$. However, (3.4.11) may be applied since $f(x)$ is continuous [and $f'(x)$ is piecewise smooth]. Noting that $f(0) = 0$, $f(L) = L$, and $(n\pi/L)b_n = 2(-1)^{n+1}$, it follows that the Fourier cosine series of df/dx is

$$\frac{df}{dx} \sim 1.$$

The constant function 1 is exactly the Fourier cosine series of df/dx since $f = x$ implies that $df/dx = 1$. Thus, the r.h.s. of (3.4.11) gives the correct expression for the Fourier cosine series of $f'(x)$ when the Fourier sine series of $f(x)$ is known, even if $f(0) \neq 0$ and/or $f(L) \neq 0$.

Method of eigenfunction expansion. Let us see how our results, concerning the conditions under which a Fourier series may be differentiated term by term, may be applied to our study of partial differential equations. We consider the heat equation (3.4.1) with zero boundary conditions at $x = 0$ and $x = L$. We will show that (3.4.2) is the correct infinite series representation of the solution of this problem. We will show this by utilizing an alternative scheme to obtain (3.4.2), known as the **method of eigenfunction expansion**, whose importance is that it may also be used when there are sources or the boundary conditions are not homogeneous (see Exercises 3.4.9–3.4.12 and Chapter 7). We begin by expanding the unknown solution $u(x, t)$ in terms of the eigenfunctions of the problem (with homogeneous boundary conditions). In this example, the eigenfunctions are $\sin n\pi x/L$, suggesting a Fourier sine series for each time:

$$u(x, t) \sim \sum_{n=1}^{\infty} B_n(t) \sin \frac{n\pi x}{L}; \tag{3.4.13}$$

the Fourier sine coefficients, B_n, will depend on time, $B_n(t)$. Equation (3.4.13) is valid if $u(x, t)$ is piecewise smooth. In fact, our physical formulation requires more, namely, that $u(x, t)$ and $\partial u/\partial x$ are continuous functions of x. Furthermore, it will be convenient to assume that $\partial^2 u/\partial x^2$ is at least piecewise smooth.

The initial condition [$u(x, 0) = f(x)$] is satisfied if

$$f(x) \sim \sum_{n=1}^{\infty} B_n(0) \sin \frac{n\pi x}{L}, \tag{3.4.14}$$

determining the Fourier sine coefficient initially

$$B_n(0) = \frac{2}{L} \int_0^L f(x) \sin \frac{n\pi x}{L} \, dx. \tag{3.4.15}$$

All that remains is to investigate whether the Fourier sine series representation of $u(x, t)$, (3.4.13), can satisfy the heat equation, $\partial u/\partial t = k\partial^2 u/\partial x^2$. To do that we must differentiate the Fourier sine series. It is here that our results concerning term-by-term differentiation are useful.

First we need to compute two derivatives with respect to x. If $u(x, t)$ is continuous, then the Fourier sine series of $u(x, t)$ can be differentiated term by term if $u(0, t) = 0$ and $u(L, t) = 0$. Since these are exactly the boundary conditions on $u(x, t)$, it follows from (3.4.13) that

$$\frac{\partial u}{\partial x} \sim \sum_{n=1}^{\infty} \frac{n\pi}{L} B_n(t) \cos \frac{n\pi x}{L}. \tag{3.4.16}$$

Since $\partial u/\partial x$ is also assumed to be continuous, an equality holds in (3.4.16). Furthermore, the Fourier cosine series of $\partial u/\partial x$ can now be term-by-term differentiated, yielding

$$\frac{\partial^2 u}{\partial x^2} \sim -\sum_{n=1}^{\infty} \left(\frac{n\pi}{L}\right)^2 B_n(t) \sin \frac{n\pi x}{L}. \tag{3.4.17}$$

Note the importance of the separation of variables solution. Sines were differentiated at the stage in which the boundary conditions occurred that allowed sines to be differentiated. Cosines occurred with no boundary condition, consistent with the fact that a Fourier cosine series does not need any subsidiary conditions in order to be differentiated. To complete the substitution of the Fourier sine series into the partial differential equation, we need only to compute $\partial u/\partial t$. If we can also term-by-term differentiate with respect to t, then

$$\frac{\partial u}{\partial t} \sim \sum_{n=1}^{\infty} \frac{dB_n}{dt} \sin \frac{n\pi x}{L}. \tag{3.4.18}$$

If this last term-by-term differentiation is justified, we see that the Fourier sine series (3.4.13) solves the partial differential equation if

$$\frac{dB_n}{dt} = -k \left(\frac{n\pi}{L}\right)^2 B_n(t). \tag{3.4.19}$$

The Fourier sine coefficient $B_n(t)$ satisfies a first-order linear differential equation with constant coefficients. The solution of (3.4.19) is

$$B_n(t) = B_n(0)e^{-(n\pi/L)^2 kt},$$

where $B_n(0)$ is given by (3.4.15). Thus, we have derived that (3.4.2) is valid, justifying the method of separation of variables.

Can we justify term-by-term differentiation with respect to the parameter t? The following theorem states the conditions under which this operation is valid:

The Fourier series of a continuous function $u(x, t)$ (depending on a parameter t)

$$u(x, t) = a_0(t) + \sum_{n=1}^{\infty} \left[a_n(t) \cos \frac{n\pi x}{L} + b_n(t) \sin \frac{n\pi x}{L} \right]$$

can be differentiated term by term with respect to the parameter t, yielding

$$\frac{\partial}{\partial t} u(x, t) \sim a_0'(t) + \sum_{n=1}^{\infty} \left[a_n'(t) \cos \frac{n\pi x}{L} + b_n'(t) \sin \frac{n\pi x}{L} \right]$$

if $\partial u / \partial t$ is piecewise smooth.

We omit its proof (see Exercise 3.4.7), which depends on the fact that

$$\frac{\partial}{\partial t} \int_{-L}^{L} g(x, t) \, dx = \int_{-L}^{L} \frac{\partial g}{\partial t} \, dx$$

is valid if g is continuous.

In summary, we have verified that the Fourier sine series is actually a solution of the heat equation satisfying the boundary conditions $u(0, t) = 0$ and $u(L, t) = 0$. Now we have two reasons for choosing a Fourier sine series for this problem. First, the method of separation of variables implies that if $u(0, t) = 0$ and $u(L, t) = 0$, then the appropriate eigenfunctions are $\sin n\pi x/L$. Second, we now see that all the differentiations of the infinite sine series are justified, where we need to assume that $u(0, t) = 0$ and $u(L, t) = 0$, exactly the physical boundary conditions.

EXERCISES 3.4

3.4.1. The integration-by-parts formula

$$\int_a^b u \frac{dv}{dx} \, dx = uv \Big|_a^b - \int_a^b v \frac{du}{dx} \, dx$$

is known to be valid for functions $u(x)$ and $v(x)$ which are continuous *and* have continuous first derivatives. However, we will assume that u, v, du/dx, and dv/dx are continuous only for $a \leqslant x \leqslant c$ and $c \leqslant x \leqslant b$; we assume that all quantities may have a jump discontinuity at $x = c$.
   *(a) Derive an expression for $\int_a^b u \, dv/dx \, dx$ in terms of $\int_a^b v \, du/dx \, dx$.
   (b) Show that this reduces to the integration-by-parts formula if u and v are continuous across $x = c$. It is *not* necessary for du/dx and dv/dx to be continuous at $x = c$!

3.4.2. Suppose that $f(x)$ and df/dx are piecewise smooth. Prove that the Fourier series of $f(x)$ can be differentiated term by term if the Fourier series of $f(x)$ is continuous.

3.4.3. Suppose that $f(x)$ is continuous [except for a jump discontinuity at $x = x_0$, $f(x_0^-) = \alpha$ and $f(x_0^+) = \beta$] and df/dx is piecewise smooth.
   *(a) Determine the Fourier sine series of df/dx in terms of the Fourier cosine series coefficients of $f(x)$.

(b) Determine the Fourier cosine series of df/dx in terms of the Fourier sine series coefficients of $f(x)$.

3.4.4. Suppose that $f(x)$ and df/dx are piecewise smooth.

(a) Prove that the Fourier sine series of a continuous function $f(x)$ can only be differentiated term by term if $f(0) = 0$ and $f(L) = 0$.

(b) Prove that the Fourier cosine series of a continuous function $f(x)$ can be differentiated term by term.

3.4.5. From the information on page 89, determine the Fourier cosine series of $\sin \pi x/L$.

3.4.6. There are some things wrong in the following demonstration. Find the mistakes and correct them.

In this problem we attempt to obtain the Fourier cosine coefficients of e^x.

$$e^x = A_0 + \sum_{n=1}^{\infty} A_n \cos \frac{n\pi x}{L}. \tag{1}$$

Differentiating yields

$$e^x - -\sum_{n=1}^{\infty} \frac{n\pi}{L} A_n \sin \frac{n\pi x}{L},$$

the Fourier sine series of e^x. Differentiating again yields

$$e^x = -\sum_{n=1}^{\infty} \left(\frac{n\pi}{L}\right)^2 A_n \cos \frac{n\pi x}{L}. \tag{2}$$

Since equations (1) and (2) give Fourier cosine series of e^x, they must be identical. Thus,

$$\left.\begin{array}{l} A_0 = 0 \\ A_n = 0 \end{array}\right\} \quad \text{(obviously wrong!)}$$

By correcting the mistakes you should be able to obtain A_0 and A_n *without* using the typical technique, that is, $A_n = 2/L \int_0^L e^x \cos n\pi x/L \, dx$.

3.4.7. Prove that the Fourier series of a continuous function $u(x, t)$ can be differentiated term by term with respect to the parameter t if $\partial u/\partial t$ is piecewise smooth.

3.4.8. Consider

$$\frac{\partial u}{\partial t} = k \frac{\partial^2 u}{\partial x^2}$$

subject to $\partial u/\partial x \, (0, t) = 0$, $\partial u/\partial x \, (L, t) = 0$, and $u(x, 0) = f(x)$. Solve in the following way. Look for the solution as a Fourier cosine series. Assume that u and $\partial u/\partial x$ are continuous and $\partial^2 u/\partial x^2$ and $\partial u/\partial t$ are piecewise smooth. Justify all differentiations of infinite series.

*3.4.9. Consider the heat equation with a *known* source $q(x, t)$:

$$\frac{\partial u}{\partial t} = k \frac{\partial^2 u}{\partial x^2} + q(x, t) \quad \text{with} \quad u(0, t) = 0 \quad \text{and} \quad u(L, t) = 0.$$

Assume that $q(x, t)$ (for each $t > 0$) is a piecewise smooth function of x. Also assume that u and $\partial u/\partial x$ are continuous functions of x (for $t > 0$) and $\partial^2 u/\partial x^2$ and $\partial u/\partial t$ are piecewise smooth. Thus,

$$u(x, t) = \sum_{n=1}^{\infty} b_n(t) \sin \frac{n\pi x}{L}.$$

What ordinary differential equation does $b_n(t)$ satisfy? Do not solve this differential equation.

3.4.10. Modify Exercise 3.4.9 if instead $\partial u/\partial x(0, t) = 0$ and $\partial u/\partial x(L, t) = 0$.

3.4.11. Consider the *nonhomogeneous* heat equation (with a steady heat source):

$$\frac{\partial u}{\partial t} = k\frac{\partial^2 u}{\partial x^2} + g(x).$$

Solve this equation with the initial condition

$$u(x, 0) = f(x)$$

and the boundary conditions

$$u(0, t) = 0 \quad \text{and} \quad u(L, t) = 0.$$

Assume that a continuous solution exists (with continuous derivatives). [*Hints:* Expand the solution as a Fourier sine series (i.e., use the method of eigenfunction expansion). Expand $g(x)$ as a Fourier sine series. Solve for the Fourier sine series of the solution. Justify all differentiations with respect to x.]

*3.4.12. Solve the following *nonhomogeneous* problem:

$$\frac{\partial u}{\partial t} = k\frac{\partial^2 u}{\partial x^2} + e^{-t} + e^{-2t}\cos\frac{3\pi x}{L} \qquad [\text{assume that } 2 \neq k(3\,\pi/L)^2]$$

subject to

$$\frac{\partial u}{\partial x}(0, t) = 0, \frac{\partial u}{\partial x}(L, t) = 0, \quad \text{and} \quad u(x, 0) = f(x).$$

Use the following method. Look for the solution as a Fourier cosine series. Justify all differentiations of infinite series (assume appropriate continuity).

3.4.13. Consider

$$\frac{\partial u}{\partial t} = k\frac{\partial^2 u}{\partial x^2}$$

subject to

$$u(0, t) = A(t), \quad u(L, t) = 0, \quad \text{and} \quad u(x, 0) = g(x).$$

Assume that $u(x, t)$ has a Fourier sine series. Determine a differential equation for the Fourier coefficients (assume appropriate continuity).

3.5 TERM-BY-TERM INTEGRATION OF FOURIER SERIES

In doing mathematical manipulations with infinite series, we must remember that some properties of finite series do not hold for infinite series. In particular, Sec. 3.4 indicated that we must be especially careful differentiating term by term an infinite Fourier series. The following theorem, however, enables us to integrate Fourier series without caution:

A Fourier series of piecewise-smooth $f(x)$ can always be integrated term by term and the result is a *convergent* infinite series that always *converges* to the integral of $f(x)$ for $-L \leq x \leq L$ (even if the original Fourier series has jump discontinuities).

Remarkably, the new series formed by term-by-term integration is continuous. However, the new series may not be a Fourier series.

To quantify this statement, let us suppose that $f(x)$ is piecewise smooth and hence has a Fourier series in the range $-L \leq x \leq L$ (not necessarily continuous):

$$f(x) \sim a_0 + \sum_{n=1}^{\infty} a_n \cos \frac{n\pi x}{L} + \sum_{n=1}^{\infty} b_n \sin \frac{n\pi x}{L}. \tag{3.5.1}$$

We will prove our claim that we can just integrate this result term by term:

$$\int_{-L}^{x} f(t) \, dt \sim a_0(x + L) + \sum_{n=1}^{\infty} \left(a_n \int_{-L}^{x} \cos \frac{n\pi t}{L} \, dt + b_n \int_{-L}^{x} \sin \frac{n\pi t}{L} \, dt \right).$$

Performing the indicated integration yields

$$\int_{-L}^{x} f(t) \, dt \sim a_0(x + L)$$

$$+ \sum_{n=1}^{\infty} \left[\frac{a_n}{n\pi/L} \sin \frac{n\pi x}{L} + \frac{b_n}{n\pi/L} \left(\cos n\pi - \cos \frac{n\pi x}{L} \right) \right]. \tag{3.5.2}$$

We will actually show that the above statement is valid with an $=$ sign. If term-by-term integration from $-L$ to x of a Fourier series is valid, then any definite integration is also valid since

$$\int_{a}^{b} = \int_{-L}^{b} - \int_{-L}^{a}.$$

Example. Term-by-term integration has some interesting applications. Recall that the Fourier sine series for $f(x) = 1$ is given by

$$1 \sim \frac{4}{\pi} \left(\sin \frac{\pi x}{L} + \frac{1}{3} \sin \frac{3\pi x}{L} + \frac{1}{5} \sin \frac{5\pi x}{L} + \cdots \right), \tag{3.5.3}$$

where $\sim$ is used since (3.5.3) is an equality only for $0 < x < L$. Integrating term by term from 0 to x results in

$$x = \frac{4L}{\pi^2} \left(1 + \frac{1}{3^2} + \frac{1}{5^2} + \cdots \right)$$

$$- \frac{4L}{\pi^2} \left(\cos \frac{\pi x}{L} + \frac{\cos 3\pi x/L}{3^2} + \frac{\cos 5\pi x/L}{5^2} + \cdots \right), \quad 0 \leq x \leq L, \tag{3.5.4}$$

where because of our theorem the $=$ sign can be used. We immediately recognize that (3.5.4) should be the Fourier cosine series of the function x. It was obtained by integrating the Fourier sine series of $f(x) = 1$. However, an infinite series of constants appears in (3.5.4); it is the constant term of the Fourier cosine series of x. In this way we can evaluate that infinite series,

$$\frac{4L}{\pi^2} \left(1 + \frac{1}{3^2} + \frac{1}{5^2} + \cdots \right) = \frac{1}{L} \int_{0}^{L} x \, dx = \frac{1}{2} L.$$

Thus, we obtain the usual form for the Fourier cosine series for x,

$$x = \frac{L}{2} - \frac{4L}{\pi^2} \left(\cos \frac{\pi x}{L} + \frac{\cos 3\pi x/L}{3^2} + \frac{\cos 5\pi x/L}{5^2} + \cdots \right), \quad 0 \leq x \leq L. \tag{3.5.5}$$

The process of deriving new series from old ones can be continued. Integrating (3.5.5) from 0 to x yields

$$\frac{x^2}{2} = \frac{L}{2}x - \frac{4L^2}{\pi^3}\left(\sin\frac{\pi x}{L} + \frac{\sin 3\pi x/L}{3^3} + \frac{\sin 5\pi x/L}{5^3} + \cdots\right). \qquad (3.5.6)$$

This example illustrates that *integrating a Fourier series term by term does not necessarily yield another Fourier series*. However, (3.5.6) can be looked at as either yielding:

1. The Fourier sine series of $x^2/2 - (L/2)x$, or
2. The Fourier sine series of $x^2/2$, where the Fourier sine series of x is needed first [see (3.3.11)].

An alternative procedure is to perform indefinite integration. In this case an arbitrary constant must be included and evaluated. For example, reconsider the Fourier sine series of $f(x) = 1$, (3.5.3). By term-by-term indefinite integration we derive the Fourier cosine series of x,

$$x = c - \frac{4L}{\pi^2}\left(\cos\frac{\pi x}{L} + \frac{\cos 3\pi x/L}{3^2} + \frac{\cos 5\pi x/L}{5^2} + \cdots\right).$$

The constant of integration is not arbitrary; it must be evaluated. Here c is again the constant term of the Fourier cosine series of x, $c = (1/L)\int_0^L x\,dx = L/2$.

Proof on integrating Fourier series. Consider

$$F(x) = \int_{-L}^{x} f(t)\,dt. \qquad (3.5.7)$$

This integral is a continuous function of x since $f(x)$ is piecewise smooth. $F(x)$ has a continuous Fourier series only if $F(L) = F(-L)$ [otherwise, remember that the periodic nature of the Fourier series implies that the Fourier series does not converge to $f(x)$ at the end points $x = \pm L$]. However, note that from the definition (3.5.7),

$$F(-L) = 0 \qquad \text{and} \qquad F(L) = \int_{-L}^{L} f(t)\,dt = 2La_0.$$

Thus, in general $F(x)$ does not have a continuous Fourier series. In Fig. 3.5.1, $F(x)$ is sketched, illustrating the fact that usually $F(-L) \neq F(L)$. However, consider the straight line connecting the point $F(-L)$ to $F(L)$, $y = a_0(x + L)$. $G(x)$, defined to be the difference between $F(x)$ and the straight line,

$$G(x) \equiv F(x) - a_0(x + L), \qquad (3.5.8)$$

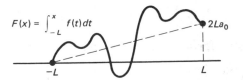

$$F(x) = \int_{-L}^{x} f(t)\,dt$$
$2La_0$
$-L$ L

Figure 3.5.1 $F(x)$ with $F(-L) \neq F(L)$.

Fourier Series   Chap. 3

will be zero at both ends, $x = \pm L$,
$$G(-L) = G(L) = 0,$$
as illustrated in Fig. 3.5.1. $G(x)$ is also continuous. Thus, $G(x)$ satisfies the properties that enable the Fourier series of $G(x)$ actually to equal $G(x)$:

$$G(x) = A_0 + \sum_{n=1}^{\infty} \left(A_n \cos \frac{n\pi x}{L} + B_n \sin \frac{n\pi x}{L} \right), \qquad (3.5.9)$$

where the $=$ sign is emphasized. These Fourier coefficients can be computed as

$$A_n = \frac{1}{L} \int_{-L}^{L} [F(x) - a_0(x + L)] \cos \frac{n\pi x}{L} \, dx \qquad (n \neq 0).$$

The x term can be dropped since it is odd (i.e., $\int_{-L}^{L} x \cos n\pi x/L \, dx = 0$). The resulting expression can be integrated by parts as follows:

$$u = F(x) - a_0 L \qquad\qquad dv = \cos \frac{n\pi x}{L} \, dx$$

$$du = \frac{dF}{dx} \, dx = f(x) \, dx \qquad v = \frac{L}{n\pi} \sin \frac{n\pi x}{L},$$

yielding

$$A_n = \frac{1}{L} \left[(F(x) - a_0 L) \frac{\sin n\pi x/L}{n\pi/L} \Big|_{-L}^{L} \right.$$
$$\left. - \frac{L}{n\pi} \int_{-L}^{L} f(x) \sin \frac{n\pi x}{L} \, dx \right] = -\frac{b_n}{n\pi/L}, \qquad (3.5.10)$$

where we have recognized that b_n is the Fourier sine coefficient of $f(x)$. In a similar manner (which we leave as an exercise) it can be shown that

$$B_n = \frac{a_n}{n\pi/L}.$$

A_0 can be calculated in a different manner (the previous method will not work). Since $G(L) = 0$ [and the Fourier series of $G(x)$ is pointwise convergent], from (3.5.9) it follows that

$$0 = A_0 + \sum_{n=1}^{\infty} A_n \cos n\pi = A_0 - \sum_{n=1}^{\infty} \frac{b_n}{n\pi/L} \cos n\pi$$

since $A_n = -b_n/(n\pi/L)$. Thus, we have shown from (3.5.9) that

$$F(x) = a_0(x + L) + \sum_{n=1}^{\infty} \left[\frac{a_n}{n\pi/L} \sin \frac{n\pi x}{L} + \frac{b_n}{n\pi/L} \left(\cos n\pi - \cos \frac{n\pi x}{L} \right) \right],$$
$$(3.5.11)$$

exactly the result of simple term-by-term integration. However, notice that (3.5.11) is *not* the Fourier series of $F(x)$, since $a_0 x$ appears. Nonetheless, (3.5.11) is valid. We have now justified term-by-term integration of Fourier series.

EXERCISES 3.5

3.5.1. Consider

$$x^2 \sim \sum_{n=1}^{\infty} a_n \sin \frac{n\pi x}{L}. \qquad (3.5.12)$$

  (a) Determine a_n from (3.5.6) and (3.3.17).
  (b) For what values of x is (3.5.12) an equality?
  *(c) Derive the Fourier cosine series for x^3 from (3.5.12). For what values of x will this be an equality?

3.5.2. (a) Using (3.3.11), obtain the Fourier cosine series of x^2.
  (b) From part (a), determine the Fourier sine series of x^3.

3.5.3. Generalize Exercise 3.5.2, in order to derive the Fourier sine series of x^m, m odd.

***3.5.4.** Suppose that $\cosh x \sim \sum_{n=1}^{\infty} b_n \sin n\pi x/L$.
  (a) Determine b_n by correctly differentiating this series twice.
  (b) Determine b_n by integrating this series twice.

3.5.5. Show that B_n in (3.5.9) satisfies $B_n = a_n/(n\pi/L)$, where a_n is defined by (3.5.1).

3.5.6. Evaluate

$$1 + \frac{1}{2^2} + \frac{1}{3^2} + \frac{1}{4^2} + \frac{1}{5^2} + \frac{1}{6^2} + \cdots$$

by evaluating (3.5.5) at $x = 0$.

***3.5.7.** Evaluate

$$1 - \frac{1}{3^3} + \frac{1}{5^3} - \frac{1}{7^3} + \cdots$$

using (3.5.6).

4

Vibrating Strings and Membranes

4.1 INTRODUCTION

At this point in our study of partial differential equations, the only physical problem we have introduced is the conduction of heat. To broaden the scope of our discussions, we now investigate the vibrations of perfectly elastic strings and membranes. We begin by formulating the governing equations for a vibrating string from physical principles. The appropriate boundary conditions will be shown to be similar in a mathematical sense to those boundary conditions for the heat equations. Examples will be solved by the method of separation of variables.

4.2 DERIVATION OF A VERTICALLY VIBRATING STRING

A vibrating string is a complicated physical system. We would like to present a simple derivation. A string vibrates only if it is tightly stretched. Consider a horizontally stretched string in its equilibrium configuration, as illustrated in Fig. 4.2.1. We imagine that the ends are tied down in some way (to be described in Sec. 4.3), maintaining the tightly stretched nature of the string. You may wish to think of stringed musical instruments as examples. We begin by tracking the motion of each particle that comprises the string. We let α be the x-coordinate of a particle when the string is in the horizontal equilibrium position. The string moves in time; it is located somewhere other than the equilibrium position at time t, as illustrated in Fig. 4.2.1. The trajectory of particle α is indicated with both horizontal and vertical components.

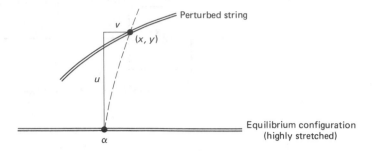

Figure 4.2.1 Vertical and horizontal displacements of a particle on a highly stretched string.

We will assume the slope of the string is small, in which case it can be shown that the horizontal displacement v can be neglected. As an approximation, **the motion is entirely vertical**, $x = \alpha$. In this situation,* the vertical displacement u depends on x and t:

$$y = u(x, t). \qquad (4.2.1)$$

Derivations including the effect of a horizontal displacement are necessarily complicated (see Weinberger [1965] and Antman [1980]).

Newton's law. We consider an infinitesimally thin segment of the string contained between x and $x + \Delta x$ (as illustrated in Fig. 4.2.2). In the unperturbed (yet stretched) horizontal position, we assume the **mass density** $\rho_0(x)$ is known. For the thin segment, the total mass is approximately $\rho_0(x) \, \Delta x$. Our object is to derive a partial differential equation describing how the displacement u changes in time. Accelerations are due to forces; we must use Newton's law. For simplicity we will analyze Newton's law for a point mass:

$$\mathbf{F} = m\mathbf{a}. \qquad (4.2.2)$$

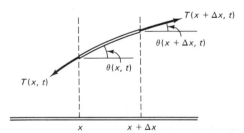

Figure 4.2.2 Stretching of a finite segment of string, illustrating the tensile forces.

We must discuss the forces acting on this segment of the string. There are body forces, which we assume act only in the vertical direction (e.g., the gravitational force), as well as forces acting on the ends of the segment of string. We assume that the string is **perfectly flexible**; it offers no resistance to bending. This means that the force exerted by the rest of the string on the end points of the segment of the string is in the direction tangent to the string. This tangential force is

* In general ($x \neq \alpha$), it is best to let $y = u(\alpha, t)$.

known as the **tension** in the string, and we denote its magnitude by $T(x, t)$. In Fig. 4.2.2 we show that the force due to the tension (exerted by the rest of the string) pulls at both ends in the direction of the tangent, trying to stretch the small segment. To obtain components of the tensile force, the angle θ between the horizon and the string is introduced. The angle depends on both the position x and time t. Furthermore, the slope of the string may be represented as either dy/dx or $\tan \theta$:

$$\frac{dy}{dx} = \tan \theta(x, t) = \frac{\partial u}{\partial x}. \tag{4.2.3}$$

The horizontal component of Newton's law prescribes the horizontal motion, which we claim is small and can be neglected. The vertical equation of motion states that the mass $\rho_0(x) \Delta x$ times the vertical component of acceleration ($\partial^2 u/\partial t^2$, where $\partial/\partial t$ is used since x is fixed for this motion) equals the vertical component of the tensile forces plus the vertical component of the body forces:

$$\rho_0(x) \Delta x \frac{\partial^2 u}{\partial t^2} = T(x + \Delta x, t) \sin \theta(x + \Delta x, t)$$
$$- T(x, t)\sin \theta(x, t) + \rho_0(x) \Delta x\, Q(x, t), \tag{4.2.4}$$

where $T(x, t)$ is the magnitude of the tensile force and $Q(x, t)$ is the vertical component of the body force per unit mass. Dividing (4.2.4) by Δx and taking the limit as $\Delta x \to 0$ yields

$$\rho_0(x) \frac{\partial^2 u}{\partial t^2} = \frac{\partial}{\partial x} [T(x, t)\sin \theta(x, t)] + \rho_0(x)Q(x, t). \tag{4.2.5}$$

For small angles θ,

$$\frac{\partial u}{\partial x} = \tan \theta = \frac{\sin \theta}{\cos \theta} \approx \sin \theta,$$

and hence (4.2.5) becomes

$$\rho_0(x) \frac{\partial^2 u}{\partial t^2} = \frac{\partial}{\partial x} \left(T \frac{\partial u}{\partial x} \right) + \rho_0(x)\, Q(x, t). \tag{4.2.6}$$

Perfectly elastic strings. The tension of a string is determined by experiments. Real strings are nearly **perfectly elastic**, by which we mean that the magnitude of the tensile force $T(x, t)$ depends only on the local stretching of the string. Since the angle θ is assumed to be small, the stretching of the string is nearly the same as for the unperturbed highly stretched horizontal string, where the tension is constant, T_0 (to be in equilibrium). Thus, **the tension $T(x, t)$ may be approximated by a constant** T_0. Consequently, the small vertical vibrations of a highly stretched string are governed by

$$\rho_0(x) \frac{\partial^2 u}{\partial t^2} = T_0 \frac{\partial^2 u}{\partial x^2} + Q(x, t)\, \rho_0(x). \tag{4.2.7}$$

One-dimensional wave equation. If the only body force per unit mass is gravity, then $Q(x, t) = -g$ in (4.2.7). In many such situations, this force is small (relative to the tensile force $\rho_0 g \ll |T_0 \partial^2 u/\partial x^2|$) and can be neglected.

Alternatively, gravity sags the string, and we can calculate the vibrations with respect to the sagged equilibrium position. In either way we are often lead to investigate (4.2.7) in the case in which $Q(x, t) = 0$,

$$\rho_0(x) \frac{\partial^2 u}{\partial t^2} = T_0 \frac{\partial^2 u}{\partial x^2} \tag{4.2.8a}$$

or

$$\frac{\partial^2 u}{\partial t^2} = c^2 \frac{\partial^2 u}{\partial x^2}, \tag{4.2.8b}$$

where $c^2 = T_0/\rho_0(x)$. Equation (4.2.8b) is called the **one-dimensional wave equation**. The notation c^2 is introduced because $T_0/\rho_0(x)$ has the dimensions of velocity squared. We will show that c is a very important velocity. For a uniform string, c is constant.

EXERCISES 4.2

4.2.1. (a) Using (4.2.7), compute the sagged equilibrium position $u_E(x)$ if $Q(x, t) = -g$. The boundary conditions are $u(0) = 0$ and $u(L) = 0$.
(b) Show that $v(x, t) = u(x, t) - u_E(x)$ satisfies (4.2.8b).

4.2.2. Show that c^2 has the dimensions of velocity squared.

4.2.3. Consider a particle whose x-coordinate (in horizontal equilibrium) is designated by α. If its vertical and horizontal displacements are u and v, respectively, determine its position x and y. Then show that

$$\frac{dy}{dx} = \frac{\partial u/\partial \alpha}{1 + \partial v/\partial \alpha}.$$

4.2.4. Derive equations for horizontal and vertical displacements without ignoring v. Assume that the string is perfectly flexible and that the tension is determined by an experimental law.

4.3 BOUNDARY CONDITIONS

The partial differential equation for a vibrating string, (4.2.7) or (4.2.8), has a second-order spatial partial derivative. We will apply one boundary condition at each end, just as we did for the one-dimensional heat equation.

The simplest boundary condition is that of a fixed end, usually fixed with zero displacement. For example, if a string is fixed (with zero displacement) at $x = L$, then

$$u(L, t) = 0. \tag{4.3.1}$$

Alternatively, we might vary an end of the string in a prescribed way:

$$u(L, t) = f(t). \tag{4.3.2}$$

Vibrating Strings and Membranes   Chap. 4

Both (4.3.1) and (4.3.2) are linear boundary conditions; (4.3.2) is nonhomogeneous, while (4.3.1) is homogeneous.

A more interesting boundary condition occurs if one end of the string is attached to a dynamical system. Let us suppose that the left end, $x = 0$, of a string is attached to a spring–mass system, as illustrated in Fig. 4.3.1. We will insist that the motion be entirely vertical. To accomplish this, we must envision the mass to be on a vertical track (possibly frictionless). The track applies a horizontal force to the mass when necessary to prevent the large horizontal component of the tensile force from turning over the spring–mass system. The string is attached to the mass so that if the position of the mass is $y(t)$, so is the position of the left end:

$$u(0, t) = y(t). \tag{4.3.3}$$

However, $y(t)$ is unknown and itself satisfies an ordinary differential equation determined from Newton's laws. We assume that the spring has unstretched length l and obeys Hooke's law with spring constant k. To make the problem even more interesting, we let the support of the spring move in some prescribed way, $y_s(t)$. Thus, the length of the spring is $y(t) - y_s(t)$ and the stretching of the spring is $y(t) - y_s(t) - l$. According to Newton's law (using Hooke's law with spring constant k),

$$m \frac{d^2 y}{dt^2} = -k(y(t) - y_s(t) - l) + \text{other forces on mass.}$$

The other vertical forces on the mass are a tensile force applied by the string $T(0, t) \sin \theta\,(0, t)$ and a force $g(t)$ representing any other external forces on the mass. Recall that we must be restricted to small angles, such that the tension is nearly constant, T_0. In that case, the vertical component is approximately $T_0\, \partial u / \partial x$:

$$T(0, t) \sin \theta\,(0, t) \approx T(0, t) \frac{\sin \theta\,(0, t)}{\cos \theta\,(0, t)} = T(0, t) \frac{\partial u}{\partial x}\,(0, t) \approx T_0 \frac{\partial u}{\partial x}(0, t),$$

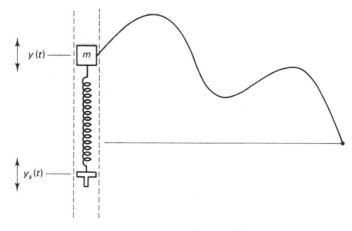

Figure 4.3.1 Spring-mass system with a variable support attached to a stretched string.

since for small angles $\cos \theta \approx 1$. In this way the boundary condition at $x = 0$ for a vibrating string attached to a spring–mass system [with a variable support $y_s(t)$ and an external force $g(t)$] is

$$m\frac{d^2}{dt^2}u\,(0, t) = -k(u(0, t) - y_s(t) - l) + T_0\frac{\partial u}{\partial x}(0, t) + g(t). \qquad (4.3.4)$$

Let us consider some special cases in which there are no external forces on the mass, $g(t) = 0$. If, in addition, the mass is sufficiently small so that the forces on the mass are in balance, then

$$T_0\frac{\partial u}{\partial x}(0, t) = k(u(0, t) - u_E(t)), \qquad (4.3.5)$$

where $u_E(t)$ is the equilibrium position of the mass, $u_E(t) = y_s(t) + l$. This form, known as the nonhomogeneous **elastic** boundary condition, is exactly analogous to Newton's law of cooling [with an external temperature of $u_E(t)$] for the heat equation. If the equilibrium position of the mass coincides with the equilibrium position of the string, $u_E(t) = 0$, the homogeneous version of the elastic boundary condition results:

$$T_0\frac{\partial u}{\partial x}(0, t) = ku(0, t). \qquad (4.3.6)$$

$\partial u/\partial x$ is proportional to u. Since for physical reasons $T_0 > 0$ and $k > 0$, the signs in (4.3.6) are prescribed. This is the same choice of signs that occurs for Newton's law of cooling. A diagram (Fig. 4.3.2) illustrates both the correct and incorrect choice of signs. This figure shows that (assuming $u = 0$ is an equilibrium position for both string and mass) if $u > 0$ at $x = 0$, then $\partial u/\partial x > 0$ in order to get a balance of vertical forces on the *massless* spring-mass system. A similar argument shows that there is an important sign change if the elastic boundary condition occurs at $x = L$:

$$T_0\frac{\partial u}{\partial x}(L, t) = -k(u(L, t) - u_E(t)), \qquad (4.3.7)$$

the same sign change we obtained for Newton's law of cooling.

For a vibrating string, another boundary condition that can be discussed is the **free end**. It is not literally free. Instead, the end is attached to a frictionless

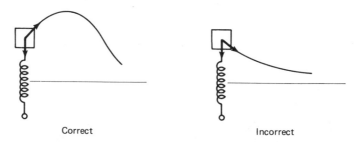

Correct   Incorrect

Figure 4.3.2 Boundary conditions for massless spring-mass system.

vertically moving track as before and is free to move up and down. There is no spring–mass system, nor external forces. However, one can obtain this boundary condition by taking the limit as $k \to 0$ of either (4.3.6) or (4.3.7):

$$T_0 \frac{\partial u}{\partial x} = 0. \tag{4.3.8}$$

This says that the vertical component of the tensile force must vanish at the end since there are no other vertical forces at the end. If the vertical component did not vanish, the end would have an infinite vertical acceleration. Boundary condition (4.3.8) is exactly analogous to the insulated boundary condition for the one-dimensional heat equation.

EXERCISES 4.3

4.3.1. If $m = 0$, which of the diagrams for the right end shown in Fig. 4.3.3 is possibly correct? *Briefly* explain. Assume that the mass can move only vertically.

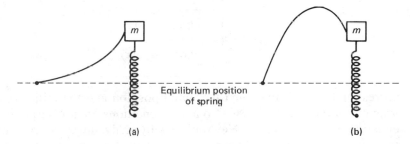

Figure 4.3.3

4.3.2. Consider two vibrating strings connected at $x = L$ to a spring-mass system on a vertical frictionless track as in Fig. 4.3.4. Assume that the spring is unstretched when the string is horizontal (the spring has a fixed support). Also suppose that there is an external force $f(t)$ acting on the mass m.
  (a) What "jump" conditions apply at $x = L$ relating the string on the left to the string on the right?
  (b) In what situations is this mathematically analogous to perfect thermal contact?

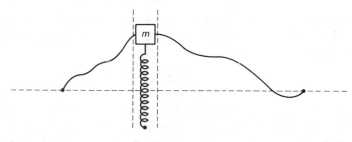

Figure 4.3.4

4.4 VIBRATING STRING WITH FIXED ENDS

In this section we solve the one-dimensional wave equation, which represents a uniform vibrating string without external forces,

$$\text{PDE:} \qquad \boxed{\frac{\partial^2 u}{\partial t^2} = c^2 \frac{\partial^2 u}{\partial x^2},} \qquad (4.4.1)$$

where $c^2 = T_0/\rho_0$, subject to the simplest homogeneous boundary conditions,

$$\text{BC:} \qquad \boxed{\begin{array}{l} u(0, t) = 0 \\ u(L, t) = 0, \end{array}} \qquad (4.4.2)$$

both ends being fixed with zero displacement. Since the partial differential equation (4.4.1) contains the second time derivative, two initial conditions are needed. We prescribe both $u(x, 0)$ and $\partial u/\partial t(x, 0)$.

$$\text{IC:} \qquad \boxed{\begin{array}{l} u(x, 0) = f(x) \\ \dfrac{\partial u}{\partial t}(x, 0) = g(x), \end{array}} \qquad (4.4.3)$$

corresponding to being given the initial position and the initial velocity of each segment of the string. These two initial conditions are not surprising as the wave equation was derived from Newton's law by analyzing each segment of the string as a particle; ordinary differential equations for particles require both initial position and velocity.

Since both the partial differential equation and the boundary conditions are linear and homogeneous, the method of separation of variables is attempted. As with the heat equation the nonhomogeneous initial conditions are put aside temporarily. We look for special product solutions of the form

$$u(x, t) = \phi(x)h(t). \qquad (4.4.4)$$

Substituting (4.4.4) into (4.4.1) yields

$$\phi(x) \frac{d^2 h}{dt^2} = c^2 h(t) \frac{d^2 \phi}{dx^2}. \qquad (4.4.5)$$

Dividing by $\phi(x)h(t)$ separates the variables, but it is more convenient to divide additionally by the constant c^2, since then the resulting eigenvalue problem will not contain the parameter c^2:

$$\frac{1}{c^2} \frac{1}{h} \frac{d^2 h}{dt^2} = \frac{1}{\phi} \frac{d^2 \phi}{dx^2} = -\lambda. \qquad (4.4.6)$$

A separation constant is introduced since $(1/c^2)(1/h)(d^2 h/dt^2)$ depends only on t and $(1/\phi)(d^2\phi/dx^2)$ depends only on x. The minus sign is inserted *purely for convenience*. With this minus sign, let us explain why we expect that $\lambda > 0$.

We need the two ordinary differential equations which follow from (4.4.6):

$$\frac{d^2h}{dt^2} = -\lambda c^2 h \tag{4.4.7}$$

and

$$\frac{d^2\phi}{dx^2} = -\lambda\phi. \tag{4.4.8}$$

The two homogeneous boundary conditions (4.4.2) show that

$$\phi(0) = \phi(L) = 0. \tag{4.4.9}$$

Thus, (4.4.8) and (4.4.9) form a boundary value problem. Instead of first reviewing the solution of (4.4.8) and (4.4.9), let us analyze the time-dependent ODE (4.4.7). If $\lambda > 0$, the general solution of (4.4.7) is a linear combination of sines and cosines,

$$h(t) = c_1 \cos c\sqrt{\lambda}\, t + c_2 \sin c\sqrt{\lambda}\, t. \tag{4.4.10}$$

If $\lambda = 0$, $h(t) = c_1 + c_2 t$, and if $\lambda < 0$, $h(t)$ is a linear combination of exponentially growing and decaying solutions in time. Since we are solving a vibrating string, it should seem more reasonable that the time-dependent solutions oscillate. This does not prove that $\lambda > 0$. Instead, it serves as an immediate motivation for choosing the minus sign in (4.4.6). Now by analyzing the boundary value problem, we may indeed determine that the eigenvalues are nonnegative.

The boundary value problem is

$$\frac{d^2\phi}{dx^2} = -\lambda\phi$$

$$\phi(0) = 0$$

$$\phi(L) = 0.$$

Although we could solve this by proceeding through three cases, we ought to recall that all the eigenvalues are positive. In fact,

$$\lambda = \left(\frac{n\pi}{L}\right)^2, \qquad n = 1, 2, 3, \ldots,$$

and the corresponding eigenfunctions are $\sin n\pi x/L$. The time-dependent part of the solution has been obtained above, (4.4.10). Thus, there are two families of product solutions: $\sin n\pi x/L \sin n\pi ct/L$ and $\sin n\pi x/L \cos n\pi ct/L$. The principle of superposition then implies that we should be able to solve the initial value problem by considering a linear combination of all product solutions:

$$u(x, t) = \sum_{n=1}^{\infty} \left(A_n \sin \frac{n\pi x}{L} \cos \frac{n\pi ct}{L} + B_n \sin \frac{n\pi x}{L} \sin \frac{n\pi ct}{L} \right). \tag{4.4.11}$$

The initial conditions (4.4.3) are satisfied if

$$f(x) = \sum_{n=1}^{\infty} A_n \sin \frac{n\pi x}{L}$$

$$g(x) = \sum_{n=1}^{\infty} B_n \frac{n\pi c}{L} \sin \frac{n\pi x}{L}. \tag{4.4.12}$$

The boundary conditions have implied that sine series are important. There are two initial conditions and two families of coefficients to be determined. From our previous work on Fourier sine series, we know that $\sin n\pi x/L$ forms an orthogonal set. A_n will be the coefficients of the Fourier sine series of $f(x)$ and $B_n n\pi c/L$ will be for the Fourier sine series of $g(x)$:

$$A_n = \frac{2}{L} \int_0^L f(x) \sin \frac{n\pi x}{L} \, dx \qquad (4.4.13)$$

$$B_n \frac{n\pi c}{L} = \frac{2}{L} \int_0^L g(x) \sin \frac{n\pi x}{L} \, dx.$$

Let us interpret these results in the context of musical stringed instruments (with fixed ends). The vertical displacement is composed of a linear combination of simple product solutions,

$$\sin \frac{n\pi x}{L} \left(A_n \cos \frac{n\pi ct}{L} + B_n \sin \frac{n\pi ct}{L} \right).$$

These are called the **normal modes** of vibration. The intensity of the sound produced depends on the amplitude,* $\sqrt{A_n^2 + B_n^2}$. The time dependence is simple harmonic with **circular frequency** (the number of oscillations in 2π units of time) equaling $n\pi c/L$, where $c = \sqrt{T_0/\rho_0}$. The sound produced consists of the superposition of these infinite number of **natural frequencies** ($n = 1, 2, \ldots$). The normal mode $n = 1$ is called the first harmonic or fundamental. In the case of a vibrating string the fundamental mode has a circular frequency of $\pi c/L$.† The larger the natural frequency, the higher the pitch of the sound produced. To produce a desired fundamental frequency, $c = \sqrt{T_0/\rho_0}$ or L can be varied. Usually, the mass density is fixed. Thus, the instrument is tuned by varying the tension T_0; the larger T_0, the higher the fundamental frequency. While playing a stringed instrument the musician can also vary the pitch by varying the effective length L, by clamping down the string. Shortening L makes the note higher. The nth normal mode is called the nth harmonic. For vibrating strings (with fixed ends) the frequencies of the higher harmonics are all integral multiples of the fundamental. It is not necessarily true for other types of musical instruments. This is thought to be pleasing to the ear.

Let us attempt to illustrate the motion associated with each normal mode. The fundamental and higher harmonics are sketched in Fig. 4.4.1. To indicate what these look like, we sketch for various values of t. At each t, each mode looks like a simple oscillation in x. The amplitude varies periodically in time. These are called **standing waves**. In all cases there is no displacement at both ends due to the boundary conditions. For the second harmonic ($n = 2$), the displacement is also zero for all time in the middle $x = L/2$. $x = L/2$ is called a **node** for the second harmonic. Similarly, there are two nodes for the third harmonic. This can be generalized; the nth harmonic has $n - 1$ nodes.‡

---

* $A_n \cos \omega t + B_n \sin \omega t = \sqrt{A_n^2 + B_n^2} \sin(\omega t + \theta)$, where $\theta = \tan^{-1} A_n/B_n$.

† Frequencies are usually measured in cycles per second, not cycles per 2π units of time. The fundamental thus has a frequency of $c/2L$, cycles per second.

‡ You can visualize experimentally this result by rapidly oscillating at the appropriate frequency

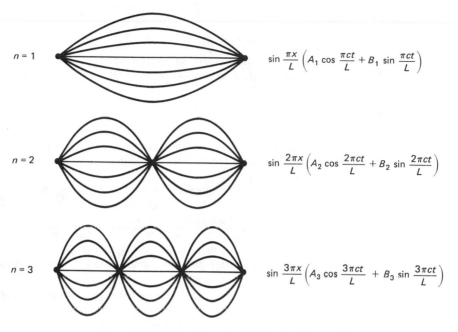

$$\sin \frac{\pi x}{L}\left(A_1 \cos \frac{\pi ct}{L} + B_1 \sin \frac{\pi ct}{L}\right)$$

$$\sin \frac{2\pi x}{L}\left(A_2 \cos \frac{2\pi ct}{L} + B_2 \sin \frac{2\pi ct}{L}\right)$$

$$\sin \frac{3\pi x}{L}\left(A_3 \cos \frac{3\pi ct}{L} + B_3 \sin \frac{3\pi ct}{L}\right)$$

Figure 4.4.1 Normal modes of vibration for a string.

It is interesting to note that the vibration corresponding to the second harmonic looks like two identical strings each with length $L/2$ vibrating in the fundamental mode, since $x = L/2$ is a node. We should find that the frequencies of vibration are identical; that is, the frequency for the fundamental ($n = 1$) with length $L/2$ should equal the frequency for the second harmonic ($n = 2$) with length L. The formula for the frequency, $\omega = n\pi c/L$, verifies this observation.

Each standing wave can be shown to be composed of two traveling waves. For example, consider the term $\sin n\pi x/L \sin n\pi ct/L$. From trigonometric identities

$$\sin \frac{n\pi x}{L} \sin \frac{n\pi ct}{L} = \underbrace{\frac{1}{2} \cos \frac{n\pi}{L}(x - ct)}_{\substack{\text{wave traveling} \\ \text{to the right} \\ \text{(with velocity } c)}} - \underbrace{\frac{1}{2} \cos \frac{n\pi}{L}(x + ct)}_{\substack{\text{wave traveling} \\ \text{to the left} \\ \text{(with velocity } -c)}}. \qquad (4.4.14)$$

In fact, since the solution (4.4.11) to the wave equation consists of a superposition of standing waves, it can be shown that this solution is a combination of just two waves (each rather complicated)—one traveling to the left at velocity $-c$ with fixed shape and the other to the right at velocity c with a different fixed shape. We are claiming that the solution to the one-dimensional wave equation

---

one end of a long rope which is tightly held at the other end. The result appears more easily for an expandable spiral telephone cord or a "slinky."

can be written as

$$u(x, t) = R(x - ct) + S(x + ct),$$

even if the boundary conditions are not fixed at $x = 0$ and $x = L$. We will show and discuss this further in the exercises and in Chapter 11.

EXERCISES 4.4

4.4.1. Consider vibrating strings of uniform density ρ_0 and tension T_0.

       *(a) What are the natural frequencies of a vibrating string of length L, fixed at both ends?

       *(b) What are the natural frequencies of a vibrating string of length H, which is fixed at $x = 0$ and "free" at the other end (i.e., $\partial u/\partial x(H, t) = 0$)? Sketch a few modes of vibration as in Fig. 4.4.1.

       (c) Show that the modes of vibration for the *odd* harmonics (i.e., $n = 1, 3, 5, \ldots$) of part (a) are identical to modes of part (b) if $H = L/2$. Verify that their natural frequencies are the same. Briefly explain using symmetry arguments.

4.4.2. In Sec. 4.2 it was shown that the displacement u of a nonuniform string satisfies

$$\rho_0 \frac{\partial^2 u}{\partial t^2} = T_0 \frac{\partial^2 u}{\partial x^2} + Q,$$

where Q represents the vertical component of the body force per unit length. If $Q = 0$, the partial differential equation is homogeneous. A slightly different homogeneous equation occurs if $Q = \alpha u$.

       (a) Show that if $\alpha < 0$, the body force is restoring (toward $u = 0$). Show that if $\alpha > 0$, the body force tends to push the string farther away from its unperturbed position, $u = 0$.

       (b) Separate variables if $\rho_0(x)$ and $\alpha(x)$, but T_0 is constant for physical reasons. Analyze the time-dependent ordinary differential equation.

      *(c) Specialize part (b) to the constant coefficient case. Solve the initial value problem if $\alpha < 0$:

$$u(0, t) = 0 \qquad u(x, 0) = 0$$

$$u(L, t) = 0 \qquad \frac{\partial u}{\partial t}(x, 0) = f(x).$$

    What are the frequencies of vibration?

4.4.3. Consider a slightly damped vibrating string that satisfies

$$\rho_0 \frac{\partial^2 u}{\partial t^2} = T_0 \frac{\partial^2 u}{\partial x^2} - \beta \frac{\partial u}{\partial t}.$$

       (a) Briefly explain why $\beta > 0$.

      *(b) Determine the solution (by separation of variables) which satisfies the boundary conditions

$$u(0, t) = 0 \quad \text{and} \quad u(L, t) = 0$$

    and the initial conditions

$$u(x, 0) = f(x) \quad \text{and} \quad \frac{\partial u}{\partial t}(x, 0) = g(x).$$

    Assume that this frictional coefficient β is relatively small ($\beta^2 < 4\pi^2 \rho_0 T_0/L^2$).

4.4.4. Redo Exercise 4.4.3(b) by the eigenfunction expansion method.

4.4.5. Redo Exercise 4.4.3(b) if $4\pi^2\rho_0 T_0/L^2 < \beta^2 < 16\pi^2\rho_0 T_0/L^2$.

4.4.6. For (4.4.1)–(4.4.3), from (4.4.11) show that
$$u(x, t) = R(x - ct) + S(x + ct),$$
where R and S are some functions.

4.4.7. If a vibrating string satisfying (4.4.1)–(4.4.3) is initially at rest, $g(x) = 0$, show that
$$u(x, t) = \tfrac{1}{2}[F(x - ct) + F(x + ct)],$$
where $F(x)$ is the odd periodic extension of $f(x)$. [*Hints:*

1. For all x, $F(x) = \displaystyle\sum_{n=1}^{\infty} A_n \sin \frac{n\pi x}{L}$.

2. $\sin a \cos b = \tfrac{1}{2}[\sin (a + b) + \sin (a - b)]$.

Comment: This result shows that *the practical difficulty of summing an infinite number of terms of a Fourier series may be avoided for the one-dimensional wave equation.*

4.4.8. If a vibrating string satisfying (4.4.1)–(4.4.3) is initially unperturbed, $f(x) = 0$, with the initial velocity given, show that
$$u(x, t) = \frac{1}{2c} \int_{x-ct}^{x+ct} G(\bar{x}) \, d\bar{x},$$
where $G(x)$ is the odd periodic extension of $g(x)$. [*Hints:*

1. For all x, $G(x) = \displaystyle\sum_{n=1}^{\infty} B_n \frac{n\pi c}{L} \sin \frac{n\pi x}{L}$.

2. $\sin a \sin b = \tfrac{1}{2}[\cos (a - b) - \cos (a + b)]$.]

See the comment after Exercise 4.4.7.

4.4.9 From (4.4.1), derive conservation of energy for a vibrating string,
$$\frac{dE}{dt} = c^2 \frac{\partial u}{\partial x} \frac{\partial u}{\partial t} \bigg|_0^L, \tag{4.4.15}$$
where the total energy E is the sum of the kinetic energy, defined by $\displaystyle\int_0^L \frac{1}{2}\left(\frac{\partial u}{\partial t}\right)^2 dx$, and the potential energy, defined by $\displaystyle\int_0^L \frac{c^2}{2}\left(\frac{\partial u}{\partial x}\right)^2 dx$.

4.4.10. What happens to the total energy E of a vibrating string (see Exercise 4.4.9):
(a) If $u(0, t) = 0$ and $u(L, t) = 0$.

(b) If $\dfrac{\partial u}{\partial x}(0, t) = 0$ and $u(L, t) = 0$.

(c) If $u(0, t) = 0$ and $\dfrac{\partial u}{\partial x}(L, t) = -\gamma u(L, t)$ with $\gamma > 0$.

(d) If $\gamma < 0$ in part (c).

4.4.11. Show that the potential and kinetic energies (defined in Exercise 4.4.9) are equal for a traveling wave, $u = R(x - ct)$.

4.4.12. Using (4.4.15), prove that the solution of (4.4.1)–(4.4.3) is unique.

4.4.13. (a) Using (4.4.15), calculate the energy of one normal mode.

  (b) Show that the total energy, when $u(x, t)$ satisfies (4.4.11), is the sum of the energies contained in each mode.

4.5 VIBRATING MEMBRANE

The heat equation in one spatial dimension is $\partial u/\partial t = k\,\partial^2 u/\partial x^2$. In two or three dimensions, the temperature satisfies $\partial u/\partial t = k\,\nabla^2 u$. In a similar way, the vibration of a string (one dimension) can be extended to the vibration of a membrane (two dimensions).

The vertical displacement of a vibrating string satisfies the one-dimensional wave equation

$$\frac{\partial^2 u}{\partial t^2} = c^2 \frac{\partial^2 u}{\partial x^2}.$$

There are important physical problems which solve

$$\frac{\partial^2 u}{\partial t^2} = c^2 \nabla^2 u, \tag{4.5.1}$$

known as the two- or three-dimensional wave equation. An example of a physical problem that satisfies a two-dimensional wave equation is the vibration of a highly stretched membrane. This can be thought of as a two-dimensional vibrating string. We will give a *brief* derivation in the manner described by Kaplan [1981], omitting some of the details we discussed for a vibrating string. We again introduce the displacement $z = u(x, y, t)$, which depends on x, y and t (as illustrated in Fig. 4.5.1). If all slopes (i.e., $\partial u/\partial x$ and $\partial u/\partial y$) are small, then as an approximation we may assume that the vibrations are entirely vertical and the tension is approximately constant. Then the mass density (mass per unit surface area), $\rho_0(x, y)$, of the membrane in the unperturbed position does not change appreciably when the membrane is perturbed.

The tensile force (per unit arc length), $\mathbf{F}_T$, is tangent to the membrane and acts along the entire edge. The direction of the tensile force (see Figure 4.5.1) is obtained by crossing the unit tangent vector to the edge, $\hat{\mathbf{t}}$, with the unit normal vector to the membrane, $\hat{\mathbf{n}}$. Since the tensile force has constant magnitude ($|\mathbf{F}_T| = T_0$), it follows that

$$\mathbf{F}_T = T_0\,\hat{\mathbf{t}} \times \hat{\mathbf{n}},$$

where the vertical component is obtained by $\mathbf{F}_T \cdot \hat{\mathbf{k}}$.

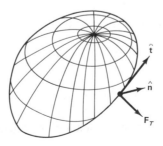

Figure 4.5.1 Perturbed stretched membrane with approximately constant tension T_0. The normal vector to the surface is $\hat{\mathbf{n}}$ and the tangent vector to the edge is $\hat{\mathbf{t}}$.

Newton's law for vertical motion must be applied to each differential section of the membrane and then summed (integrated). The sum (surface integral) of the mass ($\rho_0\, dA$) times the vertical acceleration ($\partial^2 u/\partial t^2$) equals the total (closed line integral) *vertical* tensile force (ignoring body forces)

$$\iint \rho_0 \frac{\partial^2 u}{\partial t^2}\, dA = \oint T_0\, \hat{\mathbf{t}} \times \hat{\mathbf{n}} \cdot \hat{\mathbf{k}}\, ds = \oint T_0(\hat{\mathbf{n}} \times \hat{\mathbf{k}}) \cdot \hat{\mathbf{t}}\, ds, \qquad (4.5.2)$$

where ds is the differential arc length, dA is differential surface area, and the vector triple product relation has been used ($\mathbf{A} \times \mathbf{B} \cdot \mathbf{C} = \mathbf{B} \times \mathbf{C} \cdot \mathbf{A}$). Stokes theorem ($\iint \nabla \times \mathbf{B} \cdot \hat{\mathbf{n}}\, dA = \oint \mathbf{B} \cdot \hat{\mathbf{t}}\, ds$), will be applied (for the only time in this text):

$$\iint \rho_0 \frac{\partial^2 u}{\partial t^2}\, dA = \iint T_0[\nabla \times (\hat{\mathbf{n}} \times \hat{\mathbf{k}})] \cdot \hat{\mathbf{n}}\, dA. \qquad (4.5.3)$$

Since the region is arbitrary, we derive

$$\rho_0 \frac{\partial^2 u}{\partial t^2} = T_0[\nabla \times (\hat{\mathbf{n}} \times \hat{\mathbf{k}})] \cdot \hat{\mathbf{n}}. \qquad (4.5.4)$$

A point on the membrane is described by $z = u(x, y)$. Thus, the unit normal to the vibrating membrane (using the gradient, see Appendix to Sec. 1.5) is calculated as follows:

$$\hat{\mathbf{n}} = \frac{-\dfrac{\partial u}{\partial x}\hat{\mathbf{i}} - \dfrac{\partial u}{\partial y}\hat{\mathbf{j}} + \hat{\mathbf{k}}}{\sqrt{\left(\dfrac{\partial u}{\partial x}\right)^2 + \left(\dfrac{\partial u}{\partial y}\right)^2 + 1}} \approx -\frac{\partial u}{\partial x}\hat{\mathbf{i}} - \frac{\partial u}{\partial y}\hat{\mathbf{j}} + \hat{\mathbf{k}},$$

since the partial derivatives are assumed to be small. We now begin to calculate the expression needed in (4.5.4):

$$\hat{\mathbf{n}} \times \hat{\mathbf{k}} = \begin{vmatrix} \hat{\mathbf{i}} & \hat{\mathbf{j}} & \hat{\mathbf{k}} \\[4pt] -\dfrac{\partial u}{\partial x} & -\dfrac{\partial u}{\partial y} & 1 \\[6pt] 0 & 0 & 1 \end{vmatrix} = -\frac{\partial u}{\partial y}\hat{\mathbf{i}} + \frac{\partial u}{\partial x}\hat{\mathbf{j}}.$$

Continuing, we obtain

$$\nabla \times (\hat{\mathbf{n}} \times \hat{\mathbf{k}}) = \begin{vmatrix} \hat{\mathbf{i}} & \hat{\mathbf{j}} & \hat{\mathbf{k}} \\[4pt] \dfrac{\partial}{\partial x} & \dfrac{\partial}{\partial y} & \dfrac{\partial}{\partial z} \\[6pt] -\dfrac{\partial u}{\partial y} & \dfrac{\partial u}{\partial x} & 0 \end{vmatrix} = \hat{\mathbf{k}}\left(\frac{\partial^2 u}{\partial x^2} + \frac{\partial^2 u}{\partial y^2}\right).$$

In this way we obtain the partial differential equation for a vibrating membrane,

$$\rho_0 \frac{\partial^2 u}{\partial t^2} = T_0\left(\frac{\partial^2 u}{\partial x^2} + \frac{\partial^2 u}{\partial y^2}\right).$$

Dividing by ρ_0 yields the two-dimensional wave equation

$$\frac{\partial^2 u}{\partial t^2} = c^2 \left(\frac{\partial^2 u}{\partial x^2} + \frac{\partial^2 u}{\partial y^2} \right), \tag{4.5.5}$$

where, again, $c^2 = T_0/\rho_0$. The solutions of problems for a vibrating membrane are postponed until Chapter 6.

EXERCISE 4.5

4.5.1. If a membrane satisfies an "elastic" boundary condition, show that

$$T_0 \nabla u \cdot \hat{\mathbf{n}} = -ku \tag{4.5.6}$$

if there is a restoring force per unit length proportional to the displacement.

Sturm–Liouville Eigenvalue Problems

5.1 INTRODUCTION

We have found the method of separation of variables to be quite successful in solving some homogeneous partial differential equations with homogeneous boundary conditions. In all examples we have analyzed *so far*, the boundary value problem that determines the needed eigenvalues (separation constants) has involved the simple ordinary differential equation

$$\frac{d^2\phi}{dx^2} + \lambda\phi = 0. \tag{5.1.1}$$

Explicit solutions of this equation determined the eigenvalues λ from the homogeneous boundary conditions. The principle of superposition resulted in our needing to analyze infinite series. We pursued three different cases (depending on the boundary conditions): Fourier sine series, Fourier cosine series, and Fourier series (both sines and cosines). Fortunately, we verified by explicit integration that the eigenfunctions were orthogonal. This enabled us to determine the coefficients of the infinite series from the remaining nonhomogeneous condition.

In this section we further explain and generalize these results. We show that the orthogonality of the eigenfunctions can be derived even if we cannot solve the defining differential equation in terms of elementary functions [as in (5.1.1)]. Instead, orthogonality is a direct result of the differential equation. We investigate other boundary value problems resulting from separation of variables which yield other families of orthogonal functions. These generalizations of Fourier series will not always involve sines and cosines since (5.1.1) is not necessarily appropriate.

5.2 EXAMPLES

5.2.1 Heat Flow in a Nonuniform Rod

In Sec. 1.2 we showed that the temperature u in a nonuniform rod solves the following partial differential equation:

$$c\rho \frac{\partial u}{\partial t} = \frac{\partial}{\partial x}\left(K_0 \frac{\partial u}{\partial x}\right) + Q, \qquad (5.2.1)$$

where Q represents any possible sources of heat energy. Here, in order to consider the case of a *nonuniform* rod, we allow the thermal coefficients c, ρ, K_0 to depend on x. The method of separation of variables can be applied only if (5.2.1) is linear and *homogeneous*. Usually, to make (5.2.1) homogeneous, we consider only situations without sources, $Q = 0$. However, we will be slightly more general. We will allow the heat source Q to be proportional to the temperature u,

$$Q = \alpha u, \qquad (5.2.2)$$

in which case

$$c\rho \frac{\partial u}{\partial t} = \frac{\partial}{\partial x}\left(K_0 \frac{\partial u}{\partial x}\right) + \alpha u. \qquad (5.2.3)$$

We also allow α to depend on x (but not on t), as though the specific types of sources depend on the material. Although $Q \neq 0$, (5.2.3) is still a linear and homogeneous partial differential equation. To understand the effect of this source Q, we present a plausible physical situation in which terms such as $Q = \alpha u$ might arise. Suppose that a chemical reaction generates heat (called an *exothermic* reaction) corresponding to $Q > 0$. Conceivably, this reaction could be more intense at higher temperatures. In this way the heat energy generated might be proportional to the temperature and thus $\alpha > 0$ (assuming that $u > 0$). Other types of chemical reactions (known as *endothermic*) would remove heat energy from the rod and also could be proportional to the temperature. For positive temperatures ($u > 0$), this corresponds to $\alpha < 0$. In our problem $\alpha = \alpha(x)$, and hence it is possible that $\alpha > 0$ in some parts of the rod and $\alpha < 0$ in other parts. We summarize these results by noting that if $\alpha(x) < 0$ for all x, then heat energy is being taken out of the rod, and vice versa. Later in our mathematical analysis, we will correspondingly discuss the special case $\alpha(x) < 0$.

Equation (5.2.3) is suited for the method of separation of variables if we in addition assume that there is one homogeneous boundary condition (as yet unspecified) at each end, $x = 0$ and $x = L$. We have already analyzed cases in which $\alpha = 0$ and c, ρ, K_0 are constant. In separating variables, we substitute the product form,

$$u(x, t) = \phi(x)h(t), \qquad (5.2.4)$$

into (5.2.3), which yields

$$c\rho\phi(x)\frac{dh}{dt} = h(t)\frac{d}{dx}\left(K_0\frac{d\phi}{dx}\right) + \alpha\phi(x)h(t).$$

Dividing by $\phi(x)h(t)$ does not necessarily separate variables since $c\rho$ may depend on x. However, dividing by $c\rho\phi(x)h(t)$ is always successful:

$$\frac{1}{h}\frac{dh}{dt} = \frac{1}{c\rho\phi}\frac{d}{dx}\left(K_0\frac{d\phi}{dx}\right) + \frac{\alpha}{c\rho} = -\lambda. \tag{5.2.5}$$

The separation constant $-\lambda$ has been introduced with a minus sign because in this form the time-dependent equation [following from (5.2.5)],

$$\frac{dh}{dt} = -\lambda h, \tag{5.2.6}$$

has exponentially decaying solutions if $\lambda > 0$. Solutions to (5.2.6) exponentially grow if $\lambda < 0$ (and are constant if $\lambda = 0$). Solutions exponentially growing in time are not usually encountered in physical problems. However, for problems in which $\alpha > 0$ for at least part of the rod, thermal energy is being put into the rod by the exothermic reaction, and hence it is possible for there to be some negative eigenvalues ($\lambda < 0$).

The spatial differential equation implied by separation of variables is

$$\boxed{\frac{d}{dx}\left(K_0\frac{d\phi}{dx}\right) + \alpha\phi + \lambda c\rho\phi = 0,} \tag{5.2.7}$$

which forms a boundary value problem when complemented by two homogeneous boundary conditions. This differential equation is not $d^2\phi/dx^2 + \lambda\phi = 0$. Neither does (5.2.7) have constant coefficients, because the thermal coefficients K_0, c, ρ, α are not constant. *In general, one way in which nonconstant-coefficient differential equations occur is in situations where physical properties are nonuniform.*

Note that we cannot decide on the appropriate convenient sign for the separation constant by quickly analyzing the spatial ordinary differential equation (5.2.7) with its homogeneous boundary conditions. Usually, we will be unable to solve (5.2.7) in the variable coefficient case, other than by a numerical approximate solution on the computer. Consequently, we will describe in Sec. 5.3 certain important qualitative properties of the solution of (5.2.7). Later, with a greater understanding of (5.2.7), we will return to reinvestigate heat flow in a nonuniform rod. For now, let us describe another example which yields a boundary value problem with nonconstant coefficients.

5.2.2 Circularly Symmetric Heat Flow

Nonconstant-coefficient differential equations can also arise if the physical parameters are constant. In Sec. 1.5 we showed that if the temperature u in some plane two-dimensional region is circularly symmetric (so that u only depends on time t and on the radial distance r from the origin), then u solves the linear and homogeneous partial differential equation

$$\frac{\partial u}{\partial t} = k\frac{1}{r}\frac{\partial}{\partial r}\left(r\frac{\partial u}{\partial r}\right), \tag{5.2.8}$$

under the assumption that all the thermal coefficients are constant.

We apply the method of separation of variables by seeking solutions in the form of a product:

$$u(r, t) = \phi(r)h(t).$$

Equation (5.2.8) then implies that

$$\phi(r)\frac{dh}{dt} = \frac{kh(t)}{r}\frac{d}{dr}\left(r\frac{d\phi}{dr}\right).$$

Dividing by $\phi(r)h(t)$ separates the variables, but also dividing by the constant k is convenient since it eliminates this constant from the resulting boundary value problem:

$$\frac{1}{k}\frac{1}{h}\frac{dh}{dt} = \frac{1}{r\phi}\frac{d}{dr}\left(r\frac{d\phi}{dr}\right) = -\lambda. \tag{5.2.9}$$

The two ordinary differential equations implied by (5.2.9) are

$$\frac{dh}{dt} = -\lambda kh \tag{5.2.10}$$

$$\frac{d}{dr}\left(r\frac{d\phi}{dr}\right) + \lambda r\phi = 0. \tag{5.2.11}$$

The separation constant is denoted $-\lambda$ since we expect solutions to exponentially decay in time, as is implied by (5.2.10) if $\lambda > 0$. The nonconstant coefficients in (5.2.11) are due to geometric factors introduced by the use of polar coordinates. Later in this text (Sec. 6.7) we will show that (5.2.11) can be solved using Bessel functions. However, the general discussions in the remainder of this chapter will be quite valuable in our understanding of this problem.

Let us consider the appropriate homogeneous boundary conditions for circularly symmetric heat flow in two different geometries: inside a circular annulus (as illustrated in Fig. 5.2.1a) and inside a circle (as illustrated in Fig. 5.2.1b).

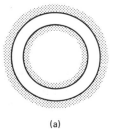

(a)

(b)

Figure 5.2.1 (a) Circular annulus; (b) circle.

In both cases we assume that all boundaries are fixed at zero temperature. For the annulus, the boundary conditions for (5.2.11) are that the temperature should be zero at the inner ($r = a$) and outer ($r = b$) concentric circular walls:

$$u(a, t) = 0 \quad \text{and} \quad u(b, t) = 0.$$

Both of these boundary conditions are exactly of the type we have already studied. However, for the circle, the same second-order differential equation (5.2.11) has only one boundary condition, $u(b, t) = 0$. Since the physical variable r ranges from $r = 0$ to $r = b$, we need a homogeneous boundary condition at $r = 0$ for mathematical reasons. (This is the same problem that occurred in studying Laplace's equation inside a cylinder. However, in that situation a nonhomogeneous condition was given at $r = b$.) On the basis of physical reasoning, we expect that the condition at $r = 0$ is that the temperature is bounded there, $|u(0, t)| < \infty$. This is an example of a **singularity condition**. It is homogeneous; it is the boundary condition that we apply at $r = 0$. Thus, we have homogeneous conditions at both $r = 0$ and $r = b$ for the circle.

5.3 STURM–LIOUVILLE EIGENVALUE PROBLEMS

5.3.1 General Classification

Differential equation. A boundary value problem consists of a linear homogeneous differential equation and corresponding linear homogeneous boundary conditions. All of the differential equations for boundary value problems that have been formulated in this text can be put in the following form:

$$\frac{d}{dx}\left(p\frac{d\phi}{dx}\right) + q\phi + \lambda\sigma\phi = 0, \tag{5.3.1}$$

where λ is the eigenvalue. Here the variable x is defined on a finite interval $a < x < b$. Four examples are:

1. *Simplest case:* $\dfrac{d^2\phi}{dx^2} + \lambda\phi = 0$; in which case, $p = 1$, $q = 0$, $\sigma = 1$.

2. *Heat flow in a nonuniform rod:* $\dfrac{d}{dx}\left(K_0\dfrac{d\phi}{dx}\right) + \alpha\phi + \lambda c\rho\phi = 0$; in which case, $p = K_0$, $q = \alpha$, $\sigma = c\rho$.

3. *Vibrations of a nonuniform string:* $T_0\dfrac{d^2\phi}{dx^2} + \alpha\phi + \lambda\rho_0\phi = 0$; in which case, $p = T_0$ (constant), $q = \alpha$, $\sigma = \rho_0$ (see Exercise 5.3.1).

4. *Circularly symmetric heat flow:* $\dfrac{d}{dr}\left(r\dfrac{d\phi}{dr}\right) + \lambda r\phi = 0$; here the independent variable $x = r$ and $p(x) = x$, $q(x) = 0$, $\sigma(x) = x$.

Many interesting results are known concerning any equation in the form (5.3.1).

Equation (5.3.1) is known as a **Sturm–Liouville differential equation**, named after two famous mathematicians active in the mid-1800s who studied it.

Boundary conditions.   The linear homogeneous boundary conditions that we have studied are of the form to follow. We also introduce some mathematical terminology:

	Heat flow	Vibrating string	Mathematical terminology
$\phi = 0$	Fixed (zero) temperature	Fixed (zero) displacement	First kind or Dirichlet condition
$\dfrac{d\phi}{dx} = 0$	Insulated	Free	Second kind or Neumann condition
$\dfrac{d\phi}{dx} = \pm\, h\phi$ $\begin{pmatrix} + \text{ left end} \\ - \text{ right end} \end{pmatrix}$	(Homogeneous) Newton's law of cooling 0° outside temperature), $h = H/K_0$, $h > 0$ (physical)	(Homogeneous) elastic boundary condition, $h = k/T_0$, $h > 0$ (physical)	Third kind or Robin condition
$\phi(-L) = \phi(L)$ $\dfrac{d\phi}{dx}(-L) = \dfrac{d\phi}{dx}(L)$	Perfect thermal contact	—	Periodicity condition (example of mixed type)
$|\phi(0)| < \infty$	Bounded temperature	—	Singularity condition

5.3.2 Regular Sturm–Liouville Eigenvalue Problem

A **regular** Sturm–Liouville eigenvalue problem consists of the Sturm–Liouville differential equation

$$\boxed{\frac{d}{dx}\left(p(x)\frac{d\phi}{dx}\right) + q(x)\phi + \lambda\sigma(x)\phi = 0} \quad a < x < b, \qquad (5.3.1)$$

subject to the boundary conditions that we have discussed (excluding periodic and singular cases):

$$\boxed{\begin{aligned} \beta_1\phi(a) + \beta_2\frac{d\phi}{dx}(a) &= 0 \\[2mm] \beta_3\phi(b) + \beta_4\frac{d\phi}{dx}(b) &= 0, \end{aligned}} \qquad (5.3.2)$$

where β_i are real. In addition, to be called regular, the coefficients p, q, and σ must be real and continuous everywhere (including the end points) and $p > 0$ and $\sigma > 0$ everywhere (also including the end points). For the regular Sturm–Liouville eigenvalue problem, many important general theorems exist. In Sec. 5.5 we will prove these results, and in Secs. 5.7 and 5.8 we will develop some more interesting examples that illustrate the significance of the general theorems.

Statement of theorems. At first let us just state (in one place) all the theorems we will discuss more fully later (and in some cases prove). For any regular Sturm–Liouville problem, all of the following theorems are valid:

---

1. All the eigenvalues λ are real.
2. There exist an infinite number of eigenvalues:
$$\lambda_1 < \lambda_2 < \cdots < \lambda_n < \lambda_{n+1} < \cdots$$
   a. There is a smallest eigenvalue, usually denoted λ_1.
   b. There is not a largest eigenvalue and $\lambda_n \to \infty$ as $n \to \infty$.
3. Corresponding to each eigenvalue λ_n, there is an eigenfunction, denoted $\phi_n(x)$ (which is unique to within an arbitrary multiplicative constant). $\phi_n(x)$ has exactly $n - 1$ zeros for $a < x < b$.
4. The eigenfunctions $\phi_n(x)$ form a "complete" set, meaning that any piecewise smooth function $f(x)$ can be represented by a generalized Fourier series of the eigenfunctions:
$$f(x) \sim \sum_{n=1}^{\infty} a_n \phi_n(x).$$

   Furthermore, this infinite series converges to $[f(x+) + f(x-)]/2$ for $a < x < b$ (if the coefficients a_n are properly chosen).
5. Eigenfunctions belonging to different eigenvalues are orthogonal relative to the weight function $\sigma(x)$. In other words,
$$\int_a^b \phi_n(x)\phi_m(x)\sigma(x)\,dx = 0 \qquad \text{if } \lambda_n \neq \lambda_m.$$
6. Any eigenvalue can be related to its eigenfunction by the **Rayleigh quotient**:
$$\lambda = \frac{-p\phi\,d\phi/dx\big|_a^b + \int_a^b [p(d\phi/dx)^2 - q\phi^2]\,dx}{\int_a^b \phi^2 \sigma\,dx},$$
   where the boundary conditions may somewhat simplify this expression.

It should be mentioned that for Sturm–Liouville eigenvalue problems that are not "regular," these theorems may be valid. An example of this is illustrated in Secs. 6.7 and 6.8.

5.3.3 Example and Illustration of Theorems

We will individually illustrate the meaning of these theorems (before proving many of them in Sec. 5.5) by referring to the simplest example of a regular Sturm–Liouville problem:

$$\frac{d^2\phi}{dx^2} + \lambda\phi = 0$$

$$\phi(0) = 0$$

$$\phi(L) = 0 \tag{5.3.3}$$

The constant-coefficient differential equation has zero boundary conditions at both ends. As we already know, the eigenvalues and corresponding eigenfunctions are

$$\lambda_n = \left(\frac{n\pi}{L}\right)^2 \quad \text{with} \quad \phi_n(x) = \sin\frac{n\pi x}{L}, \quad n = 1, 2, 3, \ldots,$$

giving rise to a Fourier sine series.

1. **Real eigenvalues.** Our theorem claims that all eigenvalues λ of a regular Sturm–Liouville problem are real. Thus, the eigenvalues of (5.3.3) should all be real. We know that the eigenvalues are $(n\pi/L)^2$, $n = 1, 2, \ldots$. However, in determining this result (see Sec. 2.3.4) we analyzed three cases: $\lambda > 0$, $\lambda = 0$, and $\lambda < 0$. We did not bother to look for complex eigenvalues because it is a relatively difficult task and we would have obtained no additional eigenvalues other than $(n\pi/L)^2$. This theorem (see Sec. 5.5 for its proof) is thus very useful. It guarantees that we do not even have to consider λ being complex.

2. **Ordering of eigenvalues.** There are an infinite number of eigenvalues for (5.3.3), namely $\lambda = (n\pi/L)^2$ for $n = 1, 2, 3, \ldots$. Sometimes we use the notation $\lambda_n = (n\pi/L)^2$. Note that there is a smallest eigenvalue, $\lambda_1 = (\pi/L)^2$, but no largest eigenvalue since $\lambda_n \to \infty$ as $n \to \infty$. Our theorem claims that this idea is valid for any regular Sturm–Liouville problem.

3. **Zeros of eigenfunctions.** For the eigenvalues of (5.3.3), $\lambda_n = (n\pi/L)^2$, the eigenfunctions are known to be $\sin n\pi x/L$. We use the notation $\phi_n(x) = \sin n\pi x/L$. The eigenfunction is unique (to within an arbitrary multiplicative constant).

An important and interesting aspect of this theorem is that we claim that for all regular Sturm–Liouville problems, the nth eigenfunction has exactly $(n - 1)$ zeros, not counting the end points. The eigenfunction ϕ_1 corresponding to the smallest eigenvalue (λ_1, $n = 1$) should have no zeros in the interior. The eigenfunction ϕ_2 corresponding to the next smallest eigenvalue (λ_2, $n = 2$) should have exactly one zero in the interior; and so on. We use our eigenvalue problem (5.3.3) to illustrate these properties. The eigenfunctions $\phi_n(x) = \sin n\pi x/L$ are sketched in Fig. 5.3.1 for $n = 1, 2, 3$. Note that the theorem is verified (since we only count zeros at interior points). $\sin \pi x/L$ has no zeros between $x = 0$ and $x = L$, $\sin 2\pi x/L$ has one zero between $x = 0$ and $x = L$, and $\sin 3\pi x/L$ has two zeros between $x = 0$ and $x = L$.

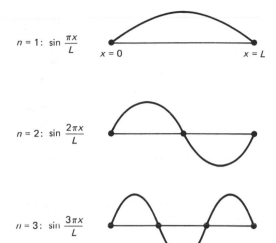

$n = 1:$ $\sin \dfrac{\pi x}{L}$

$x = 0$ $x = L$

$n = 2:$ $\sin \dfrac{2\pi x}{L}$

$n = 3:$ $\sin \dfrac{3\pi x}{L}$

Figure 5.3.1 Zeroes of eigenfunctions $\sin n\pi x/L$.

4. **Series of eigenfunctions.** According to this theorem, the eigenfunctions can always be used to represent any piecewise smooth function $f(x)$,

$$f(x) \sim \sum_{n=1}^{\infty} a_n \phi_n(x). \tag{5.3.4}$$

Thus, for our example (5.3.3),

$$f(x) \sim \sum_{n=1}^{\infty} a_n \sin \frac{n\pi x}{L}.$$

We recognize this as a Fourier sine series. We know that any piecewise smooth function can be represented by a Fourier sine series and the infinite series converges to $[f(x+) + f(x-)]/2$ for $0 < x < L$. It converges to $f(x)$ for $0 < x < L$, if $f(x)$ is continuous there. This theorem thus claims that the convergence properties of Fourier sine series are valid for all series of eigenfunctions of any regular Sturm–Liouville eigenvalue problem. Equation (5.3.4) is referred to as an expansion of $f(x)$ in terms of the eigenfunctions $\phi_n(x)$ or more simply as an eigenfunction expansion. It is also called a **generalized Fourier series of f(x)**. The coefficients a_n are called the coefficients of the eigenfunction expansion or the **generalized Fourier coefficients**. The fact that rather arbitrary functions may be represented in terms of an infinite series of eigenfunctions will enable us to solve partial differential equations by the method of separation of variables.

5. **Orthogonality of eigenfunctions.** The preceding theorem enables a function to be represented by a series of eigenfunctions, (5.3.4). Here we will show how to determine the generalized Fourier coefficients, a_n. According to the important theorem we are now describing, the eigenfunctions of any regular Sturm–Liouville eigenvalue problem will always be orthogonal. The theorem states that a weight $\sigma(x)$ must be introduced into the orthogonality relation:

$$\boxed{\int_a^b \phi_n(x)\phi_m(x)\sigma(x)\, dx = 0,} \qquad \text{if} \quad \lambda_n \neq \lambda_m. \tag{5.3.5}$$

Here $\sigma(x)$ is the possibly variable coefficient that multiplies the eigenvalue λ in the differential equation defining the eigenvalue problem. Since corresponding to each eigenvalue there is only one eigenfunction, the statement "if $\lambda_n \neq \lambda_m$" in (5.3.5) may be replaced by "if $n \neq m$." For the Fourier sine series example, the defining differential equation is $d^2\phi/dx^2 + \lambda\phi = 0$, and hence a comparison with the form of the general Sturm–Liouville problem shows that $\sigma(x) = 1$. Thus, in this case the weight is 1, and the orthogonality condition, $\int_0^L \sin n\pi x/L \sin m\pi x/L \, dx = 0$, follows if $n \neq m$, as we already know.

   As with Fourier sine series, we use the orthogonality condition to determine the generalized Fourier coefficients. In order to utilize the orthogonality condition (5.3.5), we must multiply (5.3.4) by $\phi_m(x)$ *and* $\sigma(x)$. Thus,

$$f(x)\phi_m(x)\sigma(x) = \sum_{n=1}^{\infty} a_n\phi_n(x)\phi_m(x)\sigma(x),$$

where we assume these operations on infinite series are valid, and hence introduce equal signs. Integrating from $x = a$ to $x = b$ yields

$$\int_a^b f(x)\phi_m(x)\sigma(x)\, dx = \sum_{n=1}^{\infty} a_n \int_a^b \phi_n(x)\phi_m(x)\sigma(x)\, dx.$$

Since the eigenfunctions are orthogonal [with weight $\sigma(x)$], all the integrals on the right-hand side vanish except when n reaches m:

$$\int_a^b f(x)\phi_m(x)\sigma(x)\, dx = a_m \int_a^b \phi_m^2(x)\sigma(x)\, dx.$$

The integral on the right is nonzero since the weight $\sigma(x)$ must be positive (from the definition of a regular Sturm–Liouville problem), and hence we may divide by it to determine the generalized Fourier coefficient a_m:

$$\boxed{a_m = \frac{\displaystyle\int_a^b f(x)\phi_m(x)\sigma(x)\, dx}{\displaystyle\int_a^b \phi_m^2(x)\sigma(x)\, dx}.} \tag{5.3.6}$$

In the example of a Fourier sine series, $a = 0$, $b = L$, $\phi_n(x) = \sin n\pi x/L$ and $\sigma(x) = 1$. Thus, if we recall the known integral that $\int_0^L \sin^2 n\pi x/L \, dx = L/2$, (5.3.6) reduces to the well-known formula for the coefficients of the Fourier sine series. It is not always possible to evaluate the integral in the denominator of (5.3.6) in a simple way.

   6. **Rayleigh quotient.** In Sec. 5.6 we will prove that the eigenvalue may be related to its eigenfunction in the following way:

$$\lambda = \frac{-p\phi \, d\phi/dx|_a^b + \int_a^b [p(d\phi/dx)^2 - q\phi^2] \, dx}{\int_a^b \phi^2 \sigma \, dx}, \qquad (5.3.7)$$

known as the **Rayleigh quotient**. The numerator contains integrated terms and terms evaluated at the boundaries. Since the eigenfunctions cannot be determined without knowing the eigenvalues, this expression is never used directly to determine the eigenvalues. However, interesting and significant results can be obtained from the Rayleigh quotient without solving the differential equation. Consider the Fourier sine series example (5.3.3) that we have been analyzing: $a = 0$, $b = L$, $p(x) = 1$, $q(x) = 0$, and $\sigma(x) = 1$. Since $\phi(0) = 0$ and $\phi(L) = 0$, the Rayleigh quotient implies that

$$\lambda = \frac{\int_0^L (d\phi/dx)^2 \, dx}{\int_0^L \phi^2 \, dx}. \qquad (5.3.8)$$

Although this does not determine λ since ϕ is unknown, it gives useful information. Both the numerator and the denominator are ≥ 0. Since ϕ cannot be identically zero and be called an eigenfunction, the denominator cannot be zero. Thus, $\lambda \geq 0$ follows from (5.3.8). Without solving the differential equation, we immediately conclude that there cannot be any negative eigenvalues. When we first determined eigenvalues for this problem, we worked rather hard to show that there were no negative eigenvalues (see Sec. 2.3). Now we can simply apply the Rayleigh quotient to eliminate the possibility of negative eigenvalues *for this example*. *Sometimes*, as we shall see later, we can also show that $\lambda \geq 0$ in harder problems.

Furthermore, even the possibility of $\lambda = 0$ can sometimes be analyzed using the Rayleigh quotient. For the simple problem (5.3.3) with zero boundary conditions, $\phi(0) = 0$ and $\phi(L) = 0$, let us see if it is possible for $\lambda = 0$ directly from (5.3.8). $\lambda = 0$ only if $d\phi/dx = 0$ for all x. Then, by integration, ϕ must be a constant for all x. However, from the boundary conditions [either $\phi(0) = 0$ or $\phi(L) = 0$], that constant must be zero. Thus, $\lambda = 0$ only if $\phi = 0$ everywhere. But if $\phi = 0$ everywhere, we do not call ϕ an eigenfunction. Thus, $\lambda = 0$ is not an eigenvalue *in this case*, and we have further concluded that $\lambda > 0$; all the eigenvalues must be positive. This is concluded *without* using solutions of the differential equation. The known eigenvalues in this example, $\lambda_n = (n\pi/L)^2$, $n = 1, 2, \ldots$, are clearly consistent with the conclusions from the Rayleigh quotient. Other applications of the Rayleigh quotient will appear in later sections.

EXERCISES 5.3

*5.3.1. Do Exercise 4.4.2(b). Show that the spatial differential equation may be put into Sturm–Liouville form.

5.3.2. Consider

$$\rho \frac{\partial^2 u}{\partial t^2} = T_0 \frac{\partial^2 u}{\partial x^2} + \alpha u + \beta \frac{\partial u}{\partial t}.$$

(a) Give a brief physical interpretation. What signs do α and β have to be physical?

(b) Allow ρ, α, β to be functions of x. Show that separation of variables works only if $\beta = c\rho$, where c is a constant.

(c) If $\beta = c\rho$, show that the spatial equation is a Sturm–Liouville differential equation. Solve the time equation.

*5.3.3. Consider the non-Sturm–Liouville differential equation

$$\frac{d^2\phi}{dx^2} + \alpha(x)\frac{d\phi}{dx} + [\lambda\beta(x) + \gamma(x)]\phi = 0.$$

Multiply this equation by $H(x)$. Determine $H(x)$ such that the equation may be reduced to the standard Sturm–Liouville form:

$$\frac{d}{dx}\left[p(x)\frac{d\phi}{dx}\right] + [\lambda\sigma(x) + q(x)]\phi = 0.$$

Given $\alpha(x)$, $\beta(x)$, and $\gamma(x)$, what are $p(x)$, $\sigma(x)$, and $q(x)$?

5.3.4. Consider heat flow with convection (see Exercise 1.5.2):

$$\frac{\partial u}{\partial t} = k\frac{\partial^2 u}{\partial x^2} - V_0\frac{\partial u}{\partial x}.$$

(a) Show that the spatial ordinary differential equation obtained by separation of variables is not in Sturm–Liouville form.

*(b) Solve the initial boundary value problem

$$u(0, t) = 0$$
$$u(L, t) = 0$$
$$u(x, 0) = f(x).$$

(c) Solve the initial boundary value problem

$$\frac{\partial u}{\partial x}(0, t) = 0$$

$$\frac{\partial u}{\partial x}(L, t) = 0$$

$$u(x, 0) = f(x).$$

5.3.5. For the Sturm–Liouville eigenvalue problem,

$$\frac{d^2\phi}{dx^2} + \lambda\phi = 0 \quad \text{with} \quad \frac{d\phi}{dx}(0) = 0 \quad \text{and} \quad \frac{d\phi}{dx}(L) = 0,$$

verify the following general properties:

(a) There are an infinite number of eigenvalues with a smallest but no largest.

(b) The nth eigenfunction has $n - 1$ zeros.

(c) The eigenfunctions are complete and orthogonal.

(d) What does the Rayleigh quotient say concerning negative and zero eigenvalues?

5.3.6. Redo Exercise 5.3.5 for the Sturm–Liouville eigenvalue problem

$$\frac{d^2\phi}{dx^2} + \lambda\phi = 0 \quad \text{with} \quad \frac{d\phi}{dx}(0) = 0 \quad \text{and} \quad \phi(L) = 0.$$

5.3.7. Which of statements 1–5 on page 135 are valid for the following eigenvalue problem?

$$\frac{d^2\phi}{dx^2} + \lambda\phi = 0 \qquad \text{with}$$

$$\phi(-L) = \phi(L)$$

$$\frac{d\phi}{dx}(-L) = \frac{d\phi}{dx}(L).$$

5.3.8. Show that $\lambda \geq 0$ for the eigenvalue problem

$$\frac{d^2\phi}{dx^2} + (\lambda - x^2)\phi = 0 \qquad \text{with} \qquad \frac{d\phi}{dx}(0) = 0, \quad \frac{d\phi}{dx}(1) = 0.$$

Is $\lambda = 0$ an eigenvalue?

5.3.9. Consider the eigenvalue problem

$$x^2\frac{d^2\phi}{dx^2} + x\frac{d\phi}{dx} + \lambda\phi = 0 \quad \text{with} \quad \phi(1) = 0 \quad \text{and} \quad \phi(b) = 0. \qquad (5.3.9)$$

   (a) Show that multiplying by $1/x$ puts this in the Sturm–Liouville form. (This multiplicative factor is derived in Exercise 5.3.3.)
   (b) Show that $\lambda \geq 0$.
   *(c) Since (5.3.9) is an equidimensional equation, determine all positive eigenvalues. Is $\lambda = 0$ an eigenvalue? Show that there are an infinite number of eigenvalues with a smallest, but no largest.
   (d) The eigenfunctions are orthogonal with what weight according to Sturm–Liouville theory? Verify the orthogonality using properties of integrals.
   (e) Show that the nth eigenfunction has $n - 1$ zeros.

5.3.10. Reconsider Exercise 5.3.9 with the boundary conditions

$$\frac{d\phi}{dx}(1) = 0 \qquad \text{and} \qquad \frac{d\phi}{dx}(b) = 0.$$

5.4 WORKED EXAMPLE—HEAT FLOW IN A NONUNIFORM ROD WITHOUT SOURCES

In this section we illustrate the application to partial differential equations of some of the general theorems on regular Sturm–Liouville eigenvalue problems. Consider the heat flow in a nonuniform rod (with possibly nonconstant thermal properties c, ρ, K_0) *without sources*; see Sec. 1.2 or 5.2.1. At the left end $x = 0$ the temperature is prescribed to be 0° and the right end is insulated. The initial temperature distribution is given. The mathematical formulation of this problem is:

PDE:
$$c\rho\frac{\partial u}{\partial t} = \frac{\partial}{\partial x}\left(K_0\frac{\partial u}{\partial x}\right) \qquad (5.4.1a)$$

BC:
$$u(0, t) = 0 \qquad (5.4.1b)$$
$$\frac{\partial u}{\partial x}(L, t) = 0 \qquad (5.4.1c)$$

IC:
$$u(x, 0) = f(x). \qquad (5.4.1d)$$

Since the partial differential equation and the boundary conditions are linear and homogeneous, we seek special solutions (ignoring the initial condition) in the product form:

$$u(x, t) = \phi(x)h(t). \tag{5.4.2}$$

After separation of variables (for details see Sec. 5.2.1), we find that the time part satisfies the ordinary differential equation

$$\frac{dh}{dt} = -\lambda h, \tag{5.4.3}$$

while the spatial part solves the following regular Sturm–Liouville eigenvalue problem:

$$\frac{d}{dx}\left(K_0 \frac{d\phi}{dx}\right) + \lambda c\rho\phi = 0 \tag{5.4.4a}$$

$$\phi(0) = 0 \tag{5.4.4b}$$

$$\frac{d\phi}{dx}(L) = 0. \tag{5.4.4c}$$

According to our theorems concerning Sturm–Liouville eigenvalue problems, there is an infinite sequence of eigenvalues λ_n and corresponding eigenfunctions $\phi_n(x)$. *We assume* that $\phi_n(x)$ *are known* (it might be a difficult problem to determine approximately the first few using numerical methods, but nevertheless it can be done). The time-dependent part of the differential equation is easily solved,

$$h(t) = ce^{-\lambda_n t}. \tag{5.4.5}$$

In this way we obtain an infinite sequence of product solutions of the partial differential equation

$$u(x, t) = \phi_n(x)e^{-\lambda_n t}. \tag{5.4.6}$$

According to the principle of superposition, we attempt to satisfy the initial condition with an infinite linear combination of these product solutions:

$$u(x, t) = \sum_{n=1}^{\infty} a_n \phi_n(x)e^{-\lambda_n t}. \tag{5.4.7}$$

This infinite series has the property that it solves the PDE and the homogeneous BCs. We will show that we can determine the as yet unknown constants a_n from the initial condition

$$u(x, 0) = f(x) = \sum_{n=1}^{\infty} a_n \phi_n(x). \tag{5.4.8}$$

Our theorems imply that any piecewise smooth $f(x)$ can be represented by this type of series of eigenfunctions. The coefficients a_n are the generalized Fourier coefficients of the initial condition. Furthermore, the eigenfunctions are orthogonal with a weight $\sigma(x) = c(x)\rho(x)$, determined from the physical properties of the rod:

$$\int_0^L \phi_n(x)\phi_m(x)c(x)\rho(x)\,dx = 0 \qquad \text{for } n \neq m.$$

Using these orthogonality formulas, the generalized Fourier coefficients are

$$a_n = \frac{\int_0^L f(x)\phi_n(x)c(x)\rho(x)\,dx}{\int_0^L \phi_n^2(x)c(x)\rho(x)\,dx}. \qquad (5.4.9)$$

We claim that (5.4.7) is the desired solution, with coefficients given by (5.4.9).

In order to give a minimal interpretation of the solution we should ask what happens for large t. Since the eigenvalues form an increasing sequence, each succeeding term in (5.4.7) is exponentially smaller than the preceding term for large t. Thus, for large time the solution may be accurately approximated by

$$u(x, t) \approx a_1\phi_1(x)e^{-\lambda_1 t}. \qquad (5.4.10)$$

This approximation is not very good if $a_1 = 0$, in which case (5.4.10) should begin with the first nonzero term. However, often the initial temperature $f(x)$ is nonnegative (and not identically zero). In this case, we will show from (5.4.9) that $a_1 \neq 0$:

$$a_1 = \frac{\int_0^L f(x)\phi_1(x)c(x)\rho(x)\,dx}{\int_0^L \phi_1^2(x)c(x)\rho(x)\,dx}. \qquad (5.4.11)$$

It follows that $a_1 \neq 0$, because $\phi_1(x)$ is the eigenfunction corresponding to the lowest eigenvalue and has no zeros; $\phi_1(x)$ is of one sign. Thus, if $f(x) > 0$, it follows that $a_1 \neq 0$, since $c(x)$ and $\rho(x)$ are positive physical functions. In order to sketch the solution for large fixed t, (5.4.10) shows that all that is needed is the first eigenfunction. At the very least, a numerical calculation of the first eigenfunction is easier than the computation of the first hundred.

For large time, the "shape" of the temperature distribution in space stays approximately the same in time. Its amplitude grows or decays in time depending on whether $\lambda_1 > 0$ or $\lambda_1 < 0$ (it would be constant in time if $\lambda_1 = 0$). Since this is a heat flow problem with no sources and with zero temperature at $x = 0$, we certainly expect the temperature to be exponentially decaying toward $0°$ (i.e., we expect that $\lambda_1 > 0$). Although the right end is insulated, heat energy should flow out the left end since there $u = 0$. We now prove mathematically that all

$\lambda > 0$. Since $p(x) = K_0(x)$, $q(x) = 0$, and $\sigma(x) = c(x)\rho(x)$, it follows from the Rayleigh quotient that

$$\lambda = \frac{\int_0^L K_0(x)(d\phi/dx)^2 \, dx}{\int_0^L \phi^2 c(x)\rho(x) \, dx}, \tag{5.4.12}$$

where the boundary contribution to (5.4.12) vanished due to the specific homogeneous boundary conditions, (5.4.4b) and (5.4.4c). It immediately follows from (5.4.12) that all $\lambda \geq 0$, since the thermal coefficients are positive. Furthermore, $\lambda > 0$, since ϕ = constant is not an allowable eigenfunction [because $\phi(0) = 0$]. Thus, we have shown that $\lim_{t\to\infty} u(x, t) = 0$ for this example.

EXERCISES 5.4

5.4.1. Consider

$$c\rho \frac{\partial u}{\partial t} = \frac{\partial}{\partial x}\left(K_0 \frac{\partial u}{\partial x}\right) + \alpha u,$$

where c, ρ, K_0, α are functions of x, subject to

$$u(0, t) = 0$$
$$u(L, t) = 0$$
$$u(x, 0) = f(x).$$

Assume that the appropriate eigenfunctions are known.
(a) Show that the eigenvalues are positive if $\alpha < 0$ (see Sec. 5.2.1).
(b) Solve the initial value problem.
(c) Briefly discuss $\lim_{t\to\infty} u(x, t)$.

*5.4.2. Consider

$$c\rho \frac{\partial u}{\partial t} = \frac{\partial}{\partial x}\left(K_0 \frac{\partial u}{\partial x}\right),$$

where c, ρ, K_0 are functions of x, subject to

$$\frac{\partial u}{\partial x}(0, t) = 0$$

$$\frac{\partial u}{\partial x}(L, t) = 0$$

$$u(x, 0) = f(x).$$

Assume that the appropriate eigenfunctions are known. Solve the initial value problem, briefly discussing $\lim_{t\to\infty} u(x, t)$.

*5.4.3. Solve

$$\frac{\partial u}{\partial t} = k\frac{1}{r}\frac{\partial}{\partial r}\left(r\frac{\partial u}{\partial r}\right)$$

with $u(r, 0) = f(r)$, $u(0, t)$ bounded, and $u(a, t) = 0$.

You may assume that the corresponding eigenfunctions, denoted $\phi_n(r)$, are known and are complete. (*Hint:* See Sec. 5.2.2.)

5.4.4. Consider the following boundary value problem:

$$\frac{\partial u}{\partial t} = k\frac{\partial^2 u}{\partial x^2} \quad \text{with} \quad \frac{\partial u}{\partial x}(0, t) = 0 \quad \text{and} \quad u(L, t) = 0.$$

Solve such that $u(x, 0) = \sin \pi x/L$ (initial condition). (*Hint:* If necessary, use a table of integrals.)

5.4.5. Consider

$$\rho\frac{\partial^2 u}{\partial t^2} = T_0\frac{\partial^2 u}{\partial x^2} + \alpha u,$$

where $\rho(x) > 0$, $\alpha(x) < 0$, and T_0 is constant, subject to

$$u(0, t) = 0 \qquad u(x, 0) = f(x)$$

$$u(L, t) = 0 \qquad \frac{\partial u}{\partial t}(x, 0) = g(x).$$

Assume that the appropriate eigenfunctions are known. Solve the initial value problem.

*5.4.6. Consider the vibrations of a *nonuniform* string of mass density $\rho_0(x)$. Suppose that the left end at $x = 0$ is fixed and the right end obeys the elastic boundary condition: $\partial u/\partial x = -(k/T_0)u$ at $x = L$. Suppose that the string is *initially at rest* with a known initial position $f(x)$. Solve this initial value problem. (*Hints:* Assume that the appropriate eigenvalues and corresponding eigenfunctions are known. What differential equations with what boundary conditions do they satisfy? The eigenfunctions are orthogonal with what weighting function?)

5.5 SELF-ADJOINT OPERATORS AND STURM–LIOUVILLE EIGENVALUE PROBLEMS

Introduction. In this section we prove some of the properties of regular Sturm–Liouville eigenvalue problems:

$$\frac{d}{dx}\left[p(x)\frac{d\phi}{dx}\right] + q(x)\phi + \lambda\sigma(x)\phi = 0 \tag{5.5.1a}$$

$$\beta_1\phi(a) + \beta_2\frac{d\phi}{dx}(a) = 0 \tag{5.5.1b}$$

$$\beta_3\phi(b) + \beta_4\frac{d\phi}{dx}(b) = 0, \tag{5.5.1c}$$

where β_i are real and where, on the finite interval $(a \leqslant x \leqslant b)$ p, q, σ are real continuous functions and p, σ are positive $[p(x) > 0$ and $\sigma(x) > 0]$. At times we will make some comments on the validity of our results if some of these restrictions are removed.

The proofs of three statements are somewhat difficult. We will not prove that there are an infinite number of eigenvalues. We will have to rely for understanding on the examples already presented and on some further examples developed in later sections. For Sturm–Liouville eigenvalue problems that are *not* regular, there may be no eigenvalues at all. However, in most cases of physical interest (on finite intervals) there will still be an infinite number of discrete eigenvalues. We also will not attempt to prove that any piecewise smooth function can be expanded in terms of the eigenfunctions of a regular Sturm–Liouville problem (known as the completeness property). We will not attempt to prove that each succeeding eigenfunction has one additional zero (oscillates one more time).

Linear operators. The proofs we will investigate are made easier to follow by the introduction of **operator** *notation*. Let L stand for the linear differential operator $d/dx[p(x) d/dx] + q(x)$. An operator acts on a function and yields another function. The notation means that for this L acting on the function $y(x)$,

$$L(y) \equiv \frac{d}{dx}\left[p(x)\frac{dy}{dx} \right] + q(x)y.$$

Thus, $L(y)$ is just a shorthand notation. For example, if $L \equiv d^2/dx^2 + 6$, then $L(y) = d^2y/dx^2 + 6y$ or $L(e^{2x}) = 4e^{2x} + 6e^{2x} = 10e^{2x}$.

The Sturm–Liouville differential equation is rather cumbersome to write over and over again. The use of the linear operator notation is somewhat helpful. Using the operator notation

$$L(\phi) + \lambda\sigma(x)\phi = 0, \tag{5.5.2}$$

where λ is an eigenvalue and ϕ the corresponding eigenfunction. L can operate on any function, not just an eigenfunction.

Lagrange's identity. Most of the proofs we will present concerning Sturm–Liouville eigenvalue problems are immediate consequences of an interesting and fundamental formula known as **Lagrange's identity**. For convenience, we will use the operator notation. We calculate $uL(v) - vL(u)$ where u and v are *any two functions* (not necessarily eigenfunctions). Recall that

$$L(u) = \frac{d}{dx}\left(p\frac{du}{dx} \right) + qu \quad \text{and} \quad L(v) = \frac{d}{dx}\left(p\frac{dv}{dx} \right) + qv,$$

and hence

$$uL(v) - vL(u) = u\frac{d}{dx}\left(p\frac{dv}{dx} \right) + \cancel{uqv} - v\frac{d}{dx}\left(p\frac{du}{dx} \right) - \cancel{vqu}, \tag{5.5.3}$$

where a simple cancellation has been noted. The right-hand side of (5.5.3) is manipulated to an exact differential,

$$\boxed{uL(v) - vL(u) = \frac{d}{dx}\left[p\left(u\frac{dv}{dx} - v\frac{du}{dx} \right) \right],} \tag{5.5.4}$$

known as the **differential form of Lagrange's identity**. To derive (5.5.4), we note from the product rule that

$$u\frac{d}{dx}\left(p\frac{dv}{dx} \right) = \frac{d}{dx}\left[u\left(p\frac{dv}{dx} \right) \right] - \left(p\frac{dv}{dx} \right)\frac{du}{dx},$$

and similarly

$$v\frac{d}{dx}\left(p\frac{du}{dx}\right) = \frac{d}{dx}\left[v\left(p\frac{du}{dx}\right)\right] - \left(p\frac{du}{dx}\right)\frac{dv}{dx}.$$

Equation (5.5.4) follows by subtracting these two. Later (see p. 150) we will use the differential form, (5.5.4).

Green's formula. The integral form of Lagrange's identity is also known as **Green's formula**. If follows by integrating (5.5.4):

$$\boxed{\int_a^b [uL(v) - vL(u)] \, dx = p\left(u\frac{dv}{dx} - v\frac{du}{dx}\right)\Bigg|_a^b} \qquad (5.5.5)$$

for any functions* u and v. *This is a very useful formula.*

Example. If $p = 1$ and $q = 0$ (in which case $L = d^2/dx^2$), (5.5.4) simply states that

$$u\frac{d^2v}{dx^2} - v\frac{d^2u}{dx^2} = \frac{d}{dx}\left(u\frac{dv}{dx} - v\frac{du}{dx}\right),$$

which is easily independently checked. For this example, Green's formula is

$$\int_a^b \left(u\frac{d^2v}{dx^2} - v\frac{d^2u}{dx^2}\right) dx = \left(u\frac{dv}{dx} - v\frac{du}{dx}\right)\Bigg|_a^b.$$

Self-adjointness. As an important case of Green's formula, suppose that u and v are any two functions, but with the additional restriction that the boundary terms happen to vanish,

$$p\left(u\frac{dv}{dx} - v\frac{du}{dx}\right)\Bigg|_a^b = 0.$$

Then from (5.5.5), $\int_a^b [uL(v) - vL(u)] \, dx = 0$.

Let us show how it is possible for the boundary terms to vanish. Instead of being arbitrary functions, we restrict u and v to *both* satisfy the same set of homogeneous *boundary conditions*. For example, suppose that u and v are *any* two functions that satisfy the following set of boundary conditions:

$$\phi(a) = 0$$

$$\frac{d\phi}{dx}(b) + h\phi(b) = 0.$$

Since both u and v satisfy these conditions, it follows that

$$u(a) = 0 \qquad\qquad\qquad v(a) = 0$$

$$\text{and}$$

$$\frac{du}{dx}(b) + hu(b) = 0 \qquad\qquad \frac{dv}{dx}(b) + hv(b) = 0;$$

otherwise, u and v are arbitrary. In this case, the boundary terms for Green's formula vanish:

---

* The integration requires du/dx and dv/dx to be continuous.

$$p\left(u\frac{dv}{dx} - v\frac{du}{dx}\right)\bigg|_a^b = p(b)\left[u(b)\frac{dv}{dx}(b) - v(b)\frac{du}{dx}(b)\right]$$

$$= p(b)[-u(b)hv(b) + v(b)hu(b)] = 0.$$

Thus, for any functions u and v both satisfying these homogeneous boundary conditions, we know that

$$\int_a^b [uL(v) - vL(u)]\, dx = 0.$$

In fact, we claim (see Exercise 5.5.1) that the boundary terms also vanish for any two functions u and v which both satisfy the same set of boundary conditions of the type that occur in the regular Sturm–Liouville eigenvalue problems (5.5.1b) and (5.5.1c). Thus, when discussing any regular Sturm–Liouville eigenvalue problem, we have the following theorem:

> *If u and v are any two functions satisfying the same set of homogeneous boundary conditions (of the regular Sturm–Liouville type), then*
> $$\int_a^b [uL(v) - vL(u)]\, dx = 0. \tag{5.5.6}$$

When (5.5.6) is valid, we say that the operator L (with the corresponding boundary conditions) is **self-adjoint**.*

The boundary terms also vanish in circumstances other than for boundary conditions of the regular Sturm–Liouville type. Two important further examples will be discussed briefly. The **periodic boundary condition** can be generalized (for nonconstant-coefficient operators) to

$$\phi(a) = \phi(b) \qquad \text{and} \qquad p(a)\frac{d\phi}{dx}(a) = p(b)\frac{d\phi}{dx}(b).$$

In this situation (5.5.6) also can be shown (see Exercise 5.5.1) to be valid. Another example in which the boundary terms in Green's formula vanish is the "singular" case. The singular case occurs if the coefficient of the second derivative of the differential operator is zero at an end point; for example, if $p(x) = 0$ at $x = 0$ [i.e., $p(0) = 0$]. At a singular end point, a singularity condition is imposed. The usual singularity condition at $x = 0$ is $\phi(0)$ bounded. It can also be shown that (5.5.6) is valid (see Exercise 5.5.1) if both u and v satisfy this singularity condition at $x = 0$ and any regular Sturm–Liouville type of boundary condition at $x = b$.

Orthogonal eigenfunctions. We now will show the usefulness of Green's formula. We will begin by proving the important orthogonality relationship for

---

* We usually avoid in this text an explanation of an **adjoint** operator. Here L equals its adjoint and so is called *self-adjoint*.

Sturm–Liouville eigenvalue problems. For many types of boundary conditions, **eigenfunctions corresponding to different eigenvalues are orthogonal with weight** $\sigma(\mathbf{x})$. To prove that statement, let λ_n and λ_m be eigenvalues with corresponding eigenfunctions $\phi_n(x)$ and $\phi_m(x)$. Using the operator notation, the differential equations satisfied by these eigenfunctions are

$$L(\phi_n) + \lambda_n \sigma(x)\phi_n = 0 \qquad (5.5.7a)$$

$$L(\phi_m) + \lambda_m \sigma(x)\phi_m = 0. \qquad (5.5.7b)$$

In addition, both ϕ_n and ϕ_m satisfy the same set of homogeneous boundary conditions. Since u and v are arbitrary functions, we may let $u = \phi_m$ and $v = \phi_n$ in Green's formula:

$$\int_a^b [\phi_m L(\phi_n) - \phi_n L(\phi_m)]\, dx = p(x)\left(\phi_m \frac{d\phi_n}{dx} - \phi_n \frac{d\phi_m}{dx}\right)\Bigg|_a^b.$$

$L(\phi_n)$ and $L(\phi_m)$ may be eliminated from (5.5.7). Thus,

$$(\lambda_m - \lambda_n)\int_a^b \phi_n \phi_m \sigma\, dx = p(x)\left(\phi_m \frac{d\phi_n}{dx} - \phi_n \frac{d\phi_m}{dx}\right)\Bigg|_a^b, \qquad (5.5.8)$$

corresponding to multiplying (5.5.7a) by ϕ_m, multiplying (5.5.7b) by ϕ_n, subtracting the two, and then integrating. We avoided these steps (especially the integration) by applying Green's formula. For many different kinds of boundary conditions (i.e., regular Sturm–Liouville types, periodic case, and the singular case), the boundary terms vanish if u and v both satisfy the same set of homogeneous boundary conditions. Since u and v are eigenfunctions, they satisfy this condition, and thus (5.5.8) implies that

$$(\lambda_m - \lambda_n)\int_a^b \phi_n \phi_m \sigma\, dx = 0. \qquad (5.5.9)$$

If $\lambda_m \neq \lambda_n$, then it immediately follows that

$$\int_a^b \phi_n \phi_m \sigma\, dx = 0. \qquad (5.5.10)$$

In other words, eigenfunctions (ϕ_n and ϕ_m) corresponding to different eigenvalues ($\lambda_n \neq \lambda_m$) are orthogonal with weight $\sigma(x)$.

Real eigenvalues. We can use the orthogonality of eigenfunctions to prove that the eigenvalues are real. Suppose that λ is a complex eigenvalue and $\phi(x)$ the corresponding eigenfunction (also allowed to be complex since the differential equation defining the eigenfunction would be complex):

$$L(\phi) + \lambda \sigma \phi = 0. \qquad (5.5.11)$$

We introduce the notation $^{-}$ for the complex conjugate (e.g., if $z = x + iy$, then $\bar{z} = x - iy$). Note that if $z = 0$, then $\bar{z} = 0$. Thus, the complex conjugate of (5.5.11) is also valid:

$$\overline{L(\phi)} + \bar{\lambda}\sigma\bar{\phi} = 0, \qquad (5.5.12)$$

assuming that the coefficient σ is real and hence $\bar{\sigma} = \sigma$. The complex conjugate of $L(\phi)$ is exactly L operating on the complex conjugate of ϕ, $\overline{L(\phi)} = L(\bar{\phi})$, since the coefficients of the linear differential operator are also real (see Exercise 5.5.7). Thus,

$$L(\overline{\phi}) + \overline{\lambda}\sigma\overline{\phi} = 0. \qquad (5.5.13)$$

If ϕ satisfies boundary conditions with real coefficients, then $\overline{\phi}$ satisfies the same boundary conditions. For example, if $d\phi/dx + h\phi = 0$ at $x = a$, then by taking complex conjugates, $d\overline{\phi}/dx + h\overline{\phi} = 0$ at $x = a$. Equation (5.5.13) and the boundary conditions show that $\overline{\phi}$ satisfies the Sturm–Liouville eigenvalue problem, but with the eigenvalue being $\overline{\lambda}$. We have thus proved the following theorem:*
If λ is a complex eigenvalue with corresponding eigenfunction ϕ, then $\overline{\lambda}$ is also an eigenvalue with corresponding eigenfunction $\overline{\phi}$.

However, we will show λ cannot be complex. As we have shown, if λ is an eigenvalue, then so too is $\overline{\lambda}$. According to our fundamental orthogonality theorem, the corresponding eigenfunctions (ϕ and $\overline{\phi}$) must be orthogonal (with weight σ). Thus, from (5.5.9),

$$(\lambda - \overline{\lambda}) \int_a^b \phi\overline{\phi}\sigma \, dx = 0. \qquad (5.5.14)$$

Since $\phi\overline{\phi} = |\phi|^2 \geq 0$ (and $\sigma > 0$), the integral in (5.5.14) is ≥ 0. In fact, the integral can equal zero only if $\phi \equiv 0$, which is prohibited since ϕ is an eigenfunction. Thus, (5.5.14) implies that $\lambda = \overline{\lambda}$, and hence λ is real; **all the eigenvalues are real**. The eigenfunctions can always be chosen to be real.

Unique eigenfunctions (regular and singular cases). We next prove that there is only one eigenfunction corresponding to an eigenvalue (except for the case of periodic boundary conditions). Suppose that there are two different eigenfunctions ϕ_1 and ϕ_2 corresponding to the same eigenvalue λ. We say λ is a "multiple" eigenvalue with multiplicity two. In this case, both

$$L(\phi_1) + \lambda\sigma\phi_1 = 0 \qquad (5.5.15)$$
$$L(\phi_2) + \lambda\sigma\phi_2 = 0.$$

Since λ is the same in both expressions,

$$\phi_2 L(\phi_1) - \phi_1 L(\phi_2) = 0. \qquad (5.5.16)$$

This can be integrated by some simple manipulations. However, we avoid this algebra by simply quoting the differential form of Lagrange's identity:

$$\phi_2 L(\phi_1) - \phi_1 L(\phi_2) = \frac{d}{dx}\left[p\left(\phi_2\frac{d\phi_1}{dx} - \phi_1\frac{d\phi_2}{dx}\right)\right].$$

From (5.5.16) it follows that

$$p\left(\phi_1\frac{d\phi_2}{dx} - \phi_2\frac{d\phi_1}{dx}\right) = \text{constant}. \qquad (5.5.17)$$

Often we can evaluate the constant from one of the boundary conditions.

---

* A "similar" type of theorem follows from the quadratic formula: For a quadratic equation with real coefficients, if λ is a complex root, then so is $\overline{\lambda}$. This also holds for any algebraic equation with real coefficients.

For example, if $d\phi/dx + h\phi = 0$ at $x = a$, a short calculation shows that the constant $= 0$. In fact, we claim (Exercise 5.5.10) that the constant also equals zero if *at least one of the boundary conditions is of the regular Sturm–Liouville type* (or of the singular type). For any of these boundary conditions, it follows that

$$\phi_1 \frac{d\phi_2}{dx} - \phi_2 \frac{d\phi_1}{dx} = 0. \tag{5.5.18}$$

This is equivalent to $d/dx(\phi_2/\phi_1) = 0$, and hence *for these boundary conditions*

$$\phi_2 = c\phi_1. \tag{5.5.19}$$

This shows that any two eigenfunctions ϕ_1 and ϕ_2 corresponding to the same eigenvalue must be an integral multiple of each other *for the boundary conditions above*. The two eigenfunctions are dependent; there is only one linearly independent eigenfunction; the eigenfunction is unique.

Nonunique eigenfunctions (periodic case). For periodic boundary conditions, we *cannot* conclude that the constant in (5.5.17) must be zero. Thus, it is possible that $\phi_2 \neq c\phi_1$ and that there *might* be two different eigenfunctions corresponding to the same eigenvalue.

For example, consider the simple eigenvalue problem with periodic boundary conditions,

$$\frac{d^2\phi}{dx^2} + \lambda\phi = 0$$

$$\phi(-L) = \phi(L) \tag{5.5.20}$$

$$\frac{d\phi}{dx}(-L) = \frac{d\phi}{dx}(L).$$

We know that the eigenvalue 0 has any constant as the unique eigenfunction. The other eigenvalues, $(n\pi/L)^2$, $n = 1, 2, \ldots$, each have two linearly independent eigenfunctions, $\sin n\pi x/L$ and $\cos n\pi x/L$. This, we know, gives rise to a Fourier series. However, (5.5.20) is not a regular Sturm–Liouville eigenvalue problem, since the boundary conditions are not of the prescribed form. Our theorem about unique eigenfunctions does not apply; we *may* have two[*] eigenfunctions corresponding to the same eigenvalue. Note that it is still *possible* to have only one eigenfunction, as occurs for $\lambda = 0$.

Nonunique eigenfunctions (Gram–Schmidt orthogonalization). We can solve for generalized Fourier coefficients (and correspondingly we are able to solve some partial differential equations) because of the orthogonality of the eigenfunctions. However, our theorem states that eigenfunctions *corresponding to different eigenvalues* are automatically orthogonal [with weight $\sigma(x)$]. For the case of periodic (or mixed-type) boundary conditions, it is possible for there to be more than one independent eigenfunction corresponding to the same eigenvalue.

---

[*] No more than two independent eigenfunctions are possible, since the differential equation is of second order.

For these multiple eigenvalues, the eigenfunctions are not automatically orthogonal to each other. In Sec. 6.5 Appendix, we will show that we always are able to construct the eigenfunctions such that they are orthogonal by a process called Gram–Schmidt orthogonalization.

EXERCISES 5.5

5.5.1. A Sturm–Liouville eigenvalue problem is called self-adjoint if
$$p\left(u\frac{dv}{dx} - v\frac{du}{dx}\right)\bigg|_a^b = 0$$
(since then $\int_a^b [uL(v) - vL(u)]\, dx = 0$) for any two functions u and v satisfying the boundary conditions. Show that the following yield self-adjoint problems.

(a) $\phi(0) = 0$ and $\phi(L) = 0$

(b) $\dfrac{d\phi}{dx}(0) = 0$ and $\phi(L) = 0$

(c) $\dfrac{d\phi}{dx}(0) - h\phi(0) = 0$ and $\dfrac{d\phi}{dx}(L) = 0$

(d) $\phi(a) = \phi(b)$ and $p(a)\dfrac{d\phi}{dx}(a) = p(b)\dfrac{d\phi}{dx}(b)$

(e) $\phi(a) = \phi(b)$ and $\dfrac{d\phi}{dx}(a) = \dfrac{d\phi}{dx}(b)$ [self-adjoint only if $p(a) = p(b)$]

(f) $\phi(L) = 0$ and [in the situation in which $p(0) = 0$]
$\phi(0)$ bounded and $\lim\limits_{x \to 0} p(x)\dfrac{d\phi}{dx} = 0$

*(g) Under what conditions is the following self-adjoint (if p is constant)?
$$\phi(L) + \alpha\phi(0) + \beta\frac{d\phi}{dx}(0) = 0$$
$$\frac{d\phi}{dx}(L) + \gamma\phi(0) + \delta\frac{d\phi}{dx}(0) = 0.$$

5.5.2. Prove that the eigenfunctions corresponding to different eigenvalues (of the following eigenvalue problem) are orthogonal:
$$\frac{d}{dx}\left[p(x)\frac{d\phi}{dx}\right] + q(x)\phi + \lambda\sigma(x)\phi = 0$$
with the boundary conditions
$$\phi(1) = 0$$
$$\phi(2) - 2\frac{d\phi}{dx}(2) = 0.$$
What is the weighting function?

5.5.3. Consider the eigenvalue problem $L(\phi) = -\lambda\sigma(x)\phi$, subject to a given set of homogeneous boundary conditions. Suppose that
$$\int_a^b [uL(v) - vL(u)]\, dx = 0$$

for all functions u and v satisfying the same set of boundary conditions. Prove that eigenfunctions corresponding to different eigenvalues are orthogonal (with what weight?).

5.5.4. Give an example of an eigenvalue problem with more than one eigenfunction corresponding to an eigenvalue.

5.5.5. Consider

$$L = \frac{d^2}{dx^2} + 6\frac{d}{dx} + 9.$$

(a) Show that $L(e^{rx}) = (r + 3)^2 e^{rx}$.

(b) Use part (a) to obtain solutions of $L(y) = 0$ (a second-order constant-coefficient differential equation).

(c) If z depends on x and a parameter r, show that

$$\frac{\partial}{\partial r}L(z) = L\left(\frac{\partial z}{\partial r}\right).$$

(d) Using part (c), evaluate $L(\partial z/\partial r)$ if $z = e^{rx}$.

(e) Obtain a second solution of $L(y) = 0$, using part (d).

5.5.6. Prove that if x is a root of a sixth-order polynomial *with real coefficients*, then $\bar{x}$ is also a root.

5.5.7. For

$$L = \frac{d}{dx}\left(p\frac{d}{dx}\right) + q$$

with p and q real, carefully show that

$$\overline{L(\phi)} = L(\bar{\phi}).$$

5.5.8. Consider a fourth-order linear differential operator,

$$L = \frac{d^4}{dx^4}.$$

(a) Show that $uL(v) - vL(u)$ is an exact differential.

(b) Evaluate $\int_0^1 [uL(v) - vL(u)]\, dx$ in terms of the boundary data for any functions u and v.

(c) Show that $\int_0^1 [uL(v) - vL(u)]\, dx = 0$ if u and v are any two functions satisfying the boundary conditions

$$\phi(0) = 0 \qquad \phi(1) = 0$$

$$\frac{d\phi}{dx}(0) = 0 \qquad \frac{d^2\phi}{dx^2}(1) = 0.$$

(d) Give another example of boundary conditions such that

$$\int_0^1 [uL(v) - vL(u)]\, dx = 0.$$

(e) For the eigenvalue problem [using the boundary conditions in part (c)]

$$\frac{d^4\phi}{dx^4} + \lambda e^x \phi = 0,$$

show that the eigenfunctions corresponding to different eigenvalues are orthogonal. What is the weighting function?

*5.5.9. For the eigenvalue problem

$$\frac{d^4\phi}{dx^4} + \lambda e^x \phi = 0$$

subject to the boundary conditions

$$\phi(0) = 0 \qquad \phi(1) = 0$$

$$\frac{d\phi}{dx}(0) = 0 \qquad \frac{d^2\phi}{dx^2}(1) = 0,$$

show that the eigenvalues are less than or equal to zero ($\lambda \leq 0$). (Don't worry; in a physical context that is exactly what is expected.) Is $\lambda = 0$ an eigenvalue?

5.5.10. **(a)** Show that (5.5.17) yields (5.5.18) if at least one of the boundary conditions is of the regular Sturm–Liouville type.

(b) Do part (a) if one boundary condition is of the singular type.

5.5 APPENDIX: MATRIX EIGENVALUE PROBLEM AND ORTHOGONALITY OF EIGENVECTORS

The matrix eigenvalue problem

$$\mathbf{A}\mathbf{x} = \lambda\mathbf{x}, \tag{5.5A.1}$$

where $\mathbf{A}$ is an $n \times n$ real matrix (with entries a_{ij}) and $\mathbf{x}$ is an n-dimensional column vector (with components x_i), has many properties similar to those of the Sturm–Liouville eigenvalue problem.

Eigenvalues and eigenvectors. For all values of λ, $\mathbf{x} = 0$ is a "trivial" solution of the homogeneous linear system (5.5A.1). We ask, for what values of λ are there nontrivial solutions? In general, (5.5A.1) can be rewritten as

$$(\mathbf{A} - \lambda\mathbf{I})\mathbf{x} = \mathbf{0}, \tag{5.5A.2}$$

where $\mathbf{I}$ is the identity matrix. According to the theory of linear equations (elementary linear algebra), a nontrivial solution exists only if

$$\det[\mathbf{A} - \lambda\mathbf{I}] = 0. \tag{5.5A.3}$$

Such values of λ are called **eigenvalues**, and the corresponding nonzero vectors $\mathbf{x}$ called **eigenvectors**.

In general, (5.5A.3) yields an nth-degree polynomial (known as the **characteristic polynomial**) which determines the eigenvalues; there will be n eigenvalues (but they may not be distinct). Corresponding to each distinct eigenvalue, there will be an eigenvector.

Example. If $\mathbf{A} = \begin{bmatrix} 2 & 1 \\ 6 & 1 \end{bmatrix}$, then the eigenvalues satisfy

$$0 = \det\begin{bmatrix} 2-\lambda & 1 \\ 6 & 1-\lambda \end{bmatrix} = (2-\lambda)(1-\lambda) - 6 = \lambda^2 - 3\lambda - 4 = (\lambda - 4)(\lambda + 1),$$

the characteristic polynomial. The eigenvalues are $\lambda = 4$ and $\lambda = -1$. For $\lambda = 4$, (5.5A.1) becomes

$$2x_1 + x_2 = 4x_1 \qquad \text{and} \qquad 6x_1 + x_2 = 4x_2,$$

or equivalently $x_2 = 2x_1$. The eigenvector $\begin{bmatrix} x_1 \\ x_2 \end{bmatrix} = x_1 \begin{bmatrix} 1 \\ 2 \end{bmatrix}$ is an arbitrary multiple of $\begin{bmatrix} 1 \\ 2 \end{bmatrix}$ for $\lambda = 4$. For $\lambda = -1$,

$$2x_1 + x_2 = -x_1 \quad \text{and} \quad 6x_1 + x_2 = -x_2,$$

and thus the eigenvector $\begin{bmatrix} x_1 \\ x_2 \end{bmatrix} = x_1 \begin{bmatrix} 1 \\ -3 \end{bmatrix}$ is an arbitrary multiple of $\begin{bmatrix} 1 \\ -3 \end{bmatrix}$.

Green's formula. The matrix **A** may be thought of as a linear operator in the same way that

$$L = \frac{d}{dx}\left(p\frac{d}{dx}\right) + q$$

is a linear differential operator. **A** operates on n-dimensional vectors producing an n-dimensional vector, while L operates on functions and yields a function. In analyzing the Sturm–Liouville eigenvalue problem, Green's formula was important:

$$\int_a^b [uL(v) - vL(u)] \, dx = p\left(u\frac{dv}{dx} - v\frac{du}{dx}\right)\Big|_a^b,$$

where u and v are arbitrary functions. Often, the boundary terms vanished. For vectors, the dot product is analogous to integration, $\mathbf{a} \cdot \mathbf{b} = \Sigma_i \, a_i b_i$, where a_i and b_i are the ith components of, respectively, **a** and **b** (see Sec. 2.3 Appendix). By direct analogy to Green's formula we would be lead to investigate $\mathbf{u} \cdot \mathbf{Av}$ and $\mathbf{v} \cdot \mathbf{Au}$, where **u** and **v** are arbitrary vectors. Instead, we analyze $\mathbf{u} \cdot \mathbf{Av}$ and $\mathbf{v} \cdot \mathbf{Bu}$, where **B** is any $n \times n$ matrix:

$$\mathbf{u} \cdot \mathbf{Av} = \sum_i \left(u_i \sum_j a_{ij}v_j\right) = \sum_i \sum_j a_{ij}u_iv_j$$

$$\mathbf{v} \cdot \mathbf{Bu} = \sum_i \left(v_i \sum_j b_{ij}u_j\right) = \sum_i \sum_j b_{ij}u_jv_i = \sum_i \sum_j b_{ji}u_iv_j,$$

where an alternative expression for $\mathbf{v} \cdot \mathbf{Bu}$ was derived by interchanging the roles of i and j. Thus,

$$\mathbf{u} \cdot \mathbf{Av} - \mathbf{v} \cdot \mathbf{Bu} = \sum_i \sum_j (a_{ij} - b_{ji})u_iv_j.$$

If we let **B** equal the transpose of **A** (i.e., $b_{ji} = a_{ij}$), whose notation is $\mathbf{B} = \mathbf{A}'$, then we have the following theorem:

$$\boxed{\mathbf{u} \cdot \mathbf{Av} - \mathbf{v} \cdot \mathbf{A'u} = 0,} \tag{5.5A.4}$$

quite analogous to Green's formula.

Self-adjointness. The difference between **A** and its transpose, $\mathbf{A}'$, in (5.5A.4) causes insurmountable difficulties for us. We will thus restrict our attention to **symmetric** matrices, in which case $\mathbf{A} = \mathbf{A}'$. *For symmetric matrices*

$$\boxed{\mathbf{u} \cdot \mathbf{Av} - \mathbf{v} \cdot \mathbf{Au} = 0,} \tag{5.5A.5}$$

and we will be able to use this result to prove the same theorems about eigenvalues and eigenvectors for matrices as we proved about Sturm–Liouville eigenvalue problems.

For symmetric matrices, eigenvectors corresponding to different eigenvalues are orthogonal. To prove this, suppose that $\mathbf{u}$ and $\mathbf{v}$ are eigenvectors corresponding to λ_1 and λ_2, respectively:

$$\mathbf{Au} = \lambda_1\mathbf{u} \quad \text{and} \quad \mathbf{Av} = \lambda_2\mathbf{v}.$$

If we directly apply (5.5A.5), then

$$(\lambda_2 - \lambda_1)\mathbf{u} \cdot \mathbf{v} = 0.$$

Thus, if $\lambda_1 \neq \lambda_2$ (different eigenvalues), the corresponding eigenfunctions are orthogonal in the sense that

$$\mathbf{u} \cdot \mathbf{v} = 0. \tag{5.5A.6}$$

We leave as an exercise the proof that the eigenvalues of a symmetric matrix are real.

Example. The eigenvalues of the real symmetric matrix $\begin{bmatrix} 6 & 2 \\ 2 & 3 \end{bmatrix}$ are determined from $(6 - \lambda)(3 - \lambda) - 4 = \lambda^2 - 9\lambda + 14 = (\lambda - 7)(\lambda - 2) = 0$. For $\lambda = 2$, the eigenvector satisfies

$$6x_1 + 2x_2 = 2x_1 \quad \text{and} \quad 2x_1 + 3x_2 = 2x_2,$$

and hence $\begin{bmatrix} x_1 \\ x_2 \end{bmatrix} = x_1\begin{bmatrix} 1 \\ -2 \end{bmatrix}$. For $\lambda = 7$, it follows that

$$6x_1 + 2x_2 = 7x_1 \quad \text{and} \quad 2x_1 + 3x_2 = 7x_2,$$

and the eigenvector is $\begin{bmatrix} x_1 \\ x_2 \end{bmatrix} = x_2\begin{bmatrix} 2 \\ 1 \end{bmatrix}$. As we have just proved for any real symmetric matrix, the eigenvectors are orthogonal, $\begin{bmatrix} 1 \\ -2 \end{bmatrix} \cdot \begin{bmatrix} 2 \\ 1 \end{bmatrix} = 2 - 2 = 0$.

Eigenvector expansions. *For real symmetric matrices* it can be shown that if an eigenvalue repeats R times, there will be R independent eigenvectors corresponding to that eigenvalue. These eigenvectors are automatically orthogonal to any eigenvectors corresponding to a different eigenvalue. The Gram–Schmidt procedure (see Sec. 6.5 Appendix) can be applied so that all R eigenvectors corresponding to the same eigenvalue can be constructed to be mutually orthogonal. In this manner, for real symmetric $n \times n$ matrices, n orthogonal eigenvectors can always be obtained. Since these vectors are orthogonal, they span the n-dimensional vector space and may be chosen as basis vectors. Any vector $\mathbf{v}$ may be represented in a series of the eigenvectors:

$$\mathbf{v} = \sum_{i=1}^{n} c_i\boldsymbol{\phi}_i, \tag{5.5A.7}$$

where ϕ_i is the ith eigenvector. For regular Sturm–Liouville eigenvalue problems, the eigenfunctions are complete, meaning that any (piecewise smooth) function can be represented in terms of an eigenfunction expansion

$$f(x) \sim \sum_{i=1}^{\infty} c_i \phi_i(x). \tag{5.5A.8}$$

This is analogous to (5.5A.7). In (5.5A.8) the Fourier coefficients c_i are determined by the orthogonality of the eigenfunctions. Similarly, the coordinates c_i in (5.5A.7) are determined by the orthogonality of the eigenvectors. We dot equation (5.5A.7) into ϕ_m:

$$\mathbf{v} \cdot \boldsymbol{\phi}_m = \sum_{i=1}^{n} c_i \boldsymbol{\phi}_i \cdot \boldsymbol{\phi}_m = c_m \boldsymbol{\phi}_m \cdot \boldsymbol{\phi}_m,$$

since $\boldsymbol{\phi}_i \cdot \boldsymbol{\phi}_m = 0$, $i \neq m$, determining c_m.

Linear systems. Sturm–Liouville eigenvalue problems arise in separating variables for partial differential equations. One way in which the matrix eigenvalue problem occurs is in "separating" a linear homogeneous system of ordinary differential equations with constant coefficients. We will be *very* brief. A linear homogeneous first-order system of differential equations may be represented by

$$\frac{d\mathbf{v}}{dt} = \mathbf{A}\mathbf{v}, \tag{5.5A.9}$$

where $\mathbf{A}$ is an $n \times n$ matrix and $\mathbf{v}$ is the desired n-dimensional vector solution. $\mathbf{v}$ usually satisfies given initial conditions, $\mathbf{v}(0) = \mathbf{v}_0$. We seek special solutions of the form of simple exponentials:

$$\mathbf{v}(t) = e^{\lambda t}\boldsymbol{\phi}, \tag{5.5A.10}$$

where $\boldsymbol{\phi}$ is a constant vector. This is analogous to seeking product solutions by the method of separation of variables. Since $d\mathbf{v}/dt = \lambda e^{\lambda t}\boldsymbol{\phi}$, it follows that

$$\mathbf{A}\boldsymbol{\phi} = \lambda \boldsymbol{\phi}. \tag{5.5A.11}$$

Thus, there exist solutions to (5.5A.9) of the form (5.5A.10) if λ is an eigenvalue of $\mathbf{A}$ and $\boldsymbol{\phi}$ is a corresponding eigenvector. We now restrict our attention to *real symmetric matrices* $\mathbf{A}$. There will always be n mutually orthogonal eigenvectors $\boldsymbol{\phi}_i$. We have obtained n special solutions to the linear homogeneous system (5.5A.9). A *principle of superposition* exists, and hence a linear combination of these solutions also satisfies (5.5A.9):

$$\mathbf{v} = \sum_{i=1}^{n} c_i e^{\lambda_i t} \boldsymbol{\phi}_i. \tag{5.5A.12}$$

We attempt to determine c_i so that (5.5A.12) satisfies the initial conditions, $v(0) = v_0$:

$$v_0 = \sum_{i=1}^{n} c_i \phi_i.$$

Here, the orthogonality of the eigenvectors is helpful, and thus, as before,

$$c_i = \frac{v_0 \cdot \phi_i}{\phi_i \cdot \phi_i}.$$

EXERCISES 5.5 APPENDIX

5.5A.1. Prove that the eigenvalues of real symmetric matrices are real.

5.5A.2. (a) Show that the matrix

$$A = \begin{bmatrix} 1 & 0 \\ 2 & 1 \end{bmatrix}$$

has only one independent eigenvector.

(b) Show that the matrix

$$A = \begin{bmatrix} 1 & 0 \\ 0 & 1 \end{bmatrix}$$

has two independent eigenvectors.

5.5A.3. Consider the eigenvectors of the matrix

$$A = \begin{bmatrix} 6 & 4 \\ 1 & 3 \end{bmatrix}.$$

(a) Show that the eigenvectors are not orthogonal.

(b) If the "dot product" of two vectors is defined as follows,

$$a \cdot b = \tfrac{1}{4} a_1 b_1 + a_2 b_2,$$

show that the eigenvectors are orthogonal with this dot product.

5.5A.4. Solve $dv/dt = Av$ using matrix methods if

*(a) $A = \begin{bmatrix} 6 & 2 \\ 2 & 3 \end{bmatrix}$ $v(0) = \begin{bmatrix} 1 \\ 2 \end{bmatrix}$

(b) $A = \begin{bmatrix} -1 & 2 \\ 2 & 4 \end{bmatrix}$ $v(0) = \begin{bmatrix} 2 \\ 3 \end{bmatrix}$

5.5A.5. Show that the eigenvalues are real and the eigenvectors orthogonal:

(a) $A = \begin{bmatrix} 2 & 1 \\ 1 & -4 \end{bmatrix}$

*(b) $A = \begin{bmatrix} 3 & 1-i \\ 1+i & 1 \end{bmatrix}$ (see Exercise 5.5A.6)

5.5A.6. For a matrix A whose entries are complex numbers, the complex conjugate of the transpose is denoted by A^H. For matrices in which $A^H = A$ (called **Hermitian**):

(a) Prove that the eigenvalues are real.

(b) Prove that eigenvectors corresponding to different eigenvalues are orthogonal (in the sense that $\phi_i \cdot \overline{\phi_m} = 0$, where ‾ denotes the complex conjugate).

5.6 RAYLEIGH QUOTIENT

The Rayleigh quotient can be derived from the Sturm–Liouville differential equation,

$$\frac{d}{dx}\left[p(x)\frac{d\phi}{dx}\right] + q(x)\phi + \lambda\sigma(x)\phi = 0, \qquad (5.6.1)$$

by multiplying (5.6.1) by ϕ and integrating:

$$\int_a^b \left[\phi\frac{d}{dx}\left(p\frac{d\phi}{dx}\right) + q\phi^2\right] dx + \lambda \int_a^b \phi^2\sigma \, dx = 0.$$

Since $\int_a^b \phi^2\sigma \, dx > 0$, we can solve for λ:

$$\lambda = \frac{-\int_a^b [\phi \, d/dx(p \, d\phi/dx) + q\phi^2] \, dx}{\int_a^b \phi^2\sigma \, dx}. \qquad (5.6.2)$$

Integration by parts ($\int u \, dv = uv - \int v \, du$, where $u = \phi$, $dv = d/dx \, (p \, d\phi/dx) \, dx$ and hence $du = d\phi/dx \, dx$, $v = p \, d\phi/dx$) yields an expression involving the function ϕ evaluated at the boundary:

$$\lambda = \frac{-p\phi \, d\phi/dx\Big|_a^b + \int_a^b [p(d\phi/dx)^2 - q\phi^2] \, dx}{\int_a^b \phi^2\sigma \, dx}, \qquad (5.6.3)$$

known as the **Rayleigh quotient**. In Secs. 5.3 and 5.4 we have indicated some applications of this result. Further discussion will be given in Sec. 5.7.

Nonnegative eigenvalues. Often in physical problems, the sign of λ is quite important. As shown in Sec. 5.2.1, $dh/dt + \lambda h = 0$ in certain heat flow problems. Thus, positive λ corresponds to exponential decay in time, while negative λ corresponds to exponential growth. On the other hand, in certain vibration problems (see Sec. 5.7), $d^2h/dt^2 = -\lambda h$. There, only positive λ corresponds to the "usually" expected oscillations. Thus, in both types of problems we often expect $\lambda \geq 0$:

The Rayleigh quotient (5.6.3) directly proves that $\lambda \geq 0$ if:

(a) $-p\phi\dfrac{d\phi}{dx}\bigg|_a^b \geq 0,$ and $\qquad\qquad (5.6.4)$

(b) $q \leq 0$.

We claim that both (a) and (b) are physically reasonable conditions for nonnegative λ. Consider the boundary constraint, $-p\phi \, d\phi/dx\big|_a^b \geq 0$. The simplest types of homogeneous boundary conditions, $\phi = 0$ and $d\phi/dx = 0$, do not contribute to this boundary term, satisfying (a). The condition $d\phi/dx = h\phi$ (for the physical cases of Newton's law of cooling or the elastic boundary condition) has $h > 0$ at the left end, $x = a$. Thus, it will have a positive contribution at $x = a$. The sign switch at the right end, which occurs for this type of boundary condition, will also cause a positive contribution. The periodic boundary condition [e.g., $\phi(a) = \phi(b)$ and $p(a)\, d\phi/dx(a) = p(b)\, d\phi/dx(b)$] as well as the singularity condition [$\phi(a)$ bounded, if $p(a) = 0$] also do not contribute. Thus, in all these cases $-p\phi\, d\phi/dx|_a^b \geq 0$.

The source constraint $q \leq 0$ also has a meaning in physical problems. For heat flow problems, $q \leq 0$ corresponds ($q = \alpha$, $Q = \alpha u$) to an energy-absorbing (endothermic) reaction, while for vibration problems $q \leq 0$ corresponds ($q = \alpha$, $Q = \alpha u$) to a restoring force.

Minimization principle. The Rayleigh quotient cannot be used to explicitly determine the eigenvalue (since ϕ is unknown). Nonetheless, it can be quite useful in estimating the eigenvalues. This is because of the following theorem: **The minimum value of the Rayleigh quotient for all continuous functions satisfying the boundary conditions** (but not necessarily the differential equation) **is the lowest eigenvalue**:

$$\lambda_1 = \min \frac{-pu\, du/dx|_a^b + \int_a^b [p(du/dx)^2 - qu^2]\, dx}{\int_a^b u^2 \sigma\, dx}, \tag{5.6.5}$$

where λ_1 represents the smallest eigenvalue. The minimization includes all continuous functions that satisfy the boundary conditions. The minimum is obtained only for $u = \phi_1(x)$, the lowest eigenfunction. For example, the lowest eigenvalue is important in heat flow problems (see Sec. 5.4).

Trial functions. Before proving (5.6.5), we will indicate how (5.6.5) is applied to obtain bounds on the lowest eigenvalue. Equation (5.6.5) is difficult to apply directly since we do not know how to minimize over all functions. However, let u_T be *any* continuous function satisfying the boundary conditions; u_T is known as a **trial function**. We compute the Rayleigh quotient of this trial function, $RQ[u_T]$:

$$\lambda_1 \leq RQ[u_T] = \frac{-pu_T\, du_T/dx|_a^b + \int_a^b [p(du_T/dx)^2 - qu_T^2]\, dx}{\int_a^b u_T^2 \sigma\, dx}. \tag{5.6.6}$$

We have noted that λ_1 must be less than or equal to the quotient since λ_1 is the minimum of the ratio for all functions. Equation (5.6.6) gives an **upper bound** for the lowest eigenvalue.

Example. Consider the well-known eigenvalue problem,

$$\frac{d^2\phi}{dx^2} + \lambda\phi = 0$$

$$\phi(0) = 0$$

$$\phi(1) = 0.$$

We already know that $\lambda = n^2\pi^2$ ($L = 1$), and hence the lowest eigenvalue is $\lambda_1 = \pi^2$. For this problem, the Rayleigh quotient simplifies, and (5.6.6) becomes

$$\lambda_1 \leq \frac{\int_0^1 (du_T/dx)^2 \, dx}{\int_0^1 u_T^2 \, dx}. \tag{5.6.7}$$

Trial functions must be continuous and satisfy the homogeneous boundary conditions, in this case, $u_T(0) = 0$ and $u_T(1) = 0$. In addition, we claim that the closer the trial function is to the actual eigenfunction, the more accurate is the bound of the lowest eigenvalue. Thus, we also choose trial functions with no zeros in the interior, since we already know theoretically that the lowest eigenfunction does not have a zero. We will compute the Rayleigh quotient for the three trial functions sketched in Fig. 5.6.1. For

$$u_T = \begin{cases} x, & x < \frac{1}{2} \\ 1 - x, & x > \frac{1}{2}, \end{cases}$$

(5.6.7) becomes

$$\lambda_1 \leq \frac{\int_0^{1/2} dx + \int_{1/2}^1 dx}{\int_0^{1/2} x^2 \, dx + \int_{1/2}^1 (1 - x)^2 \, dx} = \frac{1}{\frac{1}{24} + \frac{1}{24}} = 12,$$

a fair upper bound for the exact answer π^2 ($\pi^2 \approx 9.8696 \ldots$). For $u_T = x - x^2$, (5.6.7) becomes

$$\lambda_1 \leq \frac{\int_0^1 (1 - 2x)^2 \, dx}{\int_0^1 (x - x^2)^2 \, dx} = \frac{\int_0^1 (1 - 4x + 4x^2) \, dx}{\int_0^1 (x^2 - 2x^3 + x^4) \, dx} = \frac{1 - 2 + \frac{4}{3}}{\frac{1}{3} - \frac{1}{2} + \frac{1}{5}} = 10,$$

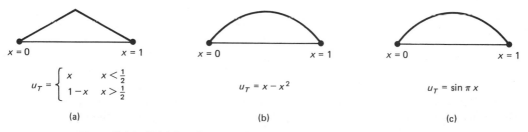

$$u_T = \begin{cases} x & x < \frac{1}{2} \\ 1 - x & x > \frac{1}{2} \end{cases}$$

(a)

$$u_T = x - x^2$$

(b)

$$u_T = \sin \pi x$$

(c)

Figure 5.6.1   Trial functions: continuous, satisfy the boundary conditions, and are of one sign.

a more accurate bound. Since $u_T = \sin \pi x$ is the actual lowest eigenfunction, the Rayleigh quotient for this trial function will exactly equal the lowest eigenvalue. Other applications of the Rayleigh quotient will be shown in later sections.

Proof. It is usual to prove the minimization property of the Rayleigh quotient using a more advanced branch of applied mathematics known as the calculus of variations. We do not have the space here to develop that material properly. Instead, we will give a proof based on eigenfunction expansions. We again calculate the Rayleigh quotient (5.6.3) for any function u that is *continuous* and *satisfies the homogeneous boundary conditions*. In this derivation, the equivalent form of the Rayleigh quotient, (5.6.2), is more useful:

$$RQ[u] = \frac{-\int_a^b uL(u)\, dx}{\int_a^b u^2 \sigma\, dx}, \qquad (5.6.8)$$

where the operator notation is quite helpful. We expand the rather arbitrary function u in terms of the (usually unknown) eigenfunctions $\phi_n(x)$:

$$u = \sum_{n=1}^{\infty} a_n \phi_n(x). \qquad (5.6.9)$$

L is a linear differential operator. We expect that

$$L(u) = \sum_{n=1}^{\infty} a_n L(\phi_n(x)), \qquad (5.6.10)$$

since this is valid for finite series. In Chapter 7 we show that (5.6.10) is valid *if u is continuous and satisfies the same homogeneous boundary conditions as the eigenfunctions $\phi_n(x)$.* Here, ϕ_n are eigenfunctions, and hence $L(\phi_n) = -\lambda_n \sigma \phi_n$. Thus, (5.6.10) becomes

$$L(u) = -\sum_{n=1}^{\infty} a_n \lambda_n \sigma \phi_n, \qquad (5.6.11)$$

which can be thought of as the eigenfunction expansion of $L(u)$. If (5.6.11) and (5.6.9) are substituted into (5.6.8) and different dummy summation indices are utilized for the product of two infinite series, we obtain

$$RQ[u] = \frac{\int_a^b \left(\sum_{m=1}^{\infty} \sum_{n=1}^{\infty} a_m a_n \lambda_n \phi_n \phi_m \sigma \right) dx}{\int_a^b \left(\sum_{m=1}^{\infty} \sum_{n=1}^{\infty} a_m a_n \phi_n \phi_m \sigma \right) dx}. \qquad (5.6.12)$$

We now do the integration in (5.6.12) before the summation. We recall that the eigenfunctions are orthogonal ($\int_a^b \phi_n \phi_m \sigma\, dx = 0$ if $n \neq m$), which implies that (5.6.12) becomes

$$RQ[u] = \dfrac{\displaystyle\sum_{n=1}^{\infty} a_n^2 \lambda_n \int_a^b \phi_n^2 \sigma \, dx}{\displaystyle\sum_{n=1}^{\infty} a_n^2 \int_a^b \phi_n^2 \sigma \, dx}. \qquad (5.6.13)$$

This is an exact expression for the Rayleigh quotient in terms of the generalized Fourier coefficients a_n of u. We denote λ_1 as the lowest eigenvalue ($\lambda_1 < \lambda_n$ for $n > 1$). Thus,

$$RQ[u] \geq \dfrac{\lambda_1 \displaystyle\sum_{n=1}^{\infty} a_n^2 \int_a^b \phi_n^2 \sigma \, dx}{\displaystyle\sum_{n-1}^{\infty} a_n^2 \int_a^b \phi_n^2 \sigma \, dx} = \lambda_1. \qquad (5.6.14)$$

Furthermore, the equality in (5.6.14) holds only if $a_n = 0$ for $n > 1$ (i.e., only if $u = a_1\phi_1$). We have showed that the smallest value of the Rayleigh quotient is the lowest eigenvalue λ_1. Moreover, the Rayleigh quotient is minimized only when $u = a_1\phi_1$ (i.e., when u is the lowest eigenfunction).

We thus have a minimization theorem for the lowest eigenvalue λ_1. We can ask if there are corresponding theorems for the higher eigenvalues. Interesting generalizations immediately follow from (5.6.13). If we insist that $a_1 = 0$, then

$$RQ[u] = \dfrac{\displaystyle\sum_{n=2}^{\infty} a_n^2 \lambda_n \int_a^b \phi_n^2 \sigma \, dx}{\displaystyle\sum_{n=2}^{\infty} a_n^2 \int_a^b \phi_n^2 \sigma \, dx}. \qquad (5.6.15)$$

This means that in addition we are restricting our function u to be orthogonal to ϕ_1, since $a_1 = \int_a^b u\phi_1\sigma \, dx / \int_a^b \phi_1^2\sigma \, dx$. We now proceed in a similar way. Since $\lambda_2 < \lambda_n$ for $n > 2$, it follows that

$$RQ[u] \geq \lambda_2,$$

and furthermore the equality holds only if $a_n = 0$ for $n > 2$ [i.e., $u = a_2\phi_2(x)$] since $a_1 = 0$ already. We have just proved the following theorem: The minimum value for all continuous functions $u(x)$ *that are orthogonal to the lowest eigenfunction* and satisfy the boundary conditions is the next-to-lowest eigenvalue. Further generalizations also follow directly from (5.6.13).

EXERCISES 5.6

5.6.1. Use the Rayleigh quotient to obtain a (reasonably accurate) upper bound for the lowest eigenvalue of

(a) $\dfrac{d^2\phi}{dx^2} + (\lambda - x^2)\phi = 0$ with $\dfrac{d\phi}{dx}(0) = 0$ and $\phi(1) = 0$

(b) $\dfrac{d^2\phi}{dx^2} + (\lambda - x)\phi = 0$ with $\dfrac{d\phi}{dx}(0) = 0$ and $\dfrac{d\phi}{dx}(1) + 2\phi(1) = 0$

*(c) $\dfrac{d^2\phi}{dx^2} + \lambda\phi = 0$ with $\phi(0) = 0$ and $\dfrac{d\phi}{dx}(1) + \phi(1) = 0$ (See Exercise 5.8.10.)

5.6.2. Consider the eigenvalue problem

$$\frac{d^2\phi}{dx^2} + (\lambda - x^2)\phi = 0$$

subject to $\dfrac{d\phi}{dx}(0) = 0$ and $\dfrac{d\phi}{dx}(1) = 0$. Show that $\lambda > 0$ (be sure to show that $\lambda \neq 0$).

5.6.3. Prove that (5.6.10) is valid in the following way. Assume $L(u)/\sigma$ is piecewise smooth so that

$$\frac{L(u)}{\sigma} = \sum_{n=1}^{\infty} b_n\phi_n(x).$$

Determine b_n. [*Hint:* Using Green's formula (5.5.5), show that $b_n = -a_n\lambda_n$ if u and du/dx are continuous and if u satisfies the same homogeneous boundary conditions as the eigenfunctions $\phi_n(x)$.]

5.7 WORKED EXAMPLE—VIBRATIONS OF A NONUNIFORM STRING

Some additional applications of the Rayleigh quotient are best illustrated in a physical problem. Consider the vibrations of a nonuniform string [constant tension T_0, but variable mass density $\rho(x)$] without sources ($Q = 0$); see Sec. 4.2. We assume that both ends are fixed with zero displacement. The mathematical equations for the initial value problem are

PDE:	$\rho\dfrac{\partial^2 u}{\partial t^2} = T_0\dfrac{\partial^2 u}{\partial x^2}$	(5.7.1a)
BC:	$u(0, t) = 0$	(5.7.1b)
	$u(L, t) = 0$	(5.7.1c)
IC:	$u(x, 0) = f(x)$	(5.7.1d)
	$\dfrac{\partial u}{\partial t}(x, 0) = g(x)$	(5.7.1e)

Again since the partial differential equation and the boundary conditions are linear and homogeneous, we are able to apply the method of separation of variables. We look for product solutions:

$$u(x, t) = \phi(x)h(t), \tag{5.7.2}$$

ignoring the nonzero initial conditions. It can be shown that $h(t)$ satisfies

$$\frac{d^2h}{dt^2} = -\lambda h, \tag{5.7.3}$$

while the spatial part solves the following regular Sturm–Liouville eigenvalue problem:

$$T_0\frac{d^2\phi}{dx^2} + \lambda\rho(x)\phi = 0$$

$$\phi(0) = 0 \tag{5.7.4}$$

$$\phi(L) = 0.$$

Usually, we presume that the infinite sequence of eigenvalues λ_n and corresponding eigenfunctions $\phi_n(x)$ are known. However, in order to analyze (5.7.3), it is necessary to know something about λ. From physical reasoning, we certainly expect $\lambda > 0$, since we expect oscillations, but we will show that the Rayleigh quotient easily guarantees that $\lambda > 0$. For (5.7.4), the Rayleigh quotient (5.6.3) becomes

$$\lambda = \frac{T_0\int_L^0 (d\phi/dx)^2\, dx}{\int_L^0 \phi^2\rho(x)\, dx}. \tag{5.7.5}$$

Clearly, $\lambda \geqslant 0$ (and as before it is impossible for $\lambda = 0$, in this case). Thus, $\lambda > 0$.

We now are assured that the solution of (5.7.3) is a linear combination of $\sin \sqrt{\lambda}\, t$ and $\cos \sqrt{\lambda}\, t$. There are two families of product solutions of the partial differential equation, $\sin \sqrt{\lambda_n}\, t\, \phi_n(x)$ and $\cos \sqrt{\lambda_n}\, t\, \phi_n(x)$. According to the principle of superposition, the solution is

$$u(x, t) = \sum_{n=1}^{\infty} a_n \sin \sqrt{\lambda_n}\, t\, \phi_n(x) + \sum_{n=1}^{\infty} b_n \cos \sqrt{\lambda_n}\, t\, \phi_n(x). \tag{5.7.6}$$

We only need to show that the two families of coefficients can be obtained from the initial conditions:

$$f(x) = \sum_{n=1}^{\infty} b_n\phi_n(x) \quad \text{and} \quad g(x) = \sum_{n=1}^{\infty} a_n\sqrt{\lambda_n}\, \phi_n(x). \tag{5.7.7a,b}$$

Thus, b_n are the generalized Fourier coefficients of the initial position $f(x)$, while $a_n\sqrt{\lambda_n}$ are the generalized Fourier coefficients for the initial velocity $g(x)$. Thus, due to the orthogonality of the eigenfunctions [with weight $\rho(x)$], we can easily determine a_n and b_n:

$$b_n = \frac{\int_0^L f(x)\phi_n(x)\rho(x)\,dx}{\int_0^L \phi_n^2 \rho\,dx} \qquad (5.7.8a)$$

$$a_n\sqrt{\lambda_n} = \frac{\int_0^L g(x)\phi_n(x)\rho(x)\,dx}{\int_0^L \phi_n^2 \rho\,dx} \qquad (5.7.8b)$$

The Rayleigh quotient can be used to obtain additional information about the lowest eigenvalue λ_1. (Note that the lowest frequency of vibration is $\sqrt{\lambda_1}$.) We know that

$$\lambda_1 = \min \frac{T_0 \int_0^L (du/dx)^2\,dx}{\int_0^L u^2 \rho(x)\,dx}. \qquad (5.7.9)$$

We have already shown (see Sec. 5.6) how to use trial functions to obtain an upper bound on the lowest eigenvalue. This is not always convenient since the denominator in (5.7.9) depends on the mass density $\rho(x)$. Instead, we will develop another method for an upper bound. By this method we will also obtain a lower bound.

Let us suppose, as is usual, that the variable mass density has upper and lower bounds,

$$0 < \rho_{\min} \leq \rho(x) \leq \rho_{\max}.$$

For any $u(x)$ it follows that

$$\rho_{\min} \int_0^L u^2\,dx \leq \int_0^L u^2 \rho(x)\,dx \leq \rho_{\max} \int_0^L u^2\,dx.$$

Consequently, from (5.7.9),

$$\frac{T_0}{\rho_{\max}} \min \frac{\int_0^L (du/dx)^2\,dx}{\int_0^L u^2\,dx} \leq \lambda_1 \leq \frac{T_0}{\rho_{\min}} \min \frac{\int_0^L (du/dx)^2\,dx}{\int_0^L u^2\,dx}. \qquad (5.7.10)$$

We can evaluate the expressions in (5.7.10), since we recognize the minimum of $\int_0^L (du/dx)^2\,dx / \int_0^L u^2\,dx$ subject to $u(0) = 0$ and $u(L) = 0$ as the lowest eigenvalue of a different problem: namely, one with constant coefficients,

$$\frac{d^2\phi}{dx^2} + \bar{\lambda}\phi = 0$$

$$\phi(0) = 0 \qquad \text{and} \qquad \phi(L) = 0.$$

We already know that $\bar{\lambda} = (n\pi/L)^2$, and hence the lowest eigenvalue for this problem is $\bar{\lambda}_1 = (\pi/L)^2$. But the minimization property of the Rayleigh quotient implies that

$$\bar{\lambda}_1 = \min \frac{\int_0^L (du/dx)^2 \, dx}{\int_0^L u^2 \, dx}.$$

Finally, we have proved that the lowest eigenvalue of our problem with variable coefficients satisfies the following inequality:

$$\frac{T_0}{\rho_{max}} \left(\frac{\pi}{L}\right)^2 \leq \lambda_1 \leq \frac{T_0}{\rho_{min}} \left(\frac{\pi}{L}\right)^2.$$

We have obtained an upper and a lower bound for the smallest eigenvalue. By taking square roots,

$$\frac{\pi}{L} \sqrt{\frac{T_0}{\rho_{max}}} \leq \sqrt{\lambda_1} \leq \frac{\pi}{L} \sqrt{\frac{T_0}{\rho_{min}}}.$$

The physical meaning of this is clear; the lowest frequency of oscillation of a variable string lies in between the lowest frequencies of vibration of two constant density strings, one with the minimum density and the other with the maximum. Similar results concerning the higher frequencies of vibration are also valid, but are harder to prove (see Weinberger [1965], Courant and Hilbert [1953], or Birkhoff and Rota [1978]).

EXERCISES 5.7

*5.7.1. Determine an upper and a (nonzero) lower bound for the lowest frequency of vibration of a nonuniform string fixed at $x = 0$ and $x - 1$ with $c^2 = 1 + 4\alpha^2(x - \frac{1}{2})^2$.

5.7.2. Consider heat flow in a one-dimensional rod without sources with nonconstant thermal properties. Assume that the temperature is zero at $x - 0$ and $x - L$. Suppose that $c\rho_{min} \leq c\rho \leq c\rho_{max}$, and $K_{min} \leq K_0(x) \leq K_{max}$. Obtain an upper and (nonzero) lower bound on the slowest exponential rate of decay of the product solution.

5.8 BOUNDARY CONDITIONS OF THE THIRD KIND

Introduction. So far we have analyzed two general varieties of boundary value problems: very specific, easily solved ones (such as the ones that give rise to Fourier sine series, Fourier cosine series, or Fourier series) and somewhat abstract Sturm–Liouville eigenvalue problems, where our theorems guaranteed many needed properties. In one case the differential equation had constant coefficients (with simple boundary conditions), and in the other we discussed differential equations with variable coefficients.

In this section we analyze problems with a boundary condition of the third kind. It will also be easily solved (since the differential equation will still have constant coefficients). However, due to its boundary conditions, it will illustrate more convincingly the general ideas of Sturm–Liouville eigenvalue problems.

Physical examples. We consider some simple problems with constant physical parameters. Heat flow in a uniform rod satisfies

$$\frac{\partial u}{\partial t} = k\frac{\partial^2 u}{\partial x^2}, \tag{5.8.1}$$

while a uniform vibrating string solves

$$\frac{\partial^2 u}{\partial t^2} = c^2\frac{\partial^2 u}{\partial x^2}. \tag{5.8.2}$$

In either case we suppose that the left end is fixed, but the right end satisfies a homogeneous boundary condition of the third kind:

$$u(0, t) = 0 \tag{5.8.3}$$

$$\frac{\partial u}{\partial x}(L, t) = -hu(L, t). \tag{5.8.4}$$

Recall that, for heat conduction (5.8.4) corresponds to Newton's law of cooling if $h > 0$, and for the vibrating string problem, (5.8.4) corresponds to a restoring force if $h > 0$, the so-called elastic boundary condition. We note that usually in physical problems $h \geq 0$. However, for mathematical reasons we will investigate both cases with $h < 0$ and $h \geq 0$. If $h < 0$, the vibrating string has a destabilizing force at the right end, while for the heat flow problem, thermal energy is being constantly put into the rod through the right end.

Sturm–Liouville eigenvalue problem. After separation of variables,

$$u(x, t) = G(t)\phi(x), \tag{5.8.5}$$

the time part satisfies the following ordinary differential equations:

$$\text{heat flow:} \qquad \frac{dG}{dt} = -\lambda kG \tag{5.8.6a}$$

$$\text{vibrating string:} \quad \frac{d^2 G}{dt^2} = -\lambda c^2 G. \tag{5.8.6b}$$

We wish to concentrate on the effect of the third type of boundary condition, (5.8.4). For either physical problem, the spatial part, $\phi(x)$, satisfies the following regular Sturm–Liouville eigenvalue problem:

$$\frac{d^2\phi}{dx^2} + \lambda\phi = 0 \tag{5.8.7a}$$

$$\phi(0) = 0 \tag{5.8.7b}$$

$$\frac{d\phi}{dx}(L) + h\phi(L) = 0, \tag{5.8.7c}$$

where h is a given fixed constant. If $h \geq 0$, this is what we call the "physical" case, while if $h < 0$ we call it the "nonphysical" case. Although the differential equation (5.8.7a) has constant coefficients, the boundary conditions will give rise to some new ideas. For the moment we ignore certain aspects of our theory of Sturm–Liouville eigenvalue problems (except for the fact that the eigenvalues are real). In solving (5.8.7a) we must consider three distinct cases: $\lambda > 0$, $\lambda < 0$, and $\lambda = 0$. This will be especially important when we analyze the nonphysical case $h < 0$.

Positive eigenvalues. If $\lambda > 0$, the solution of the differential equation is a linear combination of sines and cosines:

$$\phi(x) = c_1 \cos \sqrt{\lambda}\, x + c_2 \sin \sqrt{\lambda}\, x. \tag{5.8.8}$$

The boundary condition $\phi(0) = 0$ implies that $0 = c_1$, and hence

$$\boxed{\phi(x) = c_2 \sin \sqrt{\lambda}\, x.} \tag{5.8.9}$$

Clearly, sine functions are needed to satisfy the zero condition at $x = 0$. We will need the first derivative,

$$\frac{d\phi}{dx} = c_2 \sqrt{\lambda} \cos \sqrt{\lambda}\, x.$$

Thus, the boundary condition of the third kind, (5.8.7c), implies that

$$c_2(\sqrt{\lambda} \cos \sqrt{\lambda}\, L + h \sin \sqrt{\lambda}\, L) = 0. \tag{5.8.10}$$

If $c_2 = 0$, (5.8.9) shows that $\phi \equiv 0$, which cannot be an eigenfunction. Thus, eigenvalues exist for $\lambda > 0$ for all values of λ that satisfy

$$\sqrt{\lambda} \cos \sqrt{\lambda}\, L + h \sin \sqrt{\lambda}\, L = 0. \tag{5.8.11}$$

The more elementary case $h = 0$ will be analyzed later. Equation (5.8.11) is a transcendental equation for the positive eigenvalues λ (if $h \neq 0$). In order to solve (5.8.11), it is convenient to divide by $\cos \sqrt{\lambda}\, L$ to obtain an expression for $\tan \sqrt{\lambda}\, L$:

$$\boxed{\tan \sqrt{\lambda}\, L = -\frac{\sqrt{\lambda}}{h}.} \tag{5.8.12}$$

We are allowed to divide by $\cos \sqrt{\lambda}\, L$ because it is not zero [if $\cos \sqrt{\lambda}\, L = 0$, then $\sin \sqrt{\lambda}\, L \neq 0$ and (5.8.11) would not be satisfied]. We could have obtained an expression for cotangent rather than tangent by dividing (5.8.11) by $\sin \sqrt{\lambda}\, L$, but we are presuming that the reader feels more comfortable with the tangent function.

Graphical technique ($\lambda > 0$). Equation (5.8.12) is a transcendental equation. We cannot solve it exactly. However, let us describe a *graphical technique* to obtain information about the eigenvalues. In order to graph the solution of a transcendental equation, we introduce an artificial coordinate z. Let

$$z = \tan \sqrt{\lambda} L \qquad (5.8.13a)$$

and thus also

$$z = -\frac{\sqrt{\lambda}}{h}. \qquad (5.8.13b)$$

Now the simultaneous solution of (5.8.13a) and (5.8.13b) (i.e., their points of intersection) corresponds to solutions of (5.8.12). Equation (5.8.13a) is a pure tangent function (not compressed) as a function of $\sqrt{\lambda} L$, where $\sqrt{\lambda} L > 0$ since $\lambda > 0$. We sketch (5.8.13a) in Fig. 5.8.1. We note that the tangent function is periodic with period π; it is zero at $\sqrt{\lambda} L = 0$, π, 2π, etc.; and it approaches $\pm\infty$ as $\sqrt{\lambda} L$ approaches $\pi/2$, $3\pi/2$, $5\pi/2$, etc. We will intersect the tangent function with (5.8.13b). Since we are sketching our curves as functions of $\sqrt{\lambda} L$, we will express (5.8.13b) as a function of $\sqrt{\lambda} L$. This is easily done by multiplying numerator and denominator of (5.8.13b) by L:

$$z = -\frac{\sqrt{\lambda} L}{hL}. \qquad (5.8.13c)$$

As a function of $\sqrt{\lambda} L$, (5.8.13c) is a straight line with slope $-1/hL$. However, this line is sketched quite differently depending on whether $h > 0$ (physical case) or $h < 0$ (nonphysical case).

Positive eigenvalues (physical case, $h > 0$). The intersection of the two curves is sketched in Fig. 5.8.1 for the physical case ($h > 0$). There are an infinite number of intersections; each corresponds to a positive eigenvalue. (We exclude $\sqrt{\lambda} L = 0$ since we have assumed throughout that $\lambda > 0$.) The eigenfunctions are $\phi = \sin \sqrt{\lambda} x$, where the allowable eigenvalues are determined graphically.

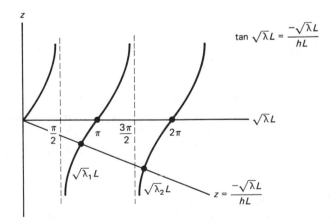

Figure 5.8.1 Graphical determination of positive eigenvalues ($h > 0$).

We cannot determine these eigenvalues exactly. However, we know from Fig. 5.8.1 that

$$\frac{\pi}{2} < \sqrt{\lambda_1}\, L < \pi \qquad (5.8.14a)$$

$$\frac{3\pi}{2} < \sqrt{\lambda_2}\, L < 2\pi, \qquad (5.8.14b)$$

and so on. It is interesting to note that as n increases, the intersecting points more closely approach the position of the vertical portions of the tangent function. We thus are able to obtain the following approximate (asymptotic) formula for the eigenvalues

$$\sqrt{\lambda_n}\, L \sim (n - \tfrac{1}{2})\pi \qquad (5.8.14c)$$

as $n \to \infty$. This becomes more and more accurate as $n \to \infty$. An asymptotic formula for the large eigenvalues similar to (5.8.14c) exists even for cases where the differential equation cannot be easily solved. We will discuss this in Sec. 5.9.

To obtain accurate values, a numerical method such as Newton's method (as often described in elementary calculus texts) can be used. A practical scheme is to use Newton's numerical method for the first few roots, until you reach a root whose solution is reasonably close to the asymptotic formula, (5.8.14c) (or improvements to this elementary asymptotic formula). Then, for larger roots, the asymptotic formula (5.8.14c) is accurate enough.

Positive eigenvalues (nonphysical case, $h < 0$). The nonphysical case ($h < 0$) also will be a good illustration of various general ideas concerning Sturm–Liouville eigenvalue problems. If $h < 0$, positive eigenvalues again are determined by graphically sketching (5.8.12), $\tan \sqrt{\lambda}\, L = -\sqrt{\lambda}/h$. The straight line (here with positive slope) must intersect the tangent function. It intersects the "first branch" of the tangent function only if the slope of the straight line is greater than 1 (see Fig. 5.8.2a). We are using the property of the tangent function that its slope is 1 at $x = 0$ and its slope increases along the first branch. Thus, if

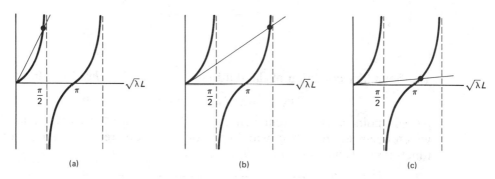

Figure 5.8.2 Graphical determination of positive eigenvalues: (a) $0 > hL > -1$; (b) $hL = -1$; (c) $hL < -1$

$h < 0$ (the nonphysical case), there are two major subcases ($-1/hL > 1$ and $0 < -1/hL < 1$) and a minor subcase ($-1/hL = 1$). We sketch these three cases in Fig. 5.8.2. In each of these three figures, there are an infinite number of intersections, corresponding to an infinite number of positive eigenvalues. The eigenfunctions are again $\sin \sqrt{\lambda} \, x$.

In these cases, the graphical solutions also show that the large eigenvalues are approximately located at the singularities of the tangent function. Equation (5.8.14c) is again asymptotic; the larger is n, the more accurate is (5.8.14c).

Zero eigenvalue. Is $\lambda = 0$ an eigenvalue for (5.8.7)? Equation (5.8.8) is *not* the general solution of (5.8.7a) if $\lambda = 0$. Instead,

$$\phi = c_1 + c_2 x; \tag{5.8.15a}$$

the eigenfunction must be a straight line. The boundary condition $\phi(0) = 0$ makes $c_1 = 0$, insisting that the straight line goes through the origin,

$$\phi = c_2 x. \tag{5.8.15b}$$

Finally, $d\phi/dx(L) + h\phi(L) = 0$ implies that

$$c_2(1 + hL) = 0. \tag{5.8.16}$$

If $hL \neq -1$ (including all physical situations, $hL > 0$), it follows that $c_2 = 0$, $\phi = 0$, and thus $\lambda = 0$ is not an eigenvalue. However, if $hL = -1$, then from (5.8.16) c_2 is arbitrary, and $\lambda = 0$ is an eigenvalue with eigenfunction x.

Negative eigenvalues. We do not expect any negative eigenvalues in the physical situations [see (5.8.6)]. If $\lambda < 0$, we introduce $s = -\lambda$, so that $s > 0$. Then (5.8.7a) becomes

$$\frac{d^2\phi}{dx^2} = s\phi. \tag{5.8.17}$$

The zero boundary condition at $x = 0$ suggests that it is more convenient to express the general solution of (5.8.17) in terms of the hyperbolic functions:

$$\phi = c_1 \cosh \sqrt{s} \, x + c_2 \sinh \sqrt{s} \, x. \tag{5.8.18}$$

Only the hyperbolic sines are needed, since $\phi(0) = 0$ implies that $c_1 = 0$:

$$\phi = c_2 \sinh \sqrt{s} \, x$$
$$\frac{d\phi}{dx} = c_2\sqrt{s} \cosh \sqrt{s} \, x. \tag{5.8.19}$$

The boundary condition of the third kind, $d\phi/dx(L) + h\phi(L) = 0$, implies that

$$c_2 (\sqrt{s} \cosh \sqrt{s} \, L + h \sinh \sqrt{s} \, L) = 0. \tag{5.8.20}$$

At this point it is apparent that the analysis for $\lambda < 0$ directly parallels that which occurred for $\lambda > 0$ (with hyperbolic functions replacing the trigonometric functions). Thus, since $c_2 \neq 0$,

$$\boxed{\tanh \sqrt{s} \, L = -\frac{\sqrt{s}}{h} = -\frac{\sqrt{s} \, L}{hL}.} \tag{5.8.21}$$

Graphical solution for negative eigenvalues. Negative eigenvalues are determined by the graphical solution of transcendental equation (5.8.21). Here properties of the hyperbolic tangent function are quite important. tanh is sketched as a function of $\sqrt{s}\,L$ in Fig. 5.8.3. Let us note some properties of the tanh function that follow from its definition,

$$\tanh x = \frac{\sinh x}{\cosh x} = \frac{e^x - e^{-x}}{e^x + e^{-x}}.$$

As $\sqrt{s}\,L \to \infty$, $\tanh \sqrt{s}\,L$ asymptotes to 1. We will also need to note that the slope* of tanh equals 1 at $\sqrt{s}\,L = 0$ and decreases toward zero as $\sqrt{s}\,L \to \infty$. This function must be intersected with the straight line implied by the r.h.s. of (5.8.21). The same four cases appear, as is sketched in Fig. 5.8.3. In physical situations ($h > 0$), there are no intersections with $\sqrt{s}\,L > 0$; there are no negative eigenvalues in the physical situations ($h > 0$). All the eigenvalues are nonnegative. However, if $hL < -1$ (and only in these situations), then there is exactly one intersection; there is one negative eigenvalue (if $hL < -1$). If we denote the intersection by $s = s_1$, the negative eigenvalue is $\lambda = -s_1$, and the corresponding eigenfunction is $\phi = \sinh \sqrt{s_1}\,x$. In nonphysical situations, there are a finite number of negative eigenvalues (one if $hL < 1$, none otherwise).

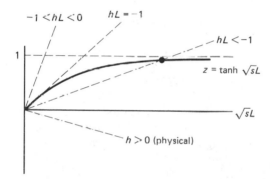

Figure 5.8.3 Graphical determination of negative eigenvalues.

Special case $h = 0$. Although if $h = 0$, the boundary conditions are not of the third kind, the eigenvalues and eigenfunctions are still of interest. If $h = 0$, then all eigenvalues are positive [see (5.8.16) and (5.8.20)] and easily explicitly determined from (5.8.11):

$$\lambda = \left[\frac{(n - 1/2)\pi}{L} \right]^2, \qquad n = 1, 2, 3, \ldots .$$

The eigenfunctions are $\sin \sqrt{\lambda}\,x$.

Summary. We have shown there to be five somewhat different cases depending on the value of the parameter h in the boundary condition. Table 5.8.1 summarizes the eigenvalues and eigenfunctions for these cases.

* $d/dx \tanh x = \operatorname{sech}^2 x = 1/\cosh^2 x$.

TABLE 5.8.1 EIGENFUNCTIONS FOR (5.8.7)

		$\lambda > 0$	$\lambda = 0$	$\lambda < 0$
Physical	$h > 0$	$\sin \sqrt{\lambda}\, x$		
	$h = 0$	$\sin \sqrt{\lambda}\, x$		
Nonphysical	$-1 < hL < 0$	$\sin \sqrt{\lambda}\, x$		
	$hL = -1$	$\sin \sqrt{\lambda}\, x$	x	
	$hL < -1$	$\sin \sqrt{\lambda}\, x$		$\sinh \sqrt{s_1}\, x$

In some sense there actually are only three cases: If $-1 < hL$, all the eigenvalues are positive; if $hL = -1$, there are no negative eigenvalues, but zero is an eigenvalue; and if $hL < -1$, there are still an infinite number of positive eigenvalues, but there is also one negative one.

Rayleigh quotient. We have shown by explicitly solving the eigenvalue problem,

$$\frac{d^2\phi}{dx^2} + \lambda\phi = 0 \tag{5.8.22a}$$

$$\phi(0) = 0 \tag{5.8.22b}$$

$$\frac{d\phi}{dx}(L) + h\phi(L) = 0, \tag{5.8.22c}$$

that in physical problems ($h \geq 0$) all the eigenvalues are positive, while in nonphysical problems ($h < 0$) there may or may not be negative eigenvalues. We will show that the Rayleigh quotient is consistent with this result:

$$\lambda = \frac{-p\phi\, d\phi/dx|_a^b + \int_a^b \left[p\left(\frac{d\phi}{dx}\right)^2 - q\phi^2 \right] dx}{\int_a^b \phi^2\sigma\, dx} = \frac{h\phi^2(L) + \int_0^L \left(\frac{d\phi}{dx}\right)^2 dx}{\int_0^L \phi^2\, dx} \tag{5.8.23}$$

since from (5.8.22a), $p(x) = 1$, $\sigma(x) = 1$, $q(x) = 0$, and $a = 0$, $b = L$, and where the boundary conditions (5.8.22b) and (5.8.22c) have been utilized to simplify the boundary terms in the Rayleigh quotient. If $h \geq 0$ (the physical cases) it readily follows from (5.8.23) that the eigenvalues must be positive, exactly what we concluded by doing the explicit calculations. However, if $h < 0$ (nonphysical case), the numerator of the Rayleigh quotient contains a negative term $h\phi^2(L)$ and a positive term $\int_0^L (d\phi/dx)^2\, dx$. It is impossible to make any conclusions concerning the sign of λ. Thus, it may be possible to have negative eigenvalues if $h < 0$. However, we are unable to conclude that there must be negative eigenvalues. A negative eigenvalue occurs only when $|h\phi^2(L)| > \int_0^L (d\phi/dx)^2\, dx$. From the Rayleigh quotient we cannot determine when this happens. It is only from an explicit calculation that we know that a negative eigenvalue occurs only if $hL < -1$.

Zeros of eigenfunctions. The Sturm–Liouville eigenvalue problem that we have been discussing in this section,

$$\frac{d^2\phi}{dx^2} + \lambda\phi = 0 \qquad \begin{aligned} \phi(0) &= 0 \\ \frac{d\phi}{dx}(L) + h\phi(L) &= 0 \end{aligned} \qquad (5.8.24)$$

is a good example for illustrating the general theorem concerning the zeros of the eigenfunctions. The theorem states that the eigenfunction corresponding to the lowest eigenvalue has no zeros in the interior. More generally, the nth eigenfunction has $n - 1$ zeros.

There are five cases of (5.8.24) worthy of discussion: $h > 0$, $h = 0$, $-1 < hL < 0$, $hL = -1$, $hL < -1$. However, the line of reasoning used in investigating the zeros of the eigenfunctions is quite similar in all cases. For that reason we will analyze only one case ($hL < -1$) and leave the others for the exercises. In this case ($hL < -1$) there is one negative eigenvalue (with corresponding eigenfunction $\sinh \sqrt{s_1}\, x$) and an infinite number of positive eigenvalues (with corresponding eigenfunctions $\sin \sqrt{\lambda}\, x$). We will need to analyze carefully the positive eigenvalues and so we reproduce Fig. 5.8.2c (as Fig. 5.8.4), used for the graphical determination of the eigenvalues in $hL < -1$. We designate the intersections starting from λ_n, $n = 2$, since the lowest eigenvalue is negative, $\lambda_1 = s_1$. Graphically, we are able to obtain bounds for these eigenvalues:

$$\pi < \sqrt{\lambda_2}\, L < \frac{3\pi}{2} \qquad (5.8.25a)$$

$$2\pi < \sqrt{\lambda_3}\, L < \frac{5\pi}{2}, \qquad (5.8.25b)$$

which is easily generalized

$$(n - 1)\pi < \sqrt{\lambda_n}\, L < (n - 1/2)\pi, \qquad n \geqslant 2. \qquad (5.8.25c)$$

Let us investigate zeros of the eigenfunctions. The lowest eigenfunction is $\sinh \sqrt{s_1}\, x$. Since the hyperbolic sine function is never zero (except at the end $x = 0$), we have verified one part of the theorem—that the eigenfunction

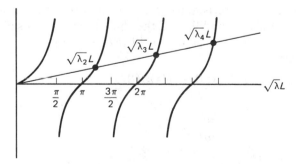

Figure 5.8.4 Positive eigenvalues ($hL < -1$).

corresponding to the lowest eigenvalue does not have a zero in the interior. The other eigenfunctions are $\sin \sqrt{\lambda_n} x$, sketched in Fig. 5.8.5. In this figure the end point $x = 0$ is clearly marked, but $x = L$ depends on λ. For example, for λ_3, the end point $x = L$ occurs at $\sqrt{\lambda_3}\, L$, which is sketched in Fig. 5.8.5 due to (5.8.25b). As x varies from 0 to L the eigenfunction is sketched in Fig. 5.8.5 up to the dashed line. This eigenfunction has two zeros ($\sqrt{\lambda_3}\, x = \pi$ and 2π). This reasoning can be used for any of these eigenfunctions. Thus, the number of zeros for the nth eigenfunction corresponding to λ_n is $n - 1$, exactly as the general theorem specifies. Our theorem does *not* state that the eigenfunction corresponding to the lowest *positive* eigenvalue has no zeros. Instead, **the eigenfunction corresponding to the lowest eigenvalue has no zeros.** To repeat, in this example the lowest eigenvalue is negative and *its* corresponding eigenfunction has no zeros.

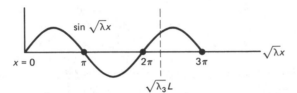

Figure 5.8.5 Zeroes of the eigenfunctions $\sin \sqrt{\lambda}\, x$.

Heat flow with a nonphysical boundary condition. To understand further the boundary condition of the third kind, let us complete the investigation of one example. We consider heat flow in a uniform rod:

PDE:	$\dfrac{\partial u}{\partial t} = k\dfrac{\partial^2 u}{\partial x^2}$	(5.8.26a)
BC:	$u(0,\, t) = 0$	(5.8.26b)
	$\dfrac{\partial u}{\partial x}(L,\, t) = -hu(L,\, t)$	(5.8.26c)
IC:	$u(x,\, 0) = f(x)$	(5.8.26d)

We assume that the temperature is zero at $x = 0$, and that the "nonphysical" case ($h < 0$) of the boundary condition of the third kind is imposed at $x = L$. Thermal energy flows into the rod at $x = L$ [if $u(L,\, t) > 0$].

Separating variables,

$$u(x,\, t) = \phi(x)G(t), \qquad (5.8.27)$$

yields

$$\frac{dG}{dt} = -\lambda k G \qquad (5.8.28)$$

$$\frac{d^2\phi}{dx^2} + \lambda\phi = 0 \qquad (5.8.29a)$$

$$\phi(0) = 0 \qquad (5.8.29b)$$

$$\frac{d\phi}{dx}(L) + h\phi(L) = 0. \qquad (5.8.29c)$$

The time part is an exponential, $G = ce^{-\lambda kt}$. Here, we only consider the case in which

$$hL < -1.$$

Then there exists one negative eigenvalue ($\lambda_1 = -s_1$), with corresponding eigenfunction $\sinh \sqrt{s_1}\, x$, where s_1 is determined as the unique solution of $\tanh \sqrt{s}\, L = -\sqrt{s}/h$. *The time part exponentially grows.* All the other eigenvalues λ_n are positive. For these the eigenfunctions are $\sin \sqrt{\lambda}\, x$ (where $\tan \sqrt{\lambda}\, x = -\sqrt{\lambda}/h$ has an infinite number of solutions), while the corresponding time-dependent part exponentially decays being proportional to $e^{-\lambda kt}$. The forms of the product solutions are $\sin \sqrt{\lambda}\, x\, e^{-\lambda kt}$ and $\sinh \sqrt{s_1}\, x\, e^{s_1 kt}$. Here, the somewhat "abstract" notation may be considered more convenient; the product solutions are $\phi_n(x)e^{-\lambda_n kt}$, where the eigenfunctions are

$$\phi_n(x) = \begin{cases} \sinh \sqrt{s_1}\, x & n = 1 \\ \sin \sqrt{\lambda_n}\, x & n > 1. \end{cases}$$

According to the principle of superposition, we attempt to satisfy the initial value problem with a linear combination of *all* possible product solutions:

$$u(x, t) = \sum_{n=1}^{\infty} a_n \phi_n(x) e^{\lambda_n kt}.$$

The initial condition, $u(x, 0) = f(x)$, implies that

$$f(x) = \sum_{n=1}^{\infty} a_n \phi_n(x).$$

Since the coefficient $\sigma(x) = 1$ in (5.8.29), the eigenfunctions $\phi_n(x)$ are orthogonal with weight 1. Thus, we know that the generalized Fourier coefficients of the initial condition $f(x)$ are:

$$a_n = \frac{\int_0^L f(x)\phi_n(x)\, dx}{\int_0^L \phi_n^2\, dx} = \begin{cases} \int_0^L f(x) \sinh \sqrt{s_1}\, x\, dx \Big/ \int_0^L \sinh^2 \sqrt{s_1}\, x\, dx & n = 1 \\ \int_0^L f(x) \sin \sqrt{\lambda_n}\, x\, dx \Big/ \int_0^L \sin^2 \sqrt{\lambda_n}\, x\, dx & n \geq 2. \end{cases}$$

In particular, we could show $\int_0^L \sin^2 \sqrt{\lambda_n}\, x\, dx \neq L/2$. Perhaps we should emphasize one additional point. We have utilized the theorem that states that eigenfunctions corresponding to different eigenvalues are orthogonal; it is guaranteed that $\int_0^L \sin \sqrt{\lambda_n} \, x \sin \sqrt{\lambda_m} \, x \, dx = 0 \ (n \neq m)$ and $\int_0^L \sin \sqrt{\lambda_n} \, x \sinh \sqrt{s_1} \, x \, dx = 0$. We do not need to verify these by integration (although it can be done).

Other problems with boundary conditions of the third kind appear in the exercises.

EXERCISES 5.8

5.8.1. Consider

$$\frac{\partial u}{\partial t} = k \frac{\partial^2 u}{\partial x^2}$$

subject to $u(0, t) = 0$, $\dfrac{\partial u}{\partial x}(L, t) = -hu(L, t)$, and $u(x, 0) = f(x)$.

(a) Solve if $hL > -1$.
(b) Solve if $hL = -1$.

5.8.2. Consider the eigenvalue problem (5.8.7). Show that the nth eigenfunction has $n - 1$ zeros in the interior if
(a) $h > 0$ (b) $h = 0$
*(c) $-1 < hL < 0$ (d) $hL = -1$

5.8.3. Consider the eigenvalue problem

$$\frac{d^2 \phi}{dx^2} + \lambda \phi = 0,$$

subject to $\dfrac{d\phi}{dx}(0) = 0$ and $\dfrac{d\phi}{dx}(L) + h\phi(L) = 0$ with $h > 0$.

(a) Prove that $\lambda > 0$ (without solving the differential equation).
*(b) Determine all eigenvalues graphically. Obtain upper and lower bounds. Estimate the large eigenvalues.
(c) Show that the nth eigenfunction has $n - 1$ zeros in the interior.

5.8.4. Redo Exercise 5.8.3 parts (b) and (c) only if $h < 0$.

5.8.5. Consider

$$\frac{\partial u}{\partial t} = k \frac{\partial^2 u}{\partial x^2}$$

with $\dfrac{\partial u}{\partial x}(0, t) = 0$, $\dfrac{\partial u}{\partial x}(L, t) = -hu(L, t)$, and $u(x, 0) = f(x)$.

(a) Solve if $h > 0$.
(b) Solve if $h < 0$.

5.8.6. Consider (with $h > 0$)

$$\frac{\partial^2 u}{\partial t^2} = c^2 \frac{\partial^2 u}{\partial x^2}$$

$$\frac{\partial u}{\partial x}(0, t) - hu(0, t) = 0 \qquad\qquad u(x, 0) = f(x)$$

$$\frac{\partial u}{\partial x}(L, t) = 0 \qquad\qquad \frac{\partial u}{\partial t}(x, 0) = g(x).$$

(a) Show that there are an infinite number of different frequencies of oscillation.

(b) Estimate the large frequencies of oscillation.

(c) Solve the initial value problem.

*5.8.7. Consider the eigenvalue problem

$$\frac{d^2\phi}{dx^2} + \lambda\phi = 0, \text{ subject to } \phi(0) = 0 \text{ and } \phi(\pi) - 2\frac{d\phi}{dx}(0) = 0.$$

(a) Show that usually

$$\int_0^\pi \left(u\frac{d^2v}{dx^2} - v\frac{d^2u}{dx^2} \right) dx \neq 0$$

for any two functions u and v satisfying these homogeneous boundary conditions.

(b) Determine all positive eigenvalues.

(c) Determine all negative eigenvalues.

(d) Is $\lambda = 0$ an eigenvalue?

(e) Is it possible that there are other eigenvalues besides those determined in parts (b) through (d)? *Briefly* explain.

5.8.8. Consider the boundary value problem

$$\phi(0) - \frac{d\phi}{dx}(0) = 0$$

$$\frac{d^2\phi}{dx^2} + \lambda\phi = 0 \qquad \text{with}$$

$$\phi(1) + \frac{d\phi}{dx}(1) = 0.$$

(a) Using the Rayleigh quotient, show that $\lambda \geq 0$. Why is $\lambda > 0$?

(b) Prove that eigenfunctions corresponding to different eigenvalues are orthogonal.

*(c) Show that

$$\tan \sqrt{\lambda} = \frac{2\sqrt{\lambda}}{\lambda - 1}.$$

Determine the eigenvalues graphically. Estimate the large eigenvalues.

(d) Solve

$$\frac{\partial u}{\partial t} = k\frac{\partial^2 u}{\partial x^2}$$

with

$$u(0, t) - \frac{\partial u}{\partial x}(0, t) = 0$$

$$u(1, t) + \frac{\partial u}{\partial x}(1, t) = 0$$

$$u(x, 0) = f(x).$$

You may call the relevant eigenfunctions $\phi_n(x)$ and assume that they are known.

5.8.9. Consider the eigenvalue problem

$$\frac{d^2\phi}{dx^2} + \lambda\phi = 0 \text{ with } \phi(0) = \frac{d\phi}{dx}(0) \text{ and } \phi(1) = \beta\frac{d\phi}{dx}(1).$$

For what values (if any) of β is $\lambda = 0$ an eigenvalue?

5.8.10. Consider the special case of the eigenvalue problem of Sec. 5.8:

$$\frac{d^2\phi}{dx^2} + \lambda\phi = 0 \text{ with } \phi(0) = 0 \text{ and } \frac{d\phi}{dx}(1) + \phi(1) = 0.$$

*(a) Determine the lowest eigenvalue to at least two or three significant figures using tables or a calculator.

*(b) Determine the lowest eigenvalue using a root finding algorithm (e.g., Newton's method) on a computer.

(c) Compare either part (a) or (b) to the bound obtained using the Rayleigh quotient [see Exercise 5.6.1(c)].

5.8.11. Determine all negative eigenvalues for

$$\frac{d^2\phi}{dx^2} + 5\phi = -\lambda\phi \text{ with } \phi(0) = 0 \text{ and } \phi(\pi) = 0.$$

5.8.12. Consider $\partial^2 u/\partial t^2 = c^2 \, \partial^2 u/\partial x^2$ with the boundary conditions

$$u = 0 \qquad\qquad \text{at } x = 0$$

$$m\frac{\partial^2 u}{\partial t^2} = -T_0\frac{\partial u}{\partial x} - ku \qquad \text{at } x = L.$$

(a) Give a brief physical interpretation of the boundary conditions.

(b) Show how to determine the frequencies of oscillation. Estimate the large frequencies of oscillation.

(c) *Without* attempting to use the Rayleigh quotient, explicitly determine if there are any separated solutions that do not oscillate in time. (*Hint:* There are none.)

(d) Show that the boundary condition is *not* self-adjoint; that is, show

$$\int_0^L \left(u_n\frac{d^2 u_m}{dx^2} - u_m\frac{d^2 u_n}{dx^2} \right) dx \neq 0$$

even when u_n and u_m are eigenfunctions corresponding to different eigenvalues.

*5.8.13. Simplify $\int_0^L \sin^2 \sqrt{\lambda} \, x \, dx$ when λ is given by (5.8.12).

5.9 LARGE EIGENVALUES (ASYMPTOTIC BEHAVIOR)

For the variable coefficient case, the eigenvalues for the Sturm–Liouville differential equation,

$$\frac{d}{dx}\left[p(x)\frac{d\phi}{dx} \right] + [\lambda\sigma(x) + q(x)]\phi = 0, \qquad (5.9.1)$$

usually must be calculated numerically. We know that there will be an infinite number of eigenvalues with no largest one. Thus, there will be an infinite sequence of large eigenvalues. In this section we state and explain reasonably good approximations to these large eigenvalues and corresponding eigenfunctions. Thus, numerical solutions will be needed only for the first few eigenvalues and eigenfunctions.

A careful derivation with adequate explanations of the asymptotic method would be lengthy. Nonetheless, some motivation for our result will be presented. We begin by attempting to approximate solutions of the differential equation (5.9.1) if the unknown eigenvalue λ is large ($\lambda \gg 1$). Interpreting (5.9.1) as a spring–mass system (x is time, ϕ is position) with time-varying parameters is helpful. Then (5.9.1) has a large restoring force $[-\lambda\sigma(x)\phi]$ such that we expect the solution to have rapid oscillation in x. Alternatively, we know that eigenfunctions corresponding to large eigenvalues have many zeros. Since the solution oscillates rapidly, over a few periods (each small) the variable coefficients are approximately constant. Thus, near any point x_0, the differential equation may be approximated crudely by one with constant coefficients:

$$p(x_0)\frac{d^2\phi}{dx^2} + \lambda\sigma(x_0)\phi \approx 0, \tag{5.9.2}$$

since in addition $\lambda\sigma(x) \gg q(x)$. According to (5.9.2) the solution is expected to oscillate with "local" spatial (circular) frequency

$$\text{frequency} = \sqrt{\frac{\lambda\sigma(x_0)}{p(x_0)}}. \tag{5.9.3}$$

This frequency is large ($\lambda \gg 1$), and thus the period is small, as assumed. The frequency (and period) depends on x, but it varies slowly; that is, over a few periods (a short distance) the period hardly changes. After many periods, the frequency (and period) may change appreciably. This slowly varying period will be illustrated in Fig. 5.9.1.

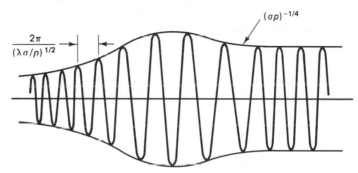

Figure 5.9.1 Liouville-Green asymptotic solution of differential equation showing rapid oscillation (or, equivalently, relatively slowly varying amplitude).

From (5.9.2) one might expect the amplitude of oscillation to be constant. However, (5.9.2) is only an approximation. Instead, we should expect the amplitude to be approximately constant over each period. Thus, both the amplitude and frequency are slowly varying:

$$\phi(x) = A(x)\cos\psi(x), \tag{5.9.4}$$

where sines can also be used. The appropriate asymptotic formula for the phase $\psi(x)$ can be obtained using the ideas we have just outlined. Since the period is small, only the values of x near any x_0 are needed to understand the oscillation implied by (5.9.4). Using the Taylor series of $\psi(x)$, we obtain

$$\phi(x) = A(x)\cos[\psi(x_0) + (x - x_0)\psi'(x_0) + \cdots]. \tag{5.9.5}$$

This is an oscillation with local frequency $\psi'(x_0)$. Thus, **the derivative of the phase is the local frequency**. From (5.9.2) we have motivated that the local frequency should be $[\lambda\sigma(x_0)/p(x_0)]^{1/2}$. Thus, we expect

$$\psi'(x_0) = \lambda^{1/2}\left[\frac{\sigma(x_0)}{p(x_0)}\right]^{1/2}. \tag{5.9.6}$$

This reasoning turns out to determine the phase correctly,

$$\psi(x) = \lambda^{1/2} \int^x \left[\frac{\sigma(x_0)}{p(x_0)} \right]^{1/2} dx_0. \tag{5.9.7}$$

Note that the phase does *not* equal the frequency times x (unless the frequency is constant).

Precise asymptotic techniques* beyond the scope of this text determine the slowly varying amplitude. It is known that two independent solutions of the differential equation can be approximated accurately (if λ is large) by

$$\phi(x) \approx (\sigma p)^{-1/4} \exp \left[\pm i\lambda^{1/2} \int^x \left(\frac{\sigma}{p} \right)^{1/2} dx_0 \right], \tag{5.9.8}$$

where sines and cosines may be used instead. A rough sketch of these solutions (using sines or cosines) is given in Fig. 5.9.1. The solution oscillates rapidly. The **envelope** of the wave is the slowly varying function $(\sigma p)^{-1/4}$, indicating the relatively slow amplitude variation. The local frequency is $(\lambda \sigma / p)^{1/2}$, corresponding to the period $2\pi/(\lambda \sigma/p)^{1/2}$.

To determine the large eigenvalues, we must apply the boundary conditions to the general solution (5.9.8). For example, if $\phi(0) = 0$, then

$$\phi(x) = (\sigma p)^{-1/4} \sin \left(\lambda^{1/2} \int_0^x \left(\frac{\sigma}{p} \right)^{1/2} dx_0 \right) + \cdots \tag{5.9.9}$$

The second boundary condition, for example, $\phi(L) = 0$, determines the eigenvalues

$$0 = \sin \left(\lambda^{1/2} \int_0^L \left(\frac{\sigma}{p} \right)^{1/2} dx_0 \right) + \cdots.$$

Thus, we derive the asymptotic formula for the large eigenvalues $\lambda^{1/2} \int_0^L (\sigma/p)^{1/2} dx_0 \approx n\pi$, or equivalently

$$\lambda \sim \left[n\pi \bigg/ \int_0^L \left(\frac{\sigma}{p} \right)^{1/2} dx_0 \right]^2, \tag{5.9.10}$$

valid if n is large. Often, this formula is reasonably accurate even when n is not very large. The eigenfunctions are given approximately by (5.9.9), where (5.9.10) should be used. Note that $q(x)$ does not appear in these asymptotic formulas; $q(x)$ does not affect the eigenvalue to *leading* order. However, more accurate formulas exist which take $q(x)$ into account.

---

* These results can be derived by various ways, such as the W.K.B.(J.) method (which should be called the Liouville–Green method) or the method of multiple scales. References for these asymptotic techniques include books by Bender and Orszag [1978], Kevorkian and Cole [1981], Nayfeh [1973], and Olver [1974].

Example. Consider the eigenvalue problem

$$\frac{d^2\phi}{dx^2} + \lambda(1 + x)\phi = 0$$

$$\phi(0) = 0$$
$$\phi(1) = 0.$$

Here $p(x) = 1$, $\sigma(x) = 1 + x$, $q(x) = 0$, $L = 1$. Our asymptotic formula for the eigenvalues is

$$\lambda \sim \left[\frac{n\pi}{\int_0^1 (1 + x_0)^{1/2}\, dx_0}\right]^2 = \frac{n^2\pi^2}{[\frac{2}{3}(1 + x_0)^{3/2}|_0^1]^2} = \frac{n^2\pi^2}{\frac{4}{9}(2^{3/2} - 1)^2}. \qquad (5.9.11)$$

In Table 5.9.1 we compare numerical results (using an accurate numerical scheme on the computer) with the asymptotic formula. Equation (5.9.11) is even a reasonable approximation if $n = 1$. The percent or relative error of the asymptotic formula improves as n increases. However, the error stays about the same (though small). There are improvements to (5.9.10) which account for the approximately constant error.

TABLE 5.9.1 EIGENVALUES λ_n

n	Numerical answer* (assumed accurate)	Asymptotic formula (5.9.11)	Error
1	6.548395	6.642429	0.094034
2	26.464937	26.569718	0.104781
3	59.674174	59.781865	0.107691
4	106.170023	106.278872	0.108849
5	165.951321	166.060737	0.109416
6	239.0177275	239.1274615	0.109734
7	325.369115	325.479045	0.109930

* Courtesy of E. C. Gartland, Jr.

EXERCISES 5.9

5.9.1. Estimate (to leading order) the large eigenvalues and corresponding eigenfunctions for

$$\frac{d}{dx}\left(p(x)\frac{d\phi}{dx}\right) + [\lambda\sigma(x) + q(x)]\phi = 0$$

if the boundary conditions are

(a) $\dfrac{d\phi}{dx}(0) = 0$ and $\dfrac{d\phi}{dx}(L) = 0$

*(b) $\phi(0) = 0$ and $\dfrac{d\phi}{dx}(L) = 0$

(c) $\phi(0) = 0$ and $\dfrac{d\phi}{dx}(L) + h\phi(L) = 0$

5.9.2. Consider

$$\frac{d^2\phi}{dx^2} + \lambda(1 + x)\phi = 0$$

subject to $\phi(0) = 0$ and $\phi(1) = 0$. Roughly sketch the eigenfunctions for λ large. Take into account amplitude and period variations.

5.9.3. Consider for $\lambda \gg 1$

$$\frac{d^2\phi}{dx^2} + [\lambda\sigma(x) + q(x)]\phi = 0.$$

*(a) Substitute

$$\phi = A(x) \exp\left[i\lambda^{1/2} \int_0^x \sigma^{1/2}(x_0)\ dx_0 \right].$$

Determine a differential equation for $A(x)$.

(b) Let $A(x) = A_0(x) + \lambda^{-1/2}A_1(x) + \cdots$. Solve for $A_0(x)$ and $A_1(x)$. Verify (5.9.8).

(c) Suppose that $\phi(0) = 0$. Use $A_1(x)$ to improve (5.9.9).

(d) Use part (c) to improve (5.9.10) if $\phi(L) = 0$.

*(e) Obtain a recursion formula for $A_n(x)$.

5.10 APPROXIMATION PROPERTIES

In many practical problems of solving partial differential equations by separation of variables, it is impossible to actually compute and work with an infinite number of terms of the infinite series. It is more usual to use a finite number of terms.* In this section, we briefly discuss the use of a finite number of terms of generalized Fourier series.

We have claimed that any piecewise smooth function $f(x)$ can be represented by a generalized Fourier series of the eigenfunctions,

$$f(x) \sim \sum_{n=1}^{\infty} a_n\phi_n(x). \tag{5.10.1}$$

Due to the orthogonality [with weight $\sigma(x)$] of the eigenfunctions, the generalized Fourier coefficients can be easily determined:

$$a_n = \frac{\int_a^b f(x)\phi_n(x)\sigma(x)\ dx}{\int_a^b \phi_n^2\sigma\ dx}. \tag{5.10.2}$$

However, suppose that we can only use the first M eigenfunctions to approximate a function $f(x)$,

$$f(x) \approx \sum_{n=1}^{M} \alpha_n\phi_n(x). \tag{5.10.3}$$

---

* Often, for numerical answers to problems in partial differential equations you may be better off using direct numerical methods.

What should the coefficients α_n be? Perhaps if we use a finite number of terms, there would be a better way to approximate $f(x)$ than by using the generalized Fourier coefficients, (5.10.2). We will pick these new coefficients α_n so that $\sum_{n=1}^{M} \alpha_n \phi_n(x)$ is the "best" approximation to $f(x)$. There are many ways to define the "best," but we will show a way that is particularly useful. In general, the coefficients α_n will depend on M. For example, suppose that we choose $M = 10$ and calculate $\alpha_1, \ldots, \alpha_{10}$ so that (5.10.3) is "best" in some way. After this calculation, we may decide that the approximation in (5.10.3) is not good enough, so we may wish to include more terms, for example, $M = 11$. We would then have to recalculate all 11 coefficients that make (5.10.3) "best" with $M = 11$. We will show that there is a way to define best such that the coefficients α_n do *not* depend on M; that is, in going from $M = 10$ to $M = 11$ only one additional coefficient need be computed, namely α_{11}.

Mean-square deviation. We define best approximation as the approximation with the least error. However, error can be defined in many different ways. The difference between $f(x)$ and its approximation $\sum_{n=1}^{M} \alpha_n \phi_n(x)$ depends on x. It is possible for $f(x) - \sum_{n=1}^{M} \alpha_n \phi_n(x)$ to be positive in some regions and negative in others. One possible measure of the error is the maximum of the deviation over the entire interval: $\max |f(x) - \sum_{n=1}^{M} \alpha_n \phi_n(x)|$. This is a reasonable definition of the error, but it is rarely used, since it is very difficult to choose the α_n to minimize this maximum deviation. Instead, we usually define the **error** to be the **mean-square deviation**,

$$E \equiv \int_a^b \left[f(x) - \sum_{n=1}^{M} \alpha_n \phi_n(x) \right]^2 \sigma(x) \, dx. \qquad (5.10.4)$$

Here a large penalty is paid for the deviation being large on even a small interval. We introduce a weight factor $\sigma(x)$ in our definition of the error because we will show it is easy to minimize this error *only with the weight $\sigma(x)$*. $\sigma(x)$ is the same function that appears in the differential equation defining the eigenfunctions $\phi_n(x)$; the weight appearing in the error is the same weight as needed for the orthogonality of the eigenfunctions.

The error, defined by (5.10.4), is a function of the coefficients $\alpha_1, \alpha_2, \ldots, \alpha_M$. To minimize a function of M variables, we usually use the first-derivative condition. We insist that the first partial derivative with respect to each α_i is zero:

$$\frac{\partial E}{\partial \alpha_i} = 0, \qquad i = 1, 2, \ldots, M.$$

We calculate each partial derivative and set it equal to zero:

$$0 = \frac{\partial E}{\partial \alpha_i} = -2 \int_a^b \left[f(x) - \sum_{n=1}^{M} \alpha_n \phi_n(x) \right] \phi_i(x) \sigma(x) \, dx, \qquad i = 1, 2, \ldots, M, \qquad (5.10.5)$$

where we have used the fact that $\partial/\partial\alpha_i \, (\Sigma_{n=1}^{M} \, \alpha_n\phi_n(x)) = \phi_i(x)$. There are M equations, (5.10.5), for the M unknowns. This would be rather difficult to solve, except for the fact that the eigenfunctions are orthogonal with the same weight $\sigma(x)$ that appears in (5.10.5). Thus, (5.10.5) becomes

$$\int_a^b f(x)\phi_i(x)\sigma(x)\,dx = \alpha_i \int_a^b \phi_i^2(x)\sigma(x)\,dx.$$

The ith equation can be easily solved for α_i. In fact, $\alpha_i = a_i$ [see (5.10.2)]; all first partial derivatives are zero if the coefficients are chosen to be the generalized Fourier coefficients. We should still show that this actually minimizes the error (not just a local critical point, where all first partial derivatives vanish). We in fact will show that **the best approximation** (*in the mean-square sense using the first M eigenfunctions*) **occurs when the coefficients are chosen to be the generalized Fourier coefficients**: In this way (1) the coefficients are easy to determine, and (2) the coefficients are independent of M.

Proof. To prove that the error E is actually minimized, we will not use partial derivatives. Instead, our derivation proceeds by expanding the square deviation in (5.10.4):

$$E = \int_a^b \left(f^2 - 2\sum_{n=1}^{M} \alpha_n f\phi_n + \sum_{n=1}^{M}\sum_{l=1}^{M} \alpha_n\alpha_l\phi_n\phi_l \right)\sigma\,dx. \qquad (5.10.6)$$

Some simplification again occurs due to the orthogonality of the eigenfunctions:

$$E = \int_a^b \left(f^2 - 2\sum_{n=1}^{M} \alpha_n f\phi_n + \sum_{n=1}^{M} \alpha_n^2\phi_n^2 \right)\sigma\,dx. \qquad (5.10.7)$$

Each α_n appears quadratically:

$$E = \sum_{n=1}^{M} \left[\alpha_n^2 \int_a^b \phi_n^2\sigma\,dx - 2\alpha_n \int_a^b f\phi_n\sigma\,dx \right] + \int_a^b f^2\sigma\,dx, \qquad (5.10.8)$$

and this suggests completing the square

$$E = \sum_{n=1}^{M} \left[\int_a^b \phi_n^2\sigma\,dx \left(\alpha_n - \frac{\int_a^b f\phi_n\sigma\,dx}{\int_a^b \phi_n^2\sigma\,dx} \right)^2 - \frac{\left(\int_a^b f\phi_n\sigma\,dx \right)^2}{\int_a^b \phi_n^2\sigma\,dx} \right] + \int_a^b f^2\sigma\,dx. \quad (5.10.9)$$

The only term that depends on the unknowns α_n appears in a nonnegative way. The minimum occurs only if that first term vanishes, determining the best coefficients

$$\alpha_n = \frac{\int_a^b f\phi_n\sigma\,dx}{\int_a^b \phi_n^2\sigma\,dx}, \qquad (5.10.10)$$

the same result as obtained using the simpler first derivative condition.

Error. In this way (5.10.9) shows that the minimal error is

$$E = \int_a^b f^2 \sigma \, dx - \sum_{n=1}^M \alpha_n^2 \int_a^b \phi_n^2 \sigma \, dx, \qquad (5.10.11)$$

where (5.10.10) has been used. Equation (5.10.11) shows that as M increases, the error decreases. Thus, we can think of a generalized Fourier series as an approximation scheme. **The more terms in the truncated series that are used, the better the approximation.**

Example. For a Fourier sine series, where $\sigma(x) = 1$, $\phi_n(x) = \sin n\pi x/L$ and $\int_0^L \sin^2 n\pi x/L \, dx = L/2$, it follows that

$$E = \int_0^L f^2 \, dx - \frac{L}{2} \sum_{n=1}^M \alpha_n^2. \qquad (5.10.12)$$

Bessel's inequality and Parseval's equality. Since $E \geqslant 0$ [see (5.10.4)], it follows from (5.10.11) that

$$\int_a^b f^2 \sigma \, dx \geqslant \sum_{n=1}^M \alpha_n^2 \int_a^b \phi_n^2 \sigma \, dx, \qquad (5.10.13)$$

known as **Bessel's inequality**. More importantly, we claim that for any Sturm–Liouville eigenvalue problem, the eigenfunction expansion of $f(x)$ converges in the mean to $f(x)$, by which we mean [see (5.10.4)] that

$$\lim_{M \to \infty} E = 0;$$

the mean-square deviation vanishes as $M \to \infty$. This shows **Parseval's equality**:

$$\int_a^b f^2 \sigma \, dx = \sum_{n=1}^\infty \alpha_n^2 \int_a^b \phi_n^2 \sigma \, dx. \qquad (5.10.14)$$

Parseval's equality, (5.10.14), is a generalization of the Pythagorean theorem. For a *right* triangle, $c^2 = a^2 + b^2$. This has an interpretation for vectors. If $\mathbf{v} = a\hat{\mathbf{i}} + b\hat{\mathbf{j}}$, then $\mathbf{v} \cdot \mathbf{v} = |\mathbf{v}|^2 = a^2 + b^2$. Here a and b are components of $\mathbf{v}$ in an *orthogonal* basis of *unit* vectors. Here we represent the function $f(x)$ in terms of our orthogonal eigenfunctions

$$f(x) = \sum_{n=1}^\infty a_n \phi_n(x).$$

If we introduce eigenfunctions with *unit* length, then

$$f(x) = \sum_{n=1}^\infty \frac{a_n l \phi_n(x)}{l},$$

where l is the length of $\phi_n(x)$:

$$l^2 = \int_a^b \phi_n^2 \, \sigma \, dx.$$

Parseval's equality simply states that the length of f squared, $\int_a^b f^2 \sigma \, dx$, equals the sum of squares of the components of f (using an orthogonal basis of functions of unit length), $(a_n l)^2 = a_n^2 \int_a^b \phi_n^2 \sigma \, dx$.

EXERCISES 5.10

5.10.1. Consider the Fourier sine series for $f(x) = 1$ on the interval $0 \leq x \leq L$. How many terms in the series should be kept so that the mean-square error is 1% of $\int_0^L f^2 \sigma \, dx$?

5.10.2. Obtain a formula for an infinite series using Parseval's equality applied to the:
   (a) Fourier sine series of $f(x) = 1$ on the interval $0 \leq x \leq L$
   *(b) Fourier cosine series of $f(x) = x$ on the interval $0 \leq x \leq L$
   (c) Fourier sine series of $f(x) = x$ on the interval $0 \leq x \leq L$

5.10.3. Consider any function $f(x)$ defined for $a \leq x \leq b$. Approximate this function by a constant. Show that the best such constant (in the mean-square sense, i.e., minimizing the mean-square deviation) is the constant equal to the average of $f(x)$ over the interval $a \leq x \leq b$.

5.10.4. (a) Using Parseval's equality, express the error in terms of the tail of a series.
   (b) Redo part (a) for a Fourier sine series on the interval $0 \leq x \leq L$.
   (c) If $f(x)$ is piecewise smooth, estimate the tail in part (b). (*Hint:* Use integration by parts.)

5.10.5. Show that if

$$L(f) = \frac{d}{dx}\left(p\frac{df}{dx}\right) + qf,$$

then

$$-\int_a^b fL(f) \, dx = -pf\frac{df}{dx}\bigg|_a^b + \int_a^b \left[p\left(\frac{df}{dx}\right)^2 - qf^2\right] dx$$

if f and df/dx are continuous.

5.10.6. Assuming that the operations of summation and integration can be interchanged, show that if

$$f = \sum \alpha_n \phi_n \quad \text{and} \quad g = \sum \beta_n \phi_n,$$

then for normalized eigenfunctions

$$\int_a^b fg\sigma \, dx = \sum_{n=1}^\infty \alpha_n \beta_n,$$

a generalization of Parseval's equality.

5.10.7. Using Exercises 5.10.5 and 5.10.6, prove that

$$\sum_{n=1}^\infty \lambda_n \alpha_n^2 = -pf\frac{df}{dx}\bigg|_a^b + \int_a^b \left[p\left(\frac{df}{dx}\right)^2 - qf^2\right] dx. \qquad (5.10.15)$$

[*Hint:* Let $g = L(f)$, assuming that term-by-term differentiation is justified.]

5.10.8. According to Schwarz's inequality (proved in Exercise 2.3.10), the absolute value of the *pointwise error* satisfies:

$$\left| f(x) - \sum_{n=1}^{M} \alpha_n \phi_n \right| = \left| \sum_{n=M+1}^{\infty} \alpha_n \phi_n \right| \leq \left\{ \sum_{n=M+1}^{\infty} |\lambda_n| \alpha_n^2 \right\}^{1/2} \left\{ \sum_{n=M+1}^{\infty} \frac{\phi_n^2}{|\lambda_n|} \right\}^{1/2} \tag{5.10.16}$$

Furthermore, Chapter 8 introduces a Green's function $G(x, x_0)$ which is shown to satisfy:

$$\sum_{n=1}^{\infty} \frac{\phi_n^2}{\lambda_n} = -G(x, x). \tag{5.10.17}$$

Using (5.10.15), (5.10.16), and (5.10.17), derive an upper bound for the pointwise error (in cases in which the generalized Fourier series is pointwise convergent). Examples and further discussion of this are given by Weinberger [1965].

Partial Differential Equations with at Least Three Independent Variables

6.1 INTRODUCTION

In our discussion of partial differential equations, we have solved many problems by the method of separation of variables, but all involved only two independent variables:

$$\frac{\partial^2 u}{\partial x^2} + \frac{\partial^2 u}{\partial y^2} = 0$$

$$\frac{\partial u}{\partial t} = k\frac{\partial^2 u}{\partial x^2} \qquad\qquad \frac{\partial^2 u}{\partial t^2} = c^2\frac{\partial^2 u}{\partial x^2}$$

$$c\rho\frac{\partial u}{\partial t} = \frac{\partial}{\partial x}\left(K_0\frac{\partial u}{\partial x}\right) \qquad\qquad \rho\frac{\partial^2 u}{\partial t^2} = T_0\frac{\partial^2 u}{\partial x^2}.$$

In this chapter we show how to apply the method of separation of variables to problems with more than two independent variables.

In particular, we discuss techniques to analyze the heat equation (with constant thermal properties) in two and three dimensions,

$$\frac{\partial u}{\partial t} = k\left(\frac{\partial^2 u}{\partial x^2} + \frac{\partial^2 u}{\partial y^2}\right) \qquad \text{(two dimensions)} \qquad\qquad (6.1.1a)$$

$$\frac{\partial u}{\partial t} = k\left(\frac{\partial^2 u}{\partial x^2} + \frac{\partial^2 u}{\partial y^2} + \frac{\partial^2 u}{\partial z^2}\right) \qquad \text{(three dimensions)} \qquad\qquad (6.1.1b)$$

for various physical regions with various boundary conditions. Also of interest will be the steady-state heat equation, Laplace's equation, in three dimensions,

$$\frac{\partial^2 u}{\partial x^2} + \frac{\partial^2 u}{\partial y^2} + \frac{\partial^2 u}{\partial z^2} = 0.$$

In all these problems, the partial differential equation has at least three independent variables. Other physical problems, not related to the flow of thermal energy, may also involve more than two independent variables. For example, the vertical displacement u of a vibrating membrane satisfies the two-dimensional wave equation

$$\frac{\partial^2 u}{\partial t^2} = c^2\left(\frac{\partial^2 u}{\partial x^2} + \frac{\partial^2 u}{\partial y^2}\right).$$

It should also be mentioned that in acoustics, the perturbed pressure u satisfies the three-dimensional wave equation

$$\frac{\partial^2 u}{\partial t^2} = c^2\left(\frac{\partial^2 u}{\partial x^2} + \frac{\partial^2 u}{\partial y^2} + \frac{\partial^2 u}{\partial z^2}\right).$$

We will discuss and analyze some of these problems.

6.2 SEPARATION OF THE TIME VARIABLE

We will show that similar methods can be applied to a variety of problems. We will begin by discussing the vibrations of a membrane of any shape, and follow that with some analysis for the conduction of heat in any two- or three-dimensional region.

6.2.1 Vibrating Membrane—Any Shape

Let us consider the displacement u of a vibrating membrane of any shape. Later (Secs. 6.3 and 6.7) we will specialize our result to rectangular and circular membranes. The displacement $u(x, y, t)$ satisfies the two-dimensional wave equation,

$$\boxed{\frac{\partial^2 u}{\partial t^2} = c^2\left(\frac{\partial^2 u}{\partial x^2} + \frac{\partial^2 u}{\partial y^2}\right).} \qquad (6.2.1)$$

The initial conditions will be

$$u(x, y, 0) = \alpha(x, y) \qquad (6.2.2a)$$

$$\frac{\partial u}{\partial t}(x, y, 0) = \beta(x, y), \qquad (6.2.2b)$$

but as usual they will be ignored at first when separating variables. A homogeneous boundary condition will be given along the entire boundary. $u = 0$ on the boundary is the most common condition. However, it is possible, for example, for the displacement to be zero on only part of the boundary and for the rest of the boundary to be "free." There are many other possible boundary conditions.

Let us now apply the method of separation of variables. We begin by showing that the time variable can be separated out from the problem for a membrane of any shape by seeking product solutions of the following form:

$$u(x, y, t) = h(t)\phi(x, y).$$

(6.2.3)

Here $\phi(x, y)$ is an as yet unknown function of the two variables x and y. We do not (at this time) specify further $\phi(x, y)$ since we might expect different results in different geometries or with different boundary conditions. Later, we will show that for rectangular membranes $\phi(x, y) = F(x)G(y)$, while for circular membranes $\phi(x, y) = F(r)G(\theta)$; that is, the form of further separation depends on the geometry. It is for this reason that we begin by analyzing the general form (6.2.3). In fact, for most regions that are *not* geometrically as simple as rectangles and circles, $\phi(x, y)$ cannot be separated further. If (6.2.3) is substituted into the equation for a vibrating membrane, (6.2.1), then the result is

$$\phi(x, y)\frac{d^2h}{dt^2} = c^2 h(t)\left(\frac{\partial^2 \phi}{\partial x^2} + \frac{\partial^2 \phi}{\partial y^2}\right).$$

(6.2.4)

We will attempt to proceed as we did when there were only two independent variables. Time can be separated from (6.2.4) by dividing by $h(t)\phi(x, y)$ (and an additional division by the constant c^2 is convenient):

$$\frac{1}{c^2}\frac{1}{h}\frac{d^2h}{dt^2} = \frac{1}{\phi}\left(\frac{\partial^2 \phi}{\partial x^2} + \frac{\partial^2 \phi}{\partial y^2}\right) = -\lambda.$$

(6.2.5)

The left-hand side of the first equation is only a function of time, while the right-hand side is only a function of space (x and y). Thus, the two (as before) must equal a separation constant. Again, we must decide what notation is convenient for the separation constant, λ or $-\lambda$. A quick glance at the resulting ordinary differential equation for $h(t)$ shows that $-\lambda$ is more convenient (as will be explained). We thus obtain two equations, but unlike the case of two independent variables, one of the equations is itself still a partial differential equation:

$$\frac{d^2h}{dt^2} = -\lambda c^2 h$$

(6.2.6a)

$$\frac{\partial^2 \phi}{\partial x^2} + \frac{\partial^2 \phi}{\partial y^2} = -\lambda\phi.$$

(6.2.6b)

The notation $-\lambda$ for the separation constant was chosen because the time-dependent differential equation (6.2.6a) has oscillatory solutions if $\lambda > 0$. If $\lambda > 0$, then h is a linear combination of $\sin c\sqrt{\lambda}\, t$ and $\cos c\sqrt{\lambda}\, t$; it oscillates with frequency $c\sqrt{\lambda}$. The values of λ determine the natural frequencies of oscillation of a vibrating membrane. However, we are not guaranteed that $\lambda > 0$. To show that $\lambda > 0$, we must analyze the resulting eigenvalue problem, (6.2.6b), where ϕ is subject to a homogeneous boundary condition along the entire boundary (e.g., $\phi = 0$ on the boundary). Here the eigenvalue problem itself involves a linear homogeneous partial differential equation. Shortly, we will show that $\lambda > 0$ by introducing a Rayleigh quotient applicable to (6.2.6b). Before analyzing (6.2.6b), we will show that it arises in other contexts.

6.2.2 Heat Conduction—Any Region

We will analyze the flow of thermal energy in any two-dimensional region. We begin by seeking product solutions of the form

$$u(x, y, t) = h(t)\phi(x, y) \qquad (6.2.7)$$

for the two-dimensional heat equation, assuming constant thermal properties and no sources, (6.1.1a). By substituting (6.2.7) into (6.1.1a) and after dividing by $kh(t)\phi(x, y)$, we obtain

$$\frac{1}{k}\frac{1}{h}\frac{dh}{dt} = \frac{1}{\phi}\left(\frac{\partial^2\phi}{\partial x^2} + \frac{\partial^2\phi}{\partial y^2}\right). \qquad (6.2.8)$$

A separation constant in the form $-\lambda$ is introduced so that the time-dependent part of the product solution exponentially decays (if $\lambda > 0$) as expected, rather than exponentially grows. Then, the two equations are

$$\frac{dh}{dt} = -\lambda k h \qquad (6.2.9a)$$

$$\frac{\partial^2\phi}{\partial x^2} + \frac{\partial^2\phi}{\partial y^2} = -\lambda\phi. \qquad (6.2.9b)$$

The eigenvalue λ relates to the decay rate of the time-dependent part. The eigenvalue λ is determined by the boundary value problem, again consisting of the partial differential equation (6.2.9b) with a corresponding boundary condition on the entire boundary of the region.

For heat flow in any three-dimensional region, (6.1.1a) must be modified by the inclusion of a $k\, \partial^2 u/\partial z^2$ term. A product solution,

$$u(x, y, z, t) = h(t)\phi(x, y, z), \qquad (6.2.10)$$

may still be sought, and after separating variables, we obtain equations similar to (6.2.9),

$$\frac{dh}{dt} = -\lambda k h \qquad (6.2.11a)$$

$$\frac{\partial^2\phi}{\partial x^2} + \frac{\partial^2\phi}{\partial y^2} + \frac{\partial^2\phi}{\partial z^2} = -\lambda\phi. \qquad (6.2.11b)$$

The eigenvalue λ is determined by finding those values of λ for which nontrival solutions of (6.2.11b) exist, subject to a homogeneous boundary condition on the entire boundary.

6.2.3 Summary

In situations described in this section the spatial part $\phi(x, y)$ or $\phi(x, y, z)$ of the solution of the partial differential equation satisfies the eigenvalue problem consisting of the partial differential equation,

$$\boxed{\nabla^2 \phi = -\lambda \phi,} \tag{6.2.12}$$

with ϕ satisfying appropriate homogeneous boundary conditions, which may be of the form [see (1.5.12) and (4.5.6)]

$$\alpha \phi + \beta \nabla \phi \cdot \hat{n} = 0, \tag{6.2.13}$$

where α and β can depend on x, y, and z. If $\beta = 0$, (6.2.13) is the fixed boundary condition. If $\alpha = 0$, (6.2.13) is the insulated or free boundary condition. If both $\alpha \neq 0$ and $\beta \neq 0$, then (6.2.13) is the higher dimensional version of Newton's law of cooling or the elastic boundary condition. In Sec. 6.4 we will describe general results for this two- or three-dimensional eigenvalue problem, similar to our theorems concerning the general one-dimensional Sturm–Liouville eigenvalue problem. However, first we will describe the solution of a simple two-dimensional eigenvalue problem in a situation in which $\phi(x, y)$ may be further separated, producing two one-dimensional eigenvalue problems.

EXERCISES 6.2

6.2.1. For a vibrating membrane of any shape which satisfies (6.2.1), show that (6.2.12) results after separating time.

6.2.2. For heat conduction in any two-dimensional region which satisfies (6.1.1a), show that (6.2.12) results after separating time.

6.2.3. (a) Obtain product solutions, $\phi = f(x)g(y)$, of (6.2.12) that satisfy $\phi = 0$ on the four sides of a rectangle. (*Hint:* If necessary, see Sec. 6.3.)

    (b) Using part (a), solve the initial value problem for a vibrating rectangular membrane (fixed on all sides).

    (c) Using part (a), solve the initial value problem for the two-dimensional heat equation with zero temperature on all sides.

6.3 VIBRATING RECTANGULAR MEMBRANE

In this section we analyze the vibrations of a rectangular membrane, as sketched in Fig. 6.3.1. The vertical displacement $u(x, y, t)$ of the membrane satisfies the two-dimensional wave equation,

$$\frac{\partial^2 u}{\partial t^2} = c^2 \left(\frac{\partial^2 u}{\partial x^2} + \frac{\partial^2 u}{\partial y^2} \right). \tag{6.3.1}$$

y = H

y = 0

x = 0   x = L   **Figure 6.3.1** Rectangular membrane.

We suppose that the boundary is given such that all four sides are fixed with zero displacement:

$$u(0, y, t) = 0 \qquad u(x, 0, t) = 0 \qquad (6.3.2\text{a,b})$$
$$u(L, y, t) = 0 \qquad u(x, H, t) = 0. \qquad (6.3.2\text{c,d})$$

We ask what is the displacement of the membrane at time t if the initial position and velocity are given:

$$u(x, y, 0) = \alpha(x, y) \qquad (6.3.3\text{a})$$

$$\frac{\partial u}{\partial t}(x, y, 0) = \beta(x, y). \qquad (6.3.3\text{b})$$

As we indicated in Sec. 6.2.1, since the partial differential equation and the boundary conditions are linear and homogeneous, we apply the method of separation of variables. First, we separate only the time variable by seeking product solutions in the form

$$u(x, y, t) = h(t)\phi(x, y). \qquad (6.3.4)$$

According to our earlier calculation, we are able to introduce a separation constant $-\lambda$, and the following two equations result:

$$\frac{d^2h}{dt^2} = -\lambda c^2 h \qquad (6.3.5\text{a})$$

$$\frac{\partial^2 \phi}{\partial x^2} + \frac{\partial^2 \phi}{\partial y^2} = -\lambda \phi. \qquad (6.3.5\text{b})$$

We will show that $\lambda > 0$, in which case $h(t)$ is a linear combination of $\sin c\sqrt{\lambda}\, t$ and $\cos c\sqrt{\lambda}\, t$. The homogeneous boundary conditions imply that the eigenvalue problem is

$$\frac{\partial^2 \phi}{\partial x^2} + \frac{\partial^2 \phi}{\partial y^2} = -\lambda \phi \qquad (6.3.6\text{a})$$

$\phi(0, y) = 0 \qquad \phi(x, 0) = 0$	$(6.3.6\text{b,c})$
$\phi(L, y) = 0 \qquad \phi(x, H) = 0;$	$(6.3.6\text{d,e})$

that is, $\phi = 0$ along the entire boundary. We call (6.3.6) a two-dimensional eigenvalue problem.

The eigenvalue problem itself is a linear homogeneous PDE in two independent variables with homogeneous boundary conditions. As such (since the boundaries are simple), we can expect that (6.3.6) can be solved by separation of variables in Cartesian coordinates. In other words, we look for product solutions of (6.3.6) in the form

$$\phi(x, y) = f(x)g(y). \tag{6.3.7}$$

Before beginning our calculations, let us note that it follows from (6.3.4) that our assumption (6.3.7) is equivalent to

$$u(x, y, t) = f(x)g(y)h(t), \tag{6.3.8}$$

a product of functions of *each* independent variable. We claim, as we show in an appendix to this section, that we could obtain the same result by substituting (6.3.8) into the wave equation (6.3.1) as we now obtain by substituting (6.3.7) into the two-dimensional eigenvalue problem (6.3.6a):

$$g(y)\frac{d^2f}{dx^2} + f(x)\frac{d^2g}{dy^2} = -\lambda f(x)g(y). \tag{6.3.9}$$

The x and y parts may be separated by dividing (6.3.9) by $f(x)g(y)$ and rearranging terms:

$$\frac{1}{f}\frac{d^2f}{dx^2} = -\lambda - \frac{1}{g}\frac{d^2g}{dy^2} = -\mu. \tag{6.3.10}$$

Since the first expression is only a function of x, while the second is only a function of y, we introduce a *second* separation constant. We choose it to be $-\mu$, so that the easily solved equation, $d^2f/dx^2 = -\mu f$ has oscillatory solutions (as expected) if $\mu > 0$. Two ordinary differential equations result from separation of variables of a partial differential equation with two independent variables:

$$\frac{d^2f}{dx^2} = -\mu f \tag{6.3.11a}$$

$$\frac{d^2g}{dy^2} = -(\lambda - \mu)g. \tag{6.3.11b}$$

Equations (6.3.11) contain *two* separation constants λ and μ, both of which must be determined. In addition $h(t)$ solves an ordinary differential equation:

$$\frac{d^2h}{dt^2} = -\lambda c^2 h. \tag{6.3.11c}$$

When we separate variables for a partial differential equation in three variables, $u(x, y, t) = f(x)g(y)h(t)$, we obtain three ordinary differential equations, one a function of each independent coordinate. However, there will be only two separation constants.

To determine the separation constants, we need to use the homogeneous boundary conditions, (6.3.6b)–(6.3.6e). The product form (6.3.3) then implies that

$$\begin{aligned} f(0) &= 0 \quad \text{and} \quad f(L) = 0 \\ g(0) &= 0 \quad \text{and} \quad g(H) = 0. \end{aligned} \tag{6.3.12}$$

Of our three ordinary differential equations, only two will be eigenvalue problems. There are homogeneous boundary conditions in x and y. Thus,

$$\frac{d^2f}{dx^2} = -\mu f$$

$$f(0) = 0 \quad \text{and} \quad f(L) = 0$$

(6.3.13)

is a Sturm–Liouville eigenvalue problem in the x-variable, where μ is the eigenvalue and $f(x)$ is the eigenfunction. Similarly, the y-dependent problem is a regular Sturm–Liouville eigenvalue problem:

$$\frac{d^2g}{dy^2} = -(\lambda - \mu)g$$

$$g(0) = 0 \quad \text{and} \quad g(H) = 0.$$

(6.3.14)

Here $\lambda - \mu$ is the eigenvalue and $g(y)$ the corresponding eigenfunction.

Not only are both (6.3.13) and (6.3.14) Sturm–Liouville eigenvalue problems, but they are both ones we should be quite familiar with. Without going through the well-known details, the eigenvalues are

$$\mu_n = \left(\frac{n\pi}{L}\right)^2, \quad n = 1, 2, 3, \ldots,$$

(6.3.15)

and the corresponding eigenfunctions are

$$f_n(x) = \sin \frac{n\pi x}{L}.$$

(6.3.16)

This determines the allowable values of the separation constant μ_n.

For each value of μ_n, (6.3.14) is still an eigenvalue problem. There are an infinite number of eigenvalues λ for each n. Thus, it should be double subscripted, λ_{nm}. In fact, from (6.3.14) the eigenvalues are

$$\lambda_{nm} \quad \mu_n = \left(\frac{m\pi}{H}\right)^2, \quad m = 1, 2, 3, \ldots,$$

(6.3.17)

where we *must* use a different index to represent the various y-eigenvalues (for each value of n). The corresponding y-eigenfunction is

$$g_{nm}(y) = \sin \frac{m\pi y}{H}.$$

(6.3.18)

The separation constant λ_{nm} now can be determined from (6.3.17):

$$\lambda_{nm} - \mu_n + \left(\frac{m\pi}{H}\right)^2 = \left(\frac{n\pi}{L}\right)^2 + \left(\frac{m\pi}{H}\right)^2,$$

(6.3.19)

where $n = 1, 2, 3, \ldots$ and $m = 1, 2, 3, \ldots$. The two-dimensional eigenvalue problem (6.3.6a) has eigenvalues λ_{nm} given by (6.3.19) and eigenfunctions given by the product of the two one-dimensional eigenfunctions. Using the notation $\phi_{nm}(x, y)$ for the two-dimensional eigenfunction corresponding to the eigenvalue

λ_{nm}, we have

$$\phi_{nm}(x, y) = \sin\frac{n\pi x}{L}\sin\frac{m\pi y}{H}, \qquad \begin{array}{l} n = 1, 2, 3, \ldots \\ m = 1, 2, 3, \ldots \end{array} \qquad (6.3.20)$$

Note how easily the homogeneous boundary conditions are satisfied.

From (6.3.19) we have explicitly shown that all the eigenvalues are positive (for this problem). Thus, the time-dependent part of the product solutions are (as previously guessed) $\sin c\sqrt{\lambda_{nm}}\, t$ and $\cos c\sqrt{\lambda_{nm}}\, t$, oscillations with natural frequencies $c\sqrt{\lambda_{nm}} = c\sqrt{(n\pi/L)^2 + (m\pi/H)^2}$, $n = 1, 2, 3, \ldots$ and $m = 1, 2, 3, \ldots$. In considering the displacement u, we have obtained two doubly infinite families of product solutions: $\sin n\pi x/L \sin m\pi y/H \sin c\sqrt{\lambda_{nm}}\, t$ and $\sin n\pi x/L \sin m\pi y/H \cos c\sqrt{\lambda_{nm}}\, t$. As with the vibrating string, each of these special product solutions is known as a mode of vibration. We sketch in Fig. 6.3.2 a representation of some of these modes. In each we sketch level contours of displacement in dotted lines at a fixed t. As time varies the shape stays the same, only the amplitude varies periodically. Each mode is a standing wave. Curves along which the displacement is always zero in a mode are called **nodal curves** and are sketched in solid lines. Cells are apparent wih neighboring cells always being out of phase; that is, when one cell has a positive displacement the neighbor has negative displacement (as represented by the + and − signs).

The principle of superposition implies that we should consider a linear combination of all possible product solutions. Thus, we must include both families, summing over both n and m,

$$u(x, y, t) = \sum_{m=1}^{\infty}\sum_{n=1}^{\infty} A_{nm}\sin\frac{n\pi x}{L}\sin\frac{m\pi y}{H}\cos c\sqrt{\lambda_{nm}}\, t$$
$$+ \sum_{m=1}^{\infty}\sum_{n=1}^{\infty} B_{nm}\sin\frac{n\pi x}{L}\sin\frac{m\pi y}{H}\sin c\sqrt{\lambda_{nm}}\, t. \qquad (6.3.21)$$

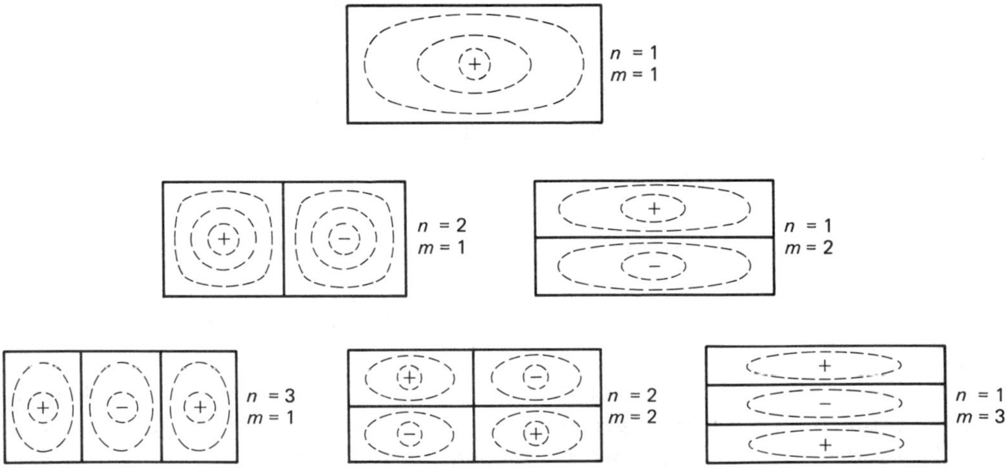

Figure 6.3.2 Nodal curves for modes of a vibrating rectangular membrane.

The two families of coefficients A_{nm} and B_{nm} hopefully will be determined from the two initial conditions. For example, $u(x, y, 0) = \alpha(x, y)$ implies that

$$\alpha(x, y) = \sum_{m=1}^{\infty} \left(\sum_{n=1}^{\infty} A_{nm} \sin \frac{n\pi x}{L} \right) \sin \frac{m\pi y}{H}. \tag{6.3.22}$$

The series in (6.3.22) is an example of what is called a **double Fourier series**. Instead of discussing the theory, we show one method to calculate A_{nm} from (6.3.22). (In Sec. 6.4 we will discuss a simpler way.) For *fixed* x, we note that $\sum_{n=1}^{\infty} A_{nm} \sin n\pi x/L$ depends only on m; furthermore, it must be the coefficients of the Fourier sine series in y of $\alpha(x, y)$ over $0 < y < H$. From our theory of Fourier sine series, we therefore know that we may easily determine the coefficients:

$$\sum_{n=1}^{\infty} A_{nm} \sin \frac{n\pi x}{L} = \frac{2}{H} \int_0^H \alpha(x, y) \sin \frac{m\pi y}{H} \, dy, \tag{6.3.23}$$

for each m. Equation (6.3.23) is valid for all x; the right-hand side is a function of x (not y, because y is integrated from 0 to H). For each m, the left-hand side is a Fourier sine series in x; in fact, it is the Fourier sine series of the right-hand side, $2/H \int_0^H \alpha(x, y) \sin m\pi y/H \, dy$. The coefficients of this Fourier sine series in x are easily determined:

$$\boxed{A_{nm} = \frac{2}{L} \int_0^L \left[\frac{2}{H} \int_0^H \alpha(x, y) \sin \frac{m\pi y}{H} \, dy \right] \sin \frac{n\pi x}{L} \, dx.} \tag{6.3.24}$$

This may be simplified to one double integral over the entire rectangular region, rather than two iterated one-dimensional integrals. In this manner we have determined one set of coefficients from one of the initial conditions.

The other coefficients B_{nm} can be determined in a similar way: in particular from (6.3.21), $\partial u/\partial t(x, y, 0) = \beta(x, y)$, which implies that

$$\beta(x, y) = \sum_{n=1}^{\infty} \sum_{m=1}^{\infty} c\sqrt{\lambda_{nm}} B_{nm} \sin \frac{n\pi x}{L} \sin \frac{m\pi y}{H}. \tag{6.3.25}$$

Thus, again using a Fourier sine series in y and a Fourier sine series in x, we obtain

$$\boxed{c\sqrt{\lambda_{nm}} B_{nm} = \frac{4}{LH} \int_0^L \int_0^H \beta(x, y) \sin \frac{m\pi y}{H} \sin \frac{n\pi x}{L} \, dy \, dx.} \tag{6.3.26}$$

The solution of our initial value problem is the doubly infinite series given by (6.3.21), where the coefficients are determined by (6.3.24) and (6.3.26).

We have shown that when all three independent variables separate for a partial differential equation, there results three ordinary differential equations, two of which are eigenvalue problems. In general, for a partial differential equation in N variables that completely separates, there will be N ordinary differential equations, $N - 1$ of which are one-dimensional eigenvalue problems (to determine the $N - 1$ separation constants). We have already shown this for $N = 3$ (this section) and $N = 2$.

EXERCISES 6.3

6.3.1. Consider the heat equation in a two-dimensional rectangular region, $0 < x < L$, $0 < y < H$,

$$\frac{\partial u}{\partial t} = k\left(\frac{\partial^2 u}{\partial x^2} + \frac{\partial^2 u}{\partial y^2}\right)$$

subject to the initial condition

$$u(x, y, 0) = f(x, y).$$

Solve the initial value problem and analyze the temperature as $t \to \infty$ if the boundary conditions are:

*(a) $u(0, y, t) = 0$ $u(x, 0, t) = 0$ **(b)** $\frac{\partial u}{\partial x}(0, y, t) = 0$ $\frac{\partial u}{\partial y}(x, 0, t) = 0$

     $u(L, y, t) = 0$ $u(x, H, t) = 0$ $\frac{\partial u}{\partial x}(L, y, t) = 0$ $\frac{\partial u}{\partial y}(x, H, t) = 0$

*(c) $\frac{\partial u}{\partial x}(0, y, t) = 0$ $u(x, 0, t) = 0$ **(d)** $u(0, y, t) = 0$ $\frac{\partial u}{\partial y}(x, 0, t) = 0$

     $\frac{\partial u}{\partial x}(L, y, t) = 0$ $u(x, H, t) = 0$ $\frac{\partial u}{\partial x}(L, y, t) = 0$ $\frac{\partial u}{\partial y}(x, II, t) = 0$

(e) $u(0, y, t) = 0$ $u(x, 0, t) = 0$

     $u(L, y, t) = 0$ $\frac{\partial u}{\partial y}(x, H, t) + hu(x, H, t) = 0$ $(h > 0)$.

6.3.2. Consider the heat equation in a three-dimensional box-shaped region, $0 < x < L$, $0 < y < H$, $0 < z < W$,

$$\frac{\partial u}{\partial t} = k\left(\frac{\partial^2 u}{\partial x^2} + \frac{\partial^2 u}{\partial y^2} + \frac{\partial^2 u}{\partial z^2}\right)$$

subject to the initial condition

$$u(x, y, z, 0) = f(x, y, z).$$

Solve the initial value problem and analyze the temperature as $t \to \infty$ if the boundary conditions are

(a) $u(0, y, z, t) = 0$ $\frac{\partial u}{\partial y}(x, 0, z, t) = 0$ $\frac{\partial u}{\partial z}(x, y, 0, t) = 0$

     $u(L, y, z, t) = 0$ $\frac{\partial u}{\partial y}(x, H, z, t) = 0$ $u(x, y, W, t) = 0$

*(b) $\frac{\partial u}{\partial x}(0, y, z, t) = 0$ $\frac{\partial u}{\partial y}(x, 0, z, t) = 0$ $\frac{\partial u}{\partial z}(x, y, 0, t) = 0$

     $\frac{\partial u}{\partial x}(L, y, z, t) = 0$ $\frac{\partial u}{\partial y}(x, H, z, t) = 0$ $\frac{\partial u}{\partial z}(x, y, W, t) = 0$.

6.3.3. Solve

$$\frac{\partial u}{\partial t} = k_1 \frac{\partial^2 u}{\partial x^2} + k_2 \frac{\partial^2 u}{\partial y^2}$$

on a rectangle $(0 < x < L, 0 < y < H)$ subject to

$$u(x, y, 0) = f(x, y)$$

$$u(0, y, t) = 0 \qquad \frac{\partial u}{\partial y}(x, 0, t) = 0$$

$$u(L, y, t) = 0 \qquad \frac{\partial u}{\partial y}(x, H, t) = 0.$$

6.3.4. Consider the wave equation for a vibrating rectangular membrane $(0 < x < L, 0 < y < H)$

$$\frac{\partial^2 u}{\partial t^2} = c^2 \left(\frac{\partial^2 u}{\partial x^2} + \frac{\partial^2 u}{\partial y^2} \right)$$

subject to the initial conditions

$$u(x, y, 0) = 0 \qquad \text{and} \qquad \frac{\partial u}{\partial t}(x, y, 0) = f(x, y).$$

Solve the initial value problem if

(a) $u(0, y, t) = 0 \qquad \frac{\partial u}{\partial y}(x, 0, t) = 0$

$\qquad u(L, y, t) = 0 \qquad \frac{\partial u}{\partial y}(x, H, t) = 0$

*(b) $\frac{\partial u}{\partial x}(0, y, t) = 0 \qquad \frac{\partial u}{\partial y}(x, 0, t) = 0$

$\qquad \frac{\partial u}{\partial x}(L, y, t) = 0 \qquad \frac{\partial u}{\partial y}(x, H, t) = 0.$

6.3.5. Consider

$$\frac{\partial^2 u}{\partial t^2} = c^2 \left(\frac{\partial^2 u}{\partial x^2} + \frac{\partial^2 u}{\partial y^2} \right) - k \frac{\partial u}{\partial t} \qquad \text{with } k > 0.$$

(a) Give a *brief* physical interpretation of this equation.
(b) Suppose that $u(x, y, t) = f(x)g(y)h(t)$. What ordinary differential equations are satisfied by f, g, and h?

6.3.6. Consider Laplace's equation

$$\nabla^2 u = \frac{\partial^2 u}{\partial x^2} + \frac{\partial^2 u}{\partial y^2} + \frac{\partial^2 u}{\partial z^2} = 0$$

in a right cylinder whose base is arbitrarily shaped (see Fig. 6.3.3). The top is $z = H$ and the bottom is $z = 0$. Assume that

$$\frac{\partial}{\partial z} u(x, y, 0) = 0$$

$$u(x, y, H) = f(x, y)$$

and $u = 0$ on the "lateral" sides.
(a) Separate the z-variable in general.
*(b) Solve for $u(x, y, z)$ if the region is a rectangular box, $0 < x < L, 0 < y < W, 0 < z < H$.

6.3.7. If possible, solve Laplace's equation,

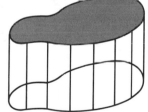

Figure 6.3.3

$$\nabla^2 u = \frac{\partial^2 u}{\partial x^2} + \frac{\partial^2 u}{\partial y^2} + \frac{\partial^2 u}{\partial z^2} = 0,$$

in a rectangular-shaped region, $0 < x < L, 0 < y < W, 0 < z < H$, subject to the boundary conditions

(a) $\frac{\partial u}{\partial x}(0, y, z) = 0, \qquad u(x, 0\ z) = 0, \qquad u(x, y, 0) = f(x, y)$

$\qquad \frac{\partial u}{\partial x}(L, y, z) = 0, \qquad u(x, W, z) = 0, \qquad u(x, y, H) = 0$

(b) $u(0, y, z) = 0,$ $u(x, 0, z) = 0,$ $u(x, y, 0) = 0$

       $u(L, y, z) = 0,$ $u(x, W, z) = f(x, z),$ $u(x, y, H) = 0$

*(c) $\dfrac{\partial u}{\partial x}(0, y, z) = 0,$ $\dfrac{\partial u}{\partial y}(x, 0, z) = 0,$ $\dfrac{\partial u}{\partial z}(x, y, 0) = 0$

       $\dfrac{\partial u}{\partial x}(L, y, z) = f(y, z),$ $\dfrac{\partial u}{\partial y}(x, W, z) = 0,$ $\dfrac{\partial u}{\partial z}(x, y, H) = 0$

*(d) $\dfrac{\partial u}{\partial x}(0, y, z) = 0,$ $\dfrac{\partial u}{\partial y}(x, 0, z) = 0,$ $\dfrac{\partial u}{\partial z}(x, y, 0) = 0$

       $u(L, y, z) = g(y, z),$ $\dfrac{\partial u}{\partial y}(x, W, z) = 0,$ $\dfrac{\partial u}{\partial z}(x, y, H) = 0.$

6.3 APPENDIX: OUTLINE OF ALTERNATIVE METHOD TO SEPARATE VARIABLES

An alternative (and equivalent) method to separate variables for

$$\frac{\partial^2 u}{\partial t^2} = c^2 \left(\frac{\partial^2 u}{\partial x^2} + \frac{\partial^2 u}{\partial y^2} \right) \tag{6.3A.1}$$

is to directly assume product solutions of the form

$$u(x, y, t) = f(x)g(y)h(t). \tag{6.3A.2}$$

By substituting (6.3A.2) into (6.3A.1), dividing by $c^2 f(x)g(y)h(t)$, we obtain

$$\frac{1}{c^2} \frac{1}{h} \frac{d^2 h}{dt^2} = \frac{1}{f} \frac{d^2 f}{dx^2} + \frac{1}{g} \frac{d^2 g}{dy^2} = -\lambda, \tag{6.3A.3}$$

after introducing a separation constant $-\lambda$. This shows that

$$\boxed{\frac{d^2 h}{dt^2} = -\lambda c^2 h.} \tag{6.3A.4}$$

Equation (6.3A.3) can be separated further,

$$\frac{1}{f} \frac{d^2 f}{dx^2} = -\lambda - \frac{1}{g} \frac{d^2 g}{dy^2} = -\mu, \tag{6.3A.5}$$

enabling a second separation constant $-\mu$ to be introduced:

$$\boxed{\frac{d^2 f}{dx^2} = -\mu f} \tag{6.3A.6a}$$

$$\boxed{\frac{d^2 g}{dy^2} = -(\lambda - \mu)g.} \tag{6.3A.6b}$$

In this way we have derived the same three ordinary differential equations (with two separation constants).

6.4 STATEMENTS AND ILLUSTRATIONS OF THEOREMS FOR THE EIGENVALUE PROBLEM $\nabla^2\phi + \lambda\phi = 0$

In solving the heat equation and the wave equation in any two- or three-dimensional region R (with constant physical properties, such as density), we have shown that the spatial part $\phi(x, y, z)$ of product form solutions $u(x, y, z, t) = \phi(x, y, z)h(t)$ satisfies the following multidimensional eigenvalue problem:

$$\boxed{\nabla^2\phi + \lambda\phi = 0,}$$

(6.4.1)

with

$$a\phi + b\,\nabla\phi \cdot \hat{\mathbf{n}} = 0$$

(6.4.2)

on the entire boundary. Here a and b can depend on x, y, and z. Equation (6.4.1) is known as the **Helmholtz equation**.

Equation (6.4.1) can be generalized to

$$\nabla \cdot (p\,\nabla\phi) + q\phi + \lambda\sigma\phi = 0,$$

(6.4.3)

where p, q, and σ are functions of x, y, and z. This eigenvalue problem [with boundary condition (6.4.2)] is directly analogous to the one-dimensional regular Sturm–Liouville eigenvalue problem. We prefer to deal with a somewhat simpler case, (6.4.1), corresponding to $p = \sigma = 1$ and $q = 0$. We will state and prove results for (6.4.1). We leave the discussion of (6.4.3) to some exercises (in Sec. 6.5).

Only for very simple geometries (for examples: rectangles, see Sec. 6.3, or circles, see Sec. 6.7) can even (6.4.1) be solved explicitly. In other situations, we may have to rely on numerical treatments. However, certain general properties of (6.4.1) are quite useful, all analogous to results we understand for the one-dimensional Sturm–Liouville problem. The reasons for the analogy will be discussed in the next section. We begin by simply stating the theorems for the two-dimensional case of (6.4.1) and (6.4.2):

---

1. All the eigenvalues are real.
2. There exist an infinite number of eigenvalues. There is a smallest eigenvalue, but no largest one.
3. Corresponding to an eigenvalue, there *may* be many eigenfunctions (unlike regular Sturm–Liouville eigenvalue problems).
4. The eigenfunctions $\phi(x, y)$ form a "complete" set, meaning that any piecewise smooth function $f(x, y)$ can be represented by a generalized Fourier series of the eigenfunctions:

$$f(x, y) \sim \sum_\lambda a_\lambda\phi_\lambda(x, y).$$

(6.4.4)

Here $\sum_\lambda a_\lambda\phi_\lambda$ means a linear combination of all the eigenfunctions. The series converges in the mean if the coefficients a_λ are chosen correctly.

---

5. Eigenfunctions belonging to different eigenvalues (λ_1 and λ_2) are orthogonal relative to the weight σ ($\sigma = 1$) over the entire region R. Mathematically,

$$\iint_R \phi_{\lambda_1}\phi_{\lambda_2}\, dx\, dy = 0 \qquad \text{if } \lambda_1 \neq \lambda_2, \tag{6.4.5}$$

where $\iint_R dx\, dy$ represents an integral over the entire region R. Furthermore, different eigenfunctions belonging to the same eigenvalue can be made orthogonal by the Gram–Schmidt process (see Sec. 6.5). Thus, we may assume that (6.4.5) is valid even if $\lambda_1 = \lambda_2$ as long as ϕ_{λ_1} is independent of ϕ_{λ_2}.

6. An eigenvalue λ can be related to the eigenfunction by the Rayleigh quotient:

$$\lambda = \frac{-\oint \phi\, \nabla\phi \cdot \hat{\mathbf{n}}\, ds + \iint_R (\nabla\phi)^2\, dx\, dy}{\iint_R \phi^2\, dx\, dy}. \tag{6.4.6}$$

The boundary conditions often simplify the boundary integral.

Here $\hat{\mathbf{n}}$ is a unit outward normal and $\oint ds$ is a closed line integral over the entire boundary of the plane two-dimensional region, where ds is the differential arc length. The three-dimensional result is nearly identical; $\iint$ must be replaced by $\iiint$ and the boundary line integral $\oint ds$ must be replaced by the boundary surface integral $\oiint dS$, where dS is the differential surface area.

Example. We will prove some of these statements in Sec. 6.5. To understand their meaning, we will show how the example of Sec. 6.3 illustrates most of these theorems. For the vibrations of a rectangular ($0 < x < L, 0 < y < H$) membrane with fixed zero boundary conditions, we have shown that the relevant eigenvalue problem is

$$
\begin{array}{|c|}
\hline
\nabla^2\phi + \lambda\phi = 0 \\
\hline
\phi(0, y) = 0 \qquad \phi(x, 0) = 0 \\
\hline
\phi(L, y) = 0 \qquad \phi(x, H) = 0. \\
\hline
\end{array}
\tag{6.4.7}
$$

We have determined that the eigenvalues and corresponding eigenfunctions are

$$
\boxed{\lambda_{nm} = \left(\frac{n\pi}{L}\right)^2 + \left(\frac{m\pi}{H}\right)^2} \quad
\begin{array}{l}
n = 1, 2, 3, \ldots \\
m = 1, 2, 3, \ldots
\end{array}
$$

with $\qquad \boxed{\phi_{nm}(x, y) = \sin\dfrac{n\pi x}{L} \sin\dfrac{m\pi y}{H}.} \tag{6.4.8}$

1. **Real eigenvalues.** In our calculation of the eigenvalues for (6.4.7) we *assumed* that the eigenfunctions existed in a product form. Under that assumption, (6.4.8) showed the eigenvalues to be real. Our theorem guarantees the eigenvalues will always be real.

2. **Ordering of eigenvalues.** There is a doubly infinite set of eigenvalues for (6.4.7), namely $\lambda_{nm} = (n\pi/L)^2 + (m\pi/H)^2$ for $n = 1, 2, 3, \ldots$ and $m = 1, 2, 3, \ldots$. There is a smallest eigenvalue, $\lambda_{11} = (\pi/L)^2 + (\pi/H)^2$, but no largest eigenvalue.

3. **Multiple eigenvalues.** For $\nabla^2\phi + \lambda\phi = 0$, our theorem states that, in general, it is possible for there to be more than one eigenfunction corresponding to the same eigenvalue. To illustrate this, consider (6.4.7) in the case in which $L = 2H$. Then

$$\lambda_{nm} = \frac{\pi^2}{4H^2}(n^2 + 4m^2) \tag{6.4.9a}$$

with

$$\phi_{nm} = \sin\frac{n\pi x}{2H}\sin\frac{m\pi y}{H}. \tag{6.4.9b}$$

We note that it is possible to have different eigenfunctions corresponding to the same eigenvalue. For example, $n = 4$, $m = 1$ and $n = 2$, $m = 2$, yield the same eigenvalue:

$$\lambda_{41} = \lambda_{22} = \frac{\pi^2}{4H^2}20.$$

For $n = 4$, $m = 1$, the eigenfunction is $\phi_{41} = \sin 4\pi x/2H \sin \pi y/H$, while for $n = 2$, $m = 2$, $\phi_{22} = \sin 2\pi x/2H \sin 2\pi y/H$. The nodal curves for these eigenfunctions are sketched in Fig. 6.4.1. They are different eigenfunctions with the same eigenvalue, $\lambda = (\pi^2/4H^2)20$. It is not surprising that the eigenvalue is the same, since a membrane vibrating in these modes has cells of the same dimensions: one $H \times H/2$ and the other $H/2 \times H$. By symmetry they will have the same natural frequency (and hence the same eigenvalue since the natural frequency is $c\sqrt{\lambda}$). In fact, in general by symmetry [as well as by formula (6.4.9)] $\lambda_{(2n)m} = \lambda_{(2m)n}$.

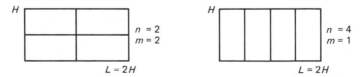

Figure 6.4.1 Nodal curves for eigenfuctions with the same eigenvalue (symmetric).

However, it is also possible for more than one eigenfunction to occur for reasons having nothing to do with symmetry. For example, $n = 1$, $m = 4$ and $n = 7$, $m = 2$ yield the same eigenvalue: $\lambda_{14} = \lambda_{72} = (\pi^2/4H^2)65$. The corresponding eigenfunctions are $\phi_{14} = \sin \pi x/2H \sin 4\pi y/H$ and $\phi_{72} = \sin 7\pi x/2H \sin 2\pi y/H$, which are sketched in Fig. 6.4.2. It is only coincidental that both of these shapes vibrate with the same frequency. In these situations, it is possible for two

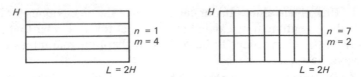

Figure 6.4.2 Nodal curves for eigenfunctions with the same eigenvalue (asymmetric).

eigenfunctions to correspond to the same eigenvalue. We can find situations with even more multiplicities (or degeneracies). Since $\lambda_{14} = \lambda_{72} = (\pi^2/4H^2)65$, it is also true that $\lambda_{28} = \lambda_{(14)4} = (\pi^2/4H^2)260$. However, by symmetry $\lambda_{28} = \lambda_{(16)1}$ and $\lambda_{(14)4} = \lambda_{87}$. Thus,

$$\lambda_{28} = \lambda_{(16)1} = \lambda_{(14)4} = \lambda_{87} = \left(\frac{\pi^2}{4H^2}\right)260.$$

Here there are four eigenfunctions corresponding to the same eigenvalue.

4a. **Series of eigenfunctions.** According to this theorem, (6.4.4), the eigenfunctions of $\nabla^2\phi + \lambda\phi = 0$ can always be used to represent any piecewise smooth function $f(x, y)$. In our illustrative example, (6.4.7), Σ_λ becomes a double sum,

$$f(x, y) \sim \sum_{n=1}^{\infty} \sum_{m=1}^{\infty} a_{nm} \sin \frac{n\pi x}{L} \sin \frac{m\pi y}{H}. \tag{6.4.10}$$

5. **Orthogonality of eigenfunctions.** We will show that the multidimensional orthogonality of the eigenfunctions, as expressed by (6.4.5) for any two different eigenfunctions, can be used to determine the generalized Fourier coefficients in (6.4.4).* We will do this in exactly the way we did for one-dimensional Sturm–Liouville eigenfunction expansions. We simply multiply (6.4.4) by ϕ_{λ_i} and integrate over the entire region R:

$$\iint_R f\phi_{\lambda_i} \, dx \, dy = \sum_\lambda a_\lambda \iint_R \phi_\lambda \phi_{\lambda_i} \, dx \, dy. \tag{6.4.11}$$

Since the eigenfunctions are all orthogonal to each other (with weight 1 because $\nabla^2\phi + \lambda\phi = 0$), it follows that

$$\iint_R f\phi_{\lambda_i} \, dx \, dy = a_{\lambda_i} \iint_R \phi_{\lambda_i}^2 \, dx \, dy, \tag{6.4.12}$$

or, equivalently,

$$a_{\lambda_i} = \frac{\displaystyle\iint_R f\phi_{\lambda_i} \, dx \, dy}{\displaystyle\iint_R \phi_{\lambda_i}^2 \, dx \, dy}. \tag{6.4.13}$$

---

* If there is more than one eigenfunction corresponding to the same eigenvalue, then we assume that the eigenfunctions have been made orthogonal (if necessary by the Gram–Schmidt process).

There is no difficulty in forming (6.4.13) from (6.4.12) since the denominator of (6.4.13) is necessarily positive.

For the special case that occurs for a rectangle with fixed zero boundary conditions, (6.4.7), the generalized Fourier coefficients a_{nm} are given by (6.4.13):

$$a_{nm} = \frac{\int_0^H \int_0^L f(x, y) \sin n\pi x/L \sin m\pi y/H \, dx \, dy}{\int_0^H \int_0^L \sin^2 n\pi x/L \sin^2 m\pi y/H \, dx \, dy}. \tag{6.4.14}$$

The integral in the denominator may be easily shown to equal $(L/2)(H/2)$ by calculating two one-dimensional integrals. In this way we rederive (6.3.24). In this case, (6.4.10), the generalized Fourier coefficient a_{nm} can be evaluated in two equivalent ways:

1. Using one two-dimensional orthogonality formula for the eigenfunctions of $\nabla^2\phi + \lambda\phi = 0$.
2. Using two one-dimensional orthogonality formulas.

4b. Convergence. As with any Sturm–Liouville eigenvalue problem (see Sec. 5.10), a *finite* series of the eigenfunctions of $\nabla^2\phi + \lambda\phi = 0$ may be used to approximate a function $f(x, y)$. In particular, we could show that if we measure error in the mean-square sense,

$$E = \iint_R (f - \sum_\lambda a_\lambda\phi_\lambda)^2 \, dx \, dy, \tag{6.4.15}$$

with weight function 1, then this mean-square error is minimized by the coefficients a_λ being chosen by (6.4.13), the generalized Fourier coefficients. It is known that the approximation improves as the number of terms increases. Furthermore, $E \to 0$ as all the eigenfunctions are included. We say that the series $\sum_\lambda a_\lambda\phi_\lambda$ **converges in the mean** to f.

EXERCISES 6.4

6.4.1. Consider the eigenvalue problem

$$\nabla^2\phi + \lambda\phi = 0$$

$$\frac{\partial\phi}{\partial x}(0, y) = 0 \qquad \phi(x, 0) = 0$$

$$\frac{\partial\phi}{\partial x}(L, y) = 0 \qquad \phi(x, H) = 0.$$

*(a) Show that there is a doubly infinite set of eigenvalues.
(b) If $L = H$, show that most eigenvalues have more than one eigenfunction.
(c) Derive that the eigenfunctions are orthogonal in a two-dimensional sense using two one-dimensional orthogonality relations.

6.4.2. Without using the explicit solution of (6.4.7), show that $\lambda \geq 0$ from the Rayleigh quotient, (6.4.6).

6.4.3. If necessary, see Sec. 6.5:

    (a) Derive that $\iint (u\nabla^2 v - v\nabla^2 u)\ dx\ dy = \oint (u\nabla v - v\nabla u)\cdot\hat{\mathbf{n}}\ ds$.

    (b) From part (a), derive (6.4.5).

6.4.4. If necessary, see Sec. 6.6. Derive (6.4.6). [*Hint:* Multiply (6.4.1) by ϕ and integrate.]

6.5 SELF-ADJOINT OPERATORS AND MULTIDIMENSIONAL EIGENVALUE PROBLEMS

Introduction. In this section we prove some of the properties of the multidimensional eigenvalue problem:

$$\boxed{\nabla^2\phi + \lambda\phi = 0} \tag{6.5.1a}$$

with

$$\boxed{\beta_1\phi + \beta_2\,\nabla\phi\cdot\hat{\mathbf{n}} = 0} \tag{6.5.1b}$$

on the entire boundary. Here β_1 and β_2 are real functions of the location in space. As with Sturm–Liouville eigenvalue problems, we will simply assume that there are an infinite number of eigenvalues for (6.5.1) and that the resulting set of eigenfunctions is complete. Proofs of these statements are difficult and beyond the intent of this text. The proofs for various other properties of the multi-dimensional eigenvalue problem are quite similar to corresponding proofs for the one-dimensional Sturm–Liouville eigenvalue problem. We let

$$L \equiv \nabla^2, \tag{6.5.2}$$

in which case the notation for the multidimensional eigenvalue problem becomes

$$L(\phi) + \lambda\phi = 0. \tag{6.5.3}$$

By comparing (6.5.3) to (5.5.2), we notice that the weight function for this multidimensional problem is expected to be 1.

Multidimensional Green's Formula. The proofs for the one-dimensional Sturm–Liouville eigenvalue problem depended on $uL(v) - vL(u)$ being an exact differential (known as Lagrange's identity). Its integrated form (known as Green's formula) was also needed. Similar identities will be derived for the Laplacian operator, $L = \nabla^2$, a multidimensional analog of the Sturm–Liouville differential operator. We will calculate $uL(v) - vL(u) = u\,\nabla^2 v - v\,\nabla^2 u$. We recall that $\nabla^2 u = \nabla\cdot(\nabla u)$ and $\nabla\cdot(a\mathbf{B}) = a\,\nabla\cdot\mathbf{B} + \nabla a\cdot\mathbf{B}$ (where a is a scalar and $\mathbf{B}$ a vector). Thus,

$$\begin{aligned}\nabla\cdot(u\,\nabla v) &= u\,\nabla^2 v + \nabla u\cdot\nabla v\\ \nabla\cdot(v\,\nabla u) &= v\,\nabla^2 u + \nabla v\cdot\nabla u.\end{aligned} \tag{6.5.4}$$

By subtracting these,

$$\boxed{u\,\nabla^2 v - v\,\nabla^2 u = \nabla\cdot(u\,\nabla v - v\,\nabla u).} \tag{6.5.5}$$

The differential form, (6.5.5), is the multidimensional version of Lagrange's identity, (5.5.4). Instead of integrating from a to b as we did in one-dimensional problems, we integrate over the entire two-dimensional region

$$\iint_R (u\,\nabla^2 v - v\,\nabla^2 u)\,dxdy = \iint_R \nabla \cdot (u\,\nabla v - v\,\nabla u)\,dx\,dy.$$

The right-hand side is in the correct form to apply the divergence theorem (recall that $\iint_R \nabla \cdot \mathbf{A}\,dx\,dy = \oint \mathbf{A} \cdot \hat{\mathbf{n}}\,ds$). Thus,

$$\boxed{\iint_R (u\,\nabla^2 v - v\,\nabla^2 u)\,dx\,dy = \oint (u\,\nabla v - v\,\nabla u) \cdot \hat{\mathbf{n}}\,ds.} \qquad (6.5.6)$$

Equation (6.5.6) is analogous to Green's formula, (5.5.5). It is known as **Green's second identity**,* but we will just refer to it as **Green's formula**.

We have shown that $L = \nabla^2$ is a multidimensional **self-adjoint** operator in the following sense:

> If u and v are any two functions, such that
>
> $$\oint (u\,\nabla v - v\,\nabla u) \cdot \hat{\mathbf{n}}\,ds = 0, \qquad (6.5.7a)$$
>
> then
>
> $$\iint_R [uL(v) - vL(u)]\,dx\,dy = 0, \qquad (6.5.7b)$$
>
> where $L = \nabla^2$.

In many problems, prescribed homogeneous boundary conditions will cause the boundary term to vanish. For example, (6.5.7b) is valid if u and v both vanish on the boundary. Again for three-dimensional problems, $\iint$ must be replaced by $\iiint$ and $\oint$ must be replaced by $\oiint$.

Orthogonality of the eigenfunctions. As with the one-dimensional Sturm–Liouville eigenvalue problem, we can prove a number of theorems directly from Green's formula (6.5.6). To show eigenfunctions corresponding to different eigenvalues are orthogonal, we consider two eigenfunctions ϕ_1 and ϕ_2, corresponding to the eigenvalues λ_1 and λ_2:

$$\begin{aligned}
\nabla^2\phi_1 + \lambda_1\phi_1 = 0 \quad &\text{or} \quad L(\phi_1) + \lambda_1\phi_1 = 0 \\
\nabla^2\phi_2 + \lambda_2\phi_2 = 0 \quad &\text{or} \quad L(\phi_2) + \lambda_2\phi_2 = 0.
\end{aligned} \qquad (6.5.8)$$

If both ϕ_1 and ϕ_2 satisfy the same homogeneous boundary conditions, then (6.5.7a) is satisfied, in which case (6.5.7b) follows. Thus

---

* *Green's first identity* arises from integrating (6.5.4) [rather than (6.5.5)] with $v = u$ and applying the divergence theorem.

$$\iint_R (-\phi_1 \lambda_2 \phi_2 + \phi_2 \lambda_1 \phi_1) \, dx \, dy = (\lambda_1 - \lambda_2) \iint_R \phi_1 \phi_2 \, dx \, dy = 0.$$

If $\lambda_1 \neq \lambda_2$, then

$$\iint_R \phi_1 \phi_2 \, dx \, dy = 0, \tag{6.5.9}$$

which means that eigenfunctions corresponding to different eigenvalues are orthogonal (in a multidimensional sense with weight 1). If two or more eigenfunctions correspond to the same eigenvalue, they can be made orthogonal to each other (as well as all other eigenfunctions) by a procedure shown in the appendix of this section. This is known as the Gram–Schmidt method.

We can now prove that the eigenvalues will be real. The proof is left for an exercise since the proof is identical to that used for the one-dimensional Sturm–Liouville problem (see Sec. 5.5).

EXERCISES 6.5

6.5.1. The vertical displacement of a nonuniform membrane satisfies

$$\frac{\partial^2 u}{\partial t^2} = c^2 \left(\frac{\partial^2 u}{\partial x^2} + \frac{\partial^2 u}{\partial y^2} \right),$$

where c depends on x and y. Suppose that $u = 0$ on the boundary of an irregularly shaped membrane.
   (a) Show that the time variable can be separated by assuming that
$$u(x, y, t) = \phi(x, y)h(t).$$
   Show that $\phi(x, y)$ satisfies the eigenvalue problem
$$\nabla^2 \phi + \lambda \sigma(x, y) \phi = 0 \qquad \text{with } \phi = 0 \text{ on the boundary.} \tag{6.5.10}$$
   What is $\sigma(x, y)$?
   (b) If the eigenvalues are known (and $\lambda > 0$), determine the frequencies of vibration.

6.5.2. See Exercise 6.5.1. Consider the two-dimensional eigenvalue problem (6.5.10).
   (a) Prove that the eigenfunctions belonging to different eigenvalues are orthogonal (with what weight?).
   (b) Prove that all the eigenvalues are real.
   (c) Do Exercise 6.6.1.

6.5.3. Redo Exercise 6.5.2 if the boundary condition is instead
   (a) $\nabla \phi \cdot \hat{\mathbf{n}} = 0$ on the boundary.
   (b) $\nabla \phi \cdot \hat{\mathbf{n}} + h(x, y)\phi = 0$ on the boundary.
   (c) $\phi = 0$ on part of the boundary and $\nabla \phi \cdot \hat{\mathbf{n}} = 0$ on the rest of the boundary.

6.5.4. Consider the heat equation in three dimensions with no sources but with nonconstant thermal properties

$$c\rho \frac{\partial u}{\partial t} = \nabla \cdot (K_0 \nabla u),$$

where $c\rho$ and K_0 are functions of x, y, and z. Assume that $u = 0$ on the boundary.

Show that the time variable can be separated by assuming that
$$u(x, y, z, t) = \phi(x, y, z)h(t).$$
Show that $\phi(x, y, z)$ satisfies the eigenvalue problem
$$\left.\begin{array}{c} \nabla \cdot (p \nabla\phi) + \lambda\sigma(x, y, z)\phi = 0 \\ \phi = 0 \text{ on the boundary.} \end{array}\right\} \tag{6.5.11}$$
What are $\sigma(x, y, z)$ and $p(x, y, z)$?

6.5.5. See Exercise 6.5.4. Consider the three-dimensional eigenvalue problem (6.5.11).
  (a) Prove that the eigenfunctions belonging to different eigenvalues are orthogonal (with what weight?).
  (b) Prove that all the eigenvalues are real.
  (c) Do Exercise 6.6.3.

6.5.6. *Derive* an expression for
$$\iint [uL(v) - vL(u)]\, dx\, dy$$
over a two-dimensional region R, where
$$L = \nabla^2 + q(x, y) \qquad [\text{i.e., } L(u) = \nabla^2 u + q(x, y)u].$$

6.5.7. Consider Laplace's equation $\nabla^2 u = 0$ in a three-dimensional region R (where u is the temperature). Suppose that the heat flux is given on the boundary (not necessarily a constant).
  (a) Explain *physically* why $\oiint \nabla u \cdot \hat{n}\, dS = 0$.
  (b) Show this mathematically.

6.5.8. Suppose that in a three-dimensional region R
$$\nabla^2 \phi = f(x, y, z)$$
with f given and $\nabla\phi \cdot \hat{n} = 0$ on the boundary.
  (a) Show mathematically that (if there is a solution)
$$\iiint_R f\, dx\, dy\, dz = 0.$$
  (b) Briefly explain physically (using the heat flow model) why condition (a) must hold for a solution. What happens in a heat flow problem if
$$\iiint_R f\, dx\, dy\, dz > 0?$$

6.5.9. Show that the boundary term (6.5.7a) vanishes if both u and v satisfy (6.5.1b):
  (a) Assume that $\beta_2 \neq 0$.
  (b) Assume $\beta_2 = 0$ for part of the boundary.

6.5 APPENDIX: GRAM–SCHMIDT METHOD

We wish to show in general that eigenfunctions *corresponding to the same eigenvalue* can be made orthogonal. The process is known as **Gram–Schmidt orthogonalization**. Let us suppose that $\phi_1, \phi_2, \ldots, \phi_n$ are independent eigenfunctions corresponding to the *same eigenvalue*. We will form a set of n-independent eigenfunctions denoted $\psi_1, \psi_2, \ldots, \psi_n$ which are mutually orthogonal, even if $\phi_1, \ldots, \phi_n$ are not. Let $\psi_1 = \phi_1$ be any one eigenfunction. Any linear combination of the eigenfunctions is also an eigenfunction (since they satisfy the *same* linear homogeneous differential equation and boundary conditions). Thus, $\psi_2 = \phi_2 + c\psi_1$ is also an eigenfunction (automatically independent of ψ_1), where c is an arbitrary constant. We choose c so that $\psi_2 = \phi_2 + c\psi_1$ is orthogonal to ψ_1: $\iint_R \psi_1 \psi_2 \, dx \, dy = 0$ becomes

$$\iint_R (\phi_2 + c\psi_1)\psi_1 \, dx \, dy = 0.$$

c is uniquely determined:

$$c = \frac{-\displaystyle\iint_R \phi_2 \psi_1 \, dx \, dy}{\displaystyle\iint_R \psi_1^2 \, dx \, dy}. \tag{6.5A.1}$$

Since there may be more than two eigenfunctions corresponding to the same eigenvalue, we continue this process.

A third eigenfunction is $\psi_3 = \phi_3 + c_1\psi_1 + c_2\psi_2$, where we choose c_1 and c_2 so that ψ_3 is orthogonal to the previous two: $\iint_R \psi_3 \binom{\psi_1}{\psi_2} \, dx \, dy = 0$. Thus,

$$\iint_R (\phi_3 + c_1\psi_1 + c_2\psi_2)\binom{\psi_1}{\psi_2} \, dx \, dy = 0.$$

However, ψ_2 is already orthogonal to ψ_1, and hence

$$\iint_R \phi_3\psi_1 \, dx \, dy + c_1 \iint_R \psi_1^2 \, dx \, dy = 0$$

$$\iint_R \phi_3\psi_2 \, dx \, dy + c_2 \iint_R \psi_2^2 \, dx \, dy = 0,$$

easily determining the two constants. This process can be used to determine n orthogonal eigenfunctions. In general,

$$\psi_j = \phi_j - \sum_{i=1}^{j-1} \left(\frac{\displaystyle\iint_R \phi_j \psi_i \, dx \, dy}{\displaystyle\iint_R \psi_i^2 \, dx \, dy} \right) \psi_i.$$

We have shown that even in the case of a multiple eigenvalue, we are always able to restrict our attention to orthogonal eigenfunctions, if necessary by this Gram–Schmidt construction.

6.6 RAYLEIGH QUOTIENT

In Sec. 5.6 we obtained the Rayleigh quotient, for the one-dimensional Sturm–Liouville eigenvalue problem. The result was obtained by multiplying the differential equation by ϕ, integrating over the entire region, solving for λ, and simplifying using integration by parts. We will derive a similar result for

$$\nabla^2 \phi + \lambda \phi = 0. \tag{6.6.1}$$

We proceed as before by multiplying (6.6.1) by ϕ. Integrating over the entire two-dimensional region and solving for λ yields

$$\lambda = \frac{-\iint_R \phi \, \nabla^2 \phi \, dx \, dy}{\iint_R \phi^2 \, dx \, dy}. \tag{6.6.2}$$

Next, we want to generalize integration by parts to multidimensional functions. Integration by parts is based on the product rule for the derivative, $d/dx\,(fg) = f\,dg/dx + g\,df/dx$. Instead of using the derivative, we use a product rule for the divergence, $\nabla \cdot (f\mathbf{g}) = f\nabla \cdot \mathbf{g} + \mathbf{g} \cdot \nabla f$. Letting $f = \phi$ and $\mathbf{g} = \nabla \phi$, it follows that $\nabla \cdot (\phi \, \nabla\phi) = \phi\nabla \cdot (\nabla\phi) + \nabla\phi \cdot \nabla\phi$. Since $\nabla \cdot (\nabla\phi) = \nabla^2\phi$ and $\nabla\phi \cdot \nabla\phi = |\nabla\phi|^2$,

$$\phi \, \nabla^2\phi = \nabla \cdot (\phi \, \nabla\phi) - |\nabla\phi|^2. \tag{6.6.3}$$

Using (6.6.3), (6.6.2) yields an alternative expression for the eigenvalue,

$$\lambda = \frac{-\iint_R \nabla \cdot (\phi \, \nabla\phi) \, dx \, dy + \iint_R |\nabla\phi|^2 \, dx \, dy}{\iint_R \phi^2 \, dx \, dy}. \tag{6.6.4}$$

Now we use (again) the divergence theorem to evaluate the first integral in the numerator of (6.6.4). Since $\iint_R \nabla \cdot \mathbf{A} \, dx \, dy = \oint \mathbf{A} \cdot \hat{\mathbf{n}} \, ds$, it follows that

$$\boxed{\lambda = \frac{-\oint \phi \, \nabla\phi \cdot \hat{\mathbf{n}} \, ds + \iint_R |\nabla\phi|^2 \, dx \, dy}{\iint_R \phi^2 \, dx \, dy},} \tag{6.6.5}$$

known as the **Rayleigh quotient**. This is quite similar to the Rayleigh quotient for Sturm–Liouville eigenvalue problems. Note that there is a boundary contribution for each: $-p\phi \, d\phi/dx\big|_a^b$ for (5.6.3) and $-\oint \phi \, \nabla\phi \cdot \hat{\mathbf{n}} \, ds$ for (6.6.5).

Example. We consider any region in which the boundary condition is $\phi = 0$ on the entire boundary. Then $\oint \phi \, \nabla\phi \cdot \hat{\mathbf{n}} \, ds = 0$, and hence from (6.6.5),

$\lambda \geqslant 0$. If $\lambda = 0$, then (6.6.5) implies that

$$0 = \iint_R |\nabla \phi|^2 \, dx \, dy. \tag{6.6.6}$$

Thus,

$$\nabla \phi = \frac{\partial \phi}{\partial x} \hat{\mathbf{i}} + \frac{\partial \phi}{\partial y} \hat{\mathbf{j}} = \mathbf{0}, \tag{6.6.7}$$

everywhere. From (6.6.7) it follows that $\partial \phi / \partial x = 0$ and $\partial \phi / \partial y = 0$ everywhere. Thus, ϕ is a constant everywhere, but since $\phi = 0$ on the boundary, $\phi = 0$ everywhere. $\phi = 0$ everywhere is not an eigenfunction, and thus we have shown that $\lambda = 0$ is not an eigenvalue. In conclusion $\lambda > 0$.

EXERCISES 6.6

6.6.1. See Exercise 6.5.1. Consider the two-dimensional eigenvalue problem with $\sigma > 0$

$$\nabla^2 \phi + \lambda \sigma(x, y)\phi = 0 \qquad \text{with } \phi = 0 \text{ on the boundary.}$$

(a) Prove that $\lambda \geqslant 0$.
(b) Is $\lambda = 0$ an eigenvalue, and if so, what is the eigenfunction?

6.6.2. Redo Exercise 6.6.1 if the boundary condition is instead
(a) $\nabla \phi \cdot \hat{\mathbf{n}} = 0$ on the boundary.
(b) $\nabla \phi \cdot \hat{\mathbf{n}} + h(x, y)\phi = 0$ on the boundary.
(c) $\phi = 0$ on part of the boundary and $\nabla \phi \cdot \hat{\mathbf{n}} = 0$ on the rest of the boundary.

6.6.3. Redo Exercise 6.6.1 if the differential equation is

$$\nabla \cdot (p \, \nabla \phi) + \lambda \sigma(x, y, z)\phi = 0$$

with boundary condition
(a) $\phi = 0$ on the boundary.
(b) $\nabla \phi \cdot \hat{\mathbf{n}} = 0$ on the boundary.

6.6.4. (a) If $\nabla^2 \phi = 0$ with $\phi = 0$ on the boundary, prove that $\phi = 0$ everywhere. (*Hint:* Use the fact that $\lambda = 0$ is not an eigenvalue for $\nabla^2 \phi = -\lambda \phi$.)
(b) Prove that there cannot be two different solutions of the problem

$$\nabla^2 u = f(x, y, z)$$

subject to the given boundary condition $u = g(x, y, z)$ on the boundary. [*Hint:* Consider $u_1 - u_2$ and use part (a).]

6.7 VIBRATING CIRCULAR MEMBRANE AND BESSEL FUNCTIONS

6.7.1 Introduction

An interesting application of both one-dimensional (Sturm–Liouville) and multi-dimensional eigenvalue problems occurs when considering the vibrations of a circular membrane. The vertical displacement u satisfies the two-dimensional wave equation,

$$\text{PDE:} \qquad \boxed{\frac{\partial^2 u}{\partial t^2} = c^2 \, \nabla^2 u.} \qquad\qquad (6.7.1)$$

The geometry suggests that we use polar coordinates, in which case $u = u(r, \theta, t)$. We assume that the membrane has zero displacement at the circular boundary, $r = a$:

$$\text{BC:} \qquad \boxed{u(a, \theta, t) = 0.} \qquad\qquad (6.7.2)$$

The initial position and velocity are given:

$$\text{IC:} \qquad \boxed{\begin{aligned} u(r, \theta, 0) &= \alpha(r, \theta) \\ \frac{\partial u}{\partial t}(r, \theta, 0) &= \beta(r, \theta). \end{aligned}} \qquad\qquad \begin{aligned} &(6.7.3\text{a}) \\[6pt] &(6.7.3\text{b}) \end{aligned}$$

6.7.2 Separation of Variables

We first separate out the time variable by seeking product solutions,

$$u(r, \theta, t) = \phi(r, \theta)h(t). \qquad\qquad (6.7.4)$$

Then, as shown earlier, $h(t)$ satisfies

$$\frac{d^2 h}{dt^2} = -\lambda c^2 h, \qquad\qquad (6.7.5)$$

where λ is a separation constant. From (6.7.5), the natural frequencies of vibration are $c\sqrt{\lambda}$ (if $\lambda > 0$). In addition, $\phi(r, \theta)$ satisfies the two-dimensional eigenvalue problem

$$\boxed{\nabla^2 \phi + \lambda\phi = 0,} \qquad\qquad (6.7.6)$$

with $\phi = 0$ on the entire boundary, $r = a$:

$$\phi(a, \theta) = 0. \qquad\qquad (6.7.7)$$

We will attempt to obtain solutions of (6.7.6) in the product form appropriate for polar coordinates,

$$\phi(r, \theta) = f(r)g(\theta), \qquad\qquad (6.7.8)$$

since for the circular membrane $0 < r < a$, $-\pi < \theta < \pi$. This is equivalent to originally seeking solutions to the wave equation in the form of a product of functions of each independent variable, $u(r, \theta, t) = f(r)g(\theta)h(t)$. We substitute

(6.7.8) into (6.7.6); in polar coordinates $\nabla^2\phi = 1/r\,\partial/\partial r(r\,\partial\phi/\partial r) + 1/r^2\,\partial^2\phi/\partial\theta^2$, and thus $\nabla^2\phi + \lambda\phi = 0$ becomes

$$\frac{g(\theta)}{r}\frac{d}{dr}\left(r\frac{df}{dr}\right) + \frac{f(r)}{r^2}\frac{d^2g}{d\theta^2} + \lambda f(r)g(\theta) = 0. \tag{6.7.9}$$

r and θ may be separated by multiplying by r^2 and dividing by $f(r)g(\theta)$:

$$-\frac{1}{g}\frac{d^2g}{d\theta^2} = \frac{r}{f}\frac{d}{dr}\left(r\frac{df}{dr}\right) + \lambda r^2 = \mu. \tag{6.7.10}$$

We introduce a second separation constant in the form μ because our experience with circular regions (see Secs. 2.4.2 and 2.5.2) suggests that $g(\theta)$ must oscillate in order to satisfy the periodic conditions in θ. Our three differential equations, with two separation constants, are thus

$$\frac{d^2h}{dt^2} = -\lambda c^2 h \tag{6.7.11a}$$

$$\frac{d^2g}{d\theta^2} = -\mu g \tag{6.7.11b}$$

$$r\frac{d}{dr}\left(r\frac{df}{dr}\right) + (\lambda r^2 - \mu)f = 0. \tag{6.7.11c}$$

Two of these equations must be eigenvalue problems. However, ignoring the initial conditions, the only given boundary condition is $f(a) = 0$, which follows from $u(a, \theta, t) = 0$ or $\phi(a, \theta) = 0$. We must remember that $-\pi < \theta < \pi$ and $0 < r < a$. Thus, both θ and r are defined over finite intervals. As such there should be boundary conditions at both ends. The periodic nature of the solution in θ implies that

$$g(-\pi) = g(\pi) \tag{6.7.12a}$$

$$\frac{dg}{d\theta}(-\pi) = \frac{dg}{d\theta}(\pi). \tag{6.7.12b}$$

We already have a condition at $r = a$. Polar coordinates are singular at $r = 0$; a singularity condition must be introduced there. Since the displacement must be finite, we conclude that

$$|f(0)| < \infty.$$

6.7.3 Eigenvalue Problems (One-Dimensional)

After separating variables, we have obtained two eigenvalue problems. We are quite familiar with the θ-eigenvalue problem, (6.7.11b) with (6.7.12). Although it is not a regular Sturm–Liouville problem due to the periodic boundary conditions, we know that the eigenvalues are

$$\mu_m = m^2, \qquad m = 0, 1, 2, \ldots. \tag{6.7.13}$$

The corresponding eigenfunctions are both

$$g(\theta) = \sin m\theta \quad \text{and} \quad g(\theta) = \cos m\theta, \tag{6.7.14}$$

although for $m = 0$ this reduces to one eigenfunction (not two as for $m \neq 0$). This eigenvalue problem generates a full Fourier series in θ, as we already know. m is the number of crests in the θ-direction.

For each integral value of m, (6.7.11c) helps to define an eigenvalue problem for λ:

$$r \frac{d}{dr}\left(r \frac{df}{dr}\right) + (\lambda r^2 - m^2)f = 0 \tag{6.7.15a}$$

$$f(a) = 0 \tag{6.7.15b}$$

$$|f(0)| < \infty. \tag{6.7.15c}$$

Since (6.7.15a) has nonconstant coefficients it is not surprising that (6.7.15a) is somewhat difficult to analyze. Equation (6.7.15a) can be put in the Sturm–Liouville form by dividing it by r:

$$\frac{d}{dr}\left(r \frac{df}{dr}\right) + \left(\lambda r - \frac{m^2}{r}\right)f = 0, \tag{6.7.16}$$

or $Lf + \lambda rf = 0$, where $L = d/dr \, (r \, d/dr) - m^2/r$. By comparison to the general Sturm–Liouville differential equation,

$$\frac{d}{dx}\left[p(x)\frac{d\phi}{dx}\right] + q\phi + \lambda\sigma\phi = 0,$$

with independent variable r, we have that $x = r$, $p(r) = r$, $\sigma(r) = r$, and $q(r) = -m^2/r$. Our problem is not a regular Sturm–Liouville problem due to the behavior at the origin ($r = 0$):

1. The boundary condition at $r = 0$, (6.7.15c), is not of the correct form.
2. $p(r) = 0$ and $\sigma(r) = 0$ at $r = 0$ (and hence is not positive everywhere).
3. $q(r)$ approaches ∞ as $r \to 0$ [and hence $q(r)$ is not continuous] for $m \neq 0$.

However, we claim that all the statements concerning regular Sturm–Liouville problems are still valid for this important singular Sturm–Liouville problem. To begin with there are an infinite number of eigenvalues (for each m). We designate the eigenvalues as λ_{mn}, where $m = 0, 1, 2, \ldots$ and $n = 1, 2, \ldots$, and the eigenfunctions $f_{mn}(r)$. *For each fixed m*, these eigenfunctions are orthogonal with weight r [see (6.7.16)], since it can be shown that the boundary terms vanish in Green's formula (see Exercise 5.5.1). Thus,

$$\int_0^a f_{mn_1} f_{mn_2} r \, dr = 0 \qquad \text{for } n_1 \neq n_2. \qquad (6.7.17)$$

Shortly, we will state more explicit facts about these eigenfunctions.

6.7.4 Bessel's Differential Equation

The r-dependent separation of variables solution satisfies a ''singular'' Sturm–Liouville differential equation, (6.7.16). An alternative form is obtained by using the product rule of differentiation and by multiplying by r:

$$\boxed{r^2 \frac{d^2f}{dr^2} + r \frac{df}{dr} + (\lambda r^2 - m^2)f = 0.} \qquad (6.7.18)$$

There is some additional analysis of (6.7.18) that can be performed. Equation (6.7.18) contains two parameters, m and λ. We already know that m is an integer, but the allowable values of λ are as yet unknown. It would be quite tedious to numerically solve (6.7.18) for various values of λ (for different integral values of m). Instead, we might notice that the simple scaling transformation,

$$\boxed{z = \sqrt{\lambda}\, r,} \qquad (6.7.19)$$

removes the dependence of the differential equation on λ:

$$\boxed{z^2 \frac{d^2f}{dz^2} + z \frac{df}{dz} + (z^2 - m^2)f = 0.} \qquad (6.7.20)$$

We note that the change of variables (6.7.19) may be performed since we showed in Sec. 6.6 from the multidimensional Rayleigh quotient that $\lambda > 0$* (for $\nabla^2 \phi + \lambda \phi = 0$) anytime $\phi = 0$ on the entire boundary, as it is here. We can also show that $\lambda > 0$ for this problem using the one-dimensional Rayleigh quotient based on (6.7.15) (see Exercise 6.7.13). Equation (6.7.20) has the advantage of not depending on λ; less work is necessary to compute solutions of (6.7.20) than of (6.7.18). However, we have gained more than that since (6.7.20) has been investigated for over 150 years. It is now known as **Bessel's differential equation of order** m.

6.7.5 Singular Points and Bessel's Differential Equation

In this subsection we *briefly* develop some of the properties of Bessel's differential equation. Equation (6.7.20) is a second-order linear differential equation with variable coefficients. We will not be able to obtain an exact closed-form solution

---

* In other problems if $\lambda = 0$, then the transformation (6.7.19) is invalid. However, (6.7.19) is unnecessary for $\lambda = 0$ since in this case (6.7.18) becomes an equidimensional equation and can be solved (as in Sec. 2.5.2).

of (6.7.20) *involving elementary functions.* To analyze a differential equation, one of the first things we should do is search for any special values of z that might cause some difficulties. $z = 0$ is a singular point of (6.7.20).

Perhaps we should define a singular point of a differential equation. We refer to the standard form:

$$\frac{d^2f}{dz^2} + a(z)\frac{df}{dz} + b(z)f = 0.$$

If $a(z)$ and $b(z)$ and all their derivatives are finite at $z = z_0$, then $z = z_0$ is called an ordinary point. Otherwise, $z = z_0$ is a singular point. For Bessel's differential equation, $a(z) = 1/z$ and $b(z) = 1 - m^2/z^2$. All finite z^* except $z = 0$ are ordinary points. $z = 0$ is a singular point [since, for example, $a(0)$ does not exist].

In the neighborhood of any ordinary point, it is known from the theory of differential equations that all solutions of the differential equation are well behaved [i.e., $f(z)$ and all its derivatives exist at any ordinary point]. We thus are guaranteed that all solutions of Bessel's differential equation are well behaved at every finite point except possibly at $z = 0$. The only difficulty can occur in the neighborhood of $z = 0$. We will investigate the expected behavior of solutions of Bessel's differential equation in the neighborhood of $z = 0$. We will describe a crude (but important) approximation. If z is very close to 0, then we should expect that z^2f in Bessel's differential equation can be ignored, since it is much smaller than m^2f.† We do not ignore $z^2\, d^2f/dz^2$ or $z\, df/dz$ because although z is small, it is possible that derivatives of f are large enough so that $z\, df/dz$ is as large as $-m^2f$. Dropping z^2f yields

$$z^2\frac{d^2f}{dz^2} + z\frac{df}{dz} - m^2f \approx 0, \qquad (6.7.21)$$

a valid approximation near $z = 0$. The advantage of this *approximation* is that (6.7.21) is exactly solvable, since it is an equidimensional (also known as a Cauchy or Euler) equation (see Sec. 2.5.2). Equation (6.7.21) can be solved by seeking solutions in the form

$$f \approx z^s. \qquad (6.7.22)$$

By substituting (6.7.22) into (6.7.21) we obtain a quadratic equation for s,

$$s(s - 1) + s - m^2 = 0, \qquad (6.7.23)$$

known as the **indicial equation**. Thus, $s^2 = m^2$, and the two roots (indices) are $s = \pm m$. If $m \neq 0$ (in which case we assume $m > 0$), then we obtain two independent approximate solutions,

$$f \approx z^m \quad \text{and} \quad f \approx z^{-m} \quad (m > 0). \qquad (6.7.24)$$

---

* With an appropriate definition, it can be shown that $z = \infty$ is *not* an ordinary point for Bessel's differential equation.

† Even if $m = 0$, we still claim that z^2f can be neglected near $z = 0$ and the result will give a reasonable approximation.

However, if $m = 0$, we only obtain one independent solution $f \approx z^0 = 1$. A second solution is easily derived from (6.7.21). If $m = 0$,

$$z^2 \frac{d^2f}{dz^2} + z \frac{df}{dz} \approx 0 \quad \text{or} \quad z \frac{d}{dz}\left(z \frac{df}{dz}\right) \approx 0.$$

Thus, $z\, df/dz$ is constant and, in addition to $f \approx 1$, it is also possible for $f \approx \ln z$. In summary, for $m = 0$, two independent solutions have the expected behavior near $z = 0$,

$$f \approx 1 \quad \text{and} \quad f \approx \ln z \quad (m = 0). \tag{6.7.25}$$

The general solution of Bessel's differential equation will be a linear combination of two independent solutions, satisfying (6.7.24) if $m \neq 0$ and (6.7.25) if $m = 0$. We have only obtained the expected approximate behavior near $z = 0$. More will be discussed in the next subsection. Because of the singular point at $z = 0$, it is possible for solutions not to be well behaved at $z = 0$. We see from (6.7.24) and (6.7.25) that independent solutions of Bessel's differential equation can be chosen such that one is well behaved at $z = 0$ and one solution is not well behaved at $z = 0$ [note that for one solution $\lim_{z \to 0} f(z) = \pm\infty$].

6.7.6 Bessel Functions and Their Asymptotic Properties (near z = 0)

We continue to discuss Bessel's differential equation of order m,

$$z^2 \frac{d^2f}{dz^2} + z \frac{df}{dz} + (z^2 - m^2)f = 0. \tag{6.7.26}$$

Near $z = 0$, there are well-behaved and singular solutions. Different values of m yield a different differential equation. Its corresponding solution will depend on m. We introduce the standard notation for a *well-behaved solution* of (6.7.26), $J_m(z)$, called the **Bessel function of the first kind of order m**. In a similar vein, we introduce the notation for a singular solution of Bessel's differential equation, $Y_m(z)$, called the **Bessel function of the second kind of order m**. You can solve a lot of problems using Bessel's differential equation by just remembering that $Y_m(z)$ *approaches $\pm\infty$ as $z \to 0$.*

The general solution of any linear homogeneous second-order differential equation is a linear combination of two independent solutions. Thus, the general solution of Bessel's differential equation (6.7.26) is

$$\boxed{f = c_1 J_m(z) + c_2 Y_m(z).} \tag{6.7.27}$$

Precise definitions of $J_m(z)$ and $Y_m(z)$ are given in Sec. 6.8. However, for our immediate purposes, we simply note that they satisfy the following asymptotic properties *for small z ($z \to 0$):*

$$J_m(z) \sim \begin{cases} 1 & m = 0 \\ \dfrac{1}{2^m m!} z^m & m > 0 \end{cases}$$

$$Y_m(z) \sim \begin{cases} \dfrac{2}{\pi} \ln z & m = 0 \\ -\dfrac{2^m(m-1)!}{\pi} z^{-m} & m > 0. \end{cases} \tag{6.7.28}$$

It should be seen that (6.7.28) is consistent with our approximate behavior, (6.7.24) and (6.7.25). We see that $J_m(z)$ is bounded as $z \to 0$, whereas $Y_m(z)$ is not.

6.7.7 Eigenvalue Problem Involving Bessel Functions

In this section we determine the eigenvalues of the singular Sturm–Liouville problem (m fixed):

$$\frac{d}{dr}\left(r\frac{df}{dr}\right) + \left(\lambda r - \frac{m^2}{r}\right)f = 0 \tag{6.7.29a}$$

$$f(a) = 0 \tag{6.7.29b}$$

$$|f(0)| < \infty. \tag{6.7.29c}$$

By the change of variables $z = \sqrt{\lambda}\, r$, (6.7.29a) becomes Bessel's differential equation,

$$z^2 \frac{d^2 f}{dz^2} + z\frac{df}{dz} + (z^2 - m^2)f = 0.$$

The general solution is a linear combination of Bessel functions, $f = c_1 J_m(z) + c_2 Y_m(z)$. The scale change implies that in terms of the radial coordinate r,

$$f = c_1 J_m(\sqrt{\lambda}\, r) + c_2 Y_m(\sqrt{\lambda}\, r). \tag{6.7.30}$$

Applying the homogeneous boundary conditions (6.7.29b) and (6.7.29c) will determine the eigenvalues. $f(0)$ must be finite. However, $Y_m(0)$ is infinite. Thus, $c_2 = 0$, implying that

$$f = c_1 J_m(\sqrt{\lambda}\, r). \tag{6.7.31}$$

Thus, the condition $f(a) = 0$ determines the eigenvalues:

$$\boxed{J_m(\sqrt{\lambda}\, a) = 0.} \qquad (6.7.32)$$

We see that $\sqrt{\lambda}\, a$ must be a zero of the Bessel function $J_m(z)$. Later in Sec. 6.8.1, we show that a Bessel function is a decaying oscillation. There are an infinite number of zeros of each Bessel function $J_m(z)$. Let z_{mn} designate the nth zero of $J_m(z)$. Then

$$\sqrt{\lambda}\, a = z_{mn} \qquad \text{or} \qquad \lambda_{mn} = \left(\frac{z_{mn}}{a}\right)^2. \qquad (6.7.33)$$

For each m, there are an infinite number of eigenvalues. (6.7.33) is analogous to $\lambda = (n\pi/L)^2$, where $n\pi$ are the zeros of $\sin x$.

Example. Consider $J_0(z)$, sketched in detail in Fig. 6.7.1. From accurate tables,* it is known that the first zero of $J_0(z)$ is $z = 2.4048255577 \ldots$. Other zeros are recorded in Fig. 6.7.1. The eigenvalues are $\lambda_{0_n} = (z_{0_n}/a)^2$.

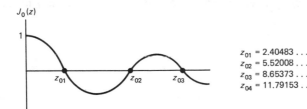

$$z_{01} = 2.40483 \ldots$$
$$z_{02} = 5.52008 \ldots$$
$$z_{03} = 8.65373 \ldots$$
$$z_{04} = 11.79153 \ldots$$

Figure 6.7.1 Sketch of $J_0(z)$ and its zeros.

Eigenfunctions.   The eigenfunctions are thus

$$J_m(\sqrt{\lambda_{mn}}\, r) = J_m\left(z_{mn}\frac{r}{a}\right), \qquad (6.7.34)$$

for $m = 0, 1, 2, \ldots, n = 1, 2, \ldots$. For each m, these are an infinite set of eigenfunctions for the singular Sturm–Liouville problem, (6.7.29). *For fixed m* they are orthogonal with weight r [as already discussed, see (6.7.17)]:

$$\int_0^a J_m(\sqrt{\lambda_{mp}}\, r)J_m(\sqrt{\lambda_{mq}}\, r)r\, dr = 0, \qquad p \neq q. \qquad (6.7.35)$$

It is known that this infinite set of eigenfunctions (m fixed) is complete. Thus, any piecewise smooth function of r can be represented by a generalized Fourier series of the eigenfunctions:

$$\alpha(r) = \sum_{n=1}^{\infty} a_n J_m(\sqrt{\lambda_{mn}}\, r), \qquad (6.7.36)$$

---

* Separate tables of the zeros are available. The *Handbook of Mathematical Functions* (Abramowitz and Stegun [1965]) is one source. Alternatively, over 700 pages are devoted to Bessel functions in *A Treatise on the Theory of Bessel Functions* by Watson [1966].

where m is fixed. This is sometimes known as a **Fourier–Bessel series.** The coefficients can be determined by the orthogonality of the Bessel functions (with weight r):

$$a_n = \frac{\int_0^a \alpha(r) J_m(\sqrt{\lambda_{mn}}\, r) r\, dr}{\int_0^a J_m^2(\sqrt{\lambda_{mn}}\, r) r\, dr}. \tag{6.7.37}$$

This illustrates the one-dimensional orthogonality of the Bessel functions. We omit the evaluation of the **normalization integrals** $\int_0^a J_m^2(\sqrt{\lambda_{mn}}\, r) r\, dr$ (e.g., see Churchill [1972] and Berg and McGregor [1966]).

6.7.8 Initial Value Problem for a Vibrating Circular Membrane

The vibrations $u(r, \theta, t)$ of a circular membrane are described by the two-dimensional wave equation, (6.7.1), with u being fixed on the boundary, (6.7.2), subject to the initial conditions (6.7.3). When we apply the method of separation of variables, we obtain four families of product solutions, $u(r, \theta, t) = f(r)g(\theta)h(t)$:

$$J_m(\sqrt{\lambda_{mn}}\, r) \left\{ \begin{array}{c} \cos m\,\theta \\ \sin m\,\theta \end{array} \right\} \left\{ \begin{array}{c} \cos c\sqrt{\lambda_{mn}}\, t \\ \sin c\sqrt{\lambda_{mn}}\, t \end{array} \right\}. \tag{6.7.38}$$

To simplify the algebra, we will assume that the membrane is initially at rest,

$$\frac{\partial u}{\partial t}(r, \theta, 0) = \beta(r, \theta) = 0.$$

Thus, the $\sin c\sqrt{\lambda_{mn}}\, t$ terms in (6.7.38) will not be necessary. Then according to the principle of superposition, we attempt to satisfy the initial value problem by considering the infinite linear combination of the remaining product solutions:

$$\begin{aligned} u(r, \theta, t) = &\sum_{m=0}^{\infty} \sum_{n=1}^{\infty} A_{mn} J_m(\sqrt{\lambda_{mn}}\, r) \cos m\theta \cos c\sqrt{\lambda_{mn}}\, t \\ &+ \sum_{m=1}^{\infty} \sum_{n=1}^{\infty} B_{mn} J_m(\sqrt{\lambda_{mn}}\, r) \sin m\theta \cos c\sqrt{\lambda_{mn}}\, t. \end{aligned} \tag{6.7.39}$$

The initial position $u(r, \theta, 0) = \alpha(r, \theta)$ implies that

$$\begin{aligned} \alpha(r, \theta) = &\sum_{m=0}^{\infty} \left(\sum_{n=1}^{\infty} A_{mn} J_m(\sqrt{\lambda_{mn}}\, r) \right) \cos m\theta \\ &+ \sum_{m=1}^{\infty} \left(\sum_{n=1}^{\infty} B_{mn} J_m(\sqrt{\lambda_{mn}}\, r) \right) \sin m\theta. \end{aligned} \tag{6.7.40}$$

By properly arranging the terms in (6.7.40), we see that this is an ordinary Fourier series in θ. Their Fourier coefficients are Fourier–Bessel series (note that m is fixed). Thus, the coefficients may be determined by the orthogonality of $J_m(\sqrt{\lambda_{mn}} \, r)$ with weight r [as in (6.7.37)]. As such we can determine the coefficients by repeated application of one-dimensional orthogonality. Two families of coefficients A_{mn} and B_{mn} (including $m = 0$) can be determined from one initial condition since the periodicity in θ yielded two eigenfunctions corresponding to each eigenvalue.

However, it is somewhat easier to determine all the coefficients using two-dimensional orthogonality. Recall that for the two-dimensional eigenvalue problem,

$$\nabla^2 \phi + \lambda \phi = 0$$

with $\phi = 0$ on the circle of radius a, the two-dimensional eigenfunctions are the doubly infinite families

$$\phi_\lambda(r, \theta) = J_m(\sqrt{\lambda_{mn}} \, r) \begin{Bmatrix} \cos m\theta \\ \sin m\theta \end{Bmatrix}.$$

Thus,

$$\alpha(r, \theta) = \sum_\lambda A_\lambda \phi_\lambda(r, \theta), \qquad (6.7.41)$$

where $\sum_\lambda$ stands for a summation over all eigenfunctions [actually two double sums, including both $\sin m\theta$ and $\cos m\theta$ as in (6.7.40)]. These eigenfunctions $\phi_\lambda(r, \theta)$ are orthogonal (in a two-dimensional sense) with weight 1. We then immediately calculate A_λ (representing both A_{mn} and B_{mn}),

$$A_\lambda = \frac{\displaystyle\iint \alpha(r, \theta)\phi_\lambda(r, \theta) \, dA}{\displaystyle\iint \phi_\lambda^2(r, \theta) \, dA}. \qquad (6.7.42)$$

Here $dA = r \, dr \, d\theta$. In two dimensions the weighting function is constant. However, for geometric reasons $dA = r \, dr \, d\theta$. Thus, **the weight r that appears in the one-dimensional orthogonality of Bessel functions is just a geometric factor.**

6.7.9 Circularly Symmetric Case

In this subsection, as an example, we consider the vibrations of a circular membrane, with $u = 0$ on the circular boundary, in the case in which the initial conditions are circularly symmetric (meaning independent of θ). We could consider this as a special case of the general problem, analyzed in Sec. 6.7.8. An alternative method, which yields the same result, is to reformulate the problem. The symmetry of the problem, including the initial conditions, suggests that the entire solution should be circularly symmetric; there should be no dependence on the angle θ. Thus,

$$u(r, t) \quad \text{and} \quad \nabla^2 u = \frac{1}{r}\frac{\partial}{\partial r}\left(r \frac{\partial u}{\partial r}\right) \quad \text{since} \quad \frac{\partial^2 u}{\partial \theta^2} = 0.$$

The mathematical formulation is thus

PDE:
$$\frac{\partial^2 u}{\partial t^2} = \frac{c^2}{r}\frac{\partial}{\partial r}\left(r\frac{\partial u}{\partial r}\right)$$
(6.7.43a)

BC:
$$u(a, t) = 0$$
(6.7.43b)

IC:
$$u(r, 0) = \alpha(r)$$
(6.7.43c)

$$\frac{\partial u}{\partial t}(r, 0) = \beta(r).$$
(6.7.43d)

We note that the partial differential equation has two independent variables; we need not study this problem in this chapter reserved for more than two independent variables. We could have analyzed this problem earlier. However, as we will see, Bessel functions are the radially dependent functions, and thus it is more natural to discuss this problem in the present part of this text.

We will apply the method of separation of variables to (6.7.43). Looking for product solutions,

$$u(r, t) = \phi(r)h(t),$$
(6.7.44)

yields

$$\frac{1}{c^2}\frac{1}{h}\frac{d^2h}{dt^2} = \frac{1}{r\phi}\frac{d}{dr}\left(r\frac{d\phi}{dr}\right) = -\lambda,$$
(6.7.45)

where $-\lambda$ is introduced because we suspect that the displacement oscillates in time. The time-dependent equation,

$$\frac{d^2h}{dt^2} = -\lambda c^2 h$$

has solutions $\sin c\sqrt{\lambda}\, t$ and $\cos c\sqrt{\lambda}\, t$ if $\lambda > 0$. The eigenvalue problem for the separation constant is

$$\frac{d}{dr}\left(r\frac{d\phi}{dr}\right) + \lambda r\phi = 0$$
(6.7.46a)

$$\phi(a) = 0$$
(6.7.46b)

$$|\phi(0)| < \infty.$$
(6.7.46c)

Since (6.7.46a) is in the form of a Sturm–Liouville problem, we immediately know that eigenfunctions corresponding to distinct eigenvalues are orthogonal with weight r.

From the Rayleigh quotient we could show that $\lambda > 0$. Thus, we may use the transformation

$$z = \sqrt{\lambda}\, r, \tag{6.7.47}$$

in which case, (6.7.46a) becomes

$$\frac{d}{dz}\left(z\,\frac{d\phi}{dz}\right) + z\phi = 0 \quad \text{or} \quad z^2\frac{d^2\phi}{dz^2} + z\frac{d\phi}{dz} + z^2\phi = 0. \tag{6.7.48}$$

We may recall that Bessel's differential equation of order m is

$$z^2\frac{d^2\phi}{dz^2} + z\frac{d\phi}{dz} + (z^2 - m^2)\phi = 0, \tag{6.7.49}$$

with solutions being Bessel functions of order m, $J_m(z)$ and $Y_m(z)$. A comparison with (6.7.49) shows that (6.7.48) is Bessel's differential equation of order 0. The general solution of (6.7.48) is thus a linear combination of the zeroth-order Bessel functions:

$$\phi = c_1 J_0(z) + c_2 Y_0(z) = c_1 J_0(\sqrt{\lambda}\, r) + c_2 Y_0(\sqrt{\lambda}\, r), \tag{6.7.50}$$

in terms of the radial variable. The singularity condition at the origin (6.7.46c) shows that $c_2 = 0$, since $Y_0(\sqrt{\lambda}\, r)$ has a logarithmic singularity at $r = 0$:

$$\phi = c_1 J_0(\sqrt{\lambda}\, r). \tag{6.7.51}$$

Finally, the eigenvalues are determined by the condition at $r = a$, (6.7.46b), in which case

$$J_0(\sqrt{\lambda}\, a) = 0. \tag{6.7.52}$$

Thus, $\sqrt{\lambda}\, a$ must be a zero of the zeroth Bessel function. We thus obtain an infinite number of eigenvalues, which we label $\lambda_1, \lambda_2, \ldots$.

We have obtained two infinite families of product solutions

$$J_0(\sqrt{\lambda_n}\, r)\sin c\sqrt{\lambda_n}\, t \quad \text{and} \quad J_0(\sqrt{\lambda_n}\, r)\cos c\sqrt{\lambda_n}\, t.$$

According to the principle of superposition, we seek solutions to our original problem, (6.7.43), in the form

$$u(r, t) = \sum_{n=1}^{\infty} a_n J_0(\sqrt{\lambda_n}\, r)\cos c\sqrt{\lambda_n}\, t + \sum_{n=1}^{\infty} b_n J_0(\sqrt{\lambda_n}\, r)\sin c\sqrt{\lambda_n}\, t. \tag{6.7.53}$$

As before, we determine the coefficients a_n and b_n from the initial conditions. $u(r, 0) = \alpha(r)$ implies that

$$\alpha(r) = \sum_{n=1}^{\infty} a_n J_0(\sqrt{\lambda_n}\, r). \tag{6.7.54}$$

The coefficients a_n are thus the Fourier–Bessel coefficients (of order 0) of $\alpha(r)$. Since $J_0(\sqrt{\lambda_n}\, r)$ forms an orthogonal set with weight r, we can easily determine a_n,

$$a_n = \frac{\displaystyle\int_0^a \alpha(r)J_0(\sqrt{\lambda_n}\, r)r\, dr}{\displaystyle\int_0^a J_0^2(\sqrt{\lambda_n}\, r)r\, dr.} \tag{6.7.55}$$

In a similar manner, the initial condition $\partial/\partial t\, u(r, 0) = \beta(r)$ determines b_n.

EXERCISES 6.7

*6.7.1. Solve as simply as possible:

$$\frac{\partial^2 u}{\partial t^2} = c^2 \nabla^2 u$$

with $u(a, \theta, t) = 0$, $u(r, \theta, 0) = 0$, and $\dfrac{\partial u}{\partial t}(r, \theta, 0) = \alpha(r) \sin 3\theta$.

6.7.2. Solve as simply as possible:

$$\frac{\partial^2 u}{\partial t^2} = c^2 \nabla^2 u \quad \text{subject to} \quad \frac{\partial u}{\partial r}(a, \theta, t) = 0$$

with initial conditions

(a) $u(r, \theta, 0) = 0$

$\dfrac{\partial u}{\partial t}(r, \theta, 0) = \beta(r) \cos 5\theta$

(b) $u(r, \theta, 0) = 0$

$\dfrac{\partial u}{\partial t}(r, \theta, 0) = \beta(r)$

(c) $u(r, \theta, 0) = \alpha(r, \theta)$

$\dfrac{\partial u}{\partial t}(r, \theta, 0) = 0$

*(d) $u(r, \theta, 0) = 0$

$\dfrac{\partial u}{\partial t}(r, \theta, 0) = \beta(r, \theta)$

6.7.3. Consider a vibrating quarter-circular membrane, $0 < r < a$, $0 < \theta < \pi/2$, with $u = 0$ on the entire boundary.
  *(a) Determine an expression for the frequencies of vibration.
  (b) Solve the initial value problem if

$$u(r, \theta, 0) = g(r, \theta)$$

$$\frac{\partial u}{\partial t}(r, \theta, 0) = 0.$$

6.7.4. Consider the displacement $u(r, \theta, t)$ of a "pie-shaped" membrane of radius a (and angle $\pi/3 = 60°$) which satisfies

$$\frac{\partial^2 u}{\partial t^2} = c^2 \nabla^2 u.$$

Assume that $\lambda > 0$. Determine the natural frequencies of oscillation if the boundary conditions are:

(a) $u(r, 0, t) = 0$ $\dfrac{\partial u}{\partial r}(a, \theta, t) = 0$
  $u(r, \pi/3, t) = 0$

(b) $u(r, 0, t) = 0$
  $u(r, \pi/3, t) = 0$ $u(a, \theta, t) = 0$

*6.7.5. Consider the displacement $u(r, \theta, t)$ of a membrane whose shape is a 90° sector of an annulus, $a < r < b$, $0 < \theta < \pi/2$, with the conditions that $u = 0$ on the entire boundary. Determine the natural frequencies of vibration.

6.7.6. Consider the circular membrane satisfying

$$\frac{\partial^2 u}{\partial t^2} = c^2 \nabla^2 u$$

subject to the boundary condition

$$u(a, \theta, t) = -\frac{\partial u}{\partial r}(a, \theta, t).$$

(a) Show that this membrane only oscillates.
(b) Obtain an expression that determines the natural frequencies.
(c) Solve the initial value problem if

$$u(r, \theta, 0) = 0$$

$$\frac{\partial u}{\partial t}(r, \theta, 0) = \alpha(r) \sin 3\theta.$$

6.7.7. Solve the heat equation

$$\frac{\partial u}{\partial t} = k \nabla^2 u$$

inside a circle of radius a with zero temperature around the entire boundary, if initially

$$u(r, \theta, 0) = f(r, \theta).$$

Briefly analyze $\lim_{t \to \infty} u(r, \theta, t)$. Compare this to what you expect to occur using physical reasoning as $t \to \infty$.

*6.7.8. Reconsider Exercise 6.7.7, but with the entire boundary insulated.

6.7.9. Solve the heat equation

$$\frac{\partial u}{\partial t} = k \nabla^2 u$$

inside a semicircle of radius a and briefly analyze the $\lim_{t \to \infty}$ if the initial conditions are

$$u(r, \theta, 0) = f(r, \theta)$$

and the boundary conditions are

(a)
$$u(r, 0, t) = 0$$
$$u(r, \pi, t) = 0$$
$$\frac{\partial u}{\partial r}(a, \theta, t) = 0$$

*(b)
$$\frac{\partial u}{\partial \theta}(r, 0, t) = 0$$
$$\frac{\partial u}{\partial \theta}(r, \pi, t) = 0$$
$$\frac{\partial u}{\partial r}(a, \theta, t) = 0$$

(c)
$$\frac{\partial u}{\partial \theta}(r, 0, t) = 0$$
$$\frac{\partial u}{\partial \theta}(r, \pi, t) = 0$$
$$u(a, \theta, t) = 0$$

(d)
$$u(r, 0, t) = 0$$
$$u(r, \pi, t) = 0$$
$$u(r, \theta, t) = 0$$

*6.7.10. Solve for $u(r, t)$ if it satisfies the circularly symmetric heat equation

$$\frac{\partial u}{\partial t} = k \frac{1}{r} \frac{\partial}{\partial r}\left(r \frac{\partial u}{\partial r}\right)$$

subject to the conditions

$$u(a, t) = 0$$
$$u(r, 0) = f(r).$$

Briefly analyze the $\lim_{t \to \infty}$.

6.7.11. Reconsider Exercise 6.7.10 with the boundary condition

$$\frac{\partial u}{\partial r}(a, t) = 0.$$

6.7.12. For the following differential equations, what is the expected approximate behavior of all solutions *near x* $= 0$?

*(a) $x^2 \dfrac{d^2y}{dx^2} + (x - 6)y = 0$ (b) $x^2 \dfrac{d^2y}{dx^2} + \left(x^2 + \dfrac{3}{16}\right)y = 0$

*(c) $x^2 \dfrac{d^2y}{dx^2} + (x + x^2) \dfrac{dy}{dx} + 4y = 0$ (d) $x^2 \dfrac{d^2y}{dx^2} + (x + x^2) \dfrac{dy}{dx} - 4y = 0$

*(e) $x^2 \dfrac{d^2y}{dx^2} - 4x \dfrac{dy}{dx} + (6 + x^3)y = 0$ (f) $x^2 \dfrac{d^2y}{dx^2} + \left(x + \dfrac{1}{4}\right)y = 0$

6.7.13. Using the one-dimensional Rayleigh quotient, show that $\lambda > 0$ as defined by (6.7.15).

6.8 MORE ON BESSEL FUNCTIONS

6.8.1 Qualitative Properties of Bessel Functions

It is helpful to have some understanding of the sketch of Bessel functions. Let us rewrite Bessel's differential equation as

$$\frac{d^2f}{dz^2} = -\left(1 - \frac{m^2}{z^2}\right)f - \frac{1}{z}\frac{df}{dz}, \tag{6.8.1}$$

in order to compare it with the equation describing the motion of a spring-mass system (unit mass, spring "constant" k and frictional coefficient c):

$$\frac{d^2y}{dt^2} = -ky - c\frac{dy}{dt}.$$

The equilibrium is $y = 0$. Thus, we might think of Bessel's differential equation as representing a time-varying frictional force ($c = 1/t$) and a time-varying "restoring" force ($k = 1 - m^2/t^2$). The latter force is a variable restoring force only for $t > m$ ($z > m$). We might expect the solutions of Bessel's differential equation to be similar to a damped oscillator (at least for $z > m$). The larger z gets, the closer the variable spring constant k approaches 1 and the more the frictional force tends to vanish. The solution should oscillate with frequency approximately 1, but should slowly decay. This is similar to an underdamped spring-mass system, but the solutions to Bessel's differential equation should decay more slowly than any exponential since the frictional force is approaching zero. Detailed numerical solutions of Bessel functions are sketched in Fig. 6.8.1,

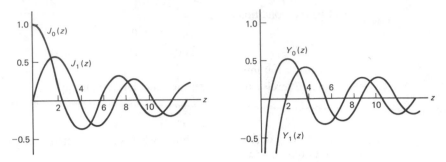

Figure 6.8.1 Sketch of various Bessel functions.

verifying these points. Note that *for small z,*

$$J_0(z) \approx 1 \qquad Y_0(z) \approx \frac{2}{\pi}\ln z$$

$$J_1(z) \approx \tfrac{1}{2}z \qquad Y_1(z) \approx -\frac{2}{\pi}z^{-1} \qquad (6.8.2)$$

$$J_2(z) \approx \tfrac{1}{8}z^2 \qquad Y_2(z) \approx -\frac{4}{\pi}z^{-2}.$$

These sketches vividly show a property worth memorizing—**Bessel functions of the first and second kind look like decaying oscillations.** In fact, it is known that $J_m(z)$ and $Y_m(z)$ may be accurately approximated for large z by simple algebraically decaying oscillations *for large z*:

$$
\boxed{
\begin{aligned}
J_m(z) &\sim \sqrt{\frac{2}{\pi z}}\cos\left(z - \frac{\pi}{4} - m\frac{\pi}{2}\right) & \text{as } z \to \infty \\[2mm]
Y_m(z) &\sim \sqrt{\frac{2}{\pi z}}\sin\left(z - \frac{\pi}{4} - m\frac{\pi}{2}\right) & \text{as } z \to \infty.
\end{aligned}
}
\qquad (6.8.3)
$$

These are known as asymptotic formulas, meaning that the approximations improve as $z \to \infty$. Earlier in Sec. 5.9 we claimed that approximation formulas similar to (6.8.3) always exist for any Sturm–Liouville problem for the large eigenvalues $\lambda \gg 1$. Here $\lambda \gg 1$ implies that $z \gg 1$ since $z = \sqrt{\lambda}\,r$ and $0 < r < a$ (as long as r is not too small).

A derivation of (6.8.3) requires facts beyond the scope of this text. However, information such as (6.8.3) is readily available from many sources.* We notice from (6.8.3) that the only difference in the approximate behavior for large z of all these Bessel functions is the precise phase shift. We also note that the

---

* A personal favorite, highly recommended to students with a serious interest in the applications of mathematics to science and engineering, is *Handbook of Mathematical Functions,* edited by M. Abramowitz and I. A. Stegun, originally published *inexpensively* by the National Bureau of Standards in 1964 and more recently reprinted by Dover in paperback.

frequency is approximately 1 (and period 2π) for large z, consistent with the comparison with a spring-mass system with vanishing friction and $k \to 1$. Furthermore, the amplitude of oscillation, $\sqrt{2/\pi z}$, decays more slowly as $z \to \infty$ than the exponential rate of decay associated with an underdamped oscillator, as previously discussed qualitatively.

6.8.2 Asymptotic Formulas for the Eigenvalues

Approximate values of the zeros of the eigenfunctions $J_m(z)$ may be obtained using these asymptotic formulas, (6.8.3). For example, for $m = 0$, for large z

$$J_0(z) \sim \sqrt{\frac{2}{\pi z}} \cos \left(z - \frac{\pi}{4} \right).$$

The zeros approximately occur when $z - \pi/4 = -\pi/2 + s\pi$, but s must be large (in order for z to be large). Thus, the large zeros are given approximately by

$$z \sim \pi(s - \tfrac{1}{4}), \tag{6.8.4}$$

for large integral s. We claim that formula (6.8.4) becomes more and more accurate as n increases. In fact, since the formula is reasonably accurate already for $n = 2$ or 3 (see Table 6.8.1), it may be unnecessary to compute the zero to a greater accuracy than is given by (6.8.4). A further indication of the accuracy of the asymptotic formula is that we see that the differences of the first few eigenvalues are already nearly π (as predicted for the large eigenvalues).

TABLE 6.8.1 ZEROS OF $J_0(z)$

n	z_{0n}	Exact	Large z formula (6.8.4)	Error	Percentage error	$z_{0n} - z_{0(n-1)}$
1	z_{01}	2.40483 . . .	2.35619	0.04864	2.0	—
2	z_{02}	5.52008 . . .	5.49779	0.02229	0.4	3.11525
3	z_{03}	8.65373 . . .	8.63938	0.01435	0.2	3.13365
4	z_{04}	11.79153 . . .	11.78097	0.01156	0.1	3.13780

6.8.3 Zeros of Bessel Functions and Nodal Curves

We have shown that the eigenfunctions are $J_m(\sqrt{\lambda_{mn}}\, r)$ where $\lambda_{mn} = (z_{mn}/a)^2$, z_{mn} being the nth zero of $J_m(z)$. Thus, the eigenfunctions are

$$J_m \left(z_{mn} \frac{r}{a} \right).$$

For example, for $m = 0$, the eigenfunctions are $J_0(z_{0n}\, r/a)$, where the sketch of $J_0(z)$ is reproduced in Fig. 6.8.2 (and the zeros are marked). As r ranges from 0 to a, the argument of the eigenfunction $J_0(z_{0n}\, r/a)$ ranges from 0 to the nth zero, z_{0n}. At $r = a$, $z = z_{0n}$, the nth zero. Thus, the nth eigenfunction has $n - 1$ zeros in the interior. Although originally stated for regular Sturm–Liouville problems, it is also valid for singular problems (if eigenfunctions exist).

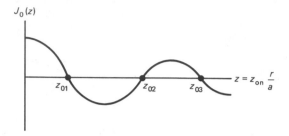

Figure 6.8.2 Sketch of $J_0(z)$ and its zeros.

The separation of variables solution of the wave equation is $u(r, \theta, t) = f(r)g(\theta)h(t)$,

$$u(r, \theta, t) = J_m\left(z_{mn}\frac{r}{a}\right)\left\{\begin{matrix}\sin m\theta \\ \cos m\theta\end{matrix}\right\}\left\{\begin{matrix}\sin c\sqrt{\lambda_{mn}}\,t \\ \cos c\sqrt{\lambda_{mn}}\,t\end{matrix}\right\}, \qquad (6.8.5)$$

known as a mode of oscillation. For each $m \neq 0$ there are four families of solutions (for $m = 0$, there are two families). Each mode oscillates with a characteristic natural frequency, $c\sqrt{\lambda_{mn}}$. At certain positions along the membrane, known as **nodal curves**, the membrane will be unperturbed for all time (for vibrating strings we called these positions nodes). The nodal curve for the $\sin m\theta$ mode is determined by

$$J_m\left(z_{mn}\frac{r}{a}\right)\sin m\theta = 0. \qquad (6.8.6)$$

The nodal curve consists of all points where $\sin m\theta = 0$ or $J_m(z_{mn} r/a) = 0$. $\sin m\theta$ is zero along $2m$ distinct rays, $\theta = s\pi/m$, $s = 1, 2, \ldots, 2m$. In order for there to be a zero of $J_m(z_{mn} r/a)$ for $0 < r < a$, $z_{mn} r/a$ must equal an earlier zero of $J_m(z)$, $z_{mn} r/a = z_{mp}$, $p = 1, 2, \ldots, n - 1$. There are thus $n - 1$ circles along which $J_m(z_{mn} r/a) = 0$ besides $r = a$. We illustrate this for $m = 3$, $n = 5$ in Fig. 6.8.3, where the nodal circles are determined from a table.

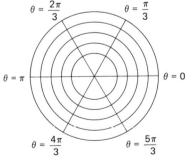

Figure 6.8.3 Nodal curves for a vibrating circular membrane.

p	z_{3p}	$r/a = z_{3p}/z_{35}$
1	6.38	0.33
2	9.76	0.50
3	13.02	0.67
4	16.22	0.84
5	19.41	—

6.8.4 Series Representation of Bessel Functions

The usual method of discussing Bessel functions relies on series solution methods for differential equations. We will obtain little useful information by pursuing this topic. However, some may find it helpful to refer to the formulas that follow.

First we review some additional results concerning series solutions around $z = 0$ for second-order linear differential equations:

$$\frac{d^2f}{dz^2} + a(z)\frac{df}{dz} + b(z)f = 0. \tag{6.8.7}$$

Recall that $z = 0$ is an ordinary point if both $a(z)$ and $b(z)$ have Taylor series around $z = 0$. In this case we are guaranteed that all solutions may be represented by a convergent Taylor series,

$$f = \sum_{n=0}^{\infty} a_n z^n = a_0 + a_1 z + a_2 z^2 + \cdots,$$

at least in some neighborhood of $z = 0$.

If $z = 0$ is not an ordinary point, then we call it a singular point (e.g., $z - 0$ is a singular point of Bessel's differential equation). If $z = 0$ is a singular point, we cannot state that all solutions have Taylor series around $z = 0$. However, if $a(z) = R(z)/z$ and $b(z) = S(z)/z^2$ with $R(z)$ and $S(z)$ having Taylor series, then we can say more about solutions of the differential equation near $z = 0$. For this case known as a **regular singular point**, the coefficients $a(z)$ and $b(z)$ can have *at worst* a simple pole and double pole, respectively. It is possible for $a(z)$ and $b(z)$ not to be that singular. For example, if $a(z) = 1 + z$ and $b(z) = (1 - z^3)/z^2$, then $z = 0$ is a regular singular point. Bessel's differential equation in the form (6.8.7) is

$$\frac{d^2f}{dz^2} + \frac{1}{z}\frac{df}{dz} + \frac{z^2 - m^2}{z^2}f = 0.$$

Here, $R(z) = 1$ and $S(z) - z^2 - m^2$; both have Taylor series around $z = 0$. Therefore, $z = 0$ is a regular singular point for Bessel's differential equation.

For a regular singular point at $z = 0$, it is known by the **method of Frobenius** that at least one solution of the differential equation is in the form

$$f = z^p \sum_{n=0}^{\infty} a_n z^n, \tag{6.8.8}$$

that is, z^p times a Taylor series, where p is one of the solutions of the quadratic indicial equation. One method to obtain the indicial equation is to substitute $f = z^p$ into the corresponding equidimensional equation that results by replacing $R(z)$ by $R(0)$ and $S(z)$ by $S(0)$. Thus,

$$p(p - 1) + R(0)p + S(0) = 0$$

is the indicial equation. If the two values of p (the roots of the indicial equation) differ by a noninteger, then two independent solutions exist in the form (6.8.8).

If the two roots of the indicial equation are identical, then only one solution is in the form (6.8.8) and the other solution is more complicated, but always involves logarithms. If the roots differ by an integer, then sometimes both solutions exist in the form (6.8.8), while other times form (6.8.8) only exists corresponding to the larger root p, and a series beginning with the smaller root p must be modified by the introduction of logarithms. Details of the method of Frobenius are presented in most elementary differential equations texts.

For Bessel's differential equation, we have shown that the indicial equation is

$$p(p - 1) + p - m^2 = 0,$$

since $R(0) = 1$ and $S(0) = -m^2$. Its roots are $\pm m$. If $m = 0$, the roots are identical. Form (6.8.8) is valid for one solution, while logarithms must enter the second solution. For $m \neq 0$ the roots of the indicial equation differ by an integer. Detailed calculations also show that logirithms must enter. The following infinite series can be verified by substitution and are often considered as definitions of $J_m(z)$ and $Y_m(z)$:

$$J_m(z) = \sum_{k=0}^{\infty} \frac{(-1)^k (z/2)^{2k+m}}{k!(k+m)!} \tag{6.8.9}$$

$$Y_m(z) = \frac{2}{\pi} \left[\left(\log \frac{z}{2} + \gamma \right) J_m(z) - \frac{1}{2} \sum_{k=0}^{m-1} \frac{(m-k-1)!(z/2)^{2k-m}}{k!} \tag{6.8.10}$$

$$+ \frac{1}{2} \sum_{k=0}^{\infty} (-1)^{k+1} [\varphi(k) + \varphi(k+m)] \frac{(z/2)^{2k+m}}{k!(m+k)!} \right]^*.$$

We have obtained these from the previously mentioned handbook edited by Abramowitz and Stegun.

EXERCISES 6.8

6.8.1. The boundary value problem for a vibrating annular membrane $1 < r < 2$ (fixed at the inner and outer radii) is

$$\frac{d}{dr} \left(r \frac{df}{dr} \right) + \left(\lambda r - \frac{m^2}{r} \right) f = 0$$

with $f(1) = 0$ and $f(2) = 0$, where $m = 0, 1, 2, \ldots$.
(a) Show that $\lambda > 0$.
*(b) Obtain an expression that determines the eigenvalues.
(c) For what value of m does the smallest eigenvalue occur?
*(d) Obtain an upper and lower bound for the smallest eigenvalue.
(e) Using a trial function, obtain an upper bound for the lowest eigenvalue.
(f) Compute approximately the lowest eigenvalue from part (b) using tables of Bessel functions. Compare to parts (d) and (e).

* 1. $\varphi(k) = 1 + \frac{1}{2} + \frac{1}{3} + \cdots + 1/k$, $\phi(0) = 0$
2. $\gamma = \lim_{k \to \infty} [\varphi(k) - \ln k] = 0.5772157 \ldots$, known as Euler's constant.
3. If $m = 0$, $\sum_{k=0}^{m-1} \equiv 0$.

6.8.2 Consider the temperature $u(r, \theta, t)$ in a quarter-circle of radius a satisfying

$$\frac{\partial u}{\partial t} = k \, \nabla^2 u$$

subject to the conditions

$$u(r, 0, t) = 0 \qquad u(a, \theta, t) = 0$$
$$u(r, \pi/2, t) = 0 \qquad u(r, \theta, 0) = G(r, \theta).$$

(a) Show that the boundary value problem is

$$\frac{d}{dr}\left(r \frac{df}{dr}\right) + \left(\lambda r - \frac{\mu}{r}\right)f = 0$$

with $f(a) = 0$ and $f(0)$ bounded.

(b) Show that $\lambda > 0$ if $\mu \geqslant 0$.

(c) Show that for each μ, the eigenfunction corresponding to the smallest eigenvalue has no zeros for $0 < r < a$.

*(d) Solve the initial value problem.

6.8.3. Reconsider Exercise 6.8.2 with the boundary conditions

$$\frac{\partial u}{\partial \theta}(r, 0, t) = 0, \qquad \frac{\partial u}{\partial \theta}\left(r, \frac{\pi}{2}, t\right) = 0 \quad \text{and} \quad u(a, \theta, t) = 0.$$

6.8.4. Consider the boundary value problem

$$\frac{d}{dr}\left(r \frac{df}{dr}\right) + \left(\lambda r - \frac{m^2}{r}\right)f = 0$$

with $f(a) = 0$ and $f(0)$ bounded. For each integral m, show that the nth eigenfunction has $n - 1$ zeros for $0 < r < a$.

6.8.5. Using the known asymptotic behavior as $z \to 0$ and as $z \to \infty$, roughly sketch for all $z > 0$

(a) $J_4(z)$ (b) $Y_1(z)$ (c) $Y_0(z)$

(d) $J_0(z)$ (e) $Y_5(z)$ (f) $J_2(z)$

6.8.6. Determine approximately the *large* frequencies of vibration of a circular membrane.

6.8.7. Consider Bessel's differential equation

$$z^2 \frac{d^2 f}{dz^2} + z \frac{df}{dz} + (z^2 - m^2)f = 0.$$

Let $f = y/z^{1/2}$. Derive that

$$\frac{d^2 y}{dz^2} + y(1 + \tfrac{1}{4} z^{-2} - m^2 z^{-2}) = 0.$$

*6.8.8. Using Exercise 6.8.7, determine exact expressions for $J_{1/2}(z)$ and $Y_{1/2}(z)$. Use and verify Eqs. (6.8.3) and (6.7.28) in this case.

6.8.9. In this exercise use the result of Exercise 6.8.7. If z is large, verify as much as possible concerning Eqs. (6.8.3).

6.8.10. In this exercise use the result of Exercise 6.8.7 in order to improve upon (6.8.3):

(a) Substitute $y = e^{iz}w(z)$ and show that

$$\frac{d^2 w}{dz^2} + 2i \frac{dw}{dz} + \frac{\gamma}{z^2}w = 0, \text{ where } \gamma = \tfrac{1}{4} - m^2.$$

(b) Substitute $w = \sum_{n=0}^{\infty} \beta_n z^{-n}$. Determine the first few terms β_n (assuming that $\beta_0 = 1$).

(c) Use part (b) to obtain an improved asymptotic solution of Bessel's differential equation. For real solutions, take real and imaginary parts.

(d) Find a recurrence formula for β_n. Show that the series diverges. (Nonetheless, a finite series is very useful.)

6.8.11. In order to "understand" the behavior of Bessel's differential equation as $z \to \infty$, let $x = 1/z$. Show that $x = 0$ is a singular point, but an irregular singular point. [The asymptotic solution of a differential equation in the neighborhood of an irregular singular point is analyzed in an unmotivated way in Exercise 6.8.10. For a more systematic presentation, see advanced texts on asymptotic or perturbation methods (such as Bender and Orszag [1978].)]

6.8.12. The lowest eigenvalue for (6.7.29) for $m = 0$ is $\lambda = (z_{01}/a)^2$. Determine a reasonably accurate upper bound by using the Rayleigh quotient with a trial function. Compare to the exact answer.

6.8.13. Explain why the nodal circles in Fig. 6.8.3 are nearly equally spaced.

6.9 LAPLACE'S EQUATION IN A CIRCULAR CYLINDER

6.9.1 Introduction

Laplace's equation,

$$\nabla^2 u = 0, \tag{6.9.1}$$

represents the steady-state heat equation (without sources). We have solved Laplace's equation in a rectangle (Sec. 2.5.1) and Laplace's equation in a circle (Sec. 2.5.2). In both cases, when variables were separated, oscillations occur in one direction, but not in the other. Laplace's equation in a rectangular box can also be solved by the method of separation of variables. As shown in some exercises in Chapter 6, the three independent variables yield two eigenvalue problems which have oscillatory solutions and solutions in one direction which are not oscillatory.

A more interesting problem is to consider Laplace's equation in a circular cylinder of radius a and height H. Using circular cylindrical coordinates,

$$x = r \cos \theta$$
$$y = r \sin \theta$$
$$z = z,$$

Laplace's equation is

$$\frac{1}{r} \frac{\partial}{\partial r} \left(r \frac{\partial u}{\partial r} \right) + \frac{1}{r^2} \frac{\partial^2 u}{\partial \theta^2} + \frac{\partial^2 u}{\partial z^2} = 0. \tag{6.9.2}$$

We prescribe u (perhaps temperature) on the entire boundary of the cylinder:

top:	$u(r, \theta, H) = \beta(r, \theta)$
bottom:	$u(r, \theta, 0) = \alpha(r, \theta)$
lateral side:	$u(a, \theta, z) = \gamma(\theta, z)$.

There are three nonhomogeneous boundary conditions. One approach is to break the problem up into the sum of three simpler problems, each solving Laplace's equation,

$$\nabla^2 u_i = 0, \qquad i = 1, 2, 3,$$

where $u = u_1 + u_2 + u_3$. This is illustrated in Fig. 6.9.1. In this way each problem satisfies two homogeneous boundary conditions, but the sum satisfies the desired nonhomogeneous conditions. We separate variables once, for all three cases, and then proceed to solve each problem individually.

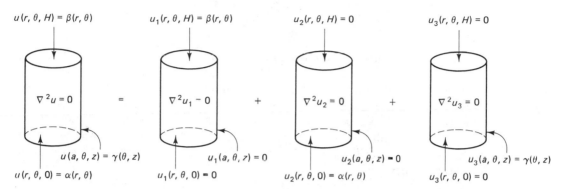

Figure 6.9.1 Laplace's equation in a circular cylinder.

6.9.2 Separation of Variables

We begin by looking for product solutions,

$$u(r, \theta, z) = f(r)g(\theta)h(z), \tag{6.9.3}$$

for Laplace's equation. Substituting (6.9.3) into (6.9.2) and dividing by $f(r)g(\theta)h(z)$ yields

$$\frac{1}{rf}\frac{d}{dr}\left(r\frac{df}{dr}\right) + \frac{1}{r^2}\frac{1}{g}\frac{d^2g}{d\theta^2} + \frac{1}{h}\frac{d^2h}{dz^2} = 0. \tag{6.9.4}$$

We immediately can separate the z-dependence, and hence

$$\frac{1}{h}\frac{d^2h}{dz^2} = \lambda. \tag{6.9.5}$$

Do we expect oscillations in z? From Fig. 6.9.1 we see that oscillations in z should be expected for the u_3-problem but not necessarily for the u_1- or u_2-problem. Perhaps $\lambda < 0$ for the u_3-problem, but not for the u_1- and u_2-problems. Thus, we do not specify λ at this time. The r and θ parts also can be separated if (6.9.4) is multiplied by r^2 (and (6.9.5) is utilized):

$$\frac{r}{f}\frac{d}{dr}\left(r\frac{df}{dr}\right) + \lambda r^2 = -\frac{1}{g}\frac{d^2g}{d\theta^2} = \mu. \tag{6.9.6}$$

A second separation constant μ is introduced, with the anticipation that $\mu > 0$ because of the expected oscillations in θ for all three problems. In fact, the implied periodic boundary conditions in θ dictate that

$$\mu = m^2, \tag{6.9.7}$$

and that $g(\theta)$ can be either $\sin m\theta$ or $\cos m\theta$, where m is a nonnegative integer, $m = 0, 1, 2, \ldots$. A Fourier series in θ will be appropriate for all these problems.

In summary, the θ-dependence is $\sin m\theta$ and $\cos m\theta$, and the remaining two differential equations are

$$\frac{d^2h}{dz^2} = \lambda h \tag{6.9.8}$$

$$r\frac{d}{dr}\left(r\frac{df}{dr}\right) + (\lambda r^2 - m^2)f = 0. \tag{6.9.9}$$

These two differential equations contain only one unspecified parameter λ. Only one will become an eigenvalue problem. The eigenvalue problem needs two homogeneous boundary conditions. Different results occur for the various problems, u_1, u_2, and u_3. For the u_3-problem, there are two homogeneous boundary conditions in z, and thus (6.9.8) will become an eigenvalue problem [and (6.9.9) will have nonoscillatory solutions]. However, for the u_1- and u_2-problems, there do not exist two homogeneous boundary conditions in z. Instead, there should be two homogeneous conditions in r. One of these is at $r = a$. The other must be a singularity condition at $r = 0$, which occurs due to the singular nature of polar (or circular cylindrical) coordinates at $r = 0$ and the singular nature of (6.9.9) at $r = 0$:

$$|f(0)| < \infty. \tag{6.9.10}$$

Thus, we will find that for the u_1- and u_2-problems, (6.9.9) will be the eigenvalue problem. The solution of (6.9.9) will oscillate, whereas the solution of (6.9.8) will not oscillate. We next describe the details of all three problems.

6.9.3 Zero Temperature on the Lateral Sides and on the Bottom or Top

The mathematical problem for u_1 is

$$\nabla^2 u_1 = 0 \tag{6.9.11a}$$

$$u_1(r, \theta, 0) = 0 \tag{6.9.11b}$$
$$u_1(r, \theta, H) = \beta(r, \theta) \tag{6.9.11c}$$
$$u_1(a, \theta, z) = 0. \tag{6.9.11d}$$

The temperature is zero on the bottom. By separation of variables, in which the nonhomogeneous condition (6.9.11c) is momentarily ignored, $u_1 = f(r)g(\theta)h(z)$. The θ-part is known to equal $\sin m\theta$ and $\cos m\theta$ (for integral $m \geq 0$). The z-dependent equation, (6.9.8), satisfies only one homogeneous condition, $h(0) = 0$. The r-dependent equation will become a boundary value problem determining the separation constant λ. The two homogeneous boundary conditions are

$$f(a) = 0 \tag{6.9.12a}$$

$$|f(0)| < \infty. \tag{6.9.12b}$$

The eigenvalue problem, (6.9.9) with (6.9.12), is one that was analyzed in Sec. 6.8. There we showed that $\lambda > 0$ (by directly using the Rayleigh quotient). Furthermore, we showed that the general solution of (6.9.9) is a linear combination of Bessel functions of order m with argument $\sqrt{\lambda}\, r$:

$$f(r) = c_1 J_m(\sqrt{\lambda}\, r) + c_2 Y_m(\sqrt{\lambda}\, r) = c_1 J_m(\sqrt{\lambda}\, r), \tag{6.9.13}$$

which has been simplified using the singularity condition, (6.9.12b).Then the homogeneous condition, (6.9.12a), determines λ:

$$\boxed{J_m(\sqrt{\lambda}\, a) = 0.} \tag{6.9.14}$$

Again $\sqrt{\lambda}\, a$ must be a zero of the mth Bessel function, and the notation λ_{mn} is used to indicate the infinite number of eigenvalues for each m. The eigenfunction $J_m(\sqrt{\lambda_{mn}}\, r)$ oscillates in r.

Since $\lambda > 0$, the solution of (6.9.8) that satisfies $h(0) = 0$ is proportional to

$$h(z) = \sinh \sqrt{\lambda}\, z. \tag{6.9.15}$$

No oscillations occur in the z-direction. There are thus two doubly infinite families of product solutions:

$$\sinh \sqrt{\lambda_{mn}}\, z\, J_m(\sqrt{\lambda_{mn}}\, r) \begin{Bmatrix} \sin m\theta \\ \cos m\theta \end{Bmatrix}, \tag{6.9.16}$$

oscillatory in r and θ, but nonoscillatory in z. The principle of superposition implies that we should consider

$$\begin{aligned}
u_1(r, \theta, z) &= \sum_{m=0}^{\infty} \sum_{n=1}^{\infty} A_{mn} \sinh \sqrt{\lambda_{mn}}\, z\, J_m(\sqrt{\lambda_{mn}}\, r) \cos m\theta \\
&\quad + \sum_{m=1}^{\infty} \sum_{n=1}^{\infty} B_{mn} \sinh \sqrt{\lambda_{mn}}\, z\, J_m(\sqrt{\lambda_{mn}}\, r) \sin m\theta.
\end{aligned} \tag{6.9.17}$$

The nonhomogeneous boundary condition, (6.9.11c), $u_1(r, \theta, H) = \beta(r, 0)$, will determine the coefficients A_{mn} and B_{mn}. It will involve a Fourier series in θ and a Fourier–Bessel series in r. Thus we can solve for A_{mn} and B_{mn} using the two one-dimensional orthogonality formulas. Alternatively, the coefficients are more easily calculated using the two-dimensional orthogonality of $J_m(\sqrt{\lambda_{mn}}\, r) \cos m\theta$ and $J_m(\sqrt{\lambda_{mn}}\, r) \sin m\theta$ (see Sec. 6.8). We omit the details.

In a similar manner, one can obtain u_2. We leave as an exercise the solution of this problem.

6.9.4 Zero Temperature on the Top and Bottom

A somewhat different mathematical problem arises if we consider the situation in which the top and bottom are held at zero temperature. The problem for u_3 is

$$\nabla^2 u_3 = 0 \tag{6.9.18a}$$

$$u_3(r, \theta, 0) = 0 \tag{6.9.18b}$$
$$u_3(r, \theta, H) = 0 \tag{6.9.18c}$$
$$u_3(a, \theta, z) = \gamma(\theta, z). \tag{6.9.18d}$$

We may again use the results of the method of separation of variables. The periodicity again implies that the θ-part will relate to a Fourier series (i.e., $\sin m\theta$ and $\cos m\theta$). However, unlike what occurred in Sec. 6.9.3, the z-equation has two homogeneous boundary conditions:

$$\frac{d^2 h}{dz^2} = \lambda h \tag{6.9.19a}$$

$$h(0) = 0 \tag{6.9.19b}$$

$$h(H) = 0. \tag{6.9.19c}$$

This is the simplest Sturm–Liouville eigenvalue problem (in a somewhat different form). In order for $h(z)$ to oscillate and satisfy (6.9.19b) and (6.9.19c), the separation constant λ must be negative. In fact, we should recognize that

$$\lambda = -\left(\frac{n\pi}{H}\right)^2 \qquad n = 1, 2, \dots \tag{6.9.20a}$$

$$h(z) = \sin \frac{n\pi z}{H}. \tag{6.9.20b}$$

The boundary conditions at top and bottom imply that we will be using an ordinary Fourier sine series in z.

We have oscillations in z and θ. The r-dependent solution should not be oscillatory; they satisfy (6.9.9), which using (6.9.20a) becomes

$$r \frac{d}{dr}\left(r \frac{df}{dr}\right) + \left(-\left(\frac{n\pi}{H}\right)^2 r^2 - m^2\right) f = 0. \tag{6.9.21}$$

A homogeneous condition, in the form of a singularity condition, exists at $r = 0$,

$$|f(0)| < \infty, \tag{6.9.22}$$

but there is no homogeneous condition at $r = a$.

Equation (6.9.21) looks similar to Bessel's differential equation, but has the wrong sign in front of the r^2 term. It cannot be changed into Bessel's differential equation using a *real* transformation. If we let

$$s = i\left(\frac{n\pi}{H}\right)r, \tag{6.9.23}$$

where $i = \sqrt{-1}$, then (6.9.21) becomes

$$s\frac{d}{ds}\left(s\frac{df}{ds}\right) + (s^2 - m^2)f = 0 \quad \text{or} \quad s^2\frac{d^2f}{ds^2} + s\frac{df}{ds} + (s^2 - m^2)f = 0.$$

We recognize this as exactly Bessel's differential equation, and thus

$$f = c_1 J_m(s) + c_2 Y_m(s) \quad \text{or} \quad f = c_1 J_m\left(i\frac{n\pi}{H}r\right) + c_2 Y_m\left(i\frac{n\pi}{H}r\right). \tag{6.9.24}$$

Therefore, the solution of (6.9.21) can be represented in terms of Bessel functions of an imaginary argument. This is not very useful since Bessel functions are not usually tabulated in this form.

Instead, we introduce a real transformation that eliminates the dependence on $n\pi/H$ of the differential equation:

$$w = \frac{n\pi}{H}r$$

Then (6.9.21) becomes

$$\boxed{w^2\frac{d^2f}{dw^2} + w\frac{df}{dw} + (-w^2 - m^2)f = 0.} \tag{6.9.25}$$

Again the wrong sign appears for this to be Bessel's differential equation. Equation (6.9.25) is a modification of Bessel's differential equation, and its solutions *which have been well tabulated* are known as **modified Bessel functions**.

Equation (6.9.25) has the same kind of singularity at $w = 0$ as Bessel's differential equation. As such, the singular behavior could be determined by the method of Frobenius.* Thus, we can specify *one solution to be well defined at $w = 0$, called the* **modified Bessel function of order m of the first kind**, denoted $I_m(w)$. Another independent solution, which is *singular at the origin, is called the* **modified Bessel function of order m of the second kind**, denoted $K_m(w)$. Both $I_m(w)$ and $K_m(w)$ are well-tabulated functions. We will need very little knowledge concerning $I_m(w)$ and $K_m(w)$. The general solution of (6.9.21) is thus

$$f = c_1 K_m\left(\frac{n\pi}{H}r\right) + c_2 I_m\left(\frac{n\pi}{H}r\right). \tag{6.9.26}$$

Since K_m is singular at $r = 0$ and I_m is not, it follows that $c_1 = 0$ and $f(r)$ is proportional to $I_m(n\pi r/H)$. We simply note that both $I_m(w)$ and $K_m(w)$ are non-oscillatory and are not zero for $w > 0$. A discussion of this and further properties are given in Sec. 6.9.5.

---

* Here it is easier to use the complex transformation (6.9.23). Then the infinite series representation for Bessel functions is valid for complex arguments, avoiding additional calculations.

There are thus two doubly infinite families of product solutions:

$$I_m\left(\frac{n\pi}{H}r\right)\sin\frac{n\pi z}{H}\cos m\theta \quad \text{and} \quad I_m\left(\frac{n\pi}{H}r\right)\sin\frac{n\pi z}{H}\sin m\theta. \quad (6.9.27)$$

These solutions are oscillatory in z and θ, but nonoscillatory in r. The principle of superposition, equivalent to a Fourier sine series in z and a Fourier series in θ, implies that

$$
\boxed{
\begin{aligned}
u_3(r, \theta, z) = \sum_{m=0}^{\infty}\sum_{n=1}^{\infty} E_{mn}I_m\left(\frac{n\pi}{H}r\right)\sin\frac{n\pi z}{H}\cos m\theta \\
+ \sum_{m=1}^{\infty}\sum_{n=1}^{\infty} F_{mn}I_m\left(\frac{n\pi}{H}r\right)\sin\frac{n\pi z}{H}\sin m\theta.
\end{aligned}
}
\qquad (6.9.28)
$$

The coefficients E_{mn} and F_{mn} can be determined [if $I_m(n\pi a/H) \neq 0$] from the nonhomogeneous equation (6.9.18d) either by two iterated one-dimensional orthogonality results or by one application of two-dimensional orthogonality. In the next section we will discuss further properties of $I_m(n\pi a/H)$, including the fact that it has no positive zeros.

In this way the solution for Laplace's equation inside a circular cylinder has been determined given any temperature distribution along the entire boundary.

6.9.5 Modified Bessel Functions

The differential equation that defines the modified Bessel functions is

$$w^2\frac{d^2f}{dw^2} + w\frac{df}{dw} + (-w^2 - m^2)f = 0. \quad (6.9.29)$$

Two independent solutions are denoted $K_m(w)$ and $I_m(w)$. The behavior in the neighborhood of the singular point $w = 0$ is determined by the roots of the indicial equation, $\pm m$, corresponding to approximate solutions near $w = 0$ of the forms $w^{\pm m}$ (for $m \neq 0$) and w^0 and $w^0 \ln w$ (for $m = 0$). We can choose the two independent solutions such that one is well behaved at $w = 0$ and the other singular.

A good understanding of these functions comes from also analyzing their behavior as $w \to \infty$. Roughly speaking for large w (6.9.29) can be rewritten as

$$\frac{d^2f}{dw^2} \approx -\frac{1}{w}\frac{df}{dw} + f. \quad (6.9.30)$$

Thinking of this as Newton's law for a particle with certain forces, the $-1/w\, df/dw$ term is a weak damping force tending to vanish as $w \to \infty$. We might expect as $w \to \infty$ that

$$\frac{d^2f}{dw^2} \approx f,$$

which suggests that the solution should be a linear combination of an exponentially growing e^w and exponentially decaying e^{-w} term. In fact, the weak damping has its effects (just as it did for ordinary Bessel functions). We state (but do not prove) a more advanced result, namely that the asymptotic behavior for large w of solutions of (6.9.29) are approximately $e^{\pm w}/w^{1/2}$. Thus, both $I_m(w)$ and $K_m(w)$ are linear combinations of these two, one exponentially growing and the other decaying.

There is only one independent linear combination which decays as $w \to \infty$. There are many combinations which grow as $w \to \infty$. We *define* $K_m(w)$ to be a solution that decays as $w \to \infty$. It must be proportional to $e^{-w}/w^{1/2}$ and it is defined *uniquely* by

$$K_m(w) \sim \sqrt{\frac{\pi}{2}} \frac{e^{-w}}{w^{1/2}}, \tag{6.9.31}$$

as $w \to \infty$. As $w \to 0$ the behavior of $K_m(w)$ will be some linear combination of the two different behaviors (e.g., w^m and w^{-m} for $m \neq 0$). In general, it will be composed of both and hence will be singular at $w = 0$. In more advanced treatments it is shown that

$$K_m(w) \sim \begin{cases} -\ln w & m = 0 \\ \frac{1}{2}(m - 1)! \, (\frac{1}{2}w)^{-m} & m \neq 0, \end{cases} \tag{6.9.32}$$

as $w \to 0$. The most important facts about this function is that the $K_m(w)$ **exponentially decays as $w \to \infty$, but is singular at $w = 0$.**

Since $K_m(w)$ is singular at $w = 0$, we would like to define a second solution $I_m(w)$ not singular at $w = 0$. $I_m(w)$ is *defined* uniquely such that

$$I_m(w) \sim \frac{1}{m!} \left(\frac{1}{2}w \right)^m, \tag{6.9.33}$$

as $w \to 0$. As $w \to \infty$, the behavior of $I_m(w)$ will be some linear combination of the two different asymptotic behaviors $(e^{\pm w}/w^{1/2})$. In general, it will be composed of both and hence is expected to exponentially grow as $w \to \infty$. In more advanced works, it is shown that

$$I_m(w) \sim \sqrt{\frac{1}{2\pi w}} e^w, \tag{6.9.34}$$

as $w \to \infty$. The most important facts about this function is that $I_m(w)$ **is well behaved at $w = 0$, but exponentially grows as $w \to \infty$.**

Some modified Bessel functions are sketched in Fig. 6.9.2. Although we have not proved it, note that both $I_m(w)$ and $K_m(w)$ are not zero for $w > 0$.

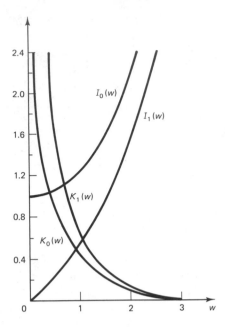

Figure 6.9.2 Various modified Bessel functions (from Abramowitz and Stegun [1965]).

EXERCISES 6.9

6.9.1. Solve Laplace's equation inside a circular cylinder subject to the boundary conditions

(a) $u(r, \theta, 0) = \alpha(r, \theta)$, $u(r, \theta, H) = 0$, $u(a, \theta, z) = 0$

*(b) $u(r, \theta, 0) = \alpha(r) \sin 7\theta$, $u(r, \theta, H) = 0$, $u(a, \theta, z) = 0$

(c) $u(r, \theta, 0) = 0$, $u(r, \theta, H) = \beta(r) \cos 3\theta$, $\dfrac{\partial u}{\partial r}(a, \theta, z) = 0$

(d) $\dfrac{\partial u}{\partial z}(r, \theta, 0) = \alpha(r) \sin 3\theta$, $\dfrac{\partial u}{\partial z}(r, \theta, H) = 0$, $\dfrac{\partial u}{\partial r}(a, \theta, z) = 0$

(e) $\dfrac{\partial u}{\partial z}(r, \theta, 0) = \alpha(r, \theta)$, $\dfrac{\partial u}{\partial z}(r, \theta, H) = 0$, $\dfrac{\partial u}{\partial r}(a, \theta, z) = 0$

   Under what condition does a solution exist?

6.9.2. Solve Laplace's equation inside a semicircular cylinder, subject to the boundary conditions

(a) $u(r, \theta, 0) = 0$, $u(r, 0, z) = 0$, $u(a, \theta, z) = 0$
   $u(r, \theta, H) = \alpha(r, \theta)$, $u(r, \pi, z) = 0$,

*(b) $u(r, \theta, 0) = 0$, $u(r, 0, z) = 0$, $u(a, \theta, z) = \beta(\theta, z)$
   $\dfrac{\partial u}{\partial z}(r, \theta, H) = 0$, $u(r, \pi, z) = 0$,

(c) $\dfrac{\partial}{\partial z}u(r, \theta, 0) = 0$, $\dfrac{\partial u}{\partial \theta}(r, 0, z) = 0$, $\dfrac{\partial u}{\partial r}(a, \theta, z) = \beta(\theta, z)$
   $\dfrac{\partial}{\partial z}u(r, \theta, H) = 0$, $\dfrac{\partial u}{\partial \theta}(r, \pi, z) = 0$,
   Under what condition does a solution exist?

(d) $u(r, \theta, 0) = 0$, $\qquad u(r, 0, z) = 0$, $\qquad u(a, \theta, z) = 0$

$\qquad u(r, \theta, H) = 0$, $\qquad \dfrac{\partial u}{\partial \theta}(r, \pi, z) = \alpha(r, z)$

6.9.3. Solve the heat equation

$$\frac{\partial u}{\partial t} = k \nabla^2 u$$

inside a quarter-circular cylinder ($0 < \theta < \pi/2$ with radius a and height H) subject to the initial condition

$$u(r, \theta, z, 0) = f(r, \theta, z).$$

Briefly explain what temperature distribution you expect to be approached as $t \to \infty$. Consider the following boundary conditions

(a) $u(r, \theta, 0) = 0$, $\qquad u(r, 0, z) = 0$, $\qquad u(a, \theta, z) = 0$
$\qquad u(r, \theta, H) = 0$, $\qquad u(r, \pi/2, z) = 0$,

*(b) $\dfrac{\partial u}{\partial z}(r, \theta, 0) = 0$, $\qquad \dfrac{\partial u}{\partial \theta}(r, 0, z) = 0$, $\qquad \dfrac{\partial u}{\partial r}(a, \theta, z) = 0$

$\qquad \dfrac{\partial u}{\partial z}(r, \theta, H) - 0$, $\qquad \dfrac{\partial u}{\partial \theta}(r, \pi/2, z) = 0$,

(c) $u(r, \theta, 0) = 0$, $\qquad \dfrac{\partial u}{\partial \theta}(r, 0, z) - 0$, $\qquad \dfrac{\partial u}{\partial r}(a, \theta, z) - 0$

$\qquad u(r, \theta, H) = 0$, $\qquad u(r, \pi/2, z) = 0$

6.9.4. Solve the heat equation

$$\frac{\partial u}{\partial t} - k \nabla^2 u$$

inside a cylinder (of radius a and height H) subject to the initial condition,

$$u(r, \theta, z, 0) = f(r, z),$$

independent of θ, if the boundary conditions are

*(a) $u(r, \theta, 0, t) = 0$, $\qquad u(r, \theta, H, t) = 0$, $\qquad u(a, \theta, z, t) = 0$

(b) $\dfrac{\partial u}{\partial z}(r, \theta, 0, t) - 0$, $\qquad \dfrac{\partial u}{\partial z}(r, \theta, H, t) - 0$, $\qquad \dfrac{\partial u}{\partial r}(a, 0, z, t) - 0$

(c) $u(r, \theta, 0, t) = 0$, $\qquad u(r, \theta, H, t) = 0$, $\qquad \dfrac{\partial u}{\partial r}(a, \theta, z, t) = 0$

6.9.5. Determine the three ordinary differential equations obtained by separation of variables for Laplace's equation in spherical coordinates

$$0 = \frac{\partial}{\partial r}\left(r^2 \frac{\partial u}{\partial r}\right) + \frac{1}{\sin \phi} \frac{\partial}{\partial \phi}\left(\sin \phi \frac{\partial u}{\partial \phi}\right) + \frac{1}{\sin^2 \phi} \frac{\partial^2 u}{\partial \theta^2}.$$

7

Nonhomogeneous Problems

7.1 INTRODUCTION

In the previous chapters we have developed only one method to solve partial differential equations, the method of separation of variables. In order to apply the method of separation of variables, the partial differential equation (with n independent variables) must be linear and homogeneous. In addition, we must be able to formulate a problem with linear and homogeneous boundary conditions for $n - 1$ variables. However, some of the most fundamental physical problems do not have homogeneous conditions.

7.2 HEAT FLOW WITH SOURCES AND NONHOMOGENEOUS BOUNDARY CONDITIONS

Time-independent boundary conditions. As an elementary example of a nonhomogeneous problem, consider the heat flow (without sources) in a uniform rod of length L with the temperature fixed at the left end at $A°$ and the right at $B°$. If the initial condition is prescribed, the mathematical problem for the temperature $u(x, t)$ is

$$\text{PDE:} \qquad \boxed{\frac{\partial u}{\partial t} = k\frac{\partial^2 u}{\partial x^2}} \qquad (7.2.1a)$$

$$\text{BC:} \qquad \boxed{\begin{aligned} u(0, t) &= A \\ u(L, t) &= B \end{aligned}} \qquad \begin{aligned} &(7.2.1b) \\ &(7.2.1c) \end{aligned}$$

$$\text{IC:} \qquad \boxed{u(x, 0) = f(x).} \qquad (7.2.1d)$$

The method of separation of variables cannot be used directly since for even this simple example the boundary conditions are not homogeneous.

Equilibrium temperature. To analyze this problem, we first obtain an equilibrium temperature distribution, $u_E(x)$. If such a temperature distribution exists, it must satisfy the steady-state (time-independent) heat equation,

$$\frac{d^2 u_E}{dx^2} = 0, \tag{7.2.2}$$

as well as the given time-independent boundary conditions,

$$u_E(0) = A \tag{7.2.3a}$$

$$u_E(L) = B. \tag{7.2.3b}$$

We ignore the initial conditions in defining an equilibrium temperature distribution. As shown in Sec. 1.5, (7.2.2) implies that the temperature distribution is linear, and the unique one that satisfies (7.2.3) can be determined geometrically or algebraically:

$$u_E(x) = A + \frac{B - A}{L}x, \tag{7.2.4}$$

which is sketched in Fig. 7.2.1. Usually $u_E(x)$ will *not* be the desired time-dependent solution, since it satisfies the initial conditions (7.2.1d) only if $f(x) = A + [(B - A)/L]x$.

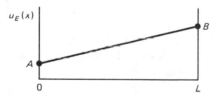

Figure 7.2.1 Equilibrium temperature distribution.

Displacement from equilibrium. For more general initial conditions, we consider the temperature displacement from the equilibrium temperature,

$$\boxed{v(x, t) \equiv u(x, t) - u_E(x).} \tag{7.2.5}$$

Instead of solving for $u(x, t)$, we will determine $v(x, t)$. Since $\partial v/\partial t = \partial u/\partial t$ and $\partial^2 v/\partial x^2 = \partial^2 u/\partial x^2$ [note that $u_E(x)$ is linear in x], it follows that $v(x, t)$ also satisfies the heat equation

$$\frac{\partial v}{\partial t} = k\frac{\partial^2 v}{\partial x^2}. \tag{7.2.6a}$$

Furthermore, both $u(x,t)$ and $u_E(x)$ equal A at $x = 0$ and equal B at $x = L$, and hence their difference is zero at $x = 0$ and at $x = L$:

$$v(0, t) = 0 \tag{7.2.6b}$$

$$v(L, t) = 0. \tag{7.2.6c}$$

Initially, $v(x, t)$ equals the difference between the given initial temperature and the equilibrium temperature,

$$v(x, 0) = f(x) - u_E(x). \tag{7.2.6d}$$

Fortunately, the mathematical problem for $v(x, t)$ is a linear homogeneous partial differential equation with linear homogeneous boundary conditions. Thus, $v(x, t)$ can be determined by the method of separation of variables. In fact, this problem is one we have encountered frequently. Hence, we note that

$$v(x, t) = \sum_{n=1}^{\infty} a_n \sin \frac{n\pi x}{L} e^{-k(n\pi/L)^2 t}, \tag{7.2.7}$$

where the initial conditions imply that

$$f(x) - u_E(x) = \sum_{n=1}^{\infty} a_n \sin \frac{n\pi x}{L}. \tag{7.2.8}$$

Thus, a_n equals the Fourier sine coefficients of $f(x) - u_E(x)$:

$$a_n = \frac{2}{L} \int_0^L [f(x) - u_E(x)] \sin \frac{n\pi x}{L} \, dx.$$

From (7.2.5) we easily obtain the desired temperature, $u(x,t) = u_E(x) + v(x,t)$. Thus,

$$\boxed{u(x,t) = u_E(x) + \sum_{n=1}^{\infty} a_n \sin \frac{n\pi x}{L} e^{-k(n\pi/L)^2 t},} \tag{7.2.9}$$

where a_n is given by (7.2.8) and $u_E(x)$ is given by (7.2.4). As $t \to \infty$, $u(x,t) \to u_E(x)$ irrespective of the initial conditions. The temperature approaches its equilibrium distribution for all initial conditions.

Steady nonhomogeneous terms. The previous method also works if there are *steady* sources of thermal energy:

PDE:	$\dfrac{\partial u}{\partial t} = k\dfrac{\partial^2 u}{\partial x^2} + Q(x)$	(7.2.10a)
BC:	$u(0, t) = A$	(7.2.10b)
	$u(L, t) = B$	(7.2.10c)
IC:	$u(x, 0) = f(x).$	(7.2.10d)

If an equilibrium solution exists (see Exercise 1.4.6 for a somewhat different example in which an equilibrium solution does *not* exist), then we determine it and again consider the displacement from equilibrium,

$$v(x, t) = u(x, t) - u_E(x).$$

We can show that $v(x, t)$ satisfies a linear homogeneous partial differential equation and linear homogeneous boundary conditions. Thus, $v(x, t)$ can be solved by the method of separation of variables.

Time-dependent nonhomogeneous terms. Unfortunately, nonhomogeneous problems are not always as easy to solve as the previous examples. In order to clarify the situation, we again consider the heat flow in a uniform rod of length L. However, we make two substantial changes. First, we introduce temperature-dependent heat sources distributed in a prescribed way throughout the rod. Thus, the temperature will solve the following nonhomogeneous partial differential equation:

$$\text{PDE:} \qquad \boxed{\frac{\partial u}{\partial t} = k\frac{\partial^2 u}{\partial x^2} + Q(x, t).} \qquad (7.2.11a)$$

Here the sources of thermal energy $Q(x, t)$ vary in space and time. In addition, we allow the temperature at the ends to vary in time. This yields time-dependent and nonhomogeneous linear boundary conditions,

$$\text{BC:} \qquad \boxed{\begin{aligned} u(0, t) &= A(t) \\ u(L, t) &= B(t), \end{aligned}} \qquad \begin{aligned} &(7.2.11b) \\ &(7.2.11c) \end{aligned}$$

instead of the time-independent ones, (7.2.1b) and (7.2.1c). Again the initial temperature distribution is prescribed:

$$\text{IC:} \qquad \boxed{u(x, 0) = f(x).} \qquad (7.2.11d)$$

The mathematical problem defined by (7.2.11) consists of a nonhomogeneous partial differential equation with nonhomogeneous boundary conditions.

Related homogeneous boundary conditions. We claim that we *cannot* always reduce this problem to a homogeneous partial differential equation with homogeneous boundary conditions, as we did for the first example of this section. Instead, we will find it quite useful to note that we can always transform our problem into one with homogeneous boundary conditions, although in general the partial differential equation will remain nonhomogeneous. We consider *any* **reference temperature distribution** $r(x, t)$ (the simpler the better) with only the property that it satisfy the given nonhomogeneous boundary conditions. In our example, this means only that

$$r(0, t) = A(t)$$
$$r(L, t) = B(t),$$

It is usually not difficult to obtain many candidates for $r(x, t)$. Perhaps the simplest choice is

$$r(x, t) = A(t) + \frac{x}{L}[B(t) - A(t)], \qquad (7.2.12)$$

although there are other possibilities.* Again the difference between the desired solution $u(x, t)$ and the chosen function $r(x, t)$ (now *not* necessarily an equilibrium solution) is employed:

$$v(x, t) \equiv u(x, t) - r(x, t). \tag{7.2.13}$$

Since both $u(x, t)$ and $r(x, t)$ satisfy the same linear (although nonhomogeneous) boundary condition at both $x = 0$ and $x = L$, it follows that $v(x, t)$ satisfies the *related* homogeneous boundary conditions:

$$v(0, t) = 0 \tag{7.2.14a}$$

$$v(L, t) = 0. \tag{7.2.14b}$$

The partial differential equation satisfied by $v(x, t)$ is derived by substituting

$$u(x, t) = v(x, t) + r(x, t)$$

into the heat equation with sources, (7.2.11a). Thus,

$$\frac{\partial v}{\partial t} = k \frac{\partial^2 v}{\partial x^2} + \left[Q(x, t) - \frac{\partial r}{\partial t} + k \frac{\partial^2 r}{\partial x^2} \right] \equiv k \frac{\partial^2 v}{\partial x^2} + \overline{Q}. \tag{7.2.14c}$$

In general, the partial differential equation for $v(x, t)$ is of the same type as for $u(x, t)$, but with a different nonhomogeneous term, since $r(x, t)$ usually does not satisfy the homogeneous heat equation. The initial condition is also usually altered:

$$v(x, 0) = f(x) - r(x, 0) = f(x) - A(0) - \frac{x}{L}[B(0) - A(0)] \equiv g(x). \tag{7.2.14d}$$

It can be seen that in general only the boundary conditions have been made homogeneous. In Sec. 7.3 we will develop a method to analyze nonhomogeneous problems with homogeneous boundary conditions.

EXERCISES 7.2

7.2.1. Solve the heat equation with time-independent sources and boundary conditions

$$\frac{\partial u}{\partial t} = k \frac{\partial^2 u}{\partial x^2} + Q(x)$$

$$u(x, 0) = f(x)$$

if an equilibrium solution exists. Analyze the limits as $t \to \infty$. If no equilibrium exists, explain why and reduce the problem to one with homogeneous boundary conditions (but do not solve). Assume

*(a) $Q(x) = 0$, $u(0, t) = A$, $\dfrac{\partial u}{\partial x}(L, t) = B$

(b) $Q(x) = 0$, $\dfrac{\partial u}{\partial x}(0, t) = 0$, $\dfrac{\partial u}{\partial x}(L, t) = B \neq 0$

(c) $Q(x) = 0$, $\dfrac{\partial u}{\partial x}(0, t) = A \neq 0$, $\dfrac{\partial u}{\partial x}(L, t) = A$

*(d) $Q(x) = k$, $u(0, t) = A$, $u(L, t) = B$

---

* Other choices for $r(x, t)$ yield equivalent solutions to the original nonhomogeneous problem.

(e) $Q(x) = k$, $\quad\quad \dfrac{\partial u}{\partial x}(0, t) = 0$, $\quad\quad \dfrac{\partial u}{\partial x}(L, t) = 0$

(f) $Q(x) = \sin \dfrac{2\pi x}{L}$, $\quad \dfrac{\partial u}{\partial x}(0, t) = 0$, $\quad\quad \dfrac{\partial u}{\partial x}(L, t) = 0$

7.2.2. Consider the heat equation with time-dependent sources and boundary conditions:

$$\frac{\partial u}{\partial t} = k \frac{\partial^2 u}{\partial x^2} + Q(x, t)$$

$$u(x, 0) = f(x).$$

Reduce the problem to one with homogeneous boundary conditions if

*(a) $\dfrac{\partial u}{\partial x}(0, t) = A(t)$ $\quad$ and $\quad$ $\dfrac{\partial u}{\partial x}(L, t) = B(t)$

(b) $u(0, t) = A(t)$ $\quad$ and $\quad$ $\dfrac{\partial u}{\partial x}(L, t) = B(t)$

*(c) $\dfrac{\partial u}{\partial x}(0, t) = A(t)$ $\quad$ and $\quad$ $u(L, t) = B(t)$

(d) $u(0, t) = 0$ $\quad$ and $\quad$ $\dfrac{\partial u}{\partial x}(L, t) + h(u(L, t) - B(t)) = 0$

(e) $\dfrac{\partial u}{\partial x}(0, t) = 0$ $\quad$ and $\quad$ $\dfrac{\partial u}{\partial x}(L, t) + h(u(L, t) - B(t)) = 0$

7.2.3. Solve the two-dimensional heat equation with circularly symmetric time-independent sources, boundary conditions, and initial conditions (inside a circle):

$$\frac{\partial u}{\partial t} = \frac{k}{r} \frac{\partial}{\partial r}\left(r \frac{\partial u}{\partial r}\right) + Q(r)$$

with

$$u(r, 0) = f(r) \quad\quad \text{and} \quad\quad u(a, t) = T.$$

7.2.4. Solve the two-dimensional heat equation with time-independent boundary conditions:

$$\frac{\partial u}{\partial t} = k\left(\frac{\partial^2 u}{\partial x^2} + \frac{\partial^2 u}{\partial y^2}\right)$$

subject to the boundary conditions

$$u(0, y, t) = 0 \quad\quad \frac{\partial}{\partial y} u(x, 0, t) = 0$$

$$u(L, y, t) = 0 \quad\quad u(x, H, t) = g(x)$$

and the initial condition

$$u(x, y, 0) = f(x, y)$$

Analyze the limit as $t \to \infty$.

7.2.5. Solve the initial value problem for a two-dimensional heat equation inside a circle (of radius a) with time-independent boundary conditions:

$$\frac{\partial u}{\partial t} = k \, \nabla^2 u$$

$$u(a, \theta, t) = g(\theta)$$

$$u(r, \theta, 0) = f(r, \theta).$$

7.2.6. Solve the wave equation with time-independent sources,

$$\frac{\partial^2 u}{\partial t^2} = c^2 \frac{\partial^2 u}{\partial x^2} + Q(x)$$

$$u(x, 0) = f(x)$$

$$\frac{\partial}{\partial t} u(x, 0) = g(x),$$

if an "equilibrium" solution exists. Analyze the behavior for large t. If no equilibrium exists, explain why and reduce the problem to one with homogeneous boundary conditions. Assume that

*(a) $Q(x) = 0$, $u(0, t) = A$, $u(L, t) = B$

(b) $Q(x) = 1$, $u(0, t) = 0$, $u(L, t) = 0$

(c) $Q(x) = 1$, $u(0, t) = A$, $u(L, t) = B$
     [*Hint:* Add problems (a) and (b).]

*(d) $Q(x) = \sin \dfrac{\pi x}{L}$, $u(0, t) = 0$, $u(L, t) = 0$

7.3 METHOD OF EIGENFUNCTION EXPANSION WITH HOMOGENEOUS BOUNDARY CONDITIONS (DIFFERENTIATING SERIES OF EIGENFUNCTIONS)

In Sec. 7.2 we showed how to introduce a problem with homogeneous boundary conditions, even if the original problem of interest has nonhomogeneous boundary conditions. For that reason we will investigate nonhomogeneous linear partial differential equations with homogeneous boundary conditions. For example, consider

PDE:
$$\frac{\partial v}{\partial t} = k \frac{\partial^2 v}{\partial x^2} + \overline{Q}(x, t) \qquad (7.3.1a)$$

BC:
$$v(0, t) = 0 \qquad (7.3.1b)$$
$$v(L, t) = 0 \qquad (7.3.1c)$$

IC:
$$v(x, 0) = g(x). \qquad (7.3.1d)$$

We will solve this problem by the **method of eigenfunction expansion**. Consider the eigenfunctions of the related homogeneous problem. The related homogeneous problem is

$$\frac{\partial u}{\partial t} = k \frac{\partial^2 u}{\partial x^2}$$

$$u(0, t) = 0 \qquad (7.3.2)$$

$$u(L, t) = 0.$$

The eigenfunctions of this related homogeneous problem satisfy

$$\frac{d^2\phi}{dx^2} + \lambda\phi = 0$$

$$\phi(0) = 0 \qquad\qquad (7.3.3)$$

$$\phi(L) = 0.$$

We know that the eigenvalues are $\lambda_n = (n\pi/L)^2$, $n = 1, 2, \ldots$, and the corresponding eigenfunctions are $\phi_n(x) = \sin n\pi x/L$. However, the eigenfunctions will be different for other problems. We do not wish to emphasize the method of eigenfunction expansion solely for this one example. Thus, we will speak in some generality. We assume that the eigenfunctions (of the related homogeneous problem) are known and we designate them $\phi_n(x)$. The eigenfunctions satisfy a Sturm–Liouville eigenvalue problem and as such they are complete (any piecewise smooth function may be expanded in a series of these eigenfunctions). **The method of eigenfunction expansion, employed to solve the nonhomogeneous problem with homogeneous boundary conditions, (7.3.1), consists in expanding the unknown solution v(x, t) in a series of the related homogeneous eigenfunctions:**

$$v(x, t) = \sum_{n=1}^{\infty} a_n(t)\phi_n(x). \qquad\qquad (7.3.4)$$

For each fixed t, $v(x, t)$ is a function of x, and hence $v(x, t)$ will have a generalized Fourier series. In our example, $\phi_n(x) = \sin n\pi x/L$, and this series is an ordinary Fourier sine series. The generalized Fourier coefficients are a_n, but the coefficients will vary as t changes. Thus, the generalized Fourier coefficients are functions of time, $a_n(t)$. At first glance expansion (7.3.4) may appear similar to what occurs in separating variables for homogeneous problems. However, (7.3.4) is substantially different. Here $a_n(t)$ are *not* the time-dependent separated solutions $e^{-k(n\pi/L)^2 t}$. Instead, $a_n(t)$ are just the generalized Fourier coefficients for $v(x, t)$. We will determine $a_n(t)$ and show that usually $a_n(t)$ is not proportional to $e^{-k(n\pi/L)^2 t}$.

Equation (7.3.4) automatically satisfies the homogeneous boundary conditions. We emphasize this by stating that both $v(x, t)$ and $\phi_n(x)$ satisfy the same homogeneous boundary conditions. The initial condition is satisfied if

$$g(x) = \sum_{n=1}^{\infty} a_n(0)\phi_n(x).$$

Due to the orthogonality of the eigenfunctions [with weight 1 in this problem because of the constant coefficient in (7.3.3)], we can determine the initial values of the generalized Fourier coefficients:

$$a_n(0) = \frac{\displaystyle\int_0^L g(x)\phi_n(x)\,dx}{\displaystyle\int_0^L \phi_n^2(x)\,dx}. \qquad\qquad (7.3.5)$$

"All" that remains is to determine $a_n(t)$ such that (7.3.4) solves the nonhomogeneous partial differential equation (7.3.1a). We will show in two different ways that $a_n(t)$ satisfies a first-order differential equation in order for (7.3.4) to satisfy (7.3.1a).

One method* to determine $a_n(t)$ is by direct substitution. This is easy to do but requires calculation of $\partial v/\partial t$ and $\partial^2 v/\partial x^2$. Since $v(x, t)$ is an infinite series, the differentiation can be a delicate process. We simply state that with some degree of generality, if v and $\partial v/\partial x$ are continuous and **if $v(x, t)$ solves the same homogeneous boundary conditions as does $\phi_n(x)$, then the necessary term-by-term differentiations can be justified.** For the cases of Fourier sine and cosine series, a more detailed investigation of the properties of the term-by-term differentiation of these series was presented in Sec. 3.4, which proved this result. For the general case, we omit a proof. However, we obtain the same solution in Sec. 7.4 by an alternative method, which thus justifies the somewhat simpler technique of the present section. We thus proceed to term-by-term differentiate $v(x, t)$:

$$\frac{\partial v}{\partial t} = \sum_{n=1}^{\infty} \frac{da_n(t)}{dt} \phi_n(x)$$

$$\frac{\partial^2 v}{\partial x^2} = \sum_{n=1}^{\infty} a_n(t) \frac{d^2\phi_n(x)}{dx^2} = -\sum_{n=1}^{\infty} a_n(t)\lambda_n\phi_n(x),$$

since $\phi_n(x)$ satisfies $d^2\phi_n/dx^2 + \lambda_n\phi_n = 0$. Substituting these results into (7.3.1a) yields

$$\sum_{n=1}^{\infty} \left[\frac{da_n}{dt} + \lambda_n k a_n \right] \phi_n(x) = \overline{Q}(x, t). \tag{7.3.6}$$

The left-hand side is the generalized Fourier series for $\overline{Q}(x, t)$. Due to the orthogonality of $\phi_n(x)$, we obtain a first-order differential equation for $a_n(t)$:

$$\frac{da_n}{dt} + \lambda_n k a_n = \frac{\displaystyle\int_0^L \overline{Q}(x, t)\phi_n(x)\, dx}{\displaystyle\int_0^L \phi_n^2(x)\, dx} \equiv \overline{q}_n(t). \tag{7.3.7}$$

The right-hand side is a known function of time (and n), namely, the Fourier coefficient of $\overline{Q}(x, t)$:

$$\overline{Q}(x, t) = \sum_{n=1}^{\infty} \overline{q}_n(t)\phi_n(x).$$

Equation (7.3.7) requires an initial condition, and sure enough $a_n(0)$ equals the generalized Fourier coefficients of the initial condition [see (7.3.5)].

Equation (7.3.7) is a nonhomogeneous linear first-order equation. Perhaps the easiest method† to solve it [unless $\overline{q}_n(t)$ is particularly simple] is to multiply

---

* A second method is discussed in Sec. 7.4.

† Another method is variation of parameters.

it by the integrating factor $e^{\lambda_n kt}$. *Thus,*

$$e^{\lambda_n kt}\left(\frac{da_n}{dt} + \lambda_n ka_n\right) = \frac{d}{dt}(a_n e^{\lambda_n kt}) = \bar{q}_n e^{\lambda_n kt}.$$

Integrating from 0 to t yields

$$a_n(t)e^{\lambda_n kt} - a_n(0) = \int_0^t \bar{q}_n(\tau)e^{\lambda_n k\tau}\, d\tau.$$

We solve for $a_n(t)$ and obtain

$$a_n(t) = a_n(0)e^{-\lambda_n kt} + e^{-\lambda_n kt}\int_0^t \bar{q}_n(\tau)e^{\lambda_n k\tau}\, d\tau. \qquad (7.3.8)$$

Note that $a_n(t)$ is in the form of a constant, $a_n(0)$, times the homogeneous solution $e^{-\lambda_n kt}$ plus a particular solution. This completes the method of eigenfunction expansions. The solution of our nonhomogeneous partial differential equation with homogeneous boundary conditions is

$$v(x, t) = \sum_{n=1}^{\infty} a_n(t)\phi_n(x),$$

where $\phi_n(x) = \sin n\pi x/L$, $\lambda_n = (n\pi/L)^2$, $a_n(t)$ is given by (7.3.8), $\bar{q}_n(\tau)$ is given by (7.3.7), and $a_n(0)$ is given by (7.3.5). The solution is rather complicated.

As a check, if the problem was homogeneous, $Q(x, t) = 0$, then the solution simplifies to

$$v(x, t) = \sum_{n=1}^{\infty} a_n(t)\phi_n(x),$$

where

$$a_n(t) = a_n(0)e^{-\lambda_n kt}$$

and $a_n(0)$ is given by (7.3.5), exactly the solution obtained by separation of variables.

Example. As an elementary example, suppose that for $0 < x < \pi$ (i.e., $L = \pi$)

$$\frac{\partial u}{\partial t} = \frac{\partial^2 u}{\partial x^2} + \sin 3x\, e^{-t} \quad \text{subject to} \quad \begin{array}{l} u(0, t) = 0 \\ u(\pi, t) = 1 \\ u(x, 0) = f(x). \end{array}$$

To make the boundary conditions homogeneous, we introduce the displacement from reference temperature distribution $v(x, , t) = u(x, t) - x/\pi$, in which case

$$\frac{\partial v}{\partial t} = \frac{\partial^2 v}{\partial x^2} + \sin 3x\, e^{-t} \quad \text{subject to} \quad \begin{array}{l} v(0, t) = 0 \\ v(\pi, t) = 0 \\ v(x, 0) = f(x) - \dfrac{x}{\pi}. \end{array}$$

The eigenfunctions are $\sin n\pi x/L = \sin nx$ (since $L = \pi$), and thus

$$v(x, t) = \sum_{n=1}^{\infty} a_n(t) \sin nx. \qquad (7.3.9)$$

This eigenfunction expansion is substituted into the PDE, yielding

$$\sum_{n=1}^{\infty} \left(\frac{da_n}{dt} + n^2 a_n \right) \sin nx = \sin 3x \, e^{-t}.$$

Thus, the unknown Fourier sine coefficients satisfy

$$\frac{da_n}{dt} + n^2 a_n = \begin{cases} 0 & n \neq 3 \\ e^{-t} & n = 3. \end{cases}$$

The solution of this does not require (7.3.8):

$$a_n(t) = \begin{cases} a_n(0)e^{-n^2 t} & n \neq 3 \\ \dfrac{1}{8}e^{-t} + [a_3(0) - \dfrac{1}{8}]e^{-9t} & n = 3, \end{cases} \tag{7.3.10a}$$

where

$$a_n(0) = \frac{2}{\pi} \int_0^\pi \left[f(x) - \frac{x}{\pi} \right] \sin nx \, dx. \tag{7.3.10b}$$

The solution to the original nonhomogeneous problem is given by $u(x, t) = v(x, t) + x/\pi$, where v satisfies (7.3.9) with $a_n(t)$ determined from (7.3.10).

EXERCISES 7.3

7.3.1. Solve the initial value problem for the heat equation with time-dependent sources

$$\frac{\partial u}{\partial t} = k \frac{\partial^2 u}{\partial x^2} + Q(x, t)$$

$$u(x, 0) = f(x)$$

subject to the following boundary conditions:

(a) $u(0, t) = 0,$ $\quad \dfrac{\partial u}{\partial x}(L, t) = 0$

(b) $u(0, t) = 0,$ $\quad u(L, t) + 2\dfrac{\partial u}{\partial x}(L, t) = 0$

*(c) $u(0, t) = A(t),$ $\quad \dfrac{\partial u}{\partial x}(L, t) = 0$

(d) $u(0, t) = A \neq 0,$ $\quad u(L, t) = 0$

(e) $\dfrac{\partial u}{\partial x}(0, t) = A(t)$ $\quad \dfrac{\partial u}{\partial x}(L, t) = B(t)$

*(f) $\dfrac{\partial u}{\partial x}(0, t) = 0,$ $\quad \dfrac{\partial u}{\partial x}(L, t) = 0$

(g) Specialize part (f) to the case $Q(x, t) = Q(x)$ (independent of t) such that $\int_0^L Q(x) \, dx \neq 0$. In this case show that there are no time-independent solutions. What happens to the time-dependent solution as $t \to \infty$? Briefly explain.

7.3.2. Consider the heat equation with a steady source

$$\frac{\partial u}{\partial t} = k\frac{\partial^2 u}{\partial x^2} + Q(x)$$

subject to the initial and boundary conditions described in this section:

$$u(0, t) = 0, \ u(L, t) = 0, \quad \text{and} \quad u(x, 0) = f(x).$$

Obtain the solution by the method of eigenfunction expansion. Show that the solution approaches a steady-state solution.

*7.3.3. Solve the initial value problem

$$c\rho\frac{\partial u}{\partial t} = \frac{\partial}{\partial x}\left(K_0\frac{\partial u}{\partial x}\right) + qu + f(x, t),$$

where c, ρ, K_0, and q are functions of x only, subject to the conditions

$$u(0, t) = 0, \ u(L, t) = 0, \quad \text{and} \quad u(x, 0) = g(x).$$

Assume that the eigenfunctions are known. $\left[\text{Hint: let } L \equiv \frac{d}{dx}\left(K_0\frac{d}{dx}\right) + q.\right]$

7.3.4. Consider

$$\frac{\partial u}{\partial t} = \frac{1}{\sigma(x)}\frac{\partial}{\partial x}\left[K_0(x)\frac{\partial u}{\partial x}\right] \qquad (K_0 > 0, \ \sigma > 0)$$

with the boundary conditions and initial conditions:

$$u(x, 0) = g(x), \qquad u(0, t) = A \quad \text{and} \quad u(L, t) = B.$$

*(a) Find a time-independent solution, $u_0(x)$.

(b) Show that $\lim_{t\to\infty} u(x, t) = f(x)$ independent of the initial conditions. [Show that $f(x) = u_0(x)$.]

*7.3.5. Solve

$$\frac{\partial u}{\partial t} = k\,\nabla^2 u + f(r, t)$$

inside the circle $(r < a)$ with $u = 0$ at $r = a$ and initially $u = 0$.

7.3.6. Solve

$$\frac{\partial u}{\partial t} = \frac{\partial^2 u}{\partial x^2} + \sin 5x\ e^{-2t}$$

subject to $u(0, t) = 1$, $u(\pi, t) = 0$, and $u(x, 0) = 0$.

*7.3.7. Solve

$$\frac{\partial u}{\partial t} = \frac{\partial^2 u}{\partial x^2}$$

subject to $u(0, t) = 0$, $u(L, t) = t$, and $u(x, 0) = 0$.

7.4 METHOD OF EIGENFUNCTION EXPANSION USING GREEN'S FORMULA (WITH OR WITHOUT HOMOGENEOUS BOUNDARY CONDITIONS)

In this section we reinvestigate problems that may have nonhomogeneous boundary conditions. We still use the method of eigenfunction expansion. For example, consider

PDE:	$\dfrac{\partial u}{\partial t} = k\dfrac{\partial^2 u}{\partial x^2} + Q(x, t)$	(7.4.1a)

BC:	$u(0, t) = A(t)$	(7.4.1b)
	$u(L, t) = B(t)$	(7.4.1c)

IC:	$u(x, 0) = f(x)$.	(7.4.1d)

The eigenfunctions of the related homogeneous problem,

$$\frac{d^2\phi_n}{dx^2} + \lambda_n\phi_n = 0 \tag{7.4.2a}$$

$$\phi_n(0) = 0 \tag{7.4.2b}$$

$$\phi_n(L) = 0, \tag{7.4.2c}$$

are known to be $\phi_n(x) = \sin n\pi x/L$, corresponding to the eigenvalues $\lambda_n = (n\pi/L)^2$. Any piecewise smooth function can be expanded in terms of these eigenfunctions. Thus, even though $u(x, t)$ satisfies nonhomogeneous boundary conditions, it is still true that

$$u(x, t) = \sum_{n=1}^{\infty} b_n(t)\phi_n(x). \tag{7.4.3}$$

Actually, the equality in (7.4.3) cannot be valid at $x = 0$ and at $x = L$ since $\phi_n(x)$ satisfies the homogeneous boundary conditions, while $u(x, t)$ does not. Nonetheless, we use the $=$ notation, where we understand that the $\sim$ notation is more proper. It is difficult to determine $b_n(t)$ by substituting (7.4.3) into (7.4.1a); the required term-by-term differentiations with respect to x are not justified since $u(x, t)$ and $\phi_n(x)$ do not satisfy the same homogeneous boundary conditions [$\partial^2 u/\partial x^2 \neq \sum_{n=1}^{\infty} b_n(t) \, d^2\phi_n/dx^2$.] However, term-by-term time derivatives are valid:

$$\frac{\partial u}{\partial t} = \sum_{n=1}^{\infty} \frac{db_n}{dt}\phi_n(x). \tag{7.4.4}$$

We will determine a first-order differential equation for $b_n(t)$. Unlike in Sec. 7.3, it will be obtained *without* calculating spatial derivatives of an infinite series of eigenfunctions. From (7.4.4) it follows that

$$\sum_{n=1}^{\infty} \frac{db_n}{dt}\phi_n(x) = k\frac{\partial^2 u}{\partial x^2} + Q(x, t),$$

and thus

$$\frac{db_n}{dt} = \frac{\int_0^L \left[k \frac{\partial^2 u}{\partial x^2} + Q(x, t) \right] \phi_n(x) \, dx}{\int_0^L \phi_n^2 \, dx}. \tag{7.4.5}$$

Equation (7.4.5) allows the partial differential equation to be satisfied. Already we note the importance of the generalized Fourier series of $Q(x, t)$:

$$Q(x, t) = \sum_{n=1}^{\infty} q_n(t)\phi_n(x), \quad \text{where} \quad q_n(t) = \frac{\int_0^L Q(x, t)\phi_n(x) \, dx}{\int_0^L \phi_n^2 \, dx}.$$

Thus, (7.4.5) simplifies:

$$\frac{db_n}{dt} = q_n(t) + \frac{\int_0^L k\frac{\partial^2 u}{\partial x^2} \phi_n(x) \, dx}{\int_0^L \phi_n^2 \, dx}. \tag{7.4.6}$$

We will show how to evaluate the integral in (7.4.6) in terms of $b_n(t)$, yielding a first-order differential equation.

If we integrate $\int_0^L \partial^2 u/\partial x^2 \, \phi_n(x) \, dx$ twice by parts, then we would obtain the desired result. However, this would be a considerable effort. There is a better way; we have already performed the required integration by parts in a more general context. Perhaps the operator notation, $L \equiv \partial^2/\partial x^2$, will help to remind you of the result we need. Using $L = \partial^2/\partial x^2$,

$$\int_0^L \frac{\partial^2 u}{\partial x^2}\phi_n(x) \, dx = \int_0^L \phi_n L(u) \, dx.$$

Now this may be simplified by employing Green's formula (derived by repeated integrations in Chapter 5). Let us restate **Green's formula**:

$$\int_0^L [uL(v) - vL(u)]dx = p\left(u\frac{dv}{dx} - v\frac{du}{dx} \right)\Big|_0^L, \tag{7.4.7}$$

where L is any Sturm–Liouville operator ($L \equiv d/dx \, (p \, d/dx) + q$). In our context, $L = \partial^2/\partial x^2$ (i.e., $p = 1$, $q = 0$). Partial derivatives may be used, since $\partial/\partial x = d/dx$ with t fixed. Thus,

$$\int_0^L \left(u\frac{\partial^2 v}{\partial x^2} - v\frac{\partial^2 u}{\partial x^2} \right) dx = \left(u\frac{\partial v}{\partial x} - v\frac{\partial u}{\partial x} \right)\Big|_0^L. \tag{7.4.8}$$

Here we let $v = \phi_n(x)$. Often both u and ϕ_n satisfy the same homogeneous boundary conditions and the right-hand side vanishes. Here $\phi_n(x) = \sin n\pi x/L$ satisfies homogeneous boundary conditions, but $u(x, t)$ does not $[u(0, t) = A(t)$ and $u(L, t) = B(t)]$. Using $d\phi_n/dx = (n\pi/L)\cos n\pi x/L$, the right-hand side of (7.4.8) simplifies to $(n\pi/L)[B(t)(-1)^n - A(t)]$. Furthermore, $\int_0^L u\, d^2\phi_n/dx^2\, dx = -\lambda_n \int_0^L u\phi_n\, dx$ since $d^2\phi_n/dx^2 + \lambda_n\phi_n = 0$. Thus, (7.4.8) becomes

$$\int_0^L \phi_n \frac{\partial^2 u}{\partial x^2} dx = -\lambda_n \int_0^L u\phi_n dx - \frac{n\pi}{L}[B(t)(-1)^n - A(t)].$$

Since $b_n(t)$ are the generalized Fourier coefficients of $u(x, t)$, we know that

$$b_n(t) = \frac{\int_0^L u\phi_n\, dx}{\int_0^L \phi_n^2\, dx}.$$

Finally, (7.4.6) reduces to a first-order differential equation for $b_n(t)$:

$$\frac{db_n}{dt} + k\lambda_n b_n = q_n(t) + \frac{k(n\pi/L)[A(t) - (-1)^n B(t)]}{\int_0^L \phi_n^2(x)\, dx}. \qquad (7.4.9)$$

The nonhomogeneous terms arise in two ways. $q_n(t)$ is due to the source terms in the PDE, while the term involving $A(t)$ and $B(t)$ is a result of the nonhomogeneous boundary conditions at $x = 0$ and $x = L$. Equation (7.4.9) is again solved by introducing the integrating factor $e^{k\lambda_n t}$. The required initial condition for $b_n(t)$ follows from the given initial condition, $u(x, 0) = f(x)$:

$$f(x) = \sum_{n=1}^{\infty} b_n(0)\phi_n(x)$$

$$b_n(0) = \frac{\int_0^L f(x)\phi_n(x)\, dx}{\int_0^L \phi_n^2\, dx}.$$

It is interesting to note that the differential equation for the coefficients $b_n(t)$ for problems with nonhomogeneous boundary conditions is quite similar to the one that occurred in the preceding section for homogeneous boundary conditions; only the nonhomogeneous term is modified.

If the boundary conditions are homogeneous, $u(0, t) = 0$ and $u(L, t) = 0$, then (7.4.9) reduces to

$$\frac{db_n}{dt} + k\lambda_n b_n = q_n(t),$$

the differential equation derived in the preceding section. Using Green's formula is an alternative procedure to derive the eigenfunction expansion. It can be used even if the boundary conditions are homogeneous. In fact, it is this derivation that justifies the differentiation of infinite series of eigenfunctions used in Sec. 7.3.

We now have two procedures to solve nonhomogeneous partial differential equations with nonhomogeneous boundary conditions. By subtracting any function that just solves the nonhomogeneous boundary conditions, we can solve a related problem with homogeneous boundary conditions by the eigenfunction expansion method. Alternatively, we can solve directly the original problem with non-homogeneous boundary conditions by the method of eigenfunction expansions. In both cases we need the eigenfunction expansion of some function $w(x, t)$:

$$w(x, t) = \sum_{n=1}^{\infty} a_n(t)\phi_n(x).$$

If $w(x, t)$ satisfies the same homogeneous boundary conditions as $\phi_n(x)$, then we claim that this series will converge reasonably fast. However, if $w(x, t)$ satisfies nonhomogeneous boundary conditions, then not only will the series not satisfy the boundary conditions (at $x = 0$ and $x = L$), but the series will converge more slowly everywhere. Thus, *the advantage of reducing a problem to homogeneous boundary conditions is that the corresponding series converges faster.*

EXERCISES 7.4

7.4.1. In these problems, do not make a reduction to homogeneous boundary conditions. Solve the initial value problem for the heat equation with time-dependent sources

$$\frac{\partial u}{\partial t} = k\frac{\partial^2 u}{\partial x^2} + Q(x, t)$$

$$u(x, 0) = f(x)$$

subject to the following boundary conditions:

(a) $u(0, t) = A(t)$, $\quad \frac{\partial u}{\partial x}(L, t) = B(t)$

*(b) $\frac{\partial u}{\partial x}(0, t) = A(t)$, $\quad \frac{\partial u}{\partial x}(L, t) = B(t)$

7.4.2. Use the method of eigenfunction expansions to solve, without reducing to homogeneous boundary conditions:

$$\frac{\partial u}{\partial t} = k\frac{\partial^2 u}{\partial x^2}$$

$$u(x, 0) = f(x) \qquad \left.\begin{array}{l} u(0, t) = A \\ u(L, t) = B \end{array}\right\} \text{constants.}$$

7.4.3. Consider

$$c(x)\rho(x)\frac{\partial u}{\partial t} = \frac{\partial}{\partial x}\left[K_0(x)\frac{\partial u}{\partial x}\right] + q(x)u + f(x, t)$$

$$u(x, 0) = g(x) \qquad u(0, t) = \alpha(t)$$

$$u(L, t) = \beta(t).$$

Assume the eigenfunctions $\phi_n(x)$ of the related homogeneous problem are known.
(a) Solve without reducing to a problem with homogeneous boundary conditions.
(b) Solve by first reducing to a problem with homogeneous boundary conditions.

7.4.4. Reconsider

$$\frac{\partial u}{\partial t} = k\frac{\partial^2 u}{\partial x^2} + Q(x, t)$$

$$u(x, 0) = f(x) \qquad u(0, t) = 0$$
$$u(L, t) = 0.$$

Assume that the solution $u(x, t)$ has the appropriate smoothness, so that it may be represented by a Fourier cosine series

$$u(x, t) = \sum_{n=0}^{\infty} c_n(t) \cos\frac{n\pi x}{L}.$$

Solve for dc_n/dt. Show that c_n satisfies a first-order nonhomogeneous ordinary differential equation, but part of the nonhomogeneous term is not known. Make a brief philosophical conclusion.

7.5 FORCED VIBRATING MEMBRANES AND RESONANCE

The method of eigenfunction expansion may also be applied to nonhomogeneous partial differential equations with more than two independent variables. An interesting example is a vibrating membrane of arbitrary shape. In our previous analysis of membranes, vibrations were caused by the initial conditions. Another mechanism that will put a membrane into motion is an external force. The linear nonhomogeneous partial differential equation that describes a vibrating membrane is

$$\frac{\partial^2 u}{\partial t^2} = c^2 \nabla^2 u + Q(x, y, t), \qquad (7.5.1a)$$

where $Q(x, y, t)$ represents a time- and spatially dependent external force. To be completely general, there should be some boundary condition along the boundary of the membrane. However, it is more usual for a vibrating membrane to be fixed with zero vertical displacement. Thus, we will specify this homogeneous boundary condition,

$$u = 0, \qquad (7.5.1b)$$

on the entire boundary. Both the initial position and initial velocity are specified:

$$u(x, y, 0) = \alpha(x, y) \qquad (7.5.1c)$$

$$\frac{\partial u}{\partial t}(x, y, 0) = \beta(x, y). \qquad (7.5.1d)$$

To use the method of eigenfunction expansion, we must assume that we "know" the eigenfunctions of the related homogeneous problem. By applying the method of separation of variables to (7.5.1a) with $Q(x, y, t) = 0$ where the boundary condition is (7.5.1b), we obtain the problem satisfied by the eigenfunctions:

$$\nabla^2 \phi = -\lambda \phi, \tag{7.5.2}$$

with $\phi = 0$ on the entire boundary. We know these eigenfunctions are complete, and that different eigenfunctions are orthogonal (in a two-dimensional sense) with weight 1. We have also shown that $\lambda > 0$. However, the specific eigenfunctions depend on the geometric shape of the region. Explicit formulas can be obtained only for certain relatively simple geometries. Recall that for a rectangle ($0 \leq x \leq L$, $0 \leq y \leq H$) the eigenvalues are $\lambda_{nm} = (n\pi/L)^2 + (m\pi/H)^2$, and the corresponding eigenfunctions are $\phi_{nm}(x, y) = \sin n\pi x/L \sin m\pi y/H$, where $n = 1, 2, 3, \ldots$ and $m = 1, 2, 3. \ldots$. Also for a circle of radius a, we have shown that the eigenvalues are $\lambda_{mn} = (z_{mn}/a)^2$, where z_{mn} are the nth zeros of the Bessel function of order m, $J_m(z_{mn}) = 0$, and the corresponding eigenfunctions are both $J_m(z_{mn} r/a) \sin m\theta$ and $J_m(z_{mn}r/a) \cos m\theta$, where $n = 1, 2, 3, \ldots$ and $m = 0, 1, 2, 3, \ldots$.

In general, we designate the related homogeneous eigenfunctions $\phi_i(x, y)$. Any (piecewise smooth) function, including the desired solution for our forced vibrating membrane, may be expressed in terms of an infinite series of these eigenfunctions. Thus,

$$u(x, y, t) = \sum_i A_i(t) \phi_i(x, y). \tag{7.5.3}$$

Here the $\sum_i$ represents a summation over *all* eigenfunctions. For membranes it will include a double sum if we are able to separate variables for $\nabla^2 \phi + \lambda \phi = 0$.

Term-by-term differentiation. We will obtain an ordinary differential equation for the time-dependent coefficients, $A_i(t)$. The differential equation will be derived in two ways: direct substitution (with the necessary differentiation of infinite series of eigenfunctions) and through use of the multidimensional Green's formula. In either approach we need to assume there are no difficulties with the term-by-term differentiation of (7.5.3) with respect to t. Thus,

$$\frac{\partial^2 u}{\partial t^2} = \sum_i \frac{d^2 A_i}{dt^2} \phi_i(x, y). \tag{7.5.4}$$

Term-by-term spatial differentiations are allowed since both u and ϕ_i solve the same homogeneous boundary conditions:

$$\nabla^2 u = \sum_i A_i(t) \nabla^2 \phi_i(x, y). \tag{7.5.5}$$

This would not be valid if $u \neq 0$ on the entire boundary. Since $\nabla^2 \phi_i = -\lambda_i \phi_i$, it follows that (7.5.1a) becomes

$$\sum_i \left(\frac{d^2 A_i}{dt^2} + c^2 \lambda_i A_i \right) \phi_i = Q(x, y, t). \tag{7.5.6}$$

If we expand $Q(x, y, t)$ in terms of these same eigenfunctions,

$$Q(x, y, t) = \sum_i q_i(t)\phi_i(x, y), \quad \text{where} \quad q_i(t) = \frac{\iint Q\phi_i \, dx \, dy}{\iint \phi_i^2 \, dx \, dy}, \quad (7.5.7)$$

then

$$\boxed{\frac{d^2 A_i}{dt^2} + c^2 \lambda_i A_i = q_i(t).} \quad (7.5.8)$$

Thus, A_i solves a linear nonhomogeneous second-order differential equation.

Green's formula. An alternative way to derive (7.5.8) is to use Green's formula. We begin this derivation by determining d^2A_i/dt^2 directly from (7.5.4), only using the two-dimensional orthogonality of $\phi_i(x, y)$ (with weight 1):

$$\frac{d^2 A_i}{dt^2} = \frac{\iint \frac{\partial^2 u}{\partial t^2} \phi_i \, dx \, dy}{\iint \phi_i^2 \, dx \, dy}. \quad (7.5.9)$$

We then eliminate $\partial^2 u/\partial t^2$ from the partial differential equation (7.5.1a):

$$\frac{d^2 A_i}{dt^2} = \frac{\iint (c^2 \nabla^2 u + Q)\phi_i \, dx \, dy}{\iint \phi_i^2 \, dx \, dy}. \quad (7.5.10)$$

Recognizing the latter integral as the generalized Fourier coefficients of Q [see (7.5.7)], we have that

$$\frac{d^2 A_i}{dt^2} = q_i(t) + \frac{\iint c^2 \nabla^2 u \phi_i \, dx \, dy}{\iint \phi_i^2 \, dx \, dy}. \quad (7.5.11)$$

It is now appropriate to use the two-dimensional version of Green's formula:

$$\iint (\phi_i \nabla^2 u - u \nabla^2 \phi_i) \, dx \, dy = \oint (\phi_i \nabla u - u \nabla \phi_i) \cdot \hat{n} \, ds, \quad (7.5.12)$$

where ds represents differential arc length along the boundary and $\hat{n}$ is a unit outward normal to the boundary. In our situation u and ϕ_i satisfy homogeneous boundary conditions, and hence the boundary term in (7.5.12) vanishes:

$$\iint (\phi_i \nabla^2 u - u \nabla^2 \phi_i) \, dx \, dy = 0. \quad (7.5.13)$$

Equation (7.5.13) is perhaps best remembered as $\iint [uL(v) - vL(u)] \, dx \, dy = 0$, where $L = \nabla^2$. If the membrane did not have homogeneous boundary conditions, then (7.5.12) would be used instead of (7.5.13), as we did in Sec. 7.4. The advantage of the use of Green's formula is that we can also solve the problem if the boundary condition was nonhomogeneous. Through the use of (7.5.13),

$$\iint \phi_i \nabla^2 u \, dx \, dy = \iint u \nabla^2 \phi_i \, dx \, dy = -\lambda_i \iint u\phi_i \, dx \, dy = -\lambda A_i(t) \iint \phi_i^2 \, dx \, dy,$$

since $\nabla^2 \phi_i + \lambda_i \phi_i = 0$, and since, $A_i(t)$ is the generalized Fourier coefficient of $u(x, y, t)$:

$$A_i(t) = \frac{\iint u\phi_i \, dx \, dy}{\iint \phi_i^2 \, dx \, dy}. \tag{7.5.14}$$

Consequently, we derive from (7.5.11) that

$$\boxed{\frac{d^2 A_i}{dt^2} + c^2 \lambda_i A_i = q_i,} \tag{7.5.15}$$

the same second-order differential equation as already derived [see (7.5.8)], justifying the simpler term-by-term differentiation performed there.

Variation of parameters. We will need some facts about ordinary differential equations in order to solve (7.5.8) or (7.5.15). Equation (7.5.15) is a second-order linear nonhomogeneous differential equation with constant coefficients (since $\lambda_i c^2$ is constant). The general solution is a particular solution plus a linear combination of homogeneous solutions. In this problem the homogeneous solutions are $\sin c\sqrt{\lambda_i} \, t$ and $\cos c\sqrt{\lambda_i} \, t$, since $\lambda_i > 0$. A particular solution can always be obtained by variation of parameters. However, the method of undetermined coefficients is usually easier and should be used if $q_i(t)$ is a polynomial, exponential, sine, or cosine (or products and/or sums of these). Using the method of variation of parameters (see Sec. 8.3.2), it can be shown that the general solution of (7.5.15) is

$$\boxed{A_i(t) = c_1 \cos c\sqrt{\lambda_i} \, t + c_2 \sin c\sqrt{\lambda_i} \, t + \int_0^t q_i(\tau) \frac{\sin c\sqrt{\lambda_i}(t - \tau)}{c\sqrt{\lambda_i}} d\tau.} \tag{7.5.16}$$

Using this form, the initial conditions may be easily satisfied:

$$A_i(0) = c_1 \tag{7.5.17a}$$

$$\frac{dA_i}{dt}(0) = c_2 c\sqrt{\lambda_i}. \tag{7.5.17b}$$

From the initial conditions (7.5.1c) and (7.5.1d), it follows that

$$A_i(0) = \frac{\iint \alpha(x, y)\phi_i(x, y) \, dx \, dy}{\iint \phi_i^2 \, dx \, dy} \tag{7.5.18a}$$

$$\frac{dA_i}{dt}(0) = \frac{\iint \beta(x, y)\phi_i(x, y)\, dx\, dy}{\iint \phi_i^2\, dx\, dy}.$$ (7.5.18b)

The solution, in general, for a forced vibrating membrane is

$$u(x, y, t) = \sum_i A_i(t)\phi_i(x, y),$$

where ϕ_i is given by (7.5.2) and $A_i(t)$ is determined by (7.5.16)–(7.5.18).

If there is no external force, $Q(x, y, t) = 0$ [i.e., $q_i(t) = 0$], then $A_i(t) = c_1 \cos c\sqrt{\lambda_i}\, t + c_2 \sin c\sqrt{\lambda_i}\, t$. In this situation, the solution is

$$u(x, y, t) = \sum_i (a_i \cos c\sqrt{\lambda_i}\, t + b_i \sin c\sqrt{\lambda_i}\, t)\phi_i(x, y),$$

exactly the solution obtained by separation of variables. The natural frequencies of oscillation of a membrane are $c\sqrt{\lambda_i}$.

Period forcing.   We have just completed the analysis of the vibrations of an arbitrarily shaped membrane with arbitrary external forcing. We could easily specialize this result to rectangular and circular membranes. Instead of doing that, we continue to discuss an arbitrarily shaped membrane. However, let us suppose that the forcing function is purely oscillatory in time; specifically,

$$\boxed{Q(x, y, t) = \overline{Q}(x, y) \cos \omega t;}$$ (7.5.19)

that is, the **forcing frequency** is ω. We do not specify the spatial dependence, $\overline{Q}(x, y)$. The eigenfunction expansion of the forcing function is also needed. From (7.5.7) it follows that

$$q_i(t) = \gamma_i \cos \omega t,$$ (7.5.20)

where γ_i are constants,

$$\gamma_i \equiv \frac{\iint \overline{Q}(x, y)\phi_i(x, y)\, dx\, dy}{\iint \phi_i^2\, dx\, dy}.$$

From (7.5.15), the generalized Fourier coefficients solve the second-order differential equation,

$$\frac{d^2 A_i}{dt^2} + c^2\lambda_i A_i = \gamma_i \cos \omega t.$$ (7.5.21)

Since the r.h.s. of (7.5.21) is simple, a particular solution is more easily obtained by the method of undetermined coefficients [rather than by using the general form (7.5.16)]. Homogeneous solutions are again a linear combination of $\sin c\sqrt{\lambda_i}\, t$ and $\cos c\sqrt{\lambda_i}\, t$, representing the natural frequencies $c\sqrt{\lambda_i}$ of the membrane. The membrane is being forced at frequency ω.

The solution of (7.5.21) is not difficult. We might guess that a *particular solution* is in the form*

$$A_i(t) = B_i \cos \omega t. \qquad (7.5.22)$$

Substituting (7.5.22) into (7.5.21) shows that

$$B_i(c^2\lambda_i - \omega^2) = \gamma_i \qquad \text{or} \qquad B_i = \frac{\gamma_i}{c^2\lambda_i - \omega^2},$$

but this division is valid only if $\omega^2 \neq c^2\lambda_i$. The physical meaning of this result is that *if the forcing frequency ω is different from a natural frequency*, then a particular solution is

$$A_i(t) = \frac{\gamma_i \cos \omega t}{c^2\lambda_i - \omega^2}, \qquad (7.5.23)$$

and the general solution is

$$A_i(t) = \frac{\gamma_i \cos \omega t}{c^2\lambda_i - \omega^2} + c_1 \cos c\sqrt{\lambda_i}\, t + c_2 \sin c\sqrt{\lambda_i}\, t. \qquad (7.5.24)$$

$A_i(t)$ represents the amplitude of the mode $\phi_i(x, y)$. Each mode is composed of a vibration at its natural frequency $c\sqrt{\lambda_i}$ and a vibration at the forcing frequency ω. The closer these two frequencies are (for a given mode) the larger the amplitude of that mode.

Resonance. However, **if the forcing frequency ω is the same as one of the natural frequencies** $c\sqrt{\lambda_i}$, then a phenomenon known as **resonance occurs**. Mathematically, if $\omega^2 = c^2\lambda_i$, then for those modes [i.e., *only* those $\phi_i(x, y)$ such that $\omega^2 = c^2\lambda_i$], (7.5.22) is not the appropriate solution since the r.h.s. of (7.5.21) is a homogeneous solution. Instead, the solution is not periodic in time. The amplitude of oscillations grows proportional to t. Some algebra shows that a particular solution of (7.5.21) is

$$A_i(t) = \frac{\gamma_i}{2\omega} t \sin \omega t, \qquad (7.5.25a)$$

and hence the general solution is

$$A_i(t) = \frac{\gamma_i}{2\omega} t \sin \omega t + c_1 \cos \omega t + c_2 \sin \omega t, \qquad (7.5.25b)$$

where $\omega = c\sqrt{\lambda_i}$, for any mode that resonates. At resonance, natural modes corresponding to the forcing frequency grow in time without a bound. The other oscillating modes remain bounded. After a while, the resonating modes will dominate. Thus the spatial structure of a solution will be primarily due to the eigenfunctions of the resonant modes. The other modes are not significantly excited. We present a brief derivation of (7.5.25a), which avoids some tedious algebra. If $\omega^2 \neq c^2\lambda_i$, we obtain the general solution (7.5.23) relatively easily.

---

* If the first derivative term were present in (7.5.21), representing a frictional force, then a particular solution must include both $\cos \omega t$ *and* $\sin \omega t$.

Unfortunately, we cannot take the limit as $\omega \to c\sqrt{\lambda_i}$ since the amplitude then approaches infinity. However, from (7.5.24) we see that

$$A_i(t) = \frac{\gamma_i}{c^2 \lambda_i - \omega^2}(\cos \omega t - \cos c\sqrt{\lambda_i}\, t) \qquad (7.5.26)$$

is also an allowable solution* if $\omega^2 \neq c^2 \lambda_i$. However, (7.5.26) may have a limit as $\omega \to c^2 \lambda_i$, since $A_i(t)$ is in the form of $0/0$ as $\omega \to c\sqrt{\lambda_i}$. We calculate the limit of (7.5.26) as $\omega \to c\sqrt{\lambda_i}$ using l'Hospital's rule:

$$A_i(t) = \lim_{\omega \to c\sqrt{\lambda_i}} \frac{\gamma_i(\cos \omega t - \cos c\sqrt{\lambda_i}\, t)}{c^2 \lambda_i - \omega^2} = \lim_{\omega \to c\sqrt{\lambda_i}} \frac{-\gamma_i t \sin \omega t}{-2\omega},$$

verifying (7.5.25a).

The displacement of the resonated mode cannot grow indefinitely as (7.5.25b) suggests. The mathematics is correct, but some physical assumptions should be modified. Perhaps it is appropriate to include a frictional force, which limits the growth as is shown in an exercise. Alternatively, perhaps the mode grows to such a large amplitude that the linearization assumption, needed in a physical derivation of the two-dimensional wave equation, is no longer valid; a different partial differential equation should be appropriate for sufficiently large displacements. Perhaps the amplitude growth due to resonance would result in the snapping of the membrane (but this is not likely to happen until after the linearization assumption has been violated).

Note that we have demonstrated this result for any geometry. The introduction of the details of rectangular or circular geometries might just cloud the basic mathematical and physical phenomena.

Resonance for a vibrating membrane is similar mathematically to resonance for spring–mass systems (also without friction). In fact, resonance occurs for any mechanical system when a forcing frequency equals one of the natural frequencies. Disasters such as the infamous Tacoma Bridge collapse and various jet airplane crashes have been blamed on resonance phenomena.

EXERCISES 7.5

7.5.1. By substitution show that

$$y(t) = \frac{1}{\omega} \int_0^t f(\bar{t}) \sin \omega(t - \bar{t})\, d\bar{t}$$

is a particular solution of

$$\frac{d^2 y}{dt^2} + \omega^2 y = f(t).$$

What is the general solution? What solution satisfies the initial conditions $y(0) = y_0$ and $\frac{dy}{dt}(0) = v_0$?

* This solution corresponds to the initial conditions $A_i(0) = 0$, $dA_i/dt(0) = 0$.

7.5.2. Consider a vibrating string with time-dependent forcing:

$$\frac{\partial^2 u}{\partial t^2} = c^2 \frac{\partial^2 u}{\partial x^2} + Q(x, t)$$

$$u(0, t) = 0 \qquad u(x, 0) = f(x)$$

$$u(L, t) = 0 \qquad \frac{\partial u}{\partial t}(x, 0) = 0.$$

(a) Solve the initial value problem.
*(b) Solve the initial value problem if $Q(x, t) = g(x) \cos \omega t$. For what values of ω does resonance occur?

7.5.3. Consider a vibrating string with friction with time-periodic forcing

$$\frac{\partial^2 u}{\partial t^2} = c^2 \frac{\partial^2 u}{\partial x^2} - \beta \frac{\partial u}{\partial t} + g(x) \cos \omega t$$

$$u(0, t) = 0 \qquad u(x, 0) = f(x)$$

$$u(L, t) = 0 \qquad \frac{\partial u}{\partial t}(x, 0) = 0.$$

(a) Solve this initial value problem if β is small $(0 < \beta < 2c\pi/L)$.
(b) Compare this solution to Exercise 7.5.2(b).

7.5.4. Solve the initial value problem for a vibrating string with time-dependent forcing,

$$\frac{\partial^2 u}{\partial t^2} = c^2 \frac{\partial^2 u}{\partial x^2} + Q(x, t), \qquad u(x, 0) = f(x), \qquad \frac{\partial u}{\partial t}(x, 0) = 0,$$

subject to the following boundary conditions. Do not reduce to homogeneous boundary conditions:

(a) $u(0, t) = A(t), \qquad u(L, t) = B(t)$

(b) $u(0, t) = 0, \qquad \frac{\partial u}{\partial x}(L, t) = 0$

(c) $\frac{\partial u}{\partial x}(0, t) = A(t), \qquad u(L, t) = 0$

7.5.5. Solve the initial value problem for a membrane with time-dependent forcing and fixed boundaries ($u = 0$),

$$\frac{\partial^2 u}{\partial t^2} = c^2 \nabla^2 u + Q(x, y, t),$$

$$u(x, y, 0) = f(x, y), \qquad \frac{\partial u}{\partial t}(x, y, 0) = 0,$$

if the membrane is
(a) a rectangle $(0 < x < L, 0 < y < H)$
(b) a circle $(r < a)$
*(c) a semicircle $(0 < \theta < \pi, r < a)$
(d) a circular annulus $(a < r < b)$

7.5.6. Consider the displacement $u(r, \theta, t)$ of a forced semicircular membrane of radius α (Fig. 7.5.1) which satisfies the partial differential equation

$$\frac{1}{c^2} \frac{\partial^2 u}{\partial t^2} = \frac{1}{r} \frac{\partial}{\partial r} \left(r \frac{\partial u}{\partial r} \right) + \frac{1}{r^2} \frac{\partial^2 u}{\partial \theta^2} + g(r, \theta, t),$$

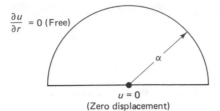

$\dfrac{\partial u}{\partial r} = 0$ (Free)

α

$u = 0$
(Zero displacement)

Figure 7.5.1

with the homogeneous boundary conditions:

$$u(r, 0, t) = 0, \quad u(r, \pi, t) = 0, \quad \text{and} \quad \frac{\partial u}{\partial r}(\alpha, \theta, t) = 0$$

and the initial conditions

$$u(r, \theta, 0) = H(r, \theta) \quad \text{and} \quad \frac{\partial u}{\partial t}(r, \theta, 0) = 0.$$

*(a) Assume that $u(r, \theta, t) = \Sigma\Sigma\, a(t)\phi(r, \theta)$, where $\phi(r, \theta)$ are the eigenfunctions of the related homogeneous problem. What initial conditions does $a(t)$ satisfy? What differential equation does $a(t)$ satisfy?

(b) What are the eigenfunctions?

(c) Solve for $u(r, \theta, t)$. (*Hint:* See Exercise 7.5.1.)

7.6 POISSON'S EQUATION

We have applied the method of eigenfunction expansion to nonhomogeneous time-dependent boundary value problems for PDEs (with or without homogeneous boundary conditions). In each case, the method of eigenfunction expansion,

$$u = \sum_i a_i(t)\phi_i,$$

yielded an *initial value problem* for the coefficients $a_i(t)$, where ϕ_i are the related homogeneous eigenfunctions satisfying, for examples,

$$\frac{d^2\phi}{dx^2} + \lambda\phi = 0 \quad \text{and} \quad \nabla^2\phi + \lambda\phi = 0.$$

Time-independent nonhomogeneous problems must be solved in a slightly different way. Consider the equilibrium temperature distribution with time-independent sources, which satisfies Poisson's equation,

$$\boxed{\nabla^2 u = Q,} \tag{7.6.1}$$

where Q is related to the sources of thermal energy. For now we do not specify the geometric region. However, we assume the temperature is specified on the entire boundary,

$$u = \alpha,$$

Nonhomogeneous Problems   Chap. 7

where α is given and usually *not* constant. This problem is nonhomogeneous in two ways; due to the forcing function Q and the boundary condition α. We can decompose the equilibrium temperature into two parts, $u = u_1 + u_2$, one u_1 due to the forcing and the other u_2 due to the boundary condition:

$$\nabla^2 u_1 = Q \qquad\qquad \nabla^2 u_2 = 0$$

$$u_1 = 0 \text{ on the boundary} \qquad u_2 = \alpha \quad \text{on the boundary.}$$

It is easily checked that $u = u_1 + u_2$ satisfies Poisson's equation and the nonhomogeneous BC. The problem for u_2 is the solution of Laplace's equation (with nonhomogeneous boundary conditions). For simple geometries this can be solved by the method of separation of variables (where in Secs. 2.5.1 and 6.9.1 we showed how homogeneous boundary conditions could be introduced).

Thus, at first in this section, we focus our attention on Poisson's equation

$$\nabla^2 u_1 = Q$$

with *homogeneous* boundary conditions ($u_1 = 0$ on the boundary). Since u_1 satisfies homogeneous BC, we should expect that the method of eigenfunction expansion is appropriate. The problem can be analyzed in two somewhat different ways: (1) we can expand the solutions in eigenfunctions of the related homogeneous problem, coming from separation of variables of $\nabla^2 u_1 = 0$ (as we did for the time-dependent problems), or (2) we can expand the solution in the eigenfunctions

$$\nabla^2 \phi + \lambda \phi = 0.$$

The two methods are different (but are related).

One-dimensional eigenfunctions. To be specific, let us consider the two-dimensional Poisson's equation in a rectangle with zero boundary conditions:

$$\boxed{\nabla^2 u_1 = Q,} \tag{7.6.2}$$

as illustrated in Fig. 7.6.1. We first describe the use of one-dimensional eigenfunctions. The related homogeneous problem, $\nabla^2 u_1 = 0$, which is Laplace's equation, can be separated (in rectangular coordinates). We may recall that the solution oscillates in one direction and is a combination of exponentials in the other direction. Thus, eigenfunctions of the related homogeneous problem (needed for the method of eigenfunction expansion) might be x-eigenfunctions or y-eigenfunctions. Since we have two homogeneous boundary conditions in *both* directions, we can use *either* x-dependent or y-dependent eigenfunctions. To be specific we use x-dependent eigenfunctions, which are $\sin n\pi x/L$ since $u_1 = 0$

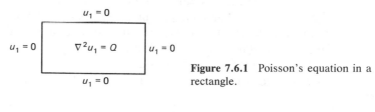

Figure 7.6.1 Poisson's equation in a rectangle.

at $x = 0$ and $x = L$. The method of eigenfunction expansion consists in expanding $u_1(x, y)$ in a series of these eigenfunctions:

$$u_1 = \sum_{n=1}^{\infty} b_n(y) \sin \frac{n\pi x}{L},$$ (7.6.3)

where the sine coefficients $b_n(y)$ are functions of y. Differentiating (7.6.3) twice with respect to y and substituting this into Poisson's equation, (7.6.2), yields

$$\sum_{n=1}^{\infty} \frac{d^2 b_n}{dy^2} \sin \frac{n\pi x}{L} + \frac{\partial^2 u_1}{\partial x^2} = Q.$$ (7.6.4a)

$\partial^2 u_1 / \partial x^2$ can be determined in two related ways (as we also showed for nonhomogeneous time-dependent problems): term-by-term differentiation with respect to x of the series (7.6.3) which is more direct or by use of Green's formula. In either way we obtain from (7.6.4a),

$$\sum_{n=1}^{\infty} \left[\frac{d^2 b_n}{dy^2} - \left(\frac{n\pi}{L} \right)^2 b_n \right] \sin \frac{n\pi x}{L} = Q,$$ (7.6.4b)

since both u_1 and $\sin n\pi x/L$ satisfy the same homogeneous boundary conditions. Thus, the Fourier sine coefficients satisfy the following second-order ordinary differential equation:

$$\frac{d^2 b_n}{dy^2} - \left(\frac{n\pi}{L} \right)^2 b_n = \frac{2}{L} \int_0^L Q(x, y) \sin \frac{n\pi x}{L} \, dx \equiv q_n(y),$$ (7.6.5)

where the right-hand side is the sine coefficient of Q,

$$Q = \sum_{n=1}^{\infty} q_n \sin \frac{n\pi x}{L}.$$ (7.6.6)

We must solve (7.6.5). Two conditions are needed. We have satisfied Poisson's equation and the boundary condition at $x = 0$ and $x = L$. The boundary condition at $y = 0$ (for all x), $u_1 = 0$, and at $y = H$ (for all x), $u_1 = 0$, imply that

$$b_n(0) = 0 \quad \text{and} \quad b_n(H) = 0.$$ (7.6.7)

Thus, the unknown coefficients in the method of eigenfunction expansion [see (7.6.5)] themselves solve a one-dimensional *nonhomogeneous boundary value problem*. Compare this result to the time-dependent nonhomogeneous PDE problems, in which the coefficients satisfied one-dimensional initial value problems. One-dimensional boundary value problems are more difficult to satisfy than initial value problems. Later we will discuss boundary value problems for ordinary differential equations. We will find different ways to solve (7.6.5) subject to the BC (7.6.7). One form of the solution we can obtain using the method of variation of parameters (see Sec. 8.3.2) is

$$b_n(y) = \sinh \frac{n\pi(H-y)}{L} \int_0^y q_n(\xi) \sinh \frac{n\pi\xi}{L} \, d\xi$$
$$+ \sinh \frac{n\pi y}{L} \int_y^H q_n(\xi) \sinh \frac{n\pi(H-\xi)}{L} \, d\xi. \tag{7.6.8}$$

Thus, we can solve Poisson's equation (with homogeneous boundary conditions) using the x-dependent related one-dimensional homogeneous eigenfunctions. Problems with nonhomogeneous boundary conditions can be solved in the same way introducing the appropriate modifications following from the use of Green's formula with nonhomogeneous conditions. In Exercise 7.6.1, the same problem is solved using the y-dependent related homogeneous eigenfunctions.

Two-dimensional eigenfunctions. A somewhat different way to solve Poisson's equation,

$$\nabla^2 u_1 = Q, \tag{7.6.9}$$

on a rectangle with zero BC is to consider the related two-dimensional eigenfunctions:

$$\nabla^2 \phi = -\lambda \phi$$

with $\phi = 0$ on the boundary. For a rectangle we know that this implies a sine series in x *and a* sine series in y:

$$\phi_{nm} = \sin \frac{n\pi x}{L} \sin \frac{m\pi y}{H}$$

$$\lambda_{nm} = \left(\frac{n\pi}{L}\right)^2 + \left(\frac{m\pi}{H}\right)^2.$$

The method of eigenfunction expansion consists of expanding the solution u_1 in terms of these two-dimensional eigenfunctions

$$u_1 = \sum_{n=1}^{\infty} \sum_{m=1}^{\infty} b_{nm} \sin \frac{n\pi x}{L} \sin \frac{m\pi y}{H}. \tag{7.6.10}$$

Here b_{nm} are constants (not a function of another variable) since u_1 only depends on x and y. The substitution of (7.6.10) into Poisson's equation (7.6.9) yields

$$\sum_{n=1}^{\infty} \sum_{m=1}^{\infty} -b_{nm}\lambda_{nm} \sin \frac{n\pi x}{L} \sin \frac{m\pi y}{H} = Q,$$

since $\nabla^2 \phi_{nm} = -\lambda_{nm}\phi_{nm}$. The Laplacian can be evaluated by term-by-term differentiation since both u_1 and ϕ_{nm} satisfy the same homogeneous boundary conditions. The eigenfunctions ϕ_{nm} are orthogonal (in a two-dimensional sense) with weight 1. Thus,

$$-b_{nm}\lambda_{nm} = \frac{\int_0^H \int_0^L Q \sin n\pi x/L \sin m\pi y/H \, dx \, dy}{\int_0^H \int_0^L \sin^2 n\pi x/L \sin^2 m\pi y/H \, dx \, dy}, \tag{7.6.11}$$

determining the b_{nm}. The expression on the r.h.s. of (7.6.11) is recognized as the generalized Fourier coefficients of Q. Dividing by λ_{nm} to solve for b_{nm} poses no difficulty since $\lambda_{nm} > 0$ (explicitly or by use of the Rayleigh quotient). It is easier to obtain the solution using the expansion in terms of two-dimensional eigenfunctions than using one-dimensional ones. However, doubly infinite series such as (7.6.10) may converge quite slowly. Numerical methods may be preferable except in simple cases. In Exercise 7.6.2 we show that the Fourier sine coefficients in y of $b_n(y)$ [see (7.6.3)] equals b_{nm} [see (7.6.10)]. This shows the equivalence of the one- and two-dimensional eigenfunction expansion approaches.

Nonhomogeneous boundary conditions (any geometry). The two-dimensional eigenfunctions can also be directly used for Poisson's equation subject to nonhomogeneous boundary conditions. It is no more difficult to indicate the solution for a rather general geometry. Suppose that

$$\boxed{\nabla^2 u = Q,} \tag{7.6.12}$$

with $u = \alpha$ on the boundary. Consider the eigenfunctions ϕ_i of $\nabla^2 \phi = -\lambda\phi$ with $\phi = 0$ on the boundary. We represent u in terms of these eigenfunctions:

$$\boxed{u = \sum_i b_i \phi_i.} \tag{7.6.13}$$

Now, it is *no longer* true that

$$\nabla^2 u = \sum_i b_i \nabla^2 \phi_i,$$

since u does not satisfy homogeneous boundary conditions. Instead, from (7.6.13) we know that

$$b_i = \frac{\iint u\phi_i \, dx \, dy}{\iint \phi_i^2 \, dx \, dy} = -\frac{1}{\lambda_i}\frac{\iint u \nabla^2\phi_i \, dx \, dy}{\iint \phi_i^2 \, dx \, dy}, \tag{7.6.14}$$

since $\nabla^2\phi_i = -\lambda_i\phi_i$. We can evaluate the numerator using Green's two-dimensional formula,

$$\iint (u \nabla^2 v - v \nabla^2 u) \, dx \, dy = \oint (u \nabla v - v \nabla u) \cdot \hat{\mathbf{n}} \, ds. \tag{7.6.15}$$

Letting $v = \phi_i$, we see

$$\iint u \nabla^2\phi_i \, dx \, dy = \iint \phi_i \nabla^2 u \, dx \, dy + \oint (u \nabla\phi_i - \phi_i \nabla u) \cdot \hat{\mathbf{n}} \, ds.$$

However, $\nabla^2 u = Q$ and on the boundary $\phi_i = 0$ and $u = \alpha$. Thus,

$$b_i = -\frac{1}{\lambda_i} \frac{\iint \phi_i Q \, dx \, dy + \oint \alpha \nabla \phi_i \cdot \hat{n} \, ds}{\iint \phi_i^2 \, dx \, dy}.$$ (7.6.16)

This is the general expression for b_i, since λ_i, ϕ_i, α, and Q are considered to be known. Again, dividing by λ_i causes no difficulty, since $\lambda_i > 0$ from the Rayleigh quotient. For problems in which $\lambda_i = 0$, see Sec. 8.4.

If u also satisfies homogeneous boundary conditions, $\alpha = 0$, then (7.6.16) becomes

$$b_i = -\frac{1}{\lambda_i} \frac{\iint \phi_i Q \, dx \, dy}{\iint \phi_i^2 \, dx \, dy},$$

agreeing with (7.6.11) in the case of a rectangular region. This shows that (7.6.10) may be term-by-term differentiated if u and ϕ satisfy the same homogeneous boundary conditions.

EXERCISES 7.6

7.6.1. Solve

$$\nabla^2 u = Q(x, y)$$

on a rectangle ($0 < x < L, 0 < y < H$) subject to

(a) $u(0, y) = 0,$ $u(x, 0) = 0$

   $u(L, y) = 0,$ $u(x, H) = 0$
   Use a Fourier sine series in y.

*(b) $u(0, y) = 0,$ $u(x, 0) = 0$

   $u(L, y) = 1,$ $u(x, H) = 0.$
   Do not reduce to homogeneous boundary conditions.

(c) Solve part (b) by first reducing to homogeneous boundary conditions.

*(d) $\dfrac{\partial u}{\partial x}(0, y) = 0,$ $\dfrac{\partial u}{\partial y}(x, 0) = 0$

   $\dfrac{\partial u}{\partial x}(L, y) = 0,$ $\dfrac{\partial u}{\partial y}(x, H) = 0.$
   In what situations are there solutions?

(e) $\dfrac{\partial u}{\partial x}(0, y) = 0,$ $u(x, 0) = 0$

   $\dfrac{\partial u}{\partial x}(L, y) = 0,$ $\dfrac{\partial u}{\partial y}(x, H) = 0.$

7.6.2. The solution of (7.6.5),

$$\frac{d^2 b_n}{dy^2} - \left(\frac{n\pi}{L}\right)^2 b_n = q_n(y),$$

subject to $b_n(0) = 0$ and $b_n(H) = 0$ is given by (7.6.8).
(a) Solve this instead by letting $b_n(y)$ equal a Fourier sine series.
(b) Show that this series is equivalent to (7.6.8).
(c) Show that this series is equivalent to the answer obtained by an expansion in the two-dimensional eigenfunctions, (7.6.10).

7.6.3. Solve (using two-dimensional eigenfunctions) $\nabla^2 u = Q(r, \theta)$ inside a circle of radius a subject to the given boundary condition. In what situations are there solutions?

*(a) $u(a, \theta) = 0$ (b) $\dfrac{\partial u}{\partial r}(a, \theta) = 0$

(c) $u(a, \theta) = f(\theta)$ (d) $\dfrac{\partial u}{\partial r}(a, \theta) = g(\theta)$

7.6.4. Solve Exercise 7.6.3 using one-dimensional eigenfunctions.

7.6.5. Consider

$$\nabla^2 u = Q(x, y)$$

inside an unspecified region with $u = 0$ on the boundary. Suppose that the eigenfunctions $\nabla^2 \phi = -\lambda\phi$ subject to $\phi = 0$ on the boundary are known. Solve for $u(x, y)$.

*7.6.6.** Solve the following example of Poisson's equation:

$$\nabla^2 u = e^{2y} \sin x$$

subject to the following boundary conditions:

$$u(0, y) = 0 \qquad u(x, 0) = 0$$
$$u(\pi, y) = 0 \qquad u(x, L) = f(x).$$

7.6.7. Solve

$$\nabla^2 u = Q(x, y, z)$$

inside a rectangular box $(0 < x < L, 0 < y < H, 0 < z < W)$ subject to $u = 0$ on the six sides.

7.6.8. Solve

$$\nabla^2 u = Q(r, \theta, z)$$

inside a circular cylinder $(0 < r < a, 0 < \theta < 2\pi, 0 < z < H)$ subject to $u = 0$ on the sides.

7.6.9. On a rectangle $(0 < x < L, 0 < y < H)$ consider

$$\nabla^2 u = Q(x, y)$$

with $\nabla u \cdot \hat{n} = 0$ on the boundary.
(a) Show that a solution exists only if $\iint Q(x, y)\, dx dy = 0$. Briefly explain using physical reasoning.
(b) Solve using the method of eigenfunction expansion. Compare to part (a). (*Hint:* $\lambda = 0$ is an eigenvalue.)
(c) If $\iint Q\, dx dy = 0$, determine the arbitary constant in the solution of part (b) by consideration of the time-dependent problem $\dfrac{\partial u}{\partial t} = k(\nabla^2 u - Q)$, subject to the initial condition $u(x, y, 0) = g(x, y)$.

7.6.10. Reconsider Exercise 7.6.9 for an arbitrary two-dimensional region.

8

Green's Functions for Time-Independent Problems

8.1 INTRODUCTION

Solutions to linear partial differential equations are nonzero due to initial conditions, nonhomogeneous boundary conditions, and forcing terms. If the partial differential equation is homogeneous and there is a set of homogeneous boundary conditions, then we usually attempt to solve the problem by the method of separation of variables. In Chapter 7 we developed the method of eigenfunction expansions to obtain solutions in cases in which there were forcing terms (and/or nonhomogeneous boundary conditions).

In this chapter, we will primarily consider problems without initial conditions (ordinary differential equations and Laplace's equation with sources). We will show that there is one function for each problem, called the Green's function, which can be used to describe the influence of both nonhomogeneous boundary conditions and forcing terms. We will develop properties of these Green's functions and show direct methods to obtain them. Time-dependent problems with initial conditions, such as the heat and wave equations, are more difficult. They will be used as motivation, but detailed study of their Green's functions will not be presented until Chapter 10.

8.2 ONE-DIMENSIONAL HEAT EQUATION

We begin by reanalyzing the one-dimensional heat equation with no sources and homogeneous boundary conditions:

$$\frac{\partial u}{\partial t} = k \frac{\partial^2 u}{\partial x^2} \tag{8.2.1a}$$

277

$$u(0, t) = 0 \tag{8.2.1b}$$

$$u(L, t) = 0 \tag{8.2.1c}$$

$$u(x, 0) = g(x). \tag{8.2.1d}$$

In Chapter 2, according to the method of separation of variables, we obtained

$$u(x, t) = \sum_{n=1}^{\infty} a_n \sin \frac{n\pi x}{L} e^{-k(n\pi/L)^2 t}, \tag{8.2.2}$$

where the initial condition implied that a_n are the coefficients of the Fourier sine series of $g(x)$,

$$g(x) = \sum_{n=1}^{\infty} a_n \sin \frac{n\pi x}{L} \tag{8.2.3a}$$

$$a_n = \frac{2}{L} \int_0^L g(x) \sin \frac{n\pi x}{L} dx. \tag{8.2.3b}$$

We examine this solution (8.2.2) more closely in order to investigate the effect of the initial condition $g(x)$. We eliminate the Fourier sine coefficients from (8.2.3b) (introducing a dummy integration variable x_0):

$$u(x, t) = \sum_{n=1}^{\infty} \left[\frac{2}{L} \int_0^L g(x_0) \sin \frac{n\pi x_0}{L} dx_0 \right] \sin \frac{n\pi x}{L} e^{-k(n\pi/L)^2 t}.$$

If we interchange the order of operations of the infinite summation and integration, we obtain

$$u(x, t) = \int_0^L g(x_0) \left(\sum_{n=1}^{\infty} \frac{2}{L} \sin \frac{n\pi x_0}{L} \sin \frac{n\pi x}{L} e^{-k(n\pi/L)^2 t} \right) dx_0. \tag{8.2.4}$$

We define the quantity in parenthesis as the **influence function** for the initial condition. It expresses the fact that the temperature at position x at time t is due to the initial temperature at x_0. To obtain the temperature $u(x, t)$, we sum (integrate) the influences of all possible initial positions.

Before further interpreting this result, it is helpful to do a similar analysis for a more general heat equation including sources, but still having homogeneous boundary conditions

$$\frac{\partial u}{\partial t} = k \frac{\partial^2 u}{\partial x^2} + Q(x, t) \tag{8.2.5a}$$

$$u(0, t) = 0 \tag{8.2.5b}$$

$$u(L, t) = 0 \tag{8.2.5c}$$

$$u(x, 0) = g(x). \tag{8.2.5d}$$

This nonhomogeneous problem is suited for the method of eigenfunction expansions,

$$u(x, t) = \sum_{n=1}^{\infty} a_n(t) \sin \frac{n\pi x}{L}. \tag{8.2.6}$$

This Fourier sine series can be differentiated term by term since both $\sin n\pi x/L$ and $u(x, t)$ solve the same homogeneous boundary conditions. Hence, $a_n(t)$ solves the following first-order differential equation:

$$\frac{da_n}{dt} + k\left(\frac{n\pi}{L}\right)^2 a_n = q_n(t) = \frac{2}{L}\int_0^L Q(x, t) \sin\frac{n\pi x}{L}\,dx, \tag{8.2.7}$$

where $q_n(t)$ are the coefficients of the Fourier sine series of $Q(x, t)$,

$$Q(x, t) = \sum_{n=1}^{\infty} q_n(t) \sin\frac{n\pi x}{L}. \tag{8.2.8}$$

The solution of (8.2.7) (using the integrating factor $e^{k(n\pi/L)^2 t}$) is

$$a_n(t) = a_n(0)e^{-k(n\pi/L)^2 t} + e^{-k(n\pi/L)^2 t}\int_0^t q_n(t_0)e^{k(n\pi/L)^2 t_0}\,dt_0. \tag{8.2.9}$$

$a_n(0)$ are the coefficients of the Fourier sine series of the initial condition, $u(x, 0) = g(x)$:

$$g(x) = \sum_{n=1}^{\infty} a_n(0) \sin\frac{n\pi x}{L} \tag{8.2.10a}$$

$$a_n(0) = \frac{2}{L}\int_0^L g(x) \sin\frac{n\pi x}{L}\,dx. \tag{8.2.10b}$$

These Fourier coefficients may be eliminated, yielding

$$u(x, t) = \sum_{n=1}^{\infty}\left[\left(\frac{2}{L}\int_0^L g(x_0) \sin\frac{n\pi x_0}{L}\,dx_0\right)e^{-k(n\pi/L)^2 t}\right.$$
$$\left. + e^{-k(n\pi/L)^2 t}\int_0^t\left(\frac{2}{L}\int_0^L Q(x_0, t_0) \sin\frac{n\pi x_0}{L}\,dx_0\right)e^{k(n\pi/L)^2 t_0}\,dt_0\right]\sin\frac{n\pi x}{L}.$$

After interchanging the order of performing the infinite summation and the integration (over both x_0 and t_0), we obtain

$$u(x, t) = \int_0^L g(x_0)\left(\sum_{n=1}^{\infty}\frac{2}{L}\sin\frac{n\pi x_0}{L}\sin\frac{n\pi x}{L}e^{-k(n\pi/L)^2 t}\right)dx_0$$
$$+ \int_0^L\int_0^t Q(x_0, t_0)\left(\sum_{n=1}^{\infty}\frac{2}{L}\sin\frac{n\pi x_0}{L}\sin\frac{n\pi x}{L}e^{-k(n\pi/L)^2(t-t_0)}\right)dt_0\,dx_0.$$

We therefore introduce the **Green's function**, $G(x, t; x_0, t_0)$,

$$G(x, t; x_0, t_0) = \sum_{n=1}^{\infty}\frac{2}{L}\sin\frac{n\pi x_0}{L}\sin\frac{n\pi x}{L}e^{-k(n\pi/L)^2(t-t_0)}. \tag{8.2.11}$$

We have shown that

$$u(x, t) = \int_0^L g(x_0)G(x, t; x_0, 0)\,dx_0$$
$$+ \int_0^L\int_0^t Q(x_0, t_0)G(x, t; x_0, t_0)\,dt_0\,dx_0. \tag{8.2.12}$$

The Green's function at $t_0 = 0$, $G(x, t; x_0, 0)$, expresses the influence of the initial temperature at x_0 on the temperature at position x and time t. In addition, $G(x, t; x_0, t_0)$ shows the influence on the temperature at the position x and time t of the forcing term $Q(x_0, t_0)$ at position x_0 and time t_0. Instead of depending on the source time t_0 and the response time t, independently, we note that the Green's function depends only on the **elapsed time** $t - t_0$:

$$G(x, t; x_0, t_0) = G(x, t - t_0; x_0, 0).$$

This occurs because the heat equation has coefficients that do not change in time; the laws of thermal physics are not changing. The Green's function exponentially decays in elapsed time $(t - t_0)$ [see (8.2.11)]. For example, this means that the influence of the source at time t_0 diminishes rapidly. It is only the most recent sources of thermal energy that are important at time t.

Equation (8.2.11) is an extremely useful representation of the Green's function if time t is large. However, for small t the series converges more slowly. In Chapter 10 we will obtain an alternative representation of the Green's function useful for small t.

In (8.2.12) we integrate over all positions x_0. The solution is the result of adding together the influences of all sources and initial temperatures. We also integrate the sources over all *past* times $0 < t_0 < t$. This is part of a **causality principle**. The temperature at time t is only due to the thermal sources that acted *before* time t. Any future sources of heat energy cannot influence the temperature now.

Among the questions we will investigate later for this and other problems are:

1. Are there more direct methods to obtain the Green's function?
2. Are there any simpler expressions for it [can we simplify (8.2.11)]?
3. Can we explain the relationships between the influence of the initial condition and the influence of the forcing terms?
4. Can we account easily for nonhomogeneous boundary conditions?

EXERCISES 8.2

8.2.1. Consider

$$\frac{\partial u}{\partial t} = k \frac{\partial^2 u}{\partial x^2} + Q(x, t)$$

$$u(x, 0) = g(x).$$

In all cases obtain formulas similar to (8.2.12) by introducing a Green's function.
(a) Use Green's formula instead of term-by-term spatial differentiation if

$$u(0, t) = 0 \quad \text{and} \quad u(L, t) = 0.$$

(b) Modify part (a) if

$$u(0, t) = A(t) \quad \text{and} \quad u(L, t) = B(t).$$

Do not reduce to a problem with homogeneous boundary conditions.

(c) Solve using any method if

$$\frac{\partial u}{\partial x}(0, t) = 0 \quad \text{and} \quad \frac{\partial u}{\partial x}(L, t) = 0.$$

*(d) Use Green's formula instead of term-by-term differentiation if

$$\frac{\partial u}{\partial x}(0, t) = A(t) \quad \text{and} \quad \frac{\partial u}{\partial x}(L, t) = B(t).$$

8.2.2. Solve by the method of eigenfunction expansion

$$c\rho \frac{\partial u}{\partial t} = \frac{\partial}{\partial x}\left(K_0 \frac{\partial u}{\partial x}\right) + Q(x, t)$$

subject to $u(0, t) = 0$, $u(L, t) = 0$, and $u(x, 0) = g(x)$, if $c\rho$ and K_0 are functions of x. Assume that the eigenfunctions are known. Obtain a formula similar to (8.2.12) by introducing a Green's function.

*8.2.3. Solve by the method of eigenfunction expansion

$$\frac{\partial^2 u}{\partial t^2} = c^2 \frac{\partial^2 u}{\partial x^2} + Q(x, t)$$

$$u(0, t) = 0 \qquad u(x, 0) = f(x)$$

$$u(L, t) = 0 \qquad \frac{\partial u}{\partial t}(x, 0) = g(x).$$

Define functions (in the simplest possible way) such that a relationship similar to (8.2.12) exists. It must be somewhat different due to the two initial conditions. (*Hint:* See Exercise 7.5.1.)

8.2.4. Modify Exercise 8.2.3 (using Green's formula if necessary) if instead

(a) $\frac{\partial u}{\partial x}(0, t) = 0$ and $\frac{\partial u}{\partial x}(L, t) = 0$

(b) $u(0, t) = A(t)$ and $u(L, t) = 0$

(c) $\frac{\partial u}{\partial x}(0, t) = 0$ and $\frac{\partial u}{\partial x}(L, t) = B(t)$

8.3 GREEN'S FUNCTIONS FOR BOUNDARY VALUE PROBLEMS FOR ORDINARY DIFFERENTIAL EQUATIONS

8.3.1 One-Dimensional Steady-State Heat Equation

Introduction. Investigating the Green's functions for the time-dependent heat equation is not an easy task. Instead, we first investigate a simpler problem. Most of the techniques discussed will be valid for more difficult problems.

We will investigate the steady-state heat equation with homogeneous boundary conditions, arising in situations in which the source term $Q(x, t) = Q(x)$ is independent of time:

$$0 = k\frac{d^2 u}{dx^2} + Q(x).$$

We prefer the form

$$\frac{d^2u}{dx^2} = f(x),$$ (8.3.1a)

in which case $f(x) = -Q(x)/k$. The boundary conditions we consider are

$$u(0) = 0 \quad \text{and} \quad u(L) = 0.$$ (8.3.1b, c)

We will solve this problem in many different ways in order to suggest methods for other harder problems.

Limit of time-dependent problem. One way (not the most obvious nor easiest) to solve (8.3.1) is to analyze our solution (8.2.12) of the time-dependent problem, obtained in the preceding section, in the special case of a steady source:

$$u(x, t) = \int_0^L g(x_0)G(x, t; x_0, 0) \, dx_0$$

$$+ \int_0^L -kf(x_0)\left(\int_0^t G(x, t; x_0, t_0) \, dt_0\right) dx_0,$$ (8.3.2)

where

$$G(x, t; x_0, t_0) = \sum_{n=1}^{\infty} \frac{2}{L} \sin\frac{n\pi x_0}{L} \sin\frac{n\pi x}{L} e^{-k(n\pi/L)^2(t-t_0)}.$$ (8.3.3)

As $t \to \infty$, $G(x, t; x_0, 0) \to 0$ such that the effect of the initial condition $u(x, 0) = g(x)$ vanishes at $t \to \infty$. However, even though $G(x, t; x_0, t_0) \to 0$ as $t \to \infty$, the steady source is still important as $t \to \infty$ since

$$\int_0^t e^{-k(n\pi/L)^2(t-t_0)} \, dt_0 = \frac{e^{-k(n\pi/L)^2(t-t_0)}}{k(n\pi/L)^2}\bigg|_{t_0=0}^{t} = \frac{1 - e^{-k(n\pi/L)^2 t}}{k(n\pi/L)^2}.$$

Thus, as $t \to \infty$,

$$u(x, t) \longrightarrow \boxed{u(x) = \int_0^L f(x_0)G(x, x_0) \, dx_0,}$$ (8.3.4)

where

$$\boxed{G(x, x_0) = -\sum_{n=1}^{\infty} \frac{2}{L} \frac{\sin n\pi x_0/L \sin n\pi x/L}{(n\pi/L)^2}.}$$ (8.3.5)

Here we obtained the steady-state temperature distribution $u(x)$ by taking the limit as $t \to \infty$ of the time-dependent problem with a steady source $Q(x) = -kf(x)$. $G(x, x_0)$ is the **influence** or **Green's function** for the steady-state problem.

The symmetry,

$$G(x, x_0) = G(x_0, x),$$

will be discussed later.

8.3.2 The Method of Variation of Parameters

There are more direct ways to obtain the solution of (8.3.1). We consider a more general nonhomogeneous problem

$$L(u) = f(x), \qquad (8.3.6)$$

defined for $a < x < b$, subject to two homogeneous boundary conditions (of the standard form discussed in Chapter 5), where L is the Sturm–Liouville operator:

$$L = \frac{d}{dx}\left(p\frac{d}{dx}\right) + q. \qquad (8.3.7)$$

For the simple steady-state heat equation of the preceding subsection, $p = 1$ and $q = 0$, so that $L = d^2/dx^2$.

Nonhomogeneous ordinary differential equations can always be solved by the **method of variation of parameters** if two* solutions of the homogeneous problem are known, $u_1(x)$ and $u_2(x)$. We briefly review this technique. In the method of variation of parameters, a particular solution of (8.3.6) is sought in the form

$$\boxed{u = v_1 \cdot u_1 + v_2 \cdot u_2,} \qquad (8.3.8)$$

where v_1 and v_2 are functions of x to be determined. The original differential equation has one unknown function, so that the extra degree of freedom allows us to assume du/dx is the same as if v_1 and v_2 were constants:

$$\frac{du}{dx} = v_1\frac{du_1}{dx} + v_2\frac{du_2}{dx}.$$

Since v_1 and v_2 are not constant, this is valid only if the other terms, arising from the variation of v_1 and v_2, vanish:

$$\frac{dv_1}{dx}u_1 + \frac{dv_2}{dx}u_2 = 0.$$

The differential equation $L(u) = f(x)$ is then satisfied if

$$\frac{dv_1}{dx}p\frac{du_1}{dx} + \frac{dv_2}{dx}p\frac{du_2}{dx} = f(x).$$

The method of variation of parameters at this stage yields two linear equations

---

* Actually, only one homogeneous solution is necessary as the method of reduction of order is a procedure for obtaining a second homogeneous solution if one is known.

for the unknowns dv_1/dx and dv_2/dx. The solution is

$$\frac{dv_1}{dx} = \frac{-fu_2}{p\left(u_1\dfrac{du_2}{dx} - u_2\dfrac{du_1}{dx}\right)} = \frac{-fu_2}{c} \qquad (8.3.9a)$$

$$\frac{dv_2}{dx} = \frac{fu_1}{p\left(u_1\dfrac{du_2}{dx} - u_2\dfrac{du_1}{dx}\right)} = \frac{fu_1}{c}, \qquad (8.3.9b)$$

where

$$c = p\left(u_1\frac{du_2}{dx} - u_2\frac{du_1}{dx}\right). \qquad (8.3.9c)$$

Using the Wronskian described shortly, we will show that c is constant. The constant c depends on the choice of homogeneous solutions u_1 and u_2. The general solution of $L(u) = f(x)$ is given by $u = u_1v_1 + u_2v_2$, where v_1 and v_2 are obtained by integrating (8.3.9).

Wronskian. We define the **Wronskian** W as

$$W = u_1\frac{du_2}{dx} - u_2\frac{du_1}{dx}.$$

It satisfies an elementary differential equation:

$$\frac{dW}{dx} = u_1\frac{d^2u_2}{dx^2} - u_2\frac{d^2u_1}{dx^2} = -\frac{dp/dx}{p}\left(u_1\frac{du_2}{dx} - u_2\frac{du_1}{dx}\right) = -\frac{dp/dx}{p}W, \quad (8.3.10)$$

where the defining differential equations for the homogeneous solutions, $L(u_1) = 0$ and $L(u_2) = 0$, have been used. Solving (8.3.10) shows that

$$W = \frac{c}{p} \quad \text{or} \quad pW = c.$$

Example   Consider the problem (8.3.1)

$$\frac{d^2u}{dx^2} = f(x) \quad \text{with} \quad u(0) = 0 \quad \text{and} \quad u(L) = 0.$$

This corresponds to the general case (8.3.6) with $p = 1$ and $q = 0$. Two homogeneous solutions of (8.3.1a) are 1 and x. However, the algebra is easier if we pick $u_1(x)$ to be a homogeneous solution satisfying one of the boundary conditions $u(0) = 0$ and $u_2(x)$ to be a homogeneous solution satisfying the other boundary condition:

$$u_1(x) = x$$

$$u_2(x) = L - x.$$

Since $p = 1$, $c = -L$ from (8.3.9c). By integrating (8.3.9), we obtain

$$v_1(x) = \frac{1}{L}\int_0^x f(x_0)\,(L - x_0)\,dx_0 + c_1$$

$$v_2(x) = -\frac{1}{L}\int_0^x f(x_0)\,x_0\,dx_0 + c_2,$$

which is needed in the method of variation of parameters ($u = u_1 v_1 + u_2 v_2$). The boundary condition $u(0) = 0$ yields $0 = c_2 L$, whereas $u(L) = 0$ yields

$$0 = \int_0^L f(x_0)(L - x_0) \, dx_0 + c_1 L,$$

so that $v_1(x) = -\dfrac{1}{L} \displaystyle\int_x^L f(x_0)(L - x_0) \, dx_0$. Thus, the solution of the nonhomogeneous boundary value problem is

$$u(x) = -\frac{x}{L} \int_x^L f(x_0)(L - x_0) \, dx_0 - \frac{L - x}{L} \int_0^x f(x_0) x_0 \, dx_0. \qquad (8.3.11)$$

We will transform (8.3.11) into the desired form,

$$u(x) = \int_0^L f(x_0) G(x, x_0) \, dx_0. \qquad (8.3.12)$$

By comparing (8.3.11) to (8.3.12), we obtain

$$G(x, x_0) = \begin{cases} -\dfrac{x(L - x_0)}{L} & x < x_0 \\ -\dfrac{x_0(L - x)}{L} & x > x_0. \end{cases} \qquad (8.3.13)$$

A sketch and interpretation of this solution will be given in Sec. 8.3.5. Although somewhat complicated, the symmetry can be seen:

$$G(x, x_0) = G(x_0, x).$$

For the steady-state heat equation we have obtained two Green's functions, (8.3.5) and (8.3.13). They appear quite different. In Exercise 8.3.1 they are shown to be the same. In particular, (8.3.13) yields a piecewise smooth function (actually continuous), and its Fourier sine series can be shown to be given by (8.3.5).

The solution also can be derived by directly integrating (8.3.1) twice:

$$u = \int_0^x \left[\int_0^{x_0} f(\bar{x}) \, d\bar{x} \right] dx_0 + c_1 x + c_2. \qquad (8.3.14)$$

In Exercise 8.3.2, you are asked to show that (8.3.13) can be obtained from (8.3.14). This can be done by interchanging the order of integration in (8.3.14) or by integrating (8.3.14) by parts.

8.3.3 The Method of Eigenfunction Expansion for Green's Functions

In Chapter 7, nonhomogeneous partial differential equations were solved by the eigenfunction expansion method. Here we show how to apply the same ideas to the general Sturm–Liouville nonhomogeneous ordinary differential equation:

$$L(u) = f(x) \tag{8.3.15}$$

subject to two homogeneous boundary conditions. We introduce a related eigenvalue problem,

$$L(\phi) = -\lambda \sigma \phi, \tag{8.3.16}$$

subject to the *same* homogeneous boundary conditions. The weight σ here can be chosen arbitrarily. However, there is usually at most one choice of $\sigma(x)$ such that the differential equation (8.3.16) is in fact well known.* We solve (8.3.15) by seeking $u(x)$ as a generalized Fourier series of the eigenfunctions

$$u(x) = \sum_{n=1}^{\infty} a_n \phi_n(x). \tag{8.3.17}$$

We can differentiate this twice term by term† since both $\phi_n(x)$ and $u(x)$ solve the same homogeneous boundary conditions:

$$\sum_{n=1}^{\infty} a_n L(\phi_n) = -\sum_{n=1}^{\infty} a_n \lambda_n \sigma \phi_n = f(x),$$

where (8.3.16) has been used. The orthogonality of the eigenfunctions (with weight σ) implies that

$$-a_n \lambda_n = \frac{\int_a^b f(x) \phi_n \, dx}{\int_a^b \phi_n^2 \sigma \, dx}. \tag{8.3.18}$$

The solution of the boundary value problem for the nonhomogeneous ordinary differential equation is thus (after interchanging summation and integration)

$$u(x) = \int_a^b f(x_0) \sum_{n=1}^{\infty} \frac{\phi_n(x) \phi_n(x_0)}{-\lambda_n \int_a^b \phi_n^2 \sigma \, dx} \, dx_0. \tag{8.3.19}$$

For this problem, the Green's function has the representation in terms of the eigenfunctions:

$$G(x, x_0) = \sum_{n=1}^{\infty} \frac{\phi_n(x) \phi_n(x_0)}{-\lambda_n \int_a^b \phi_n^2 \sigma \, dx}. \tag{8.3.20}$$

---

* For example, if $L = d^2/dx^2$, we pick $\sigma = 1$ giving trigonometric functions, but if $L = \dfrac{d}{dx}\left(x \dfrac{d}{dx}\right) - \dfrac{m^2}{x}$, we pick $\sigma = x$ so that Bessel functions occur.

† Green's formula can be used to justify this step (see Sec. 7.4).

Green's Functions for Time-Independent Problems   Chap. 8

Again the symmetry is explicitly shown. Note the appearance of the eigenvalues λ_n in the denominator. The Green's function does not exist if one of the eigenvalues is zero. This will be explained in Sec. 8.4. For now we assume that all $\lambda_n \neq 0$.

Example. For the boundary value problem,

$$\frac{d^2u}{dx^2} = f(x)$$

$$u(0) = 0 \quad \text{and} \quad u(L) = 0,$$

the related eigenvalue problem,

$$\frac{d^2\phi}{dx^2} = -\lambda\phi$$

$$\phi(0) - 0 \quad \text{and} \quad \phi(L) = 0,$$

is well known. The eigenvalues are $\lambda_n = (n\pi/L)^2$, $n = 1, 2, 3, \ldots$, and the corresponding eigenfunctions are $\sin n\pi x/L$. The Fourier sine series of $u(x)$ is given by (8.3.17). In particular,

$$u(x) = \int_0^L f(x_0)G(x, x_0)\, dx_0,$$

where the Fourier sine series of the Green's function is

$$G(x, x_0) = -\frac{2}{L} \sum_{n=1}^{\infty} \frac{\sin n\pi x/L \sin n\pi x_0/L}{(n\pi/L)^2}$$

from (8.3.20), agreeing with the answer (8.3.5) obtained by the limit as $t \to \infty$ of the time-dependent problem.

8.3.4 The Dirac Delta Function and Its Relationship to Green's Functions

We have shown that

$$u(x) = \int_0^L f(x_0)G(x, x_0)\, dx_0,$$

where we have obtained different representations of the Green's function. The Green's function shows the influence of each position x_0 of the source on the solution at x. In this section, we will find a more direct way to determine the Green's function.

Dirac delta function. Our source $f(x)$ represents a forcing of our system at all points. $f(x)$ is sketched in Fig. 8.3.1. In order to isolate the effect of each

Figure 8.3.1 Piecewise constant representation of a function.

individual point we decompose $f(x)$ into a linear combination of unit pulses of duration Δx; see Fig. 8.3.2:

$$f(x) \approx \sum_i f(x_i) \text{ (unit pulse starting at } x = x_i).$$

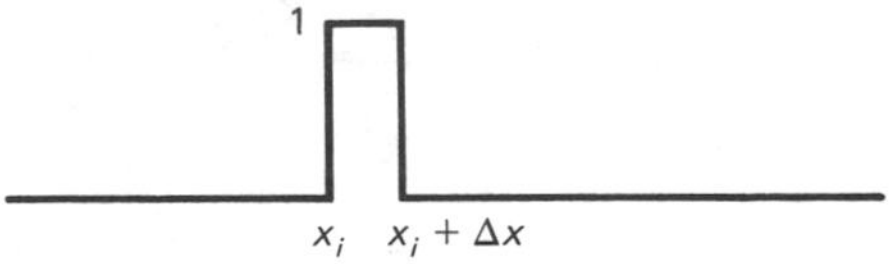

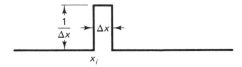

Figure 8.3.2 Pulse with unit length.

This is somewhat reminiscent of the definition of an integral. Only Δx is missing, which we introduce by multiplying and dividing by Δx:

$$f(x) = \lim_{\Delta x \to 0} \sum_i f(x_i) \frac{\text{unit pulse}}{\Delta x} \Delta x. \tag{8.3.21}$$

In this way we have motivated a rectangular pulse of width Δx and height $1/\Delta x$, sketched in Fig. 8.3.3. It has unit area. In the limit as $\Delta x \to 0$, this approaches an infinitely concentrated pulse (not really a function) $\delta(x - x_i)$ which would be zero everywhere except ∞ at $x = x_i$, still with unit area:

(a)
$$\delta(x - x_i) = \begin{cases} 0 & x \neq x_i \\ \infty & x = x_i, \end{cases} \tag{8.3.22}$$

Figure 8.3.3 Rectangular pulse with unit area.

We can think of $\delta(x - x_i)$ as a **concentrated source** or **impulsive force** at $x = x_i$. According to (8.3.21) we have

$$f(x) = \int f(x_i)\, \delta(x - x_i)\, dx_i. \tag{8.3.23}$$

Since $\delta(x - x_i)$ is not a function, we define it as an operator with the property that for any continuous $f(x)$:

(b)
$$\boxed{f(x) = \int_{-\infty}^{\infty} f(x_i)\delta(x - x_i)dx_i,} \tag{8.3.24}$$

as is suggested by (8.3.23). We call $\delta(x - x_i)$, the **Dirac delta function**.* It is so concentrated that in integrating it with any continuous function $f(x_i)$, it "sifts" out the value at $x_i = x$. The Dirac delta function may be motivated by the "limiting function" of any sequence of concentrated pulses (the shape need not be rectangular).

---

* Named after Paul Dirac, a twentieth-century mathematical physicist (1902–1984).

Other important properties of the Dirac delta function are that it has unit area:

(c)
$$1 = \int_{-\infty}^{\infty} \delta(x - x_i) \, dx_i;$$
(8.3.25)

it is an even function

(d)
$$\delta(x - x_i) = \delta(x_i - x).$$
(8.3.26)

This means that the definition (8.3.24) may be used without worrying about whether $\delta(x - x_i)$ or $\delta(x_i - x)$ appears. The Dirac delta function is also the derivative of the Heaviside unit step function $H(x - x_i)$:

$$H(x - x_i) \equiv \begin{cases} 0 & x < x_i \\ 1 & x > x_i; \end{cases}$$
(8.3.27)

(e)
$$\delta(x - x_i) = \frac{d}{dx} H(x - x_i);$$
(8.3.28a)

(f)
$$H(x - x_i) = \int_{-\infty}^{x} \delta(x_0 - x_i) \, dx_0;$$
(8.3.28b)

it has the following scaling property:

(g)
$$\delta[c(x - x_i)] = \frac{1}{|c|} \delta(x - x_i).$$
(8.3.29)

These properties are proved in the exercises.

Green's function. The solution of the nonhomogeneous problem

$$L(u) = f(x),$$
(8.3.30)

subject to two homogeneous boundary conditions is

$$u(x) = \int_{a}^{b} f(x_0) G(x, x_0) \, dx_0.$$
(8.3.31)

Here, the Green's function is the influence function for the source $f(x)$. As an example, suppose that $f(x)$ is a concentrated source at $x = x_s$, $f(x) = \delta(x - x_s)$. Then the response at x, $u(x)$, satisfies

$$u(x) = \int_{a}^{b} \delta(x_0 - x_s) G(x, x_0) \, dx_0 = G(x, x_s)$$

due to (8.3.24). This yields the fundamental interpretation of the **Green's function** $G(x, x_s)$; *it is the* **response at x due to a concentrated source at x_s**:

$$L[G(x, x_s)] = \delta(x - x_s),$$
(8.3.32)

where $G(x, x_s)$ will also satisfy the same homogeneous boundary conditions at $x = a$ and $x = b$.

As a check, let us verify that (8.3.31) satisfies (8.3.30). To satisfy (8.3.30), we must use the operator L (in the simple case, $L = d^2/dx^2$):

$$L(u) = \int_a^b f(x_0)L[G(x, x_0)] \, dx_0 = \int_a^b f(x_0) \, \delta(x - x_0) \, dx_0 = f(x),$$

where the fundamental property of both the Green's function (8.3.32) and the Dirac delta function (8.3.24) has been used.

Often (8.3.32) with two homogeneous boundary conditions is thought of as an *independent definition of the Green's function*. In this case we might want to *derive* (8.3.31), the representation of the solution of the nonhomogeneous problem in terms of the Green's function satisfying (8.3.32). The usual method to derive (8.3.31) involves Green's formula:

$$\int_a^b [uL(v) - vL(u)] \, dx = p\left(u\frac{dv}{dx} - v\frac{du}{dx}\right)\Bigg|_a^b. \tag{8.3.33}$$

If we let $v = G(x, x_0)$, then the right-hand side vanishes since both $u(x)$ and $G(x, x_0)$ satisfy the same homogeneous boundary conditions. Furthermore, from the respective differential equations, (8.3.30) and (8.3.32), it follows that

$$\int_a^b [u(x) \, \delta(x - x_0) - G(x, x_0)f(x)] \, dx = 0.$$

Thus, from the definition of the Dirac delta function

$$u(x_0) = \int_a^b f(x)G(x, x_0) \, dx.$$

If we interchange the variables x and x_0, we obtain (8.3.31):

$$u(x) = \int_a^b f(x_0)G(x_0, x) \, dx_0 = \int_a^b f(x_0)G(x, x_0) \, dx_0, \tag{8.3.34}$$

since the Green's function is known to be symmetric (8.3.13), $G(x_0, x) = G(x, x_0)$.

Maxwell's reciprocity. The symmetry of the Green's function is very important. We will prove it without using the eigenfunction expansion. Instead, we will directly use the defining differential equation (8.3.32). We again use Green's formula (8.3.33). Here we let $u = G(x, x_1)$ and $v = G(x, x_2)$. Since both satisfy the same homogeneous boundary conditions, it follows that the right-hand side is zero. In addition, $L(u) = \delta(x - x_1)$ while $L(v) = \delta(x - x_2)$, and thus

$$\int_a^b [G(x, x_1) \, \delta(x - x_2) - G(x, x_2) \, \delta(x - x_1)] \, dx = 0.$$

From the fundamental property of the Dirac delta function, it follows that

$$\boxed{G(x_1, x_2) = G(x_2, x_1),} \tag{8.3.35}$$

proving the symmetry from the differential equation defining the Green's function. This symmetry is remarkable; we call it **Maxwell's reciprocity. The response at x due to a concentrated source at x_0 is the same as the response at x_0 due to a concentrated source at x.** This is *not* physically obvious.

Jump conditions. The Green's function $G(x, x_s)$ may be determined from (8.3.32). For $x < x_s$, $G(x, x_s)$ must be a homogeneous solution satisfying the homogeneous boundary condition at $x = a$. A similar procedure is valid for $x > x_s$. Jump conditions across $x = x_s$ are determined from the singularity in (8.3.32). If $G(x, x_s)$ has a jump discontinuity at $x = x_s$, then dG/dx has a delta function singularity at $x = x_s$, and d^2G/dx^2 would be more singular than the right-hand side of (8.3.32). Thus, **the Green's function $G(x, x_s)$ is continuous at $x = x_s$.** However, dG/dx is not continuous at $x = x_s$; it has a jump discontinuity obtained by integrating (8.3.32) across $x = x_s$. We illustrate this method in the next example and leave further discussion to the exercises.

Example. Consider the solution of the steady-state heat flow problem

$$\frac{d^2u}{dx^2} = f(x) \tag{8.3.36a}$$

$$u(0) = 0 \quad \text{and} \quad u(L) = 0. \tag{8.3.36b}$$

We have shown that the solution can be represented in terms of the Green's function:

$$u(x) = \int_0^L f(x_0)G(x, x_0)\,dx_0, \tag{8.3.37}$$

where the Green's function satisfies the following problem:

$$\frac{d^2G(x, x_0)}{dx^2} = \delta(x - x_0) \tag{8.3.38a}$$

$$G(0, x_0) = 0 \quad \text{and} \quad G(L, x_0) = 0. \tag{8.3.38b}$$

One reason for defining the Green's function by the differential equation is that it gives an alternative (and often easier) way to calculate the Green's function. Here x_0 is a parameter, representing the position of a concentrated source. For $x \neq x_0$ there are no sources, and hence the steady-state heat distribution $G(x, x_0)$ must be linear ($d^2G/dx^2 = 0$):

$$G(x, x_0) = \begin{cases} a + bx & x < x_0 \\ c + dx & x > x_0, \end{cases}$$

but the constants may be different. The boundary condition at $x = 0$ applies for $x < x_0$. $G(0, x_0) = 0$ implies that $a = 0$. Similarly, $G(L, x_0) = 0$ implies that $c + dL = 0$:

$$G(x, x_0) = \begin{cases} bx & x < x_0 \\ d(x - L) & x > x_0. \end{cases}$$

This preliminary result is sketched in Fig. 8.3.4.

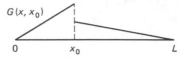

Figure 8.3.4 Green's function before application of jump conditions at $x = x_0$.

The two remaining constants are determined by two conditions at $x = x_0$. The temperature $G(x, x_0)$ must be continuous at $x = x_0$,

$$G(x_0-, x_0) = G(x_0+, x_0), \tag{8.3.39}$$

and there is a jump in the derivative of $G(x, x_0)$, most easily derived by integrating the defining differential equation (8.3.38a) from $x = x_{0-}$ to $x = x_{0+}$:

$$-\frac{dG}{dx}\bigg|_{x=x_0^+} + \frac{dG}{dx}\bigg|_{x=x_0^-} = -1. \tag{8.3.40}$$

Equation (8.3.39) implies that

$$bx_0 = d(x_0 - L),$$

while (8.3.40) yields

$$b - d = -1.$$

By solving these simultaneously, we obtain

$$d = \frac{x_0}{L} \quad \text{and} \quad b = \frac{x_0 - L}{L},$$

and thus

$$G(x, x_0) = \begin{cases} -\dfrac{x}{L}(L - x_0) & x < x_0 \\ -\dfrac{x_0}{L}(L - x) & x > x_0, \end{cases} \tag{8.3.41}$$

agreeing with (8.3.9). We sketch the Green's function in Fig. 8.3.5. The negative nature of this Green's function is due to the negative concentrated source of thermal energy, $-\delta(x - x_0)$, since $0 = d^2G/dx^2\ (x, 0) - \delta(x - x_0)$.

The symmetry of the Green's function (proved earlier) is apparent in all representations we have obtained. For example, letting $L = 1$,

$$G(x, x_0) = \begin{cases} -x(1 - x_0) & x < x_0 \\ -x_0(1 - x) & x > x_0 \end{cases} \quad \text{and} \quad G(\tfrac{1}{2}, \tfrac{1}{5}) = G(\tfrac{1}{5}, \tfrac{1}{2}) = -\tfrac{1}{10}.$$

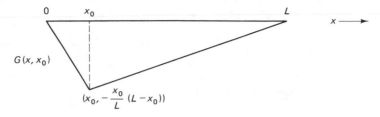

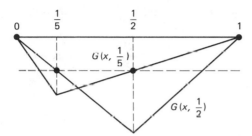

Figure 8.3.5 Green's function.

We sketch $G(x, \frac{1}{5})$ and $G(x, \frac{1}{2})$ in Fig. 8.3.6. Their equality cannot be explained by simple physical symmetries.

Figure 8.3.6 Illustration of Maxwell's reciprocity.

8.3.5 Nonhomogeneous Boundary Conditions

We have shown how to use Green's functions to solve nonhomogeneous differential equations with homogeneous boundary conditions. In this subsection we extend these ideas to include problems with nonhomogeneous boundary conditions:

$$\frac{d^2u}{dx^2} = f(x) \tag{8.3.42a}$$

$$u(0) = \alpha \quad \text{and} \quad u(L) = \beta. \tag{8.3.42b}$$

We will use the *same* Green's function as we did previously with problems with homogeneous boundary conditions:

$$\frac{d^2G}{dx^2} = \delta(x - x_0) \tag{8.3.43a}$$

$$G(0, x_0) = 0 \quad \text{and} \quad G(L, x_0) = 0; \tag{8.3.43b}$$

the Green's function always satisfies the related homogeneous boundary conditions.

To obtain the representation of the solution of (8.3.42) involving the Green's function, we again utilize Green's formula, with $v = G(x, x_0)$:

$$\int_0^L \left[u(x) \frac{d^2G(x, x_0)}{dx^2} - G(x, x_0) \frac{d^2u}{dx^2} \right] dx = u \frac{dG(x, x_0)}{dx} - G(x, x_0) \frac{du}{dx} \Big|_0^L.$$

The right-hand side now does not vanish since $u(x)$ does not satisfy homogeneous boundary conditions. Instead, using only the definitions of our problem (8.3.42) and the Green's function (8.3.43), we obtain

$$\int_0^L [u(x)\,\delta(x - x_0) - G(x, x_0)f(x)]\,dx = u(L)\left.\frac{dG(x, x_0)}{dx}\right|_{x=L} - u(0)\left.\frac{dG(x, x_0)}{dx}\right|_{x=0}.$$

We analyze this as before. Using the property of the Dirac delta function (and reversing the roles of x and x_0) and using the symmetry of the Green's function, we obtain

$$u(x) = \int_0^L f(x_0)G(x, x_0)\,dx_0 + \beta\left.\frac{dG(x, x_0)}{dx_0}\right|_{x_0=L} - \alpha\left.\frac{dG(x, x_0)}{dx_0}\right|_{x_0=0}. \qquad (8.3.44)$$

This is a representation of the solution of our nonhomogeneous problem (including nonhomogeneous boundary conditions) in terms of the standard Green's function. We must be careful in evaluating the boundary terms. In our problem, we have already shown that

$$G(x, x_0) = \begin{cases} -\dfrac{x}{L}(L - x_0) & x < x_0 \\[2mm] -\dfrac{x_0}{L}(L - x) & x > x_0. \end{cases}$$

The derivative with respect to the source position of the Green's function is thus

$$\frac{dG(x, x_0)}{dx_0} = \begin{cases} \dfrac{x}{L} & x < x_0 \\[2mm] -\left(1 - \dfrac{x}{L}\right) & x > x_0. \end{cases}$$

Evaluating this at the end points yields

$$\left.\frac{dG(x, x_0)}{dx_0}\right|_{x_0=L} = \frac{x}{L} \quad \text{and} \quad \left.\frac{dG(x, x_0)}{dx_0}\right|_{x_0=0} = -\left(1 - \frac{x}{L}\right).$$

Consequently,

$$u(x) = \int_0^L f(x_0)G(x, x_0)\,dx_0 + \beta\frac{x}{L} + \alpha\left(1 - \frac{x}{L}\right). \qquad (8.3.45)$$

The solution is the sum of a particular solution of (8.3.42a) satisfying homogeneous boundary conditions obtained earlier, $\int_0^L f(x_0)G(x, x_0)\,dx_0$, and a homogeneous solution satisfying the two required nonhomogeneous boundary conditions, $\beta(x/L) + \alpha(1 - x/L)$.

8.3.6 Summary

We have described three fundamental methods to obtain Green's functions:

1. Variation of parameters
2. Method of eigenfunction expansion
3. Using the defining differential equation for the Green's function

In addition, steady-state Green's functions can be obtained as the limit as $t \to \infty$ of the solution with steady sources. To obtain Green's functions for partial differential equations, we will discuss one important additional method. It will be described in Sec. 8.5.

EXERCISES 8.3

8.3.1. The Green's function for (8.3.1) is given explicitly by (8.3.13). The method of eigenfunction expansion yields (8.3.5). Show that the Fourier sine series of (8.3.13) yields (8.3.5).

8.3.2. (a) Derive (8.3.14).
(b) Integrate (8.3.14) by parts to derive (8.3.13).
(c) Instead of part (b), simplify the double integral in (8.3.14) by interchanging the orders of integration. Derive (8.3.13) this way.

8.3.3. Consider

$$\frac{\partial u}{\partial t} = k \frac{\partial^2 u}{\partial x^2} + Q(x, t)$$

subject to $u(0, t) = 0$, $\frac{\partial u}{\partial x}(L, t) = 0$, and $u(x, 0) = g(x)$.

(a) Solve by the method of eigenfunction expansion.
(b) Determine the Green's function for this time-dependent problem.
(c) If $Q(x, t) = Q(x)$, take the limit as $t \to \infty$ of part (b) in order to determine the Green's function for

$$\frac{d^2 u}{dx^2} = f(x) \quad \text{with} \quad u(0) = 0 \quad \text{and} \quad \frac{du}{dx}(L) = 0.$$

8.3.4. (a) Derive (8.3.25) from (8.3.24) [*Hint:* Let $f(x) - 1$.]
(b) Show that (8.3.28b) satisfies (8.3.27).
(c) Derive (8.3.26) [*Hint:* Show for any continuous $f(x)$ that

$$\int_{-\infty}^{\infty} f(x_0)\delta(x - x_0)\, dx_0 = \int_{-\infty}^{\infty} f(x_0)\delta(x_0 - x)\, dx_0$$

by letting $x_0 - x = s$ in the integral on the right.]
(d) Derive (8.3.29) [*Hint:* Evaluate $\int_{-\infty}^{\infty} f(x)\delta[c(x - x_0)]\, dx$ by making the change of variables $y = c(x - x_0)$.]

8.3.5. Consider

$$\frac{d^2 u}{dx^2} = f(x) \quad \text{with} \quad u(0) = 0 \quad \text{and} \quad \frac{du}{dx}(L) = 0.$$

*(a) Solve by direct integration.
*(b) Solve by the method of variation of parameters.
*(c) Determine $G(x, x_0)$ so that (8.3.12) is valid.
(d) Solve by the method of eigenfunction expansion. Show that $G(x, x_0)$ is given by (8.3.20).

8.3.6. Consider

$$\frac{d^2 G}{dx^2} = \delta(x - x_0) \quad \text{with} \quad G(0, x_0) = 0 \quad \text{and} \quad \frac{dG}{dx}(L, x_0) = 0.$$

*(a) Solve directly.

*(b) Graphically illustrate $G(x, x_0) = G(x_0, x)$.

(c) Compare to Exercise 8.3.5.

8.3.7. Redo Exercise 8.3.5 with the following change: $\dfrac{du}{dx}(L) + hu(L) = 0, h > 0$.

8.3.8. Redo Exercise 8.3.6 with the following change: $\dfrac{du}{dx}(L) + hu(L) = 0, h > 0$.

8.3.9. Consider

$$\frac{d^2u}{dx^2} + u = f(x) \quad \text{with} \quad u(0) = 0 \quad \text{and} \quad u(L) = 0.$$

Assume that $(n\pi/L)^2 \neq 1$ (i.e., $L \neq n\pi$ for any n).

(a) Solve by the method of variation of parameters.

*(b) Determine the Green's function so that $u(x)$ may be represented in terms of it [see (8.3.12)].

8.3.10. Solve the problem of Exercise 8.3.9 using the method of eigenfunction expansion.

8.3.11. Consider

$$\frac{d^2G}{dx^2} + G = \delta(x - x_0) \quad \text{with} \quad G(0, x_0) = 0 \quad \text{and} \quad G(L, x_0) = 0.$$

*(a) Solve for this Green's function directly. Why is it necessary to assume that $L \neq n\pi$?

(b) Show that $G(x, x_0) = G(x_0, x)$.

8.3.12. For the following problems, determine a representation of the solution in terms of the Green's function. Show that the nonhomogeneous boundary conditions can also be understood using homogeneous solutions of the differential equation:

(a) $\dfrac{d^2u}{dx^2} = f(x), u(0) = A, \dfrac{du}{dx}(L) = B.$ (See Exercise 8.3.6.).

(b) $\dfrac{d^2u}{dx^2} + u = f(x), u(0) = A, u(L) = B.$ Assume $L \neq n\pi$. (See Exercise 8.3.11).

(c) $\dfrac{d^2u}{dx^2} = f(x), u(0) = A, \dfrac{du}{dx}(L) + hu(L) = 0.$ (See Exercise 8.3.8.)

8.3.13. Consider the one-dimensional infinite space wave equation with a periodic source of frequency ω:

$$\frac{\partial^2 \phi}{\partial t^2} = c^2 \frac{\partial^2 \phi}{\partial x^2} + g(x)e^{-i\omega t}. \tag{8.3.46}$$

(a) Show that a particular solution $\phi = u(x)e^{-i\omega t}$ of (8.3.46) is obtained if u satisfies a nonhomogeneous Helmholtz equation

$$\frac{d^2u}{dx^2} + k^2u = f(x).$$

*(b) The Green's function $G(x, x_0)$ satisfies

$$\frac{d^2G}{dx^2} + k^2G = \delta(x - x_0).$$

Determine this infinite space Green's function so that the corresponding $\phi(x, t)$ is an outward propagating wave.

(c) Determine a particular solution of (8.3.46) above.

8.3.14. Consider $L(u) = f(x)$ with $L = \dfrac{d}{dx}\left(p\dfrac{d}{dx}\right) + q$. Assume that the appropriate Green's function exists. Determine the representation of $u(x)$ in terms of the Green's function if the boundary conditions are nonhomogeneous:

(a) $u(0) = \alpha$ and $u(L) = \beta$

(b) $\dfrac{du}{dx}(0) = \alpha$ and $\dfrac{du}{dx}(L) = \beta$

(c) $u(0) = \alpha$ and $\dfrac{du}{dx}(L) = \beta$

*(d) $u(0) = \alpha$ and $\dfrac{du}{dx}(L) + hu(L) = \beta$

8.3.15. Consider $L(G) = \delta(x - x_0)$ with $L = \dfrac{d}{dx}\left(p\dfrac{d}{dx}\right) + q$ subject to the boundary conditions $G(0, x_0) = 0$ and $G(L, x_0) = 0$. Introduce for all x two homogeneous solutions, y_1 and y_2, such that each solves one of the homogeneous boundary conditions:

$$L(y_1) = 0 \qquad L(y_2) = 0$$
$$y_1(0) = 0 \qquad y_2(L) = 0$$
$$\frac{dy_1}{dx}(0) = 1 \qquad \frac{dy_2}{dx}(L) = 1.$$

Even if y_1 and y_2 cannot be explicitly obtained, they can be easily calculated numerically on a computer as two *initial value problems*. Any homogeneous solution must be a linear combination of the two.

*(a) Solve for $G(x, x_0)$ in terms of $y_1(x)$ and $y_2(x)$. You may assume that $y_1(x) \neq cy_2(x)$.

(b) What goes wrong if $y_1(x) = cy_2(x)$ for all x and why?

8.3.16. Reconsider (8.3.36), whose solution we have obtained, (8.3.41). For (8.3.36) what is y_1 and y_2 in Exercise 8.3.15? Show that $G(x, x_0)$ obtained in Exercise 8.3.15 reduces to (8.3.41) for (8.3.36).

8.3.17. Consider

$$L(u) = f(x) \qquad \text{with} \qquad L = \frac{d}{dx}\left(p\frac{d}{dx}\right) + q$$

$$u(0) = 0 \qquad \text{and} \qquad u(L) = 0.$$

Introduce two homogeneous solutions, y_1 and y_2, as in Exercise 8.3.15.
(a) Determine $u(x)$ using the method of variation of parameters.
(b) Determine the Green's function from part (a).
(c) Compare to Exercise 8.3.15.

8.3.18. Reconsider Exercise 8.3.17. Determine $u(x)$ by the method of eigenfunction expansion. Show that the Green's function satisfies (8.3.20).

8.3.19. (a) If a concentrated source is placed at a node of some mode (eigenfunction), show that the amplitude of the response of that mode is zero. [*Hint:* Use the result of the method of eigenfunction expansion and recall that a node x^* of an eigenfunction means anyplace where $\phi_n(x^*) = 0$.]
(b) If the eigenfunctions are $\sin n\pi x/L$ and the source is located in the middle, $x_0 = L/2$, show that the response will have no even harmonics.

8.3.20. Derive the eigenfunction expansion of the Green's function (8.3.20) directly from the defining differential equation (8.3.36) by letting

$$G(x, x_0) = \sum_{n=1}^{\infty} a_n \phi_n(x).$$

Assume that term-by-term differentiation is justified.

*8.3.21. Solve

$$\frac{dG}{dx} = \delta(x - x_0) \qquad \text{with} \qquad G(0, x_0) = 0.$$

Show that $G(x, x_0)$ is not symmetric even though $\delta(x - x_0)$ is.

8.3.22. Solve

$$\frac{dG}{dx} + G = \delta(x - x_0) \qquad \text{with} \qquad G(0, x_0) = 0.$$

Show that $G(x, x_0)$ is not symmetric even though $\delta(x - x_0)$ is.

8.3.23. Solve

$$\frac{d^4 G}{dx^4} = \delta(x - x_0)$$

$$G(0, x_0) = 0 \qquad G(L, x_0) = 0$$

$$\frac{dG}{dx}(0, x_0) = 0 \qquad \frac{d^2 G}{dx^2}(L, x_0) = 0.$$

8.3.24. Use Exercise 8.3.23 to solve

$$\frac{d^4 u}{dx^4} = f(x)$$

$$u(0) = 0 \qquad u(L) = 0$$

$$\frac{du}{dx}(0) = 0 \qquad \frac{d^2 u}{dx^2}(L) = 0$$

(*Hint:* Exercise 5.5.8 is helpful.)

8.3.25. Use the convolution theorem for Laplace transforms to obtain particular solutions of

(a) $\dfrac{d^2 u}{dx^2} = f(x)$ (See Exercise 8.3.5.)

*(b) $\dfrac{d^4 u}{dx^4} = f(x)$ (See Exercise 8.3.24.)

8.3 APPENDIX: ESTABLISHING GREEN'S FORMULA WITH DIRAC DELTA FUNCTIONS

Green's formula is very important when analyzing Green's functions. However, our derivation of Green's formula requires integration by parts. Here we will show that Green's formula,

$$\int_a^b [uL(v) - vL(u)] \, dx = p \left(u \frac{dv}{dx} - v \frac{du}{dx} \right) \Big|_a^b, \quad \text{where } L = \frac{d}{dx}\left(p \frac{d}{dx} \right) + q \qquad (8.3A.1)$$

is valid even if v is a Green's function,

$$L(v) = \delta(x - x_0). \tag{8.3A.2}$$

We will derive (8.3A.1). We calculate the left-hand side of (8.3A.1). Since there is a singularity at $x = x_0$, we are not guaranteed that (8.3A.1) is valid. Instead, we divide the region into three parts:

$$\int_a^b = \int_a^{x_0-} + \int_{x_0-}^{x_0+} + \int_{x_0+}^b.$$

In the regions that exclude the singularity, $a \leqslant x \leqslant x_{0-}$ and $x_{0+} \leqslant x \leqslant b$, Green's formula can be used. In addition, due to the property of the Dirac delta function

$$\int_{x_0-}^{x_0+} [uL(v) - vL(u)]\, dx = \int_{x_0-}^{x_0+} [u\delta(x - x_0) - vL(u)]\, dx = u(x_0),$$

since $\int_{x_0-}^{x_0+} vL(u)\, dx = 0$. Thus, we obtain

$$\int_a^b [uL(v) - vL(u)]\, dx = p\left(u\frac{dv}{dx} - v\frac{du}{dx}\right)\bigg|_a^{x_0-} + p\left(u\frac{dv}{dx} - v\frac{du}{dx}\right)\bigg|_{x_0+}^b + u(x_0)$$

$$= p\left(u\frac{dv}{dx} - v\frac{du}{dx}\right)\bigg|_u^b + p\left(u\frac{dv}{dx} - v\frac{du}{dx}\right)\bigg|_{x_0+}^{x_0-} + u(x_0). \tag{8.3A.3}$$

Since u, du/dx, and v are continuous at $x = x_0$; it follows that

$$p\left(u\frac{dv}{dx} - v\frac{du}{dx}\right)\bigg|_{x_0+}^{x_0-} = p(x_0)u(x_0)\frac{dv}{dx}\bigg|_{x_0+}^{x_0-}.$$

However, by integrating (8.3A.2), we know that $p\,dv/dx|_{x_0-}^{x_0+} = 1$. Thus, (8.3A.1) follows from (8.3A.3). Green's formula may be utilized even if Green's functions are present.

8.4 FREDHOLM ALTERNATIVE AND MODIFIED GREEN'S FUNCTIONS

8.4.1 Introduction

If $\lambda = 0$ is an eigenvalue, then the Green's function does not exist. In order to understand the difficulty, we reexamine the nonhomogeneous problem:

$$L(u) = f(x), \tag{8.4.1}$$

subject to homogeneous boundary conditions. By the method of eigenfunction expansion, in the preceding section we obtained

$$u = \sum_{n=1}^\infty a_n\phi_n(x), \tag{8.4.2}$$

where by substitution

$$-a_n\lambda_n = \frac{\displaystyle\int_a^b f(x)\phi_n(x)\, dx}{\displaystyle\int_a^b \phi_n^2\sigma\, dx}. \tag{8.4.3}$$

If $\lambda_n = 0$ (for some n, often the lowest eigenvalue), there may not be any solutions to the nonhomogeneous boundary value problem. In particular, if $\int_a^b f(x)\phi_n(x)\, dx \neq 0$, for the eigenfunction corresponding to $\lambda_n = 0$, then (8.4.3) cannot be satisfied. This warrants further explanation.

Example. Let us consider the following simple nonhomogeneous boundary value problem:

$$\frac{d^2 u}{dx^2} = e^x \quad \text{with} \quad \frac{du}{dx}(0) = 0 \quad \text{and} \quad \frac{du}{dx}(L) = 0. \tag{8.4.4}$$

We attempt to solve (8.4.4) by integrating:

$$\frac{du}{dx} = e^x + c.$$

The two boundary conditions cannot be satisfied as they are contradictory:

$$0 = 1 + c$$
$$0 = e^L + c.$$

There is no guarantee that there are *any* solutions to a nonhomogeneous boundary value problem when $\lambda = 0$ is an eigenvalue for the related eigenvalue problem $[d^2\phi_n/dx^2 = -\lambda_n\phi_n$ with $d\phi_n/dx\,(0) = 0$ and $d\phi_n/dx\,(L) = 0]$.

In this example from one physical point of view, we are searching for an equilibrium temperature distribution. Since there are sources and the boundary conditions are of the insulated type, we know that an equilibrium temperature can only exist if there is no net input of thermal energy:

$$\int_0^L e^x\, dx = 0,$$

which is not valid. Since thermal energy is being constantly removed, there can be no equilibrium $(0 = d^2u/dx^2 - e^x)$.

Zero eigenvalue. If $\lambda = 0$ is an eigenvalue, we have shown that there may be difficulty in solving

$$L(u) = f(x), \tag{8.4.5}$$

subject to homogeneous boundary conditions. The eigenfunctions ϕ_n satisfy

$$L(\phi_n) = -\lambda_n\sigma\phi_n,$$

subject to the same homogeneous boundary conditions. Thus, if $\lambda = 0$ is an eigenvalue, the corresponding eigenfunction $\phi_h(x)$ satisfies

$$L(\phi_h) = 0 \tag{8.4.6}$$

with the same homogeneous boundary conditions. Thus, $\phi_h(x)$ is a nontrivial homogeneous solution of (8.4.5). This is important: **Nontrivial homogeneous solutions of (8.4.5) solving the same homogeneous boundary conditions are equivalent to eigenfunctions corresponding to the zero eigenvalue.** If there are no nontrivial homogeneous solutions (solving the same homogeneous boundary conditions), then $\lambda = 0$ is not an eigenvalue. If there are nontrivial homogeneous solutions, then $\lambda = 0$ is an eigenvalue.

The notion of a homogeneous solution is less confusing than can be a zero eigenvalue. For example, consider

$$\frac{d^2\phi}{dx^2} + \phi = e^x \quad \text{with} \quad \phi(0) = 0 \quad \text{and} \quad \phi(\pi) = 0. \qquad (8.4.7)$$

Are there homogeneous solutions? The answer is yes, $\phi = \sin x$. However, it may cause some confusion to say that $\lambda = 0$ is an eigenvalue (although it is true). The definition of the eigenvalues for (8.4.7) is

$$\frac{d^2\phi}{dx^2} + \phi = -\lambda\phi \quad \text{with} \quad \phi(0) = 0 \quad \text{and} \quad \phi(\pi) = 0.$$

This is best written as $d^2\phi/dx^2 + (\lambda + 1)\phi = 0$. Therefore, $\lambda + 1 = (n\pi/L)^2 = n^2$, $n = 1, 2, 3, \ldots$ and it is now clear that $\lambda = 0$ is an eigenvalue ($n = 1$).

8.4.2 Fredholm Alternative

Important conclusions can be reached from (8.4.3), obtained by the method of eigenfunction expansion. The **Fredholm alternative** summarizes these results for nonhomogeneous problems

$$L(u) = f(x), \qquad (8.4.8)$$

subject to homogeneous boundary conditions (of the self-adjoint type). Either:

1. $u = 0$ is the only homogeneous solution (i.e., $\lambda = 0$ is not an eigenvalue), in which case the nonhomogeneous problem has a unique solution, or
2. There are nontrivial homogeneous solutions $\phi_h(x)$ (i.e., $\lambda = 0$ is an eigenvalue), in which case the nonhomogeneous problem has no solutions or an infinite number of solutions.

Let us describe in more detail what occurs if $\phi_h(x)$ is a nontrivial homogeneous solution. By (8.4.3) **there are an infinite number of solutions if**

$$\int_a^b f(x)\phi_h(x)\, dx = 0, \qquad (8.4.9)$$

because the corresponding a_n is arbitrary. These nonunique solutions correspond to an arbitrary additive multiple of a homogeneous solution $\phi_h(x)$. Equation (8.4.9) corresponds to the forcing function being orthogonal to the homogeneous solution (with weight 1). If

$$\int_a^b f(x)\phi_h(x)\, dx \neq 0,$$

then the nonhomogeneous problem (with homogeneous boundary conditions) has no solutions. These results are illustrated in Table 8.4.1.

A different phrasing of the Fredholm alternative states that **for the nonhomogeneous problem (8.4.8) with homogeneous boundary conditions, solutions**

	$\displaystyle\int_a^b f(x)\phi_h(x)\,dx$	
	0	$\neq 0$
$\phi_h = 0 \ (\lambda \neq 0)$	1	Not applicable
$\phi_h \neq 0 \ (\lambda = 0)$	∞	0

exist only if the forcing function is orthogonal to all homogeneous solutions.* Note that if $u = 0$ is the only homogeneous solution, then $f(x)$ is automatically orthogonal to it (in a somewhat trivial way), and there is a solution.

Part of the Fredholm alternative can be shown without using an eigenfunction expansion. If the nonhomogeneous problem has a solution, then

$$L(u) = f(x).$$

All homogeneous solutions, $\phi_h(x)$, satisfy

$$L(\phi_h) = 0.$$

We now use Green's formula with $v = \phi_h$ and obtain

$$\int_a^b [u \cdot 0 - \phi_h f(x)]\,dx = 0 \qquad \text{or} \qquad \boxed{\int_a^b f(x)\phi_h(x)\,dx = 0,}$$

since u and ϕ_h satisfy the same homogeneous boundary conditions.

Examples. We consider three examples. First, suppose that

$$\frac{d^2u}{dx^2} = e^x \quad \text{with} \quad \frac{du}{dx}(0) = 0 \quad \text{and} \quad \frac{du}{dx}(L) = 0. \tag{8.4.10}$$

$u = 1$ is a homogeneous solution. According to the Fredholm alternative, there is a solution only if e^x is orthogonal to this homogeneous solution. Since $\int_0^L e^x \cdot 1 \, dx \neq 0$, there are no solutions of (8.4.10).

For another example, suppose that

$$\frac{d^2u}{dx^2} + 2u = e^x \quad \text{with} \quad u(0) = 0 \quad \text{and} \quad u(\pi) = 0.$$

Since there are no solutions of the corresponding homogeneous problem† (other than $u = 0$), the Fredholm alternative implies that there is a unique solution. However, to obtain that solution we must use standard techniques to solve

---

* Here the operator L is self-adjoint. For non-self-adjoint operators, the solutions exist if the forcing function is orthogonal to all solutions of the corresponding homogeneous *adjoint* problem (see Exercises 10.3.2 to 10.3.5).

† For $d^2u/dx^2 + \lambda u = 0$, with $u(0) = 0$ and $u(L) = 0$, the eigenvalues are $(n\pi/L)^2$. Here $2 \neq n^2$.

nonhomogeneous differential equations, such as the methods of undetermined coefficients, variation of parameters, or eigenfunction expansion (using sin nx).

As a more nontrivial example, we consider

$$\frac{d^2u}{dx^2} + \left(\frac{\pi}{L}\right)^2 u = \beta + x \quad \text{with} \quad u(0) = 0 \quad \text{and} \quad u(L) = 0.$$

Since $u = \sin \pi x/L$ is a solution of the homogeneous problem, the nonhomogeneous problem only has a solution if the right-hand side is orthogonal to $\sin \pi x/L$:

$$0 = \int_0^L (\beta + x) \sin \frac{\pi x}{L} \, dx.$$

This can be used to determine the only value of β for which there is a solution:

$$\beta = \frac{-\int_0^L x \sin \pi x/L \, dx}{\int_0^L \sin \pi x/L \, dx} = \frac{L}{2}.$$

However, again the Fredholm alternative cannot be used to actually obtain the solution, $u(x)$.

8.4.3 Modified Green's Functions

In this section, we will analyze

$$L(u) = f \tag{8.4.11}$$

subject to homogeneous boundary conditions when $\lambda = 0$ is an eigenvalue. If a solution to (8.4.11) exists, we will produce a particular solution of (8.4.11) by defining and constructing a modified Green's function.

If $\lambda = 0$ is not an eigenvalue, then there is a unique solution of the nonhomogeneous boundary value problem, (8.4.11), subject to homogeneous boundary conditions. In Sec. 8.3 we represented the solution using a Green's function $G(x, x_0)$ satisfying

$$L[G(x, x_0)] = \delta(x - x_0), \tag{8.4.12}$$

subject to the same homogeneous boundary conditions.

Here we analyze the case in which $\lambda = 0$ is an eigenvalue; there are nontrivial homogeneous solutions $\phi_h(x)$ of (8.4.11), $L(\phi_h) = 0$. We will assume that there are solutions of (8.4.11), that is,

$$\int_a^b f(x)\phi_h(x) \, dx = 0. \tag{8.4.13}$$

However, the Green's function defined by (8.4.12) does not exist for all x_0 since $\delta(x - x_0)$ is not orthogonal to solutions of the homogeneous problem for all x_0:

$$\int_a^b \delta(x - x_0)\phi_h(x) \, dx = \phi_h(x_0) \neq 0.$$

We need to introduce a simple comparison problem which has a solution.

$\delta(x - x_0)$ is not orthogonal to $\phi_h(x)$ because it has a "component in the direction" $\phi_h(x)$. However, there is a solution for the forcing function

$$\delta(x - x_0) + c\phi_h(x),$$

if we pick c such that this function is orthogonal to $\phi_h(x)$:

$$0 = \int_a^b \phi_h(x)[\delta(x - x_0) + c\phi_h(x)]\, dx = \phi_h(x_0) + c\int_a^b \phi_h^2(x)\, dx.$$

Thus, we introduce the **modified Green's function** $G_m(x, x_0)$ which satisfies

$$L[G_m(x, x_0)] = \delta(x - x_0) - \frac{\phi_h(x)\phi_h(x_0)}{\displaystyle\int_a^b \phi_h^2(x)\, dx}, \tag{8.4.14}$$

subject to the same homogeneous boundary conditions.

Since the right-hand side of (8.4.14) is orthogonal to $\phi_h(x)$, unfortunately there are an infinite number of solutions. In Exercise 8.4.9 it is shown that the modified Green's function can be chosen to be symmetric

$$G_m(x, x_0) = G_m(x_0, x). \tag{8.4.15}$$

If $g_m(x, x_0)$ is one symmetric modified Green's function, then the following is also a symmetric modified Green's function

$$G_m(x, x_0) = g_m(x, x_0) + \beta\phi_h(x_0)\phi_h(x)$$

for any *constant* β (independent of x and x_0). Thus, there are an infinite number of symmetric modified Green's functions. We can use any of these.

We use Green's formula to derive a representation formula for $u(x)$ using the modified Green's function. Letting $u = u(x)$ and $v = G_m(x, x_0)$, Green's formula states that

$$\int_a^b \{u(x)L[G_m(x, x_0)] - G_m(x, x_0)L[u(x)]\}\, dx = 0,$$

since both $u(x)$ and $G_m(x, x_0)$ satisfy the same homogeneous boundary conditions. The defining differential equations (8.4.11) and (8.4.14) imply that

$$\int_a^b \left\{ u(x)\left[\delta(x - x_0) - \frac{\phi_h(x)\phi_h(x_0)}{\displaystyle\int_a^b \phi_h^2(\bar{x})\, d\bar{x}} \right] - G_m(x, x_0)f(x) \right\}\, dx = 0.$$

Using the fundamental Dirac delta property (and reversing the roles of x and x_0) yields

$$u(x) = \int_a^b f(x_0)G_m(x, x_0)\, dx_0 + \frac{\phi_h(x)}{\displaystyle\int_a^b \phi_h^2(\bar{x})\, d\bar{x}} \int_a^b u(x_0)\phi_h(x_0)\, dx_0,$$

where the symmetry of $G_m(x, x_0)$ has also been utilized. The last expression is a multiple of the homogeneous solution, and thus a simple *particular* solution of (8.4.11) is

$$u(x) = \int_a^b f(x_0) G_m(x, x_0) \, dx_0, \qquad (8.4.16)$$

the same form as occurs when $\lambda = 0$ is not an eigenvalue [see (8.3.31)].

Example. The simplest example of a problem with a nontrivial homogeneous solution is

$$\frac{d^2 u}{dx^2} = f(x) \qquad (8.4.17a)$$

$$\frac{du}{dx}(0) = 0 \quad \text{and} \quad \frac{du}{dx}(L) = 0. \qquad (8.4.17b,c)$$

A constant is a homogeneous solution (eigenfunction corresponding to the zero eigenvalue). For a solution to exist, by the Fredholm alternative,* $\int_0^L f(x) \, dx = 0$. We assume $f(x)$ is of this type [e.g., $f(x) = x - L/2$]. The modified Green's function $G_m(x, x_0)$ satisfies

$$\frac{d^2 G_m}{dx^2} = \delta(x - x_0) + c \qquad (8.4.18a)$$

$$\frac{dG_m}{dx}(0) = 0 \quad \text{and} \quad \frac{dG_m}{dx}(L) = 0, \qquad (8.4.18b,c)$$

since a constant is the eigenfunction. For there to be such a modified Green's function, the r.h.s. must be orthogonal to the homogeneous solutions:

$$\int_0^L [\delta(x - x_0) + c] \, dx = 0 \qquad \text{or} \qquad c = -\frac{1}{L}.$$

We use properties of the Dirac delta function to solve (8.4.18). For $x \neq x_0$,

$$\frac{d^2 G_m}{dx^2} = -\frac{1}{L}.$$

By integration,

$$\frac{dG_m}{dx} = \begin{cases} -\dfrac{x}{L} & x < x_0 \\[2mm] -\dfrac{x}{L} + 1 & x > x_0, \end{cases} \qquad (8.4.19)$$

where the constants of integration have been chosen to satisfy the boundary conditions at $x = 0$ and $x = L$. The jump condition for the derivative $(dG_m/dx|_{x_0-}^{x_0+} = 1)$, obtained by integrating (8.4.18a), is already satisfied by (8.4.19).

---

* Physically with insulated boundaries there must be zero *net* thermal energy generated for equilibrium.

We integrate again to obtain $G_m(x, x_0)$. Assuming that $G_m(x, x_0)$ is continuous at $x = x_0$ yields

$$G_m(x, x_0) = \begin{cases} -\dfrac{1}{L}\dfrac{x^2}{2} + x_0 + c(x_0) & x < x_0 \\ -\dfrac{1}{L}\dfrac{x^2}{2} + x + c(x_0) & x > x_0. \end{cases}$$

$c(x_0)$ is an arbitrary additive constant that depends on x_0 and corresponds to an arbitrary multiple of the homogeneous solution. This is the representation of all possible modified Green's functions. Often we desire $G_m(x, x_0)$ to be symmetric. For example, $G_m(x, x_0) = G_m(x_0, x)$ for $x < x_0$ yields

$$-\frac{1}{L}\frac{x_0^2}{2} + x_0 + c(x) = -\frac{1}{L}\frac{x^2}{2} + x_0 + c(x_0)$$

or

$$c(x_0) = -\frac{1}{L}\frac{x_0^2}{2} + \beta$$

where β is an arbitrary *constant*. Thus, finally we obtain the modified Green's function,

$$G_m(x, x_0) = \begin{cases} -\dfrac{1}{L}\dfrac{(x^2 + x_0^2)}{2} + x_0 + \beta & x < x_0 \\ -\dfrac{1}{L}\dfrac{(x^2 + x_0^2)}{2} + x + \beta & x > x_0. \end{cases}$$

A solution of (8.4.17) is given by (8.4.16) with $G_m(x, x_0)$ given above.

An alternative modified Green's function. In order to solve problems with homogeneous solutions, we could introduce instead a comparison problem satisfying nonhomogeneous boundary conditions. For example, the Neumann function G_a is defined by

$$\frac{d^2 G_a}{dx^2} = \delta(x - x_0) \tag{8.4.20a}$$

$$\frac{dG_a}{dx}(0) = -c \tag{8.4.20b}$$

$$\frac{dG_a}{dx}(L) = c. \tag{8.4.20c}$$

Physically, this represents a unit negative source $-\delta(x - x_0)$ of thermal energy with heat energy flowing into both ends at the rate of c per unit time. Thus, physically there will be a solution only if $2c = 1$. This can be verified by integrating (8.4.20a) from $x = 0$ to $x = L$ or by using Green's formula. This alternate modified Green's function can be obtained in a manner similar to the previous one. In terms of this Green's function, the representation of the solution of a nonhomogeneous problem can be obtained using Green's formula (see Exercise 8.4.12).

EXERCISES 8.4

8.4.1. Consider

$$L(u) = f(x) \qquad \text{with} \qquad L = \frac{d}{dx}\left(p\frac{d}{dx}\right) + q$$

subject to two homogeneous boundary conditions. *All* homogeneous solutions ϕ_h (if they exist) satisfy $L(\phi_h) = 0$ and the same two homogeneous boundary conditions. Apply Green's formula to prove that there are no solutions u if $f(x)$ is not orthogonal (weight 1) to all $\phi_h(x)$.

8.4.2. Modify Exercise 8.4.1 if

$$L(u) = f(x)$$
$$u(0) = \alpha \qquad \text{and} \qquad u(L) = \beta$$

  *(a) Determine the condition for a solution to exist.
  (b) If this condition is satisfied, show that there are an infinite number of solutions.

8.4.3. Without determining $u(x)$ how many solutions are there of

$$\frac{d^2u}{dx^2} + \gamma u = \sin x$$

  (a) If $\gamma = 1$ and $u(0) - u(\pi) = 0$?
  *(b) If $\gamma = 1$ and $\frac{du}{dx}(0) = \frac{du}{dx}(\pi) = 0$?
  (c) If $\gamma = -1$ and $u(0) = u(\pi) = 0$?
  (d) If $\gamma = 2$ and $u(0) = u(\pi) = 0$?

8.4.4. For the following examples, obtain the general solution of the differential equation using the method of undetermined coefficients. Attempt to solve the boundary conditions, and show that the result is consistent with the Fredholm alternative:
  (a) Equation (8.4.7)
  (b) Equation (8.4.10)
  (c) Example after (8.4.10)
  (d) Second example after (8.4.10)

8.4.5. Are there any values of β for which there are solutions of

$$\frac{d^2u}{dx^2} + u = \beta + x$$

$$u(-\pi) = u(\pi) \qquad \text{and} \qquad \frac{du}{dx}(-\pi) = \frac{du}{dx}(\pi)?$$

*8.4.6. Consider

$$\frac{d^2u}{dx^2} + u = 1.$$

  (a) Find the general solution of this differential equation. Determine all solutions with $u(0) = u(\pi) = 0$. Is the Fredholm alternative consistent with your result?
  (b) Redo part (a) if $\frac{du}{dx}(0) = \frac{du}{dx}(\pi) = 0$.
  (c) Redo part (a) if $\frac{du}{dx}(-\pi) = \frac{du}{dx}(\pi)$ and $u(-\pi) = u(\pi)$.

8.4.7. Consider

$$\frac{d^2u}{dx^2} + 4u = \cos x$$

$$u(0) = u(\pi) = 0.$$

(a) Determine all solutions using the hint that a particular solution of the differential equation is in the form, $u_p = A \cos x$.

(b) Determine all solutions using the eigenfunction expansion method.

(c) Apply the Fredholm alternative. Is it consistent with parts (a) and (b)?

8.4.8. Consider

$$\frac{d^2u}{dx^2} + u = \cos x,$$

which has a particular solution of the form, $u_p = Ax \sin x$.

*(a) Suppose that $u(0) = u(\pi) = 0$. Explicitly attempt to obtain all solutions. Is your result consistent with the Fredholm alternative?

(b) Answer the same questions as in part (a) if $u(-\pi) = u(\pi)$ and $\dfrac{du}{dx}(-\pi) = \dfrac{du}{dx}(\pi)$.

8.4.9. (a) Since (8.4.14) (with homogeneous boundary conditions) is solvable, there are an infinite number of solutions. Suppose that $g_m(x, x_0)$ is one such solution that is not orthogonal to $\phi_h(x)$. Show that there is a unique modified Green's function $G_m(x, x_0)$ which is orthogonal to $\phi_h(x)$.

(b) Assume that $G_m(x, x_0)$ is the modified Green's function which is orthogonal to $\phi_h(x)$. Prove that $G_m(x, x_0)$ is symmetric. [*Hint:* Apply Green's formula with $G_m(x, x_1)$ and $G_m(x, x_2)$.]

*8.4.10. Determine the modified Green's function that is needed to solve

$$\frac{d^2u}{dx^2} + u = f(x)$$

$$u(0) = \alpha \quad \text{and} \quad u(\pi) = \beta.$$

Assume that $f(x)$ satisfies the solvability condition (see Exercise 8.4.2). Obtain a representation of the solution $u(x)$ in terms of the modified Green's function.

8.4.11. Consider

$$\frac{d^2u}{dx^2} = f(x) \quad \text{with} \quad \frac{du}{dx}(0) = 0 \quad \text{and} \quad \frac{du}{dx}(L) = 0.$$

A different modified Green's function may be defined:

$$\frac{d^2G_a}{dx^2} = \delta(x - x_0)$$

$$\frac{dG_a}{dx}(0) = 0$$

$$\frac{dG_a}{dx}(L) = c.$$

*(a) Determine c using mathematical reasoning.

*(b) Determine c using physical reasoning.

(c) Explicitly determine all possible $G_a(x, x_0)$.

*(d) Determine all symmetric $G_a(x, x_0)$.

*(e) Obtain a representation of the solution $u(x)$ using $G_a(x, x_0)$.

8.4.12. The alternate modified Green's function (Neumann function) satisfies

$$\frac{d^2 G_a}{dx^2} = \delta(x - x_0)$$

$$\frac{dG_a}{dx}(0) = -c$$

$$\frac{dG_a}{dx}(L) = c, \qquad \text{where we have shown } c = \tfrac{1}{2}.$$

(a) Determine all possible $G_a(x, x_0)$.

(b) Determine all symmetric $G_a(x, x_0)$.

(c) Determine all $G_a(x, x_0)$ which are orthogonal to $\phi_h(x)$.

(d) What relationship exists between β and γ for there to be a solution to

$$\frac{d^2 u}{dx^2} = f(x) \quad \text{with} \quad \frac{du}{dx}(0) = \beta \quad \text{and} \quad \frac{du}{dx}(L) = \gamma?$$

In this case, derive the solution $u(x)$ in terms of a Neumann function, defined above.

8.5 GREEN'S FUNCTIONS FOR POISSON'S EQUATION

8.5.1 Introduction

In Secs. 8.3 and 8.4 we discussed Green's functions for Sturm–Liouville-type ordinary differential equations, $L(u) = f$, where $L = d/dx\,(p\,d/dx) + q$. Before discussing Green's functions for time-dependent partial differential equations (such as the heat and wave equations), we will analyze Green's functions for Poisson's equation, a time-independent partial differential equation,

$$L(u) = f, \tag{8.5.1}$$

where $L = \nabla^2$, the Laplacian. At first, we will assume that u satisfies homogeneous boundary conditions. Later we will show how to use the same ideas to solve problems with nonhomogeneous boundary conditions. We will begin by assuming that the region is finite, as illustrated in Fig. 8.5.1. The extension to infinite domains will be discussed in some depth.

One-dimensional Green's functions were introduced to solve the nonhomogeneous Sturm–Liouville problem. Key relationships were provided by Green's formula. The analysis of Green's functions for Poisson's equation is quite similar. We will frequently use Green's formula for the Laplacian, either in its two- or

Figure 8.5.1 Finite two-dimensional region.

three-dimensional forms:

$$\iiint (u \, \nabla^2 v - v \, \nabla^2 u) \, dV = \oiint (u \, \nabla v - v \, \nabla u) \cdot \hat{\mathbf{n}} \, dS$$

$$\iint (u \, \nabla^2 v - v \, \nabla^2 u) \, dA = \oint (u \, \nabla v - v \, \nabla u) \cdot \hat{\mathbf{n}} \, ds.$$

We claim that these formulas are valid even for the more exotic functions we will be discussing.

8.5.2 Multidimensional Dirac Delta Function and Green's Functions

The Green's function is defined as the solution to the nonhomogeneous problem with a concentrated source, subject to homogeneous boundary conditions. We define a two-dimensional Dirac delta function as the product of two one-dimensional Dirac delta functions. If the source is concentrated at $\mathbf{x} = \mathbf{x}_0$ ($\mathbf{x} = x\hat{\mathbf{i}} + y\hat{\mathbf{j}}$, $\mathbf{x}_0 = x_0\hat{\mathbf{i}} + y_0\hat{\mathbf{j}}$), then

$$\delta(\mathbf{x} - \mathbf{x}_0) = \delta(x - x_0) \, \delta(y - y_0). \qquad (8.5.2)$$

Similar ideas hold in three dimensions. The fundamental property of this multidimensional Dirac delta function is that

$$\int_{-\infty}^{\infty} \int_{-\infty}^{\infty} f(\mathbf{x}) \, \delta(\mathbf{x} - \mathbf{x}_0) \, dA = f(\mathbf{x}_0) \qquad (8.5.3a)$$

$$\int_{-\infty}^{\infty} \int_{-\infty}^{\infty} f(x, y) \, \delta(x - x_0) \, \delta(y - y_0) \, dA = f(x_0, y_0), \qquad (8.5.3b)$$

in vector or two-dimensional component form, where $f(\mathbf{x}) = f(x, y)$. We will use the vector notation.

Green's function. In order to solve the nonhomogeneous partial differential equation,

$$\nabla^2 u = f(\mathbf{x}), \qquad (8.5.4)$$

subject to homogeneous conditions along the boundary, we introduce the Green's function $G(\mathbf{x}, \mathbf{x}_0)$ for Poisson's equation:*

$$\nabla^2 G(\mathbf{x}, \mathbf{x}_0) = \delta(\mathbf{x} - \mathbf{x}_0), \qquad (8.5.5)$$

---

* Sometimes this is called the Green's function for Laplace's equation.

subject to the *same homogeneous* boundary conditions. Here $G(\mathbf{x}, \mathbf{x}_0)$ represents the response at $\mathbf{x}$ due to a source at $\mathbf{x}_0$.

Representation formula using Green's function. Green's formula (in its two-dimensional form) with $v = G(\mathbf{x}, \mathbf{x}_0)$ becomes

$$\iint (u \, \nabla^2 G - G \, \nabla^2 u) \, dA = 0,$$

since both $u(\mathbf{x})$ and $G(\mathbf{x}_0)$ solve the same homogeneous boundary conditions such that $\oint (u \, \nabla G - G \, \nabla u) \cdot \hat{\mathbf{n}} \, ds$ vanishes. From (8.5.4) and (8.5.5), it follows that

$$u(\mathbf{x}_0) = \iint f(\mathbf{x}) G(\mathbf{x}, \mathbf{x}_0) \, dA.$$

If we reverse the role of $\mathbf{x}$ and $\mathbf{x}_0$, we obtain

$$u(\mathbf{x}) = \iint f(\mathbf{x}_0) G(\mathbf{x}_0, \mathbf{x}) \, dA_0.$$

As we will show, the Green's function is symmetric

$$\boxed{G(\mathbf{x}, \mathbf{x}_0) = G(\mathbf{x}_0, \mathbf{x}),} \tag{8.5.6}$$

and hence

$$\boxed{u(\mathbf{x}) = \iint f(\mathbf{x}_0) G(\mathbf{x}, \mathbf{x}_0) \, dA_0.} \tag{8.5.7}$$

This shows how the solution of the partial differential equation may be computed if the Green's function is known.

Symmetry. As in one-dimensional problems, to show the symmetry of the Green's function we use Green's formula with $G(\mathbf{x}, \mathbf{x}_1)$ and $G(\mathbf{x}, \mathbf{x}_2)$. Since both satisfy the same homogeneous boundary conditions, we have

$$\iint [G(\mathbf{x}, \mathbf{x}_1) \, \nabla^2 G(\mathbf{x}, \mathbf{x}_2) - G(\mathbf{x}, \mathbf{x}_2) \, \nabla^2 G(\mathbf{x}, \mathbf{x}_1)] \, dA = 0.$$

Since $\nabla^2 G(\mathbf{x}, \mathbf{x}_1) = \delta(\mathbf{x} - \mathbf{x}_1)$ and $\nabla^2 G(\mathbf{x}, \mathbf{x}_2) = \delta(\mathbf{x} - \mathbf{x}_2)$, it follows using the fundamental property of the Dirac delta function that $G(\mathbf{x}_1, \mathbf{x}_2) = G(\mathbf{x}_2, \mathbf{x}_1)$; the Green's function is symmetric.

8.5.3 Green's Functions by the Method of Eigenfunction Expansion (Multidimensional)

One method to obtain the Green's function for Poisson's equation in a *finite region* is to use an eigenfunction expansion. We consider the related eigenfunctions:

$$\nabla^2 \phi = -\lambda \phi, \tag{8.5.8}$$

subject to the same homogeneous boundary conditions. We assume that the eigenvalues and corresponding orthogonal eigenfunctions are known. Simple examples occur in rectangular and circular regions.

We attempt to solve for the Green's function $G(\mathbf{x}, \mathbf{x}_0)$ $[\nabla^2 G = \delta(\mathbf{x} - \mathbf{x}_0)]$ as an infinite series of eigenfunctions:

$$G(\mathbf{x}, \mathbf{x}_0) = \sum_\lambda a_\lambda \phi_\lambda(\mathbf{x}), \qquad (8.5.9)$$

where $\phi_\lambda(\mathbf{x})$ is the eigenfunction corresponding to the eigenvalue λ. (This series may converge very slowly.) Since both $\phi_\lambda(\mathbf{x})$ and $G(\mathbf{x}, \mathbf{x}_0)$ solve the same homogeneous boundary conditions, we expect to be able to differentiate term by term:

$$\nabla^2 G = \sum_\lambda a_\lambda \nabla^2 \phi_\lambda = -\sum_\lambda a_\lambda \lambda \phi_\lambda(\mathbf{x}).$$

This can be verified using Green's formula. Since $\nabla^2 G = \delta(\mathbf{x} - \mathbf{x}_0)$, due to the multidimensional orthogonality of $\phi_\lambda(\mathbf{x})$, it follows that

$$-a_\lambda \lambda = \frac{\displaystyle\iint \phi_\lambda(\mathbf{x}) \, \delta(\mathbf{x} - \mathbf{x}_0) \, dA}{\displaystyle\iint \phi_\lambda^2(\mathbf{x}) \, dA} = \frac{\phi_\lambda(\mathbf{x}_0)}{\displaystyle\iint \phi_\lambda^2(\mathbf{x}) \, dA}.$$

If $\lambda = 0$ is not an eigenvalue, then we can determine a_λ and the eigenfunction expansion of the Green's function,

$$\boxed{G(\mathbf{x}, \mathbf{x}_0) = \sum_\lambda \frac{\phi_\lambda(\mathbf{x})\phi_\lambda(\mathbf{x}_0)}{-\lambda \displaystyle\iint \phi_\lambda^2 \, dA}.} \qquad (8.5.10)$$

This is the natural generalization of the one-dimensional result for the Green's function corresponding to a nonhomogeneous Sturm-Liouville problem (see Sec. 8.4.10).

Example. For a rectangle, $0 < x < L$, $0 < y < H$, with boundary conditions zero on all four sides, we have shown (see Chapter 6) that the eigenvalues are $\lambda_{nm} = (n\pi/L)^2 + (m\pi/H)^2$ ($n = 1, 2, 3, \ldots$ and $m = 1, 2, 3, \ldots$) and the corresponding eigenfunctions are $\phi_\lambda(\mathbf{x}) = \sin n\pi x/L \sin m\pi y/H$. In this case the normalization constants are $\iint \phi_\lambda^2 \, dx \, dy = L/2 \cdot H/2$. The Green's function can be expanded in a series of these eigenfunctions, a Fourier sine series in x and y,

$$G(\mathbf{x}, \mathbf{x}_0) = \frac{-4}{LH} \sum_{n=1}^\infty \sum_{m=1}^\infty \frac{\sin n\pi x/L \sin m\pi y/H \sin n\pi x_0/L \sin m\pi y_0/H}{(n\pi/L)^2 + (m\pi/H)^2}.$$

Later in this section, we will obtain alternative forms of this Green's function.

Fredholm alternative. As before, difficulties occur if $\lambda = 0$ is an eigenvalue. In this case there is at least one nontrivial solution ϕ_h of Laplace's equation $\nabla^2 \phi_h = 0$ (the homogeneous equation related to Poisson's equation) satisfying homogeneous boundary conditions. It can be shown that there is a Fredholm alternative as in Sec. 8.4. There is no solution to the nonhomogeneous boundary value problem (satisfying homogeneous boundary conditions) if

$$\iint f(\mathbf{x})\phi_n(\mathbf{x})\, dA \neq 0.$$

However, if the source function $f(\mathbf{x})$ is orthogonal to *all* homogeneous solutions $\phi_h(x)$, then there are an infinite number of solutions of Poisson's equation.

Example. If the entire boundary is insulated, $\nabla\phi \cdot \hat{\mathbf{n}} = 0$, then ϕ_h equaling any constant is a nontrivial solution of $\nabla^2\phi = 0$ satisfying the boundary conditions. $\phi_h = 1$ is the eigenfunction corresponding to $\lambda = 0$. Solutions of $\nabla^2 u = f(\mathbf{x})$ then exist only if $\iint f(\mathbf{x})\, dA = 0$. Physically, for a steady-state heat equation with insulated boundaries, the *net* heat energy generated must be zero. This is just the two-dimensional version of the problem discussed in Sec. 8.4. In particular, we could introduce in some way a modified Green's function (which is also known as a **Neumann function**). We leave any discussion of this for the exercises. For the remainder of Section 8.5 we will assume that $\lambda = 0$ is not an eigenvalue.

8.5.4 Direct Solution of Green's Functions (One-Dimensional Eigenfunctions)

Green's functions can also be obtained by more direct methods. Consider the Green's function for Poisson's equation,

$$\nabla^2 G(\mathbf{x}, \mathbf{x}_0) = \delta(\mathbf{x} - \mathbf{x}_0), \tag{8.5.11}$$

inside a rectangle with zero boundary conditions, as illustrated in Fig. 8.5.2. Instead of solving for this Green's function using a series of two-dimensional eigenfunctions (see Sec. 8.5.3), we will use one-dimensional eigenfunctions, either a sine series in x or y due to the boundary conditions. Using a Fourier sine series in x,

$$G(\mathbf{x}, \mathbf{x}_0) = \sum_{n=1}^{\infty} a_n(y) \sin\frac{n\pi x}{L}. \tag{8.5.12}$$

By substituting (8.5.12) into (8.5.11), we obtain [since both $G(\mathbf{x}, \mathbf{x}_0)$ and $\sin n\pi x/L$ satisfy the same set of homogeneous boundary conditions],

$$\sum_{n=1}^{\infty}\left[\frac{d^2 a_n}{dy^2} - \left(\frac{n\pi}{L}\right)^2 a_n\right]\sin\frac{n\pi x}{L} = \delta(x - x_0)\,\delta(y - y_0)$$

or

$$\frac{d^2 a_n}{dy^2} - \left(\frac{n\pi}{L}\right)^2 a_n = \frac{2}{L}\int_0^L \delta(x - x_0)\,\delta(y - y_0)\sin\frac{n\pi x}{L}\,dx$$

$$= \frac{2}{L}\sin\frac{n\pi x_0}{L}\,\delta(y - y_0). \tag{8.5.13}$$

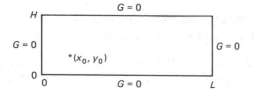

H $G = 0$

$G = 0$ $G = 0$

$*(x_0, y_0)$

0 0 $G = 0$ L

Figure 8.5.2 Green's function for Poisson's equation on a rectangle.

The boundary conditions at $y = 0$ and $y = H$ imply that the Fourier coefficients must satisfy the corresponding boundary conditions,

$$a_n(0) = 0 \quad \text{and} \quad a_n(H) = 0. \tag{8.5.14}$$

Equation (8.5.13) with boundary conditions (8.5.14) may be solved by a Fourier sine series in y; but this will yield the earlier double-sine-series analysis. On the other hand, since the nonhomogeneous term for $a_n(y)$ is a one-dimensional Dirac delta function, we may solve (8.5.13) as we have done for Green's functions. The differential equation is homogeneous if $y \neq y_0$. In addition, if we utilize the boundary conditions, we obtain

$$a_n(y) = \begin{cases} c_n \sinh \dfrac{n\pi y}{L} \sinh \dfrac{n\pi(y_0 - H)}{L} & y < y_0 \\[2ex] c_n \sinh \dfrac{n\pi(y - H)}{L} \sinh \dfrac{n\pi y_0}{L} & y > y_0, \end{cases}$$

where in this form continuity at $y = y_0$ is automatically satisfied. In addition, we integrate (8.5.13) from y_{0-} to y_{0+} to obtain the jump in the derivative:

$$\frac{da_n}{dy}\bigg|_{y_0-}^{y_0+} = \frac{2}{L} \sin \frac{n\pi x_0}{L}$$

or

$$c_n \frac{n\pi}{L}\left[\sinh \frac{n\pi y_0}{L} \cosh \frac{n\pi(y_0 - H)}{L} - \sinh \frac{n\pi(y_0 - H)}{L} \cosh \frac{n\pi y_0}{L} \right] = \frac{2}{L} \sin \frac{n\pi x_0}{L}. \tag{8.5.15}$$

Using an addition formula for hyperbolic functions, we obtain

$$c_n = \frac{2 \sin n\pi x_0/L}{n\pi \sinh n\pi H/L}.$$

This yields the Fourier sine series (in x) representation of the Green's function

$$G(\mathbf{x}, \mathbf{x}_0) = \sum_{n=1}^{\infty} \frac{2 \sin n\pi x_0/L \sin n\pi x/L}{n\pi \sinh n\pi H/L} \begin{cases} \sinh \dfrac{n\pi(y_0 - H)}{L} \sinh \dfrac{n\pi y}{L} & y < y_0 \\[2ex] \sinh \dfrac{n\pi(y - H)}{L} \sinh \dfrac{n\pi y_0}{L} & y > y_0. \end{cases} \tag{8.5.16}$$

The symmetry is exhibited explicitly. In the preceding subsection this same Green's function was represented as a double Fourier sine series in both x and y.

A third representation of this Green's function is also possible. Instead of using a Fourier sine series in x, we could have used a Fourier sine series in y. We omit the nearly identical analysis.

8.5.5 Using Green's Functions for Problems with Nonhomogeneous Boundary Conditions

As with one-dimensional problems, the *same* Green's function determined in Secs. 8.5.2 through 8.5.4, $\nabla^2 G = \delta(\mathbf{x} - \mathbf{x}_0)$ [with $G(\mathbf{x}, \mathbf{x}_0)$ satisfying homogeneous boundary conditions], may be used to solve Poisson's equation $\nabla^2 u = f(\mathbf{x})$ subject to *nonhomogeneous* boundary conditions.

For example, consider

$$\boxed{\nabla^2 u = f(\mathbf{x})} \tag{8.5.17a}$$

with

$$\boxed{u = h(\mathbf{x})} \tag{8.5.17b}$$

on the boundary. The Green's function is defined by

$$\boxed{\nabla^2 G = \delta(\mathbf{x} - \mathbf{x}_0),} \tag{8.5.18a}$$

with

$$\boxed{G(\mathbf{x}, \mathbf{x}_0) = 0} \tag{8.5.18b}$$

for $\mathbf{x}$ on the boundary ($\mathbf{x}_0$ is often not on the boundary). The Green's function satisfies the related homogeneous boundary conditions. To obtain the Green's function representation of the solution of (8.5.17), we again employ Green's formula,

$$\iint (u \, \nabla^2 G - G \, \nabla^2 u) \, dA = \oint (u \, \nabla G - G \, \nabla u) \cdot \hat{\mathbf{n}} \, ds.$$

Using the defining differential equations and the boundary conditions,

$$\iint [u(\mathbf{x}) \, \delta(\mathbf{x} - \mathbf{x}_0) - f(\mathbf{x})G(\mathbf{x}, \mathbf{x}_0)] \, dA = \oint h(\mathbf{x}) \, \nabla G \cdot \hat{\mathbf{n}} \, ds,$$

and thus,

$$u(\mathbf{x}_0) = \iint f(\mathbf{x})G(\mathbf{x}, \mathbf{x}_0) \, dA + \oint h(\mathbf{x}) \, \nabla G(\mathbf{x}, \mathbf{x}_0) \cdot \hat{\mathbf{n}} \, ds.$$

We interchange $\mathbf{x}$ and $\mathbf{x}_0$, and we use the symmetry of $G(\mathbf{x}, \mathbf{x}_0)$ to obtain

$$\boxed{u(\mathbf{x}) = \iint f(\mathbf{x}_0)G(\mathbf{x}, \mathbf{x}_0) \, dA_0 + \oint h(\mathbf{x}_0) \, \nabla_{\mathbf{x}_0} G(\mathbf{x}, \mathbf{x}_0) \cdot \hat{\mathbf{n}} \, ds_0.} \tag{8.5.19}$$

We must be especially careful with the closed line integral, representing the effect of the nonhomogeneous boundary condition. $\nabla_{\mathbf{x}_0}$ is a symbol for the gradient with respect to the position of the source,

$$\nabla_{\mathbf{x}_0} = \frac{\partial}{\partial x_0}\hat{\mathbf{i}} + \frac{\partial}{\partial y_0}\hat{\mathbf{j}}.$$

Thus, $G(\mathbf{x}, \mathbf{x}_0)$ is the influence function for the source term, while $\nabla_{\mathbf{x}_0} G(\mathbf{x}, \mathbf{x}_0) \cdot \hat{\mathbf{n}}$ is the influence function for the nonhomogeneous boundary conditions. Let us attempt to give an understanding to the influence function for the nonhomogeneous boundary conditions, $\nabla_{\mathbf{x}_0} G(\mathbf{x}, \mathbf{x}_0) \cdot \hat{\mathbf{n}}$. This is an ordinary derivative with respect to the source position in the normal direction. Using the definition of a directional derivative,

$$\nabla_{\mathbf{x}_0} G(\mathbf{x}, \mathbf{x}_0) \cdot \hat{\mathbf{n}} = \lim_{\Delta s \to 0} \frac{G(\mathbf{x}, \mathbf{x}_0 + \Delta s \hat{\mathbf{n}}) - G(\mathbf{x}, \mathbf{x}_0)}{\Delta s}.$$

This yields an interpretation of this normal derivative of the Green's function. $G(\mathbf{x}, \mathbf{x}_0 + \Delta s \hat{\mathbf{n}})/\Delta s$ is the response to a positive source of strength $1/\Delta s$ located at $\mathbf{x}_0 + \Delta s \hat{\mathbf{n}}$, while $-G(\mathbf{x}, \mathbf{x}_0)/\Delta s$ is the response to a negative source (strength $-1/\Delta s$) located at $\mathbf{x}_0$. The influence function for the nonhomogeneous boundary condition consists of two concentrated sources of opposite effects whose strength is $1/\Delta s$ and distance apart is Δs, in the limit as $\Delta s \to 0$. This is called a **dipole source**. Thus, *this nonhomogeneous boundary condition has an equivalent effect as a surface distribution of dipoles.*

8.5.6 Infinite Space Green's Functions

In Secs. 8.5.2 through 8.5.4 we obtained representations of the Green's function for Poisson's equation, $\nabla^2 G(\mathbf{x}, \mathbf{x}_0) = \delta(\mathbf{x} - \mathbf{x}_0)$. However, these representations were complicated. The resulting infinite series do not give a very good understanding of the effect at $\mathbf{x}$ of a concentrated source at $\mathbf{x}_0$. As we will show, the difficulty is caused by the presence of boundaries.

In order to obtain simpler representations, we begin by considering solving Poisson's equation

$$\nabla^2 u = f(\mathbf{x})$$

in infinite space with no boundaries. We introduce the Green's function $G(\mathbf{x}, \mathbf{x}_0)$ defined by

$$\boxed{\nabla^2 G = \delta(\mathbf{x} - \mathbf{x}_0),} \qquad (8.5.20)$$

to be valid for all $\mathbf{x}$. Since this is a model of steady-state heat flow with a concentrated source located at $\mathbf{x} = \mathbf{x}_0$ *with no boundaries*, there should be a solution that is symmetric around the source point $\mathbf{x} = \mathbf{x}_0$. Our results are somewhat different in two and three dimensions. We simultaneously solve both. We let r represent radial distance (from $\mathbf{x} = \mathbf{x}_0$) in two dimensions and ρ represent radial distance (from $\mathbf{x} = \mathbf{x}_0$) in three dimensions:

two	*three*
$\mathbf{r} = \mathbf{x} - \mathbf{x}_0$	$\boldsymbol{\rho} = \mathbf{x} - \mathbf{x}_0$
$r = |\mathbf{r}| = |\mathbf{x} - \mathbf{x}_0|$	$\rho = |\boldsymbol{\rho}| = |\mathbf{x} - \mathbf{x}_0|$
$= \sqrt{(x - x_0)^2 + (y - y_0)^2}$	$= \sqrt{(x - x_0)^2 + (y - y_0)^2 + (z - z_0)^2}.$
(8.5.21a)	(8.5.21b)

Our derivation continues with three-dimensional results in parentheses and on the right. We assume that $G(\mathbf{x}, \mathbf{x}_0)$ only depends on r (ρ):

$$G(\mathbf{x}, \mathbf{x}_0) = G(r) = G(|\mathbf{x} - \mathbf{x}_0|) \quad | \quad G(\mathbf{x}, \mathbf{x}_0) = G(\rho) = G(|\mathbf{x} - \mathbf{x}_0|).$$

Away from the source $(r \neq 0$ or $\rho \neq 0)$, the forcing function is zero $[\nabla^2 G(\mathbf{x}, \mathbf{x}_0) = 0]$. In two dimensions we look for circularly symmetric solutions of Laplace's equation for $r \neq 0$ (in three dimensions the solutions should be spherically symmetric). From our earlier work:

$$(r \neq 0) \quad \frac{1}{r}\frac{d}{dr}\left(r\frac{dG}{dr}\right) = 0 \quad \bigg| \quad (\rho \neq 0) \quad \frac{1}{\rho^2}\frac{d}{d\rho}\left(\rho^2\frac{dG}{d\rho}\right) = 0.$$

The general solution can be obtained by integration

$$G(r) = c_1 \ln r + c_2 \quad \text{(8.5.22a)} \quad \bigg| \quad G(\rho) = \frac{c_3}{\rho} + c_4. \quad \text{(8.5.22b)}$$

We will determine the constants that account for the singularity at the source. We can obtain the appropriate singularity by integrating (8.5.20) around a small circle (sphere):

$$\iint \nabla^2 G \, dA = 1 \qquad\qquad \iiint \nabla^2 G \, dV = 1$$

$$\iint \nabla \cdot (\nabla G) \, dA = \oint \nabla G \cdot \hat{n} \, ds = 1 \quad \bigg| \quad \iiint \nabla \cdot (\nabla G) \, dV = \oiint \nabla G \cdot \hat{n} \, dS = 1,$$

where the divergence theorem has been used. In two dimensions the derivative of the Green's function normal to a circle, $\nabla G \cdot \hat{n}$, is $\partial G/\partial r$ (in three dimensions $\partial G/\partial \rho$). It only depends on the radial distance [see (8.5.22)]. On the circle (sphere) the radius is constant. Thus,

$$2\pi r \frac{\partial G}{\partial r} = 1 \qquad\qquad\qquad 4\pi\rho^2 \frac{\partial G}{\partial \rho} = 1,$$

since the circumference of a circle is $2\pi r$ (the surface area of a sphere is $4\pi\rho^2$). In other problems involving infinite space Green's functions it may be necessary to consider the limit of an infinitesimally small circle (sphere). Thus, we will express the singularity condition as

$$\boxed{\lim_{r \to 0} r \frac{\partial G}{\partial r} = \frac{1}{2\pi}} \quad \text{(8.5.23a)} \quad \bigg| \quad \boxed{\lim_{\rho \to 0} \rho^2 \frac{\partial G}{\partial \rho} = \frac{1}{4\pi}.} \quad \text{(8.5.23b)}$$

From (8.5.22) and (8.5.23),

$$c_1 = \frac{1}{2\pi} \qquad\qquad\qquad\qquad c_3 = -\frac{1}{4\pi}.$$

c_2 and c_4 are arbitrary, indicating that the infinite space Green's function for Poisson's equation is determined to within an arbitrary additive constant. For convenience we let $c_2 = 0$ and $c_4 = 0$:

$$G(\mathbf{x}, \mathbf{x}_0) = \frac{1}{2\pi} \ln r \qquad (8.5.24a)$$

$$r = |\mathbf{x} - \mathbf{x}_0|$$
$$= \sqrt{(x - x_0)^2 + (y - y_0)^2}.$$

$$G(\mathbf{x}, \mathbf{x}_0) = -\frac{1}{4\pi\rho} \qquad (8.5.24b)$$

$$\rho = |\mathbf{x} - \mathbf{x}_0|$$
$$= \sqrt{(x - x_0)^2 + (y - y_0)^2 + (z - z_0)^2}.$$

Note that these are symmetric. These infinite space Green's functions are themselves singular at the concentrated source. (This does not occur in one dimension.)

In order to obtain the solution of Poisson's equation, $\nabla^2 u = f(\mathbf{x})$, *in infinite space*, using the infinite space Green's function, we need to utilize Green's formula

$$\iint (u \nabla^2 G - G \nabla^2 u) \, dA \qquad (8.5.25a)$$
$$= \oint (u \nabla G - G \nabla u) \cdot \hat{\mathbf{n}} \, ds$$

$$\iiint (u \nabla^2 G - G \nabla^2 u) \, dV \qquad (8.5.25b)$$
$$= \oiint (u \nabla G - G \nabla u) \cdot \hat{\mathbf{n}} \, dS.$$

The closed line integral $\oint$ (closed surface integral $\oiint$) represents integrating over the entire boundary. For infinite space problems with no boundaries, we must consider large circles (spheres) and take the limit as the radius approaches infinity. We would like the contribution to this closed integral "from infinity" to vanish:

$$\lim_{|\mathbf{x}| \to \infty} \oint (u \nabla G - G \nabla u) \cdot \hat{\mathbf{n}} \, ds = 0$$

$$(8.5.26a)$$

$$\lim_{|\mathbf{x}| \to \infty} \oiint (u \nabla G - G \nabla u) \cdot \hat{\mathbf{n}} \, dS = 0.$$

$$(8.5.26b)$$

In this case, using the defining differential equations, integrating (8.5.25) with the Dirac delta function, and reversing the roles of $\mathbf{x}$ and $\mathbf{x}_0$ [using the symmetry of $G(\mathbf{x}, \mathbf{x}_0)$], yields the representation formula for the solution of Poisson's equation in infinite space:

$$u(\mathbf{x}) = \iint f(\mathbf{x}_0) G(\mathbf{x}, \mathbf{x}_0) \, dA_0$$

$$(8.5.27a)$$

$$u(\mathbf{x}) = \iiint f(\mathbf{x}_0) G(\mathbf{x}, \mathbf{x}_0) \, dV_0$$

$$(8.5.27b)$$

where $G(\mathbf{x}, \mathbf{x}_0)$ is given by (8.5.24).

The condition necessary for (8.5.26) to be valid is obtained by integrating in polar (spherical) coordinates centered at $\mathbf{x} = \mathbf{x}_0$:

$$\lim_{r \to \infty} r \left(u \frac{\partial G}{\partial r} - G \frac{\partial u}{\partial r} \right) = 0 \qquad \bigg| \qquad \lim_{\rho \to \infty} \rho^2 \left(u \frac{\partial G}{\partial \rho} - G \frac{\partial u}{\partial \rho} \right) = 0,$$

since $ds = r \, d\theta$ ($dS = \rho^2 \sin \phi \, d\phi \, d\theta$; see p. 400). By substituting the known Green's functions, we obtain conditions that must be satisfied at ∞ in order for the "boundary" terms to vanish there:

$$\lim_{r \to \infty} \left(u - r \ln r \frac{\partial u}{\partial r} \right) = 0 \qquad \lim_{\rho \to \infty} \left(u + \rho \frac{\partial u}{\partial \rho} \right) = 0.$$

These are important conditions. For example, they are satisfied if $u \sim 1/r$ ($u \sim 1/\rho$) as $r \to \infty$ ($\rho \to \infty$).

8.5.7 Green's Functions for Bounded Domains Using Infinite Space Green's Functions

In this subsection we solve for the Green's function,

$$\nabla^2 G = \delta(\mathbf{x} - \mathbf{x}_0), \tag{8.5.28}$$

on a bounded *two-dimensional* domain, subject to homogeneous boundary conditions. We have already discussed some methods of solution only for simple geometries, and even these involve a considerable amount of computation. However, we now know a particular solution of (8.5.28), namely the infinite space Green's function

$$G_p(\mathbf{x}, \mathbf{x}_0) = \frac{1}{2\pi} \ln r = \frac{1}{2\pi} \ln |\mathbf{x} - \mathbf{x}_0| = \frac{1}{2\pi} \ln \sqrt{(x - x_0)^2 + (y - y_0)^2}. \tag{8.5.29}$$

Unfortunately, this infinite space Green's function will not solve the homogeneous boundary conditions. Instead, consider

$$G(\mathbf{x}, \mathbf{x}_0) = \frac{1}{2\pi} \ln |\mathbf{x} - \mathbf{x}_0| + v(\mathbf{x}, \mathbf{x}_0). \tag{8.5.30}$$

$v(\mathbf{x}, \mathbf{x}_0)$ represents the effect of the boundary. It will be a homogeneous solution, solving Laplace's equation,

$$\nabla^2 v = 0$$

subject to some nonhomogeneous boundary condition. For example, if $G = 0$ on the boundary, then $v = -(1/2\pi) \ln |\mathbf{x} - \mathbf{x}_0|$ on the boundary. $v(\mathbf{x}, \mathbf{x}_0)$ may be solved by standard methods for Laplace's equation based on separation of variables (if the geometry allows). It may be quite involved to calculate $v(\mathbf{x}, \mathbf{x}_0)$; nevertheless, this representation of the Green's function is quite important. In particular since $v(\mathbf{x}, \mathbf{x}_0)$ will be well-behaved everywhere, including $\mathbf{x} = \mathbf{x}_0$, (8.5.30) shows that **the Green's function on a finite domain will have the same singularity at source location $\mathbf{x} = \mathbf{x}_0$ as does the infinite space Green's function**. This can be explained in a somewhat physical manner. The response at a point due to a concentrated source nearby should not depend significantly on any boundaries. This technique represented by (8.5.30) removes the singularity.

8.5.8 Green's Functions for a Semi-Infinite Plane ($y > 0$) Using Infinite Space Green's Functions— the Method of Images

The infinite space Green's function can be used to obtain Green's functions for certain semi-infinite problems. Consider Poisson's equation in the two-dimensional semi-infinite region $y > 0$,

$$\boxed{\nabla^2 u = f(\mathbf{x}),} \qquad (8.5.31a)$$

subject to a nonhomogeneous condition (given temperature) on $y = 0$:

$$\boxed{u(x, 0) = h(x).} \qquad (8.5.31b)$$

The defining problem for the Green's function,
$$\nabla^2 G(\mathbf{x}, \mathbf{x}_0) = \delta(\mathbf{x} - \mathbf{x}_0), \qquad (8.5.32a)$$
satisfies the corresponding homogeneous boundary conditions,
$$G(x, 0; x_0, y_0) = 0, \qquad (8.5.32b)$$
as illustrated in Fig. 8.5.3. Here we use the notation $G(\mathbf{x}, \mathbf{x}_0) = G(x, y; x_0, y_0)$. The semi-infinite space ($y > 0$) has no sources except a concentrated source at $\mathbf{x} = \mathbf{x}_0$. The infinite space Green's function, $(1/2\pi) \ln |\mathbf{x} - \mathbf{x}_0|$, is not satisfactory since it will not be zero at $y = 0$.

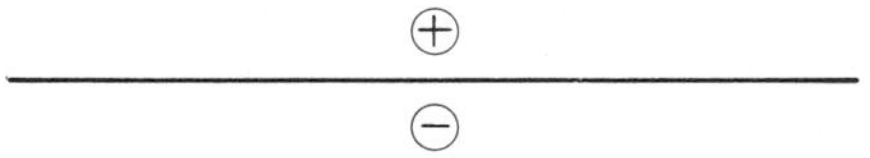

Figure 8.5.3 Image source for a semi-infinite plane.

Image source. There is a simple way to obtain a solution which is zero at $y = 0$. Consider an infinite space problem (i.e., no boundaries) with source $\delta(\mathbf{x} - \mathbf{x}_0)$ at $\mathbf{x} = \mathbf{x}_0$, and a negative image source $-\delta(\mathbf{x} - \mathbf{x}_0^*)$ at $\mathbf{x} = \mathbf{x}_0^*$ (where $\mathbf{x}_0 = x_0 \hat{\mathbf{i}} + y_0 \hat{\mathbf{j}}$ and $\mathbf{x}_0^* = x_0 \hat{\mathbf{i}} - y_0 \hat{\mathbf{j}}$):
$$\nabla^2 G = \delta(\mathbf{x} - \mathbf{x}_0) - \delta(\mathbf{x} - \mathbf{x}_0^*). \qquad (8.5.33)$$
According to the principle of superposition for nonhomogeneous problems, the response should be the sum of two individual responses:

$$G = \frac{1}{2\pi} \ln |\mathbf{x} - \mathbf{x}_0| - \frac{1}{2\pi} \ln |\mathbf{x} - \mathbf{x}_0^*|. \qquad (8.5.34)$$

By symmetry, the response at $y = 0$ due to the source at $\mathbf{x} = \mathbf{x}_0^*$ should be minus the response at $y = 0$ due to the source at $\mathbf{x} = \mathbf{x}_0$. Thus, the sum should be zero at $y = 0$ (as we will verify shortly). We call this the **method of images**. In this way, we have obtained the Green's function for Poisson's equation on a semi-infinite space ($y > 0$):

$$G(\mathbf{x}, \mathbf{x}_0) = \frac{1}{2\pi} \ln \frac{|\mathbf{x} - \mathbf{x}_0|}{|\mathbf{x} - \mathbf{x}_0^*|} = \frac{1}{4\pi} \ln \frac{(x - x_0)^2 + (y - y_0)^2}{(x - x_0)^2 + (y + y_0)^2}. \qquad (8.5.35)$$

Let us check that this is the desired solution. Equation (8.5.33), which is satisfied by (8.5.35), is not (8.5.32a). However, in the upper-half plane ($y > 0$), $\delta(\mathbf{x} - \mathbf{x}_0^*) = 0$, since $\mathbf{x} = \mathbf{x}_0^*$ is in the lower-half plane. Thus, for $y > 0$, (8.5.32a) is satisfied. Furthermore, we now show that at $y = 0$, $G(\mathbf{x}, \mathbf{x}_0) = 0$:

$$G(\mathbf{x}, \mathbf{x}_0)|_{y=0} = \frac{1}{4\pi} \ln \frac{(x - x_0)^2 + y_0^2}{(x - x_0)^2 + y_0^2} = \frac{1}{4\pi} \ln 1 = 0.$$

Solution. To solve Poisson's equation with nonhomogeneous boundary conditions, we need the solution's representation in terms of this Green's function, (8.5.35). We again use Green's formula. We need to consider a large semicircle (in the limit as the radius tends to infinity):

$$\iint (u \nabla^2 G - G \nabla^2 u) \, dA - \oint (u \nabla G - G \nabla u) \cdot \hat{\mathbf{n}} \, ds = \int_{-\infty}^{\infty} \left(G \frac{\partial u}{\partial y} - u \frac{\partial G}{\partial y} \right) \Bigg|_{y=0} dx,$$

since for the wall the outward unit normal is $\hat{\mathbf{n}} = -\hat{\mathbf{j}}$ and since the contribution at ∞ tends to vanish if $u \to 0$ sufficiently fast [in particular from p. 318 if $\lim_{r \to \infty} (u - r \ln r \, \partial u/\partial r) = 0$]. Substituting the defining differential equations and interchanging $\mathbf{x}$ and $\mathbf{x}_0$ [using the symmetry of $G(\mathbf{x}, \mathbf{x}_0)$] shows that

$$u(\mathbf{x}) = \iint f(\mathbf{x}_0) G(\mathbf{x}, \mathbf{x}_0) \, dA_0 - \int_{-\infty}^{\infty} h(x_0) \frac{\partial}{\partial y_0} G(\mathbf{x}, \mathbf{x}_0) \Bigg|_{y_0=0} dx_0, \qquad (8.5.36)$$

since $G = 0$ on $y = 0$. This can be obtained directly from (8.5.19). $G(\mathbf{x}, \mathbf{x}_0)$ from (8.5.35) is given by

$$G(\mathbf{x}, \mathbf{x}_0) = \frac{1}{4\pi} [\ln ((x - x_0)^2 + (y - y_0)^2) - \ln ((x - x_0)^2 + (y + y_0)^2)].$$

Thus,

$$\frac{\partial}{\partial y_0} G(\mathbf{x}, \mathbf{x}_0) = \frac{1}{4\pi} \left[\frac{-2(y - y_0)}{(x - x_0)^2 + (y - y_0)^2} - \frac{2(y + y_0)}{(x - x_0)^2 + (y + y_0)^2} \right].$$

Evaluating this at $y_0 = 0$ (corresponding to the source point on the boundary) yields

$$\frac{\partial}{\partial y_0} G(\mathbf{x}, \mathbf{x}_0) \Bigg|_{y_0=0} = -\frac{y/\pi}{(x - x_0)^2 + y^2}. \qquad (8.5.37)$$

This is an example of a dipole source (see Sec. 8.5.5).

Example. Consider Laplace's equation for the two-dimensional semi-infinite space ($y > 0$):

$$\nabla^2 u = 0 \qquad (8.5.38a)$$

$$u(x, 0) = h(x). \qquad (8.5.38b)$$

Equation (8.5.36) can be utilized with a zero source term. Here the solution is only due to the nonhomogeneous boundary condition. Using the normal derivative of the Green's function, (8.5.37), we obtain

$$u(x, y) = \frac{y}{\pi} \int_{-\infty}^{\infty} h(x_0) \frac{1}{(x - x_0)^2 + y^2} \, dx_0. \qquad (8.5.39)$$

The influence function for the boundary condition $h(x)$ is not the Green's function, but

$$-\frac{\partial}{\partial y_0} G(\mathbf{x}, \mathbf{x}_0)\bigg|_{y_0=0} = \frac{y}{\pi[(x - x_0)^2 + y^2]}.$$

In Chapter 9 we will obtain the same answer using Fourier transform techniques rather than using Green's functions.

Insulated boundaries. If the boundary condition for the Green's function is the insulated kind at $y = 0$, $\partial/\partial y\, G(\mathbf{x}, \mathbf{x}_0)|_{y=0} = 0$, then a positive image source must be used for $y < 0$. In this way equal sources of thermal energy are located at $\mathbf{x} = \mathbf{x}_0$ and $\mathbf{x} = \mathbf{x}_0^*$. By symmetry no heat will flow across $y = 0$, as desired. The resulting solution is obtained in an exercise.

8.5.9 Green's Functions for a Circle— The Method of Images

The Green's function for Poisson's equation for a circle of radius a (with zero boundary conditions),

$$\nabla^2 G(\mathbf{x}, \mathbf{x}_0) = \delta(\mathbf{x} - \mathbf{x}_0) \qquad (8.5.40a)$$

$$G(\mathbf{x}, \mathbf{x}_0) = 0 \qquad \text{for} \qquad |\mathbf{x}| = a, \qquad (8.5.40b)$$

rather remarkably is obtained using the method of images. The idea is that for geometric reasons there exists an image point $\mathbf{x} = \mathbf{x}_0^*$ (as sketched in Fig. 8.5.4) such that the response along the circumference of a circle is constant. Consider an infinite space Green's function corresponding to a source at $\mathbf{x} = \mathbf{x}_0$ and a negative image source at $\mathbf{x} = \mathbf{x}_0^*$, where we do not define $\mathbf{x}_0^*$ yet:

$$\nabla^2 G(\mathbf{x}, \mathbf{x}_0) = \delta(\mathbf{x} - \mathbf{x}_0) - \delta(\mathbf{x} - \mathbf{x}_0^*). \qquad (8.5.41)$$

According to the principle of superposition, the solution will be the sum of the two infinite space Green's functions. We also introduce a constant homogeneous solution of Laplace's equation so that

$$G(\mathbf{x}, \mathbf{x}_0) = \frac{1}{2\pi} \ln |\mathbf{x} - \mathbf{x}_0| - \frac{1}{2\pi} \ln |\mathbf{x} - \mathbf{x}_0^*| + c = \frac{1}{4\pi} \ln \frac{|\mathbf{x} - \mathbf{x}_0|^2}{|\mathbf{x} - \mathbf{x}_0^*|^2} + c. \qquad (8.5.42)$$

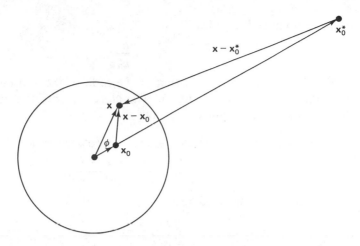

Figure 8.5.4 Green's function for Poisson's equation for a circle (image source).

We will show that there exists a point x_0^*, such that $G(x, x_0)$ given by (8.5.42) vanishes on the circle $|x| = a$. In order for this to occur,

$$|x - x_0|^2 = k\,|x - x_0^*|^2 \tag{8.5.43}$$

when $|x| = a$ (where $c = -1/4\pi \ln k$).

  We show that there is an image point x_0^* along the same radial line as the source point x_0 as illustrated in Fig. 8.5.4:

$$x_0^* = \gamma x_0. \tag{8.5.44}$$

We introduce the angle ϕ between x and x_0 (the same as the angle between x and x_0^*). Therefore,

$$
\begin{aligned}
|x - x_0|^2 &= (x - x_0) \cdot (x - x_0) = |x|^2 + |x_0|^2 - 2x \cdot x_0 \\
&= |x|^2 + |x_0|^2 - 2|x|\,|x_0|\cos\phi \\
|x - x_0^*|^2 &= (x - x_0^*) \cdot (x - x_0^*) = |x|^2 + |x_0^*|^2 - 2x \cdot x_0^* \\
&= |x|^2 + |x_0^*|^2 - 2|x|\,|x_0^*|\cos\phi,
\end{aligned}
\tag{8.5.45}
$$

otherwise known as the *law of cosines*. Equation (8.5.43) will be valid on the circle $|x| = a$ [using (8.5.44)] only if

$$a^2 + r_0^2 - 2ar_0\cos\phi = k(a^2 + \gamma^2 r_0^2 - 2a\gamma r_0\cos\phi),$$

where $r_0 = |x_0|$. This must hold for all angles ϕ, requiring γ and k to satisfy the following two equations:

$$a^2 + r_0^2 = k(a^2 + \gamma^2 r_0^2)$$

$$-2ar_0 = k(-2a\gamma r_0).$$

We obtain $k = 1/\gamma$, and thus

$$a^2 + r_0^2 = \frac{1}{\gamma}a^2 + \gamma r_0^2 \quad \text{or} \quad \gamma = \frac{a^2}{r_0^2}.$$

The image point is located at

$$\mathbf{x}_0^* = \frac{a^2}{r_0^2}\mathbf{x}_0.$$

Note that $|\mathbf{x}_0^*| = a^2/r_0$ (the product of the radii of the source and image points is the radius squared of the circle). The closer $\mathbf{x}_0$ is to the center of the circle, the farther the image point moves away.

Green's function. From $k = 1/\gamma = r_0^2/a^2$, we obtain the Green's function

$$G(\mathbf{x}, \mathbf{x}_0) = \frac{1}{4\pi} \ln\left(\frac{|\mathbf{x} - \mathbf{x}_0|^2\, a^2}{|\mathbf{x} - \mathbf{x}_0^*|^2\, r_0^2}\right).$$

Using the law of cosines, (8.5.45),

$$G(\mathbf{x}, \mathbf{x}_0) = \frac{1}{4\pi} \ln\left(\frac{a^2}{r_0^2}\frac{r^2 + r_0^2 - 2rr_0\cos\phi}{r_0^2 r^2 + r_0^{*2} - 2rr_0^*\cos\phi}\right),$$

where $r = |\mathbf{x}|$, $r_0 = |\mathbf{x}_0|$, and $r_0^* = |\mathbf{x}_0^*|$. Since $r_0^* = a^2/r_0$,

$$G(\mathbf{x}, \mathbf{x}_0) = \frac{1}{4\pi} \ln\left(\frac{a^2}{r_0^2}\frac{r^2 + r_0^2 - 2rr_0\cos\phi}{r^2 + a^4/r_0^2 - 2r\,a^2/r_0\cos\phi}\right)$$

or equivalently

$$\boxed{\;G(\mathbf{x}, \mathbf{x}_0) = \frac{1}{4\pi} \ln\left(a^2\frac{r^2 + r_0^2 - 2rr_0\cos\phi}{r^2r_0^2 + a^4 - 2rr_0a^2\cos\phi}\right),\;}\qquad (8.5.46)$$

where ϕ is the angle between $\mathbf{x}$ and $\mathbf{x}_0$ and $r = |\mathbf{x}|$ and $r_0 = |\mathbf{x}_0|$. In these forms it can be seen that on the circle $r = a$, $G(\mathbf{x}, \mathbf{x}_0) = 0$.

Solution. The solution of Poisson's equation is directly represented in terms of the Green's function. In general from (8.5.19),

$$u(\mathbf{x}) = \iint f(\mathbf{x}_0)G(\mathbf{x}, \mathbf{x}_0)\, dA_0 + \oint h(\mathbf{x}_0)\, \nabla_{\mathbf{x}_0}G(\mathbf{x}, \mathbf{x}_0)\cdot\hat{\mathbf{n}}\, ds. \qquad (8.5.47)$$

This line integral on the circular boundary can be evaluated. It is best to use polar coordinates ($ds = a\, d\theta_0$), in which case

$$\oint h(\mathbf{x}_0)\, \nabla_{\mathbf{x}_0}G(\mathbf{x}, \mathbf{x}_0)\cdot\hat{\mathbf{n}}\, ds = \int_0^{2\pi} h(\theta_0)\frac{\partial}{\partial r_0}\, G(\mathbf{x}, \mathbf{x}_0)\bigg|_{r_0 = a}\, a\, d\theta_0, \qquad (8.5.48)$$

where $r_0 = |\mathbf{x}_0|$. From (8.5.46)

$$\frac{\partial G}{\partial r_0} = \frac{1}{4\pi}\left(\frac{2r_0 - 2r\cos\phi}{r^2 + r_0^2 - 2rr_0\cos\phi} - \frac{2r^2r_0 - 2ra^2\cos\phi}{r^2r_0^2 + a^4 - 2rr_0a^2\cos\phi}\right).$$

Evaluating this for source points on the circle $r_0 = a$ yields

$$\frac{\partial G}{\partial r_0}\bigg|_{r_0 = a} = \frac{1}{4\pi}\frac{2a - 2r\cos\phi - (2r^2/a - 2r\cos\phi)}{r^2 + a^2 - 2ar\cos\phi}$$

$$= \frac{a}{2\pi}\frac{1 - (r/a)^2}{r^2 + a^2 - 2ar\cos\phi},$$

$$(8.5.49)$$

where ϕ is the angle between $\mathbf{x}$ and $\mathbf{x}_0$. If polar coordinates are used for both $\mathbf{x}$ and $\mathbf{x}_0$, then $\phi = \theta - \theta_0$ as indicated in Fig. 8.5.5.

Example. For Laplace's equation, $\nabla^2 u = 0$ [i.e., $f(\mathbf{x}) = 0$ in Poisson's equation] inside a circle with $u(x, y) = h(\theta)$ at $r = a$, we obtain from (8.5.47)–(8.5.49) in polar coordinates

$$u(r, \theta) = \frac{1}{2\pi} \int_0^{2\pi} h(\theta_0) \frac{a^2 - r^2}{r^2 + a^2 - 2ar \cos (\theta - \theta_0)} \, d\theta_0, \qquad (8.5.50)$$

known as **Poisson's formula**. Previously, we obtained a solution of Laplace's equation in this situation by the method of separation of variables (see Sec. 2.5.2). It can be shown that the infinite series solution so obtained can be summed to yield Poisson's formula (see Exercise 8.5.18).

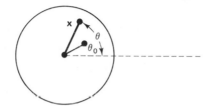

Figure 8.5.5 Polar coordinates.

EXERCISES 8.5

8.5.1. Consider (8.5.9), the eigenfunction expansion for $G(\mathbf{x}, \mathbf{x}_0)$. Assume that $\nabla^2 G$ has some eigenfunction expansion. Using Green's formula, verify that $\nabla^2 G$ may be obtained by term-by-term differentiation of (8.5.9).

8.5.2. **(a)** Solve

$$\nabla^2 u = f(x, y)$$

on a rectangle ($0 < x < L$, $0 < y < H$) with $u = 0$ on the boundary using the method of eigenfunction expansion.

(b) Write the solution in the form

$$u(\mathbf{x}) = \int_0^L \int_0^H f(\mathbf{x}_0) G(\mathbf{x}, \mathbf{x}_0) \, dx_0 \, dy_0.$$

Show that this $G(\mathbf{x}, \mathbf{x}_0)$ is the Green's function obtained previously.

8.5.3. Using the method of (multidimensional) eigenfunction expansion, determine $G(\mathbf{x}, \mathbf{x}_0)$ if

$$\nabla^2 G = \delta(\mathbf{x} - \mathbf{x}_0)$$

and

(a) on the rectangle ($0 < x < L$, $0 < y < H$)

$$\text{at } x = 0, \quad G = 0 \qquad \text{at } y = 0, \quad \frac{\partial G}{\partial y} = 0$$

$$\text{at } x = L, \quad \frac{\partial G}{\partial x} = 0 \qquad \text{at } y = H, \quad \frac{\partial G}{\partial y} = 0.$$

(b) on the rectangular-shaped box ($0 < x < L, 0 < y < H, 0 < z < W$) with $G = 0$ on the six sides.

*(c) on the semicircle ($0 < r < a, 0 < \theta < \pi$) with $G = 0$ on the entire boundary.

(d) on the quarter-circle ($0 < r < a, 0 < \theta < \pi/2$) with $G = 0$ on the straight sides and $\partial G/\partial r = 0$ at $r = a$.

*8.5.4. Consider in some three-dimensional region

$$\nabla^2 u = f$$

with $u = h(\mathbf{x})$ on the boundary. Represent $u(\mathbf{x})$ in terms of the Green's function (assumed to be known).

8.5.5. Consider inside a circle of radius a

$$\nabla^2 u = f$$

with

$$u(a, \theta) = h_1(\theta) \qquad \text{for } 0 < \theta < \pi$$

$$\frac{\partial u}{\partial r}(a, \theta) = h_2(\theta) \qquad \text{for } -\pi < \theta < 0.$$

Represent $u(r, \theta)$ in terms of the Green's function (assumed to be known).

8.5.6. Consider $\nabla^2 u = f(\mathbf{x})$ in two dimensions, satisfying homogeneous boundary conditions. Suppose that ϕ_h is a homogeneous solution,

$$\nabla^2 \phi_h = 0,$$

satisfying the same homogeneous boundary conditions. There may be more than one function ϕ_h.

(a) Show that there are no solutions $u(\mathbf{x})$ if $\iint f(\mathbf{x})\phi_h(\mathbf{x}) \, dA \neq 0$ for any $\phi_h(\mathbf{x})$.

(b) Show that there are an infinite number of solutions if $\iint f(\mathbf{x})\phi_h(\mathbf{x}) \, dA = 0$.

8.5.7. Consider in three dimensions

$$\left.\begin{array}{c} \nabla^2 u = f(\mathbf{x}) \\[6pt] \nabla u \cdot \hat{\mathbf{n}} = 0. \end{array}\right\} \tag{8.5.51}$$

with on the boundary

(a) Show that $\phi_h = 1$ is a homogeneous solution satisfying homogeneous boundary conditions.

(b) Under what condition is there a solution of (8.5.51)?

(c) What problem defines a modified Green's function for (8.5.51)? (Do *not* attempt to determine a modified Green's function.)

(d) *Assume* that the modified Green's function is symmetric. *Derive* a representation formula for $u(\mathbf{x})$ in terms of your modified Green's function.

8.5.8. Redo Exercise 8.5.7 if on the boundary $\nabla u \cdot \hat{\mathbf{n}} = h(\mathbf{x})$.

8.5.9. Using the method of (one-dimensional) eigenfunction expansion, determine $G(\mathbf{x}, \mathbf{x}_0)$ if

$$\nabla^2 G = \delta(\mathbf{x} - \mathbf{x}_0)$$

and

(a) On the rectangle ($0 < x < L, 0 < y < H$)

$$\text{at } x = 0, \quad G = 0 \qquad \text{at } y = 0, \quad \frac{\partial G}{\partial y} = 0$$

$$\text{at } x = L, \quad \frac{\partial G}{\partial x} = 0 \qquad \text{at } y = H, \quad \frac{\partial G}{\partial y} = 0.$$

Use y-dependent eigenfunctions.

*(b) On the semicircle $(0 < r < a, 0 < \theta < \pi)$ with $G = 0$ on the entire boundary. Use θ-dependent eigenfunctions.

*8.5.10. Consider the wave equation with a periodic source of frequency $\omega > 0$

$$\frac{\partial^2 \phi}{\partial t^2} = c^2 \nabla^2 \phi + g(\mathbf{x})e^{-i\omega t}.$$

Show that a particular solution at the same frequency, $\phi = u(\mathbf{x})e^{-i\omega t}$, satisfies a nonhomogeneous Helmholtz equation

$$\nabla^2 u + k^2 u = f(\mathbf{x}). \qquad \text{[What are } k^2 \text{ and } f(\mathbf{x})?\text{]}$$

The Green's function satisfies

$$\nabla^2 G + k^2 G = \delta(\mathbf{x} - \mathbf{x}_0).$$

(a) What is Green's formula for the operator $\nabla^2 + k^2$?

(b) In infinite three-dimensional space, show that

$$G = \frac{c_1 e^{ik\rho} + c_2 e^{-ik\rho}}{\rho}.$$

Choose c_1 and c_2 so that the corresponding $\phi(\mathbf{x}, t)$ is an *outward*-propagating wave.

(c) In infinite two-dimensional space, show that the Green's function is a linear combination of Bessel functions. Determine the constants so that the corresponding $\phi(\mathbf{x}, t)$ is an *outward*-propagating wave for r sufficiently large. [*Hint:* See (6.7.28) and (6.8.3).]

8.5.11. (a) Determine the Green's function for $y > 0$ (in two dimensions) for $\nabla^2 G = \delta(\mathbf{x} - \mathbf{x}_0)$ subject to $\partial G/\partial y = 0$ on $y = 0$. (*Hint:* Consider a positive source at $\mathbf{x}_0$ and a positive image source at $\mathbf{x}_0^*$.)

(b) Use part (a) to solve $\nabla^2 u = f(\mathbf{x})$ with

$$\frac{\partial u}{\partial y} = h(x) \qquad \text{at } y = 0.$$

Ignore the contribution at ∞.

8.5.12. Modify Exercise 8.5.11 if the physical region is $y < 0$.

*8.5.13. Modify Exercise 8.5.11 if the region is three-dimensional with $y > 0$. Note that $h(x)$ in part (b) becomes $h(x, z)$.

*8.5.14. Using the method of images, solve

$$\nabla^2 G = \delta(\mathbf{x} - \mathbf{x}_0)$$

in the first quadrant $(x \geq 0$ and $y \geq 0)$ with $G = 0$ on the boundaries.

8.5.15. (a) Reconsider Exercise 8.5.14 if $G = 0$ at $x = 0$ and $\partial G/\partial y = 0$ at $y = 0$.

(b) Use part (a) to solve $(x \geq 0$ and $y \geq 0)$

$$\nabla^2 u = f(x, y)$$
$$u(0, y) = g(y)$$
$$\frac{\partial u}{\partial y}(x, 0) = h(x).$$

8.5.16. (a) Using the method of images, solve

$$\nabla^2 G = \delta(\mathbf{x} - \mathbf{x}_0)$$

in the 60° wedge-shaped region $(0 < \theta < \pi/3, 0 < r < \infty)$ with $G = 0$ on the boundaries.

(b) For what other angled wedges can the method of images be used?

8.5.17. A modified Green's function $G_m(\mathbf{x}, \mathbf{x}_0)$ satisfies

$$\nabla^2 G_m = \delta(\mathbf{x} - \mathbf{x}_0) + c$$

with

$$\nabla G_m \bullet \hat{\mathbf{n}} = 0$$

on the boundary of the rectangle $(0 < x < L, 0 < y < H)$.

(a) Show that the method of eigenfunction expansion (two-dimensional) only works for $c = -1/LH$. For this c, determine $G_m(\mathbf{x}, \mathbf{x}_0)$. If possible, make $G_m(\mathbf{x}, \mathbf{x}_0)$ symmetric.

(b) Show that the method of eigenfunction expansion (one-dimensional) works only for $c = -1/LH$. For this c, determine $G_m(\mathbf{x}, \mathbf{x}_0)$. If possible, make $G_m(\mathbf{x}, \mathbf{x}_0)$ symmetric.

8.5.18. Solve $\nabla^2 u = 0$ inside a circle of radius a with $u(x, y) = h(\theta)$ at $r = a$, using the method of separation of variables. Show that

$$u(r, \theta) = \int_0^{2\pi} h(\theta_0) I(r, \theta, \theta_0) \, d\theta_0.$$

Show that the infinite series for $I(r, \theta, \theta_0)$ can be summed yielding Poisson's formula (8.5.50).

*8.5.19. Determine the Green's function $G(\mathbf{x}, \mathbf{x}_0)$ inside the semicircle $(0 < r < a, 0 < \theta < \pi)$

$$\nabla^2 G = \delta(\mathbf{x} - \mathbf{x}_0)$$

with $G = 0$ on the boundary.

8.5.20. Modify Exercise 8.5.19 if $G = 0$ on $r = a$, but $\partial G/\partial \theta = 0$ on $\theta = 0$ and $\theta = \pi$.

8.5.21. Determine the Green's function $G(\mathbf{x}, \mathbf{x}_0)$ inside the sphere of radius a

$$\nabla^2 G = \delta(\mathbf{x} - \mathbf{x}_0)$$

with $G = 0$ on the boundary.

8.5.22. Use the method of multiple images to obtain the Green's function $G(\mathbf{x}, \mathbf{x}_0)$

$$\nabla^2 G = \delta(\mathbf{x} - \mathbf{x}_0)$$

(a) on the rectangle $(0 < x < L, 0 < y < H)$ if $G = 0$ at $x = 0$ and $x = L$ and $\partial G/\partial y = 0$ at $y = 0$ and $y = H$.

(b) on the infinite strip $(0 < x < L, -\infty < y < \infty)$ if $G = 0$ at $x = 0$ and $\partial G/\partial x = 0$ at $x = L$.

*(c) on the infinite strip $(0 < x < L, -\infty < y < \infty, -\infty < z < \infty)$ if $G = 0$ at $x = 0$ and $G = 0$ at $x = L$.

(d) on the semi-infinite strip $(0 < x < L, 0 < y < \infty)$ if $G = 0$ along the boundaries.

(e) on the semi-infinite strip $(0 < x < L, -\infty < y < 0)$ if $G = 0$ at $x = 0$, $G = 0$ at $x = L$, $\partial G/\partial y = 0$ at $y = 0$.

8.5.23. Determine a particular solution of

$$\nabla^2 u = f(\mathbf{x})$$

in infinite two-dimensional space if $f(\mathbf{x}) = g(r)$, where $r = |\mathbf{x}|$:

(a) Use the infinite space Green's function (8.5.24a).

(b) Use a Green's function for the ordinary differential equation

$$\frac{1}{r}\frac{d}{dr}\left(r\frac{du}{dr}\right) = g(r).$$

(c) Compare parts (a) and (b).

8.6 PERTURBED EIGENVALUE PROBLEMS

8.6.1 Introduction

When a small change (called a **perturbation**) is made to a problem that we know how to solve, then the resulting problem may not have a simple exact solution. Here, we will develope an approximate (asymptotic) procedure to analyze perturbed eigenvalue problems. Nonhomogeneous boundary value problems with nontrivial homogeneous solutions will occur, and hence our development in Section 8.4 of the Fredholm alternative will be helpful. We begin with an elementary mathematical example before considering the more interesting case of a perturbed circular membrane.

8.6.2 Mathematical Example

The simplest example of a perturbed eigenvalue problem is

$$\frac{d^2\phi}{dx^2} + (\lambda + \varepsilon f(x))\phi = 0 \tag{8.6.1a}$$

with

$$\phi(0) = 0 \quad \text{and} \quad \phi(L) = 0. \tag{8.6.1b}$$

If $\varepsilon = 0$, this is the usual eigenvalue problem ($\lambda = (n\pi/L)^2$, $\phi = \sin n\pi x/L$). If ε is a small nonzero parameter, then the coefficient deviates from a constant by a small given amount, $\varepsilon f(x)$. It is known that the eigenvalues and eigenfunctions are well-behaved functions of ε:

$$\lambda = \lambda_0 + \varepsilon\lambda_1 + \cdots \quad \text{and} \quad \phi = \phi_0 + \varepsilon\phi_1 + \cdots \tag{8.6.2a, b}$$

This is called a **perturbation expansion**. By substituting (8.6.2) into (8.6.1a), we obtain

$$\frac{d^2}{dx^2}(\phi_0 + \varepsilon\phi_1 + \cdots) + [\lambda_0 + \varepsilon\lambda_1 + \cdots + \varepsilon f(x)](\phi_0 + \varepsilon\phi_1 + \cdots) = 0. \tag{8.6.3}$$

Equation (8.6.3) is valid for all ε. The terms without ε must equal zero (resulting from letting $\varepsilon - 0$). Thus,

$$\frac{d^2\phi_0}{dx^2} + \lambda_0\phi_0 = 0. \tag{8.6.4}$$

The boundary conditions [obtained by substituting (8.6.2b) into (8.6.1b)] are $\phi_0(0) = 0$ and $\phi_0(L) = 0$. Consequently, as expected, the *leading order* eigenvalues λ_0 and corresponding eigenfunctions ϕ_0 are the same as those of the unperturbed problem ($\varepsilon = 0$):

$$\lambda_0 = \left(\frac{n\pi}{L}\right)^2 \quad \text{and} \quad \phi_0 = \sin\frac{n\pi x}{L}, \tag{8.6.5a, b}$$

where $n = 1, 2, 3, \ldots$. A more precise notation would be $\lambda_n^{(0)}$.

The ε terms in (8.6.3) must also vanish:

$$\frac{d^2\phi_1}{dx^2} + \lambda_0\phi_1 = -f(x)\phi_0 - \lambda_1\phi_0, \tag{8.6.6a}$$

where

$$\phi_1(0) = 0 \quad \text{and} \quad \phi_1(L) = 0 \qquad (8.6.6b)$$

follows from (8.6.1b) and (8.6.2b). This is a nonhomogeneous differential equation with homogeneous boundary conditions. We note that $\phi_0 = \sin n\pi x/L$ is a nontrivial homogeneous solution satisfying the homogeneous boundary conditions. Thus, by the Fredholm alternative, there is a solution to (8.6.6) only if the right-hand side of (8.6.6a) is orthogonal to ϕ_0:

$$0 = \int_0^L f(x)\phi_0^2 \, dx + \lambda_1 \int_0^L \phi_0^2 \, dx. \qquad (8.6.7)$$

From (8.6.7) we determine the resulting perturbation of the eigenvalue:

$$\lambda_1 = -\frac{\displaystyle\int_0^L f(x)\phi_0^2 \, dx}{\displaystyle\int_0^L \phi_0^2 \, dx} = -\frac{2}{L}\int_0^L f(x)\phi_0^2 \, dx. \qquad (8.6.8)$$

Instead of using the Fredholm alternative, we could have applied the method of eigenfunction expansion to (8.6.6). In this latter way, we obtain ϕ_1 as well as λ_1 given by (8.6.8).

8.6.3 Vibrating Nearly Circular Membrane

For a physical problem involving similar ideas, consider the vibrations of a nearly circular membrane with mass impurities. We have already determined (see Section 6.7) the natural frequencies for a circular membrane with constant mass density. We want to know how these frequencies are changed due to small changes in both the density and geometry. In general, a vibrating membrane satisfies the two-dimensional wave equation

$$\frac{\partial^2 u}{\partial t^2} = c^2 \nabla^2 u, \qquad (8.6.9)$$

where $c^2 = T/\rho$ may be a function of r and θ. We assume that $u = 0$ on the boundary. By separating variables, $u(r, \theta, t) = \phi(r, \theta)h(t)$, we obtain

$$\frac{d^2 h}{dt^2} = -\lambda h \quad \text{and} \quad \nabla^2 \phi = -\frac{\lambda}{c^2}\phi. \qquad (8.6.10a, b)$$

Here, the separation constants λ are such that $\sqrt{\lambda}$ are the natural frequencies of oscillation. We know how to solve this problem (see Sec. 6.7) if c^2 is constant and the membrane is circular. However, we want to consider the case in which the constant mass density is slightly perturbed (perhaps due to a small imperfection), $\rho = \rho_0 + \varepsilon\rho_1(r, \theta)$, where the perturbation of the density $\varepsilon\rho_1(r, \theta)$ is given and ε is a very small parameter ($0 < |\varepsilon| \ll 1$). Thus

$$\frac{1}{c^2} = \frac{\rho}{T} = \frac{\rho_0 + \varepsilon\rho_1(r, \theta)}{T} = \frac{1}{c_0^2} + \varepsilon\frac{\rho_1(r, \theta)}{T},$$

where c_0 is the sound speed for a uniform membrane. The perturbed eigenvalue problem is to solve the partial differential equation

$$\nabla^2 \phi = -\lambda \left(\frac{1}{c_0^2} + \varepsilon \frac{\rho_1(r, \theta)}{T} \right) \phi, \qquad (8.6.11)$$

subject to $\phi = 0$ on the boundary:

$$\phi(a + \varepsilon g(\theta), \theta) = 0, \qquad (8.6.12)$$

since we express a perturbed circle as $r = a + \varepsilon g(\theta)$ with $g(\theta)$ given.

Boundary condition. The boundary condition (8.6.12) is somewhat difficult; we may wish to consider the simple case in which the boundary is circular ($r = a$ or $g(\theta) = 0$). In general, $\phi = 0$ along a complicated boundary, which is near to the simpler boundary $r = a$. This suggests that we utilize a Taylor series, in which case (8.6.12) may be replaced by

$$\phi + \varepsilon g(\theta) \frac{\partial \phi}{\partial r} + \frac{\varepsilon^2 g^2(\theta)}{2!} \frac{\partial^2 \phi}{\partial r^2} + \cdots = 0, \qquad (8.6.13)$$

evaluated at $r = a$.

Perturbation expansion. To solve (8.6.11) with (8.6.13), we assume the eigenvalues and eigenfunctions depend on the small parameter such that

$$\phi = \phi_0 + \varepsilon \phi_1 + \cdots \quad \text{and} \quad \lambda = \lambda_0 + \varepsilon \lambda_1 + \cdots \qquad (8.6.14a, b)$$

We substitute (8.6.14) into (8.6.11) and (8.6.13). The terms of order ε^0 are

$$\nabla^2 \phi_0 = -\frac{\lambda_0 \phi_0}{c_0^2} \qquad (8.6.15)$$

with $\phi_0 = 0$ at $r = a$. Thus, λ_0 are known unperturbed eigenvalues and ϕ_0 are the corresponding known eigenfunctions for a circular membrane with uniform density ρ_0 (see Section 6.7). We are most interested in determining λ_1, the leading order change of each eigenvalue due to the perturbed density and shape. We will determine λ_1 by considering the equations for ϕ_1 obtained by keeping only the ε terms when the perturbation expansions (8.6.14) are substituted into (8.6.11) and (8.6.13):

$$\nabla^2 \phi_1 + \frac{\lambda_0}{c_0^2} \phi_1 = -\frac{\lambda_1}{c_0^2} \phi_0 - \frac{\lambda_0}{T} \rho_1(r, \theta) \phi_0, \qquad (8.6.16a)$$

subject to the boundary condition (at $r = a$)

$$\phi_1 = -g(\theta) \frac{\partial \phi_0}{\partial r}. \qquad (8.6.16b)$$

The right-hand side of (8.6.16a) contains the known perturbation of the density ρ_1 and the unknown perturbation of each eigenvalue λ_1, whereas the right-hand side of (8.6.16b) involves the known perturbation of the shape ($r = a + \varepsilon g(\theta)$).

Compatibility condition. Boundary value problem (8.6.16) is a nonhomogeneous partial differential equation with nonhomogeneous boundary conditions. Most importantly, there is a nontrivial homogeneous solution, $\phi_{1_h} = \phi_0$; the leading order eigenfunction satisfies the *corresponding homogeneous* partial differential equation and the *homogeneous* boundary conditions [see (8.6.15)]. Thus, there is a solution to (8.6.16) only if the compatibility equation is valid. This is most easily obtained using Green's formula (with $u = \phi_0$ and $v = \phi_1$):

$$\iint [\phi_0 L(\phi_1) - \phi_1 L(\phi_0)] \, dA = \oint (\phi_0 \nabla \phi_1 - \phi_1 \nabla \phi_0) \cdot \hat{\mathbf{n}} \, ds. \qquad (8.6.17)$$

The right-hand side of (8.6.17) does not vanish since ϕ_1 does not satisfy homogeneous boundary conditions. Using (8.6.15) and (8.6.16) yields the solvability condition

$$-\iint \phi_0 \left(\frac{\lambda_1}{c_0^2} \phi_0 + \frac{\lambda_0}{T} \rho_1(r, \theta) \phi_0 \right) dA = \oint g(\theta) \frac{\partial \phi_0}{\partial r} \nabla \phi_0 \cdot \hat{\mathbf{n}} \, ds. \qquad (8.6.18)$$

We easily determine the perturbed eigenvalue λ_1 from (8.6.18):

$$\lambda_1 = \frac{\dfrac{\lambda_0}{T} \iint \rho_1(r, \theta) \phi_0^2 \, r \, dr d\theta + \displaystyle\int_0^{2\pi} g(\theta) \left(\dfrac{\partial \phi_0}{\partial r} \right)^2 a \, d\theta}{-\dfrac{1}{c_0^2} \iint \phi_0^2 \, r \, dr d\theta}, \qquad (8.6.19)$$

using $\nabla \phi_0 \cdot \hat{\mathbf{n}} = \partial \phi_0 / \partial r$ (evaluated at $r = a$), $dA = r \, dr d\theta$, and $ds = a \, d\theta$. The eigenvalues decrease if the density is increased ($\rho_1 > 0$) or if the membrane is enlarged [$g(\theta) > 0$].

As is elaborated on in the exercises, this result is valid only if there is one eigenfunction ϕ_0 corresponding to the eigenvalue λ_0. In fact, for a circular membrane, usually there are two eigenfunctions corresponding to each eigenvalue (from $\sin m\theta$ and $\cos m\theta$). Both must be considered.

If we were interested in ϕ_1, it could now be obtained from (8.6.16a) using the method of eigenfunction expansion. However, in many applications it is the perturbed eigenvalues (here frequencies) that are of greater importance.

Fredholm alternative. If $g(\theta) = 0$, then ϕ_0 and ϕ_1 satisfy the same set of homogeneous boundary conditions. Then (8.6.18) is equivalent to the Fredholm alternative; that is, solutions exist to (8.6.16) if and only if the right-hand side of (8.6.16) is orthogonal to the homogeneous solution ϕ_0. Equation (8.6.18) shows the appropriate modification for nonhomogeneous boundary conditions.

EXERCISES 8.6

8.6.1. Consider the perturbed eigenvalue problem (8.6.1a). Determine the perturbations of the eigenvalue λ_1 if

(a) $\dfrac{d\phi}{dx}(0) = 0$ and $\dfrac{d\phi}{dx}(L) = 0$

(b) $\phi(0) = 0$ and $\dfrac{d\phi}{dx}(L) = 0$

8.6.2. Reconsider Exercise 8.6.1. Determine the perturbations of the eigenvalues λ_1 and the eigenfunctions ϕ_1 using the method of eigenfunction expansion:

(a) $\dfrac{d\phi}{dx}(0) = 0$ and $\dfrac{d\phi}{dx}(L) = 0$

(b) $\phi(0) = 0$ and $\dfrac{d\phi}{dx}(L) = 0$

(c) $\phi(0) = 0$ and $\phi(L) = 0$

8.6.3. Reconsider Exercise 8.6.1 subject to the periodic boundary conditions $\phi(-L) = \phi(L)$ and $d\phi/dx(-L) = d\phi/dx(L)$. For $n \neq 0$, the eigenvalue problem is **degenerate** if $\varepsilon = 0$; that is, there is more than one eigenfunction ($\sin n\pi x/L$ and $\cos n\pi x/L$) corresponding to the same eigenvalue. *Determine the perturbed eigenvalues λ_1. Show that the eigenvalue **splits**. This means that if $\varepsilon \neq 0$, there is one eigenfunction for each eigenvalue, but as $\varepsilon \to 0$, two eigenvalues will approach each other* (**coalesce**), yielding eigenvalues with two eigenfunctions. [*Hint:* It is necessary to consider a linear combination of both eigenfunctions ($\varepsilon = 0$). For each eigenvalue, *determine the specific combination of these eigenfunctions that is the unique eigenfunction when $\varepsilon \neq 0$.*]

8.6.4. Reconsider Exercise 8.6.1 subject to the boundary conditions $\phi(0) = 0$ and $\phi(L) = 0$. Do additional calculations to obtain λ_2. Insist that the eigenfunctions are normalized, $\int_0^L \phi^2 \, dx = 1$.

8.6.5. Consider the *nonlinearly* perturbed eigenvalue problem:

$$\frac{d^2\phi}{dx^2} + \lambda\phi = \varepsilon\phi^3$$

with $\phi(0) = 0$ and $\phi(L) = 0$. Determine the perturbation of the eigenvalue λ_1. Since the problem is nonlinear, the amplitude is important. Assume $\int_0^L \phi^2 \, dx = a^2$. Sketch a as a function of λ.

8.6.6. Consider a vibrating string with approximately uniform tension T and mass density $\rho_0 + \varepsilon\rho_1(x)$ subject to fixed boundary conditions. Determine the changes in the natural frequencies induced by the mass variation.

8.6.7. Consider a uniform membrane of fixed shape with known frequencies and known natural modes of vibration. Suppose the mass density is perturbed. Determine how the frequencies are perturbed. You may assume there is only one mode of vibration for each frequency.

8.6.8. For a circular membrane, determine the change in the natural frequencies of the circularly symmetric ($m = 0$) eigenfunctions due to small mass and shape variations.

8.6.9. For noncircularly symmetric eigenfunctions ($m \neq 0$), (8.6.16) is valid with $\phi_0 = a\phi_0^{(1)} + b\phi_0^{(2)}$, where $\phi_0^{(1)}$ and $\phi_0^{(2)}$ are two mutually orthogonal eigenfunctions corresponding to the same eigenvalue λ_0. Here a and b are arbitrary constants.
(a) Determine a homogeneous linear system of equations for a and b derived from the fact that ϕ_1 has two homogeneous solutions $\phi_0^{(1)}$ and $\phi_0^{(2)}$. This will be the compatability condition for (8.6.16).
(b) Solve the linear system of part (a) to determine the perturbed frequencies and the corresponding natural modes of vibration.

8.7 SUMMARY

We have calculated a few examples of time-independent Green's functions by some different techniques:

1. Limit of time-dependent problem
2. Variation of parameter (ordinary differential equation only)
3. Eigenfunction expansion of Green's function
4. Direct solution of differential equation defining Green's function
5. Using infinite space Green's function:
   a. Removing singularity
   b. Method of images

For time-independent problems, perhaps the best techniques are based on infinite space Green's functions. We will find the same to be true for time-dependent problems. In that case we will need to discuss more techniques to solve partial differential equations in an infinite domain. For that reason in Chapter 9, we investigate solutions of homogeneous partial differential equations on infinite domains using Fourier transforms. Then in Chapter 10 we return to nonhomogeneous problems for time-dependent partial differential equations using Green's functions.

Infinite Domain Problems—
Fourier Transform Solutions
of Partial Differential Equations

9.1 INTRODUCTION

Most of the partial differential equations that we have analyzed previously were defined on finite regions (e.g., heat flow in a finite one-dimensional rod or in enclosed two- or three-dimensional regions). The solutions we obtained depended on conditions at these boundaries. In this chapter we analyze problems that extend indefinitely in at least one direction. Physical problems never are infinite, but by introducing a mathematical model with infinite extent, we are able to determine behavior of problems in situations in which the influence of actual boundaries is expected to be negligible. We will solve problems with infinite or semi-infinite extent by generalizing the method of separation of variables.

9.2 HEAT EQUATION ON AN INFINITE DOMAIN

We begin by considering heat conduction in one dimension, unimpeded by any boundaries. For the simplest case with constant thermal properties and no sources, the temperature $u(x, t)$ satisfies the heat equation,

$$\frac{\partial u}{\partial t} = k\frac{\partial^2 u}{\partial x^2}, \tag{9.2.1}$$

defined for all x, $-\infty < x < \infty$. We impose an initial condition,

$$u(x, 0) = f(x). \tag{9.2.2}$$

We would like to use (9.2.1) to predict the future temperature.

For problems in a finite region, boundary conditions are needed at both ends (usually $x = 0$ and $x = L$). Frequently, problems on an infinite domain $(-\infty < x < \infty)$ seem to be posed without any boundary conditions. However, usually there are physical conditions at $\pm\infty$, even if they are not stated as such. In the simplest case, suppose that the initial temperature distribution $f(x)$ approaches 0 as $x \to \pm\infty$. This means that *initially* for all x sufficiently large, the temperature is approximately 0. Physically, *for all time* the temperature approaches 0 as $x \to \pm\infty$:

$$u(-\infty, t) = 0 \quad \text{and} \quad u(\infty, t) = 0.$$

In this way, our problem has homogeneous "boundary" conditions.

Separation of variables. We separate variables as before

$$u(x, t) = \phi(x)h(t), \tag{9.2.3}$$

so that

$$\frac{1}{kh}\frac{dh}{dt} = \frac{1}{\phi}\frac{d^2\phi}{dx^2} = -\lambda.$$

This yields the same two ordinary differential equations as for a finite geometry,

$$\frac{dh}{dt} = -\lambda kh \tag{9.2.4a}$$

$$\frac{d^2\phi}{dx^2} = -\lambda\phi. \tag{9.2.4b}$$

Determining the separation constant λ is not difficult, but it is very subtle in this case; doing the obvious will be *wrong*. We would expect that the boundary conditions at $\pm\infty$ for $\phi(x)$ are $\phi(-\infty) = 0$ and $\phi(\infty) = 0$. However, we would quickly see that there are no values of λ for which there are nontrivial solutions of $d^2\phi/dx^2 = -\lambda\phi$ which approach 0 at both $x = \pm\infty$. For example, for $\lambda > 0$, $\phi = c_1 \cos \sqrt{\lambda}x + c_2 \sin \sqrt{\lambda}x$ and these solutions do *not* approach 0 as $x \to \pm\infty$. As we will verify later, the *correct boundary condition for the separated spatial part* $\phi(x)$ *at* $x = \pm\infty$ *is different from the boundary condition for* $u(x, t)$ *at* $x = \pm\infty$. Instead, we specify that $\phi(x)$ is just bounded at $x = \pm\infty$, $|\phi(-\infty)| < \infty$ and $|\phi(\infty)| < \infty$. This is rather strange, but later we will show that although $|\phi(\pm\infty)| < \infty$, the eventual solution of the partial differential equation (after superposition) will in fact satisfy $u(\pm\infty, t) = 0$.

Eigenvalue problem. We thus claim that the boundary value problem of interest on an infinite domain is

$$\boxed{\begin{aligned} \frac{d^2\phi}{dx^2} + \lambda\phi &= 0 \\ |\phi(\pm\infty)| &< \infty. \end{aligned}}$$

$$\frac{d^2\phi}{dx^2} + \lambda\phi = 0 \tag{9.2.5a}$$

$$|\phi(\pm\infty)| < \infty. \tag{9.2.5b}$$

Let us determine those values of λ for which both $|\phi(\pm\infty)| < \infty$. If $\lambda < 0$, the solution is a linear combination of exponentially growing and exponentially decaying solutions. It is impossible for both $|\phi(\pm\infty)| < \infty$; there are no negative eigenvalues.

However, if $\lambda > 0$, then

$$\phi = c_1 \cos \sqrt{\lambda}\, x + c_2 \sin \sqrt{\lambda}\, x.$$

This solution remains bounded for all x no matter what λ is ($\lambda > 0$). Thus, all values of λ ($\lambda > 0$) are eigenvalues. Furthermore, the eigenfunctions are both sines and cosines (since both c_1 and c_2 are arbitrary). We can also verify that $\lambda = 0$ is an eigenvalue whose eigenfunction is a constant. This is very similar to a Fourier series in that both sines and cosines (including a constant) are eigenfunctions. However, in a Fourier series the eigenvalues were **discrete**, $\lambda = (n\pi/L)^2$, whereas here all nonnegative values of λ are allowable. The set of eigenvalues for a problem is sometimes referred to as the spectrum. In this case we have a **continuous spectrum**, $\lambda \geq 0$ (rather than discrete).

Superposition principle. The time-dependent ordinary differential equation is easily solved, $h = ce^{-\lambda kt}$, and thus we obtain the following product solutions:

$$\sin \sqrt{\lambda}\, x\, e^{-\lambda kt} \quad \text{and} \quad \cos \sqrt{\lambda}\, x\, e^{-\lambda kt},$$

for all $\lambda \geq 0$. The principle of superposition suggests that we can form another solution by the most general linear combination of these. Instead of summing over all $\lambda \geq 0$, we integrate:

$$u(x, t) = \int_0^\infty [c_1(\lambda) \cos \sqrt{\lambda}\, x\, e^{-\lambda kt} + c_2(\lambda) \sin \sqrt{\lambda}\, x\, e^{-\lambda kt}] d\lambda,$$

where $c_1(\lambda)$ and $c_2(\lambda)$ are arbitrary functions of λ. This is a generalized principle of superposition. It may be verified by direct computation that the integral satisfies (9.2.1). It is usual to let $\lambda = \omega^2$, so that

$$\boxed{u(x, t) = \int_0^\infty [A(\omega) \cos \omega x\, e^{-k\omega^2 t} + B(\omega) \sin \omega x\, e^{-k\omega^2 t}]\, d\omega,} \qquad (9.2.6)$$

where $A(\omega)$ and $B(\omega)$ are arbitrary functions.* This is analogous to the solution for finite regions (with periodic boundary conditions):

$$u(x, t) = \sum_{n=0}^\infty \left[a_n \cos \frac{n\pi x}{L}\, e^{-k(n\pi/L)^2 t} + b_n \sin \frac{n\pi x}{L}\, e^{-k(n\pi/L)^2 t} \right].$$

In order to solve for the arbitrary functions $A(\omega)$ and $B(\omega)$, we must insist that (9.2.6) satisfies the initial condition $u(x, 0) = f(x)$:

$$f(x) = \int_0^\infty [A(\omega) \cos \omega x + B(\omega) \sin \omega x]\, d\omega. \qquad (9.2.7)$$

In later sections we will explain that there exist $A(\omega)$ and $B(\omega)$ such that (9.2.7) is valid for most functions $f(x)$. More importantly, we will discover how to determine $A(\omega)$ and $B(\omega)$.

---

* To be precise, note that $c_1(\lambda)\, d\lambda = c_1(\omega^2)\, 2\omega\, d\omega = A(\omega)\, d\omega$.

Complex exponentials. The x-dependent eigenfunctions were determined to be $\sin \sqrt{\lambda}\, x$ and $\cos \sqrt{\lambda}\, x$ for all $\lambda \geq 0$. Sometimes different independent solutions are utilized. One possibility is to use the complex functions $e^{i\sqrt{\lambda}x}$ and $e^{-i\sqrt{\lambda}x}$ for all $\lambda \geq 0$. If we introduce $\omega = \sqrt{\lambda}$, then the x-dependent eigenfunctions become $e^{i\omega x}$ and $e^{-i\omega x}$ for all $\omega \geq 0$. Alternatively, we may consider only* $e^{-i\omega x}$, but for all ω (including both positive and negatives). Thus, as explained further in Sec. 9.3, the product solutions are $e^{-i\omega x}e^{-k\omega^2 t}$ for all ω. The generalized principle of superposition implies that a solution of the heat equation on an infinite interval is

$$u(x, t) = \int_{-\infty}^{\infty} c(\omega)e^{-i\omega x}e^{-k\omega^2 t}\, d\omega. \tag{9.2.8}$$

This can be shown to be equivalent to (9.2.6) using Euler's formulas [see Exercise 9.2.1]. In this form, the initial condition $u(x, 0) = f(x)$ is satisfied if

$$f(x) = \int_{-\infty}^{\infty} c(\omega)e^{-i\omega x}\, d\omega. \tag{9.2.9}$$

$u(x, t)$ is real if $f(x)$ is real (see Exercises 9.2.1 and 9.2.2). We need to understand (9.2.9). In addition, we need to determine the "coefficents" $c(\omega)$. In the case of a finite region, (9.2.9) would be an infinite series (since the eigenvalues are discrete). We usually determine the coefficients by the orthogonality of the eigenfunctions. Here, the orthogonality concept is a little more complicated and is not explained until Chapter 10 (also see Exercise 9.4.18). However, in the next sections we will learn how to determine $c(\omega)$.

EXERCISES 9.2

*9.2.1. Determine complex $c(\omega)$ so that (9.2.8) is equivalent to (9.2.6) with real $A(\omega)$ and $B(\omega)$. Show that $c(-\omega) = \bar{c}(\omega)$, where the overbar denotes the complex conjugate.

9.2.2. If $c(-\omega) = \bar{c}(\omega)$ (see Exercise 9.2.1), show that $u(x, t)$ given by (9.2.8) is real.

9.3 COMPLEX FORM OF FOURIER SERIES

In solving boundary value problems for $-\infty < x < \infty$, it is usual to analyze $e^{-i\omega x}$ (for all ω) rather than $\sin \sqrt{\lambda}\, x$ and $\cos \sqrt{\lambda}\, x$ (for $\lambda \geq 0$), where $\lambda = \omega^2$. This same alternative approach can be used for problems on a finite domain.

With periodic boundary conditions, we have found the theory of Fourier series to be quite useful:

$$f(x) \sim \frac{a_0}{2} + \sum_{n=1}^{\infty}\left(a_n \cos \frac{n\pi x}{L} + b_n \sin \frac{n\pi x}{L}\right), \tag{9.3.1}$$

* It is conventional to use $e^{-i\omega x}$ rather than $e^{i\omega x}$. $|\omega|$ is the **wave number**, the number of waves in 2π distance. It is a spatial frequency.

where

$$a_n = \frac{1}{L} \int_{-L}^{L} f(x) \cos \frac{n\pi x}{L} \, dx \qquad (9.3.2a)$$

$$b_n = \frac{1}{L} \int_{-L}^{L} f(x) \sin \frac{n\pi x}{L} \, dx. \qquad (9.3.2b)$$

Here, it is convenient to have the factor $\frac{1}{2}$ in the series rather than in the integral. To introduce complex exponentials instead of sines and cosines, we use Euler's formulas

$$\cos \theta = \frac{e^{i\theta} + e^{-i\theta}}{2} \qquad \text{and} \qquad \sin \theta = \frac{e^{i\theta} - e^{-i\theta}}{2i}.$$

It follows that

$$f(x) \sim \frac{1}{2}a_0 + \frac{1}{2} \sum_{n=1}^{\infty} (a_n - ib_n)e^{in\pi x/L} + \frac{1}{2} \sum_{n=1}^{\infty} (a_n + ib_n)e^{-in\pi x/L}. \qquad (9.3.3)$$

In order to only have $e^{-in\pi x/L}$, we change the dummy index in the first summation, replacing n by $-n$. Thus,

$$f(x) \sim \frac{1}{2}a_0 + \frac{1}{2} \sum_{n=-1}^{-\infty} [a_{(-n)} - ib_{(-n)}]e^{-in\pi x/L} + \frac{1}{2} \sum_{n=1}^{\infty} (a_n + ib_n)e^{in\pi x/L}.$$

From the definition of a_n and b_n, (9.3.2), $a_{(-n)} = a_n$ and $b_{(-n)} = -b_n$. Thus, if we define

$$c_n = \frac{a_n + ib_n}{2},$$

then $f(x)$ becomes simply

$$f(x) \sim \sum_{n=-\infty}^{\infty} c_n e^{-in\pi x/L}. \qquad (9.3.4)$$

(The $n = 0$ term reduces to $a_0/2$ since $b_0 = 0$.) Equation (9.3.4) is known as the **complex form of the Fourier series of $f(x)$.**[*] It is equivalent to the usual form. It is more compact to write, but it is only used infrequently. In this form the complex Fourier coefficients are

$$c_n = \frac{1}{2}(a_n + ib_n) = \frac{1}{2L} \int_{-L}^{L} f(x) \left(\cos \frac{n\pi x}{L} + i \sin \frac{n\pi x}{L} \right) dx.$$

We immediately recognize a simplification, using Euler's formula. Thus, we derive a formula for the complex Fourier coefficients

$$c_n = \frac{1}{2L} \int_{-L}^{L} f(x) e^{in\pi x/L} \, dx. \qquad (9.3.5)$$

---

[*] As before, an equal sign appears if $f(x)$ is continuous [and periodic, $f(-L) = f(L)$]. At a jump discontinuity of $f(x)$ in the interior, the series converges to $[f(x+) + f(x-)]/2$.

Notice that the complex Fourier series representation of $f(x)$ has $e^{-in\pi x/L}$ and is summed over the discrete integers corresponding to the sum over the discrete eigenvalues. The complex Fourier coefficients, on the other hand, involve $e^{+in\pi x/L}$ and are integrated over the region of definition of $f(x)$ (with periodic boundary conditions), namely $-L \leqslant x \leqslant L$. If $f(x)$ is real, $c_{-n} = \bar{c}_n$ (see Exercise 9.3.3).

Complex orthogonality. There is an alternative way to derive the formula for the complex Fourier coefficients. Always, in the past, we have determined Fourier coefficients using the orthogonality of the eigenfunctions. A similar idea holds here. However, here the eigenfunctions $e^{-in\pi x/L}$ are complex. For complex functions the concept of orthogonality must be slightly modified. A complex function ϕ is said to be orthogonal to a complex function ψ (over an interval $a \leqslant x \leqslant b$) if $\int_a^b \bar{\phi}\psi \, dx = 0$, where $\bar{\phi}$ is the complex conjugate of ϕ. This guarantees that the length squared of a complex function f, defined by $\int_a^b \bar{f}f \, dx$, is positive (this would not have been valid for $\int_a^b ff \, dx$ since f is complex).

Using this notion of orthogonality, the eigenfunctions $e^{-in\pi x/L}$, $-\infty < n < \infty$, can be verified to form an orthogonal set because by simple integration

$$\int_{-L}^{L} \overline{(e^{-im\pi x/L})} e^{-in\pi x/L} \, dx = \begin{cases} 0 & n \neq m \\ 2L & n = m, \end{cases}$$

since

$$\overline{(e^{-im\pi x/L})} = e^{im\pi x/L}.$$

Now to determine the complex Fourier coefficients c_n, we multiply (9.3.4) by $e^{im\pi x/L}$ and integrate from $-L$ to $+L$ (assuming that the term-by-term use of these operations is valid). In this way

$$\int_{-L}^{L} f(x)e^{im\pi x/L} \, dx = \sum_{n=-\infty}^{\infty} c_n \int_{-L}^{L} e^{im\pi x/L} e^{-in\pi x/L} \, dx.$$

Using the orthogonality condition, the sum reduces to one term, $n = m$. Thus,

$$\int_{-L}^{L} f(x)e^{im\pi x/L} \, dx = 2Lc_m,$$

which explains the $1/2L$ in (9.3.5) as well as the switch of signs in the exponent.

EXERCISES 9.3

*9.3.1. Consider

$$f(x) = \begin{cases} 0 & x < x_0 \\ 1/\Delta & x_0 < x < x_0 + \Delta \\ 0 & x > x_0 + \Delta \end{cases}$$

Assume that $x_0 > -L$ and $x_0 + \Delta < L$. Determine the complex Fourier coefficients c_n.

9.3.2. Consider the eigenvalue problem

$$\frac{d^2\phi}{dx^2} + \lambda\phi = 0$$

$$\phi(-L) = \phi(L)$$

$$\frac{d\phi}{dx}(-L) = \frac{d\phi}{dx}(L).$$

Assume that λ_n is real but ϕ_n is complex. Without solving the differential equation, prove that

$$\int_{-L}^{L} \phi_n \bar{\phi}_m \, dx = 0 \qquad \text{if } \lambda_n \neq \lambda_m.$$

9.3.3. If $f(x)$ is real, show that $c_{-n} = \bar{c}_n$.

9.4 FOURIER TRANSFORM PAIR

9.4.1 Introduction

In solving boundary value problems on a finite interval ($L < x < L$, with periodic boundary conditions), we can use the complex form of a Fourier series:

$$\frac{f(x+) + f(x-)}{2} = \sum_{n=-\infty}^{\infty} c_n e^{-in\pi x/L}. \tag{9.4.1}$$

Here $f(x)$ is represented by a linear combination of all possible sinusoidal functions which are periodic with period $2L$. The Fourier coefficients were determined in Sec. 9.3,

$$c_n = \frac{1}{2L} \int_{-L}^{L} f(x) e^{in\pi x/L} \, dx. \tag{9.4.2}$$

The entire region of interest $-L < x < L$ is the domain of integration. We will extend these ideas to functions defined for $-\infty < x < \infty$, and apply it to the heat equation.

9.4.2 Fourier Integral

Motivation. The **Fourier series identity** follows by eliminating c_n (and using a dummy integration variable $\bar{x}$ to distinguish it from the spatial position x):

$$\frac{f(x+) + f(x-)}{2} = \sum_{n=-\infty}^{\infty} \left[\frac{1}{2L} \int_{-L}^{L} f(\bar{x}) e^{in\pi\bar{x}/L} \, d\bar{x} \right] e^{-in\pi x/L}. \tag{9.4.3}$$

We will show that the fundamental Fourier integral identity may be roughly defined as the limit of (9.4.3) as $L \to \infty$. In other words, functions defined for $-\infty < x < \infty$ may be thought of in some sense as periodic functions with an infinite period.

For periodic functions, $-L < x < L$, the allowable wave numbers ω (number of waves in 2π distance) are the infinite set of discrete values (see Fig. 9.4.1),

$$\omega = \frac{n\pi}{L} = 2\pi \frac{n}{2L}.$$

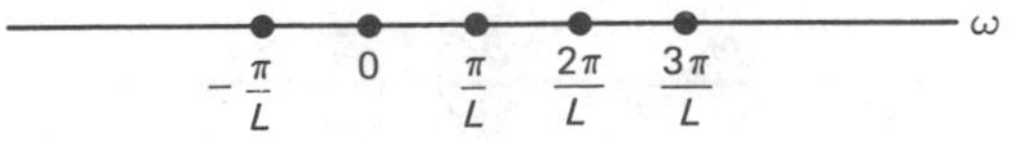

Figure 9.4.1 Discrete wave numbers.

The wave lengths are $2L/n$, integral partitions of the original region of length $2L$. The distance between successive values of the wave number is

$$\Delta\omega = \frac{(n+1)\pi}{L} - \frac{n\pi}{L} = \frac{\pi}{L};$$

they are equally spaced. From (9.4.3),

$$\frac{f(x+) + f(x-)}{2} = \sum_{n=-\infty}^{\infty} \frac{\Delta\omega}{2\pi} \int_{-L}^{L} f(\bar{x})e^{i\omega\bar{x}} \, d\bar{x} \, e^{-i\omega x}. \tag{9.4.4}$$

The values of ω are the square root of the eigenvalues. As $L \to \infty$, they become closer and closer, $\Delta\omega \to 0$. The eigenvalues approach a continuum; all possible wave numbers are allowable. The function $f(x)$ should be represented by a "sum" (which we will show becomes an integral) of waves of all possible wave lengths. Equation (9.4.4) represents a sum of rectangles (starting from $\omega = -\infty$ and going to $\omega = +\infty$) of base $\Delta\omega$ and height $(1/2\pi)$ $[\int_{-L}^{L} f(\bar{x})e^{i\omega\bar{x}} \, d\bar{x}] \, e^{-i\omega x}$. As $L \to \infty$, this height is not significantly different from

$$\frac{1}{2\pi} \int_{-\infty}^{\infty} f(\bar{x})e^{i\omega\bar{x}} \, d\bar{x} \, e^{-i\omega x}.$$

Thus, we expect as $L \to \infty$ that the areas of the rectangles approach the Riemann sum. Since $\Delta\omega \to 0$ as $L \to \infty$, (9.4.4) becomes the **Fourier integral identity**,

$$\frac{f(x+) + f(x-)}{2} = \frac{1}{2\pi} \int_{-\infty}^{\infty} \left[\int_{-\infty}^{\infty} f(\bar{x})e^{i\omega\bar{x}} \, d\bar{x} \right] e^{-i\omega x} \, d\omega. \tag{9.4.5}$$

A careful proof of this fundamental identity (see Exercise 9.4.9) closely parallels the somewhat complicated proof for the convergence of a Fourier series.

Fourier transform. We now accept (9.4.5) as fact. We next introduce $F(\omega)$ and *define* it to be the **Fourier transform of $f(x)$**:

$$F(\omega) \equiv \frac{\gamma}{2\pi} \int_{-\infty}^{\infty} f(\bar{x})e^{i\omega\bar{x}} \, d\bar{x}, \tag{9.4.6}$$

where γ is arbitrary. From (9.4.5), it then follows that

$$\frac{f(x+) + f(x-)}{2} = \frac{1}{\gamma} \int_{-\infty}^{\infty} F(\omega)e^{-i\omega x} \, d\omega. \tag{9.4.7}$$

If $f(x)$ is continuous, then $[f(x+) + f(x-)]/2 = f(x)$. Equation (9.4.7) shows that $f(x)$ is composed of waves $e^{-i\omega x}$ of all* wave numbers ω (and all wave lengths); it is known as the **Fourier integral representation of $f(x)$** or simply the **Fourier integral**. $F(\omega)$, the Fourier transform of $f(x)$, represents the amplitude of the wave with wave number ω; it is analogous to the Fourier coefficients of a Fourier series. It is determined by integrating over the entire infinite domain. Compare this to (9.4.2), where for periodic functions defined for $-L < x < L$, integration occurred only over that finite interval. Similarly, $f(x)$ may be determined from (9.4.7) if the Fourier transform $F(\omega)$ is known. $f(x)$, as determined from (9.4.7), is called the **inverse Fourier transform of $F(\omega)$**.

These relationships, (9.4.6) and (9.4.7), are quite important. They are also known as the **Fourier transform pair**. In (9.4.7) when you integrate over ω (called the **transform variable**) a function of x occurs, whereas in (9.4.6) when you integrate over x, a function of ω results. One integrand contains $e^{-i\omega x}$; the other has $e^{i\omega x}$. It is difficult to remember which is which. It hardly matters, but we must be consistent throughout. The different multiplicative factors, $1/\gamma$ in (9.4.7) and $\gamma/2\pi$ in (9.4.6), may cause some confusion. One can put any factor in front of each _as long as their product is $1/2\pi$_. We will let $\gamma = 1$ (but others may let $\gamma = \sqrt{2\pi}$, for example). We claim that (9.4.6) and (9.4.7) are valid if $f(x)$ satisfies $\int_{-\infty}^{\infty} |f(x)|dx < \infty$, in which case we say that $f(x)$ is absolutely integrable.†

An alternative notation $\mathscr{F}[f(x)]$ is sometimes used for $F(\omega)$, the Fourier transform of $f(x)$, given by (9.4.6). Similarly, the inverse Fourier transform of $F(\omega)$ is given the notation $\mathscr{F}^{-1}[F(\omega)]$.

9.4.3 Inverse Fourier Transform of a Gaussian

In Sec. 9.5, in order to complete our solution of the heat equation, we will need the inverse Fourier transform of the "bell-shaped" curve, known as a **Gaussian**,

$$G(\omega) = e^{-\alpha\omega^2},$$

sketched in Fig. 9.4.2. The function $g(x)$, whose Fourier transform is $G(\omega)$, is given by

$$g(x) = \int_{-\infty}^{\infty} G(\omega)e^{-i\omega x}\, d\omega = \int_{-\infty}^{\infty} e^{-\alpha\omega^2} e^{-i\omega x}\, d\omega, \qquad (9.4.8)$$

according to (9.4.7). By evaluating the integral in (9.4.8), we will derive shortly in the appendix to this section (see also Exercise 9.4.13) that

$$g(x) = \sqrt{\frac{\pi}{\alpha}} e^{-x^2/4\alpha} \qquad (9.4.9)$$

if $G(\omega) = e^{-\alpha\omega^2}$. As a function of x, $g(x)$ is also bell-shaped. We will have

---

* Not just wave numbers $n\pi/L$ as for periodic problems for $-L < x < L$.
† If $f(x)$ is piecewise smooth and if $f(x) \to 0$ as $x \to \pm\infty$ _sufficiently_ fast, then $f(x)$ is absolutely integrable. However, there are other kinds of functions which are absolutely integrable and hence for which the Fourier transform pair may be used.

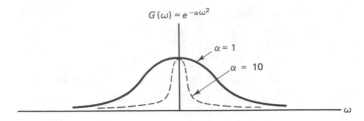

$$G(\omega) = e^{-\alpha\omega^2}$$

$\alpha = 1$

$\alpha = 10$

ω

Figure 9.4.2 Bell-shaped Gaussian.

shown the unusual result that **the inverse Fourier transform of a Gaussian is itself a Gaussian**.

This result can be used to obtain the Fourier transform $F(\omega)$ of a Gaussian $e^{-\beta x^2}$. Due to the linearity of the Fourier transform pair, the Fourier transform of $e^{-x^2/4\alpha}$ is $\sqrt{\alpha/\pi}\,e^{-\alpha\omega^2}$. Letting $\beta = 1/4\alpha$, the Fourier transform of $e^{-\beta x^2}$ is $1/\sqrt{4\beta\pi}\,e^{-\omega^2/4\beta}$. Thus, **the Fourier transform of a Gaussian is also a Gaussian**. We summarize these results in Table 9.4.1. If β is small, then $f(x)$ is a "broadly spread" Gaussian; its Fourier transform is "sharply peaked" near $\omega = 0$. On the other hand, if $f(x)$ is a narrowly peaked Gaussian function corresponding to β being large, its Fourier transform is broadly spread.

TABLE 9.4.1 FOURIER TRANSFORM OF A GAUSSIAN

$f(x) = \displaystyle\int_{-\infty}^{\infty} F(\omega)e^{-i\omega x}\,d\omega$	$F(\omega) = \dfrac{1}{2\pi}\displaystyle\int_{-\infty}^{\infty} f(x)e^{i\omega x}\,dx$
$e^{-\beta x^2}$	$\dfrac{1}{\sqrt{4\pi\beta}}\,e^{-\omega^2/4\beta}$
$\sqrt{\dfrac{\pi}{\alpha}}\,e^{-x^2/4\alpha}$	$e^{-\alpha\omega^2}$

9.4 APPENDIX: DERIVATION OF THE INVERSE FOURIER TRANSFORM OF A GAUSSIAN

The inverse Fourier transform of a Gaussian $e^{-\alpha\omega^2}$ is given by (9.4.8). By completing the square, $g(x)$ becomes

$$g(x) = \int_{-\infty}^{\infty} e^{-\alpha[\omega^2 + i(x/\alpha)\omega]}\,d\omega = \int_{-\infty}^{\infty} e^{-\alpha[\omega + i(x/2\alpha)]^2} e^{-x^2/4\alpha}\,d\omega$$

$$= e^{-x^2/4\alpha}\int_{-\infty}^{\infty} e^{-\alpha[\omega + i(x/2\alpha)]^2}\,d\omega.$$

The change of variables $s = \omega + i(x/2\alpha)$ $(ds = d\omega)$ appears to simplify the calculation,

$$g(x) = e^{-x^2/4\alpha} \int_{-\infty}^{\infty} e^{-\alpha s^2} \, ds. \tag{9.4A.1}$$

However, although (9.4A.1) is correct, we have not given the correct reasons. Actually, the change of variables $s = \omega + i(x/2\alpha)$ introduces complex numbers into the calculation. Since ω is being integrated "along the real axis" from $\omega = -\infty$ to $\omega = +\infty$, the variable s has nonzero imaginary part and does not vary along the real axis as indicated by (9.4A.1). Instead,

$$g(x) = e^{-x^2/4\alpha} \int_{-\infty + i(x/2\alpha)}^{\infty + i(x/2\alpha)} e^{-\alpha s^2} \, ds. \tag{9.4A.2}$$

The full power of the theory of complex variables is necessary to show that (9.4A.1) is equivalent to (9.4A.2).

This is not the place to attempt to teach complex variables, but a little hint of what is involved may be of interest to many readers. We sketch a complex s-plane in Fig. 9.4.3. To compute integrals from $-\infty$ to $+\infty$, we integrate from a to b (and later consider the limits as $a \to -\infty$ and $b \to +\infty$). Equation (9.4A.1) involves integrating along the real axis, while (9.4A.2) involves shifting off the real axis [with s equaling a constant imaginary part, $\text{Im}(s) = ix/2\alpha$]. According to Cauchy's theorem (from complex variables), the closed line integral is zero, $\oint e^{-\alpha s^2} \, ds = 0$, since the integrand $e^{-\alpha s^2}$ has no "singularities" inside (or on) the contour. Here, we use a rectangular contour, as sketched in Fig. 9.4.3. The closed line integral is composed of four simpler integrals, and hence

$$\int_{①} + \int_{②} + \int_{③} + \int_{④} = 0.$$

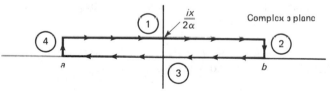

Figure 9.4.3 Closed contour integral in the complex plane.

It can be shown that in the limit as $a \to -\infty$ and $b \to +\infty$ both $\int_{②} = 0$ and $\int_{④} = 0$, since the integrand is exponentially vanishing on that path (and these paths are finite, of length $x/2\alpha$). Thus,

$$\int_{-\infty + i(x/2\alpha)}^{\infty + i(x/2\alpha)} e^{-\alpha s^2} \, ds + \int_{\infty}^{-\infty} e^{-\alpha s^2} \, ds = 0.$$

This verifies that (9.4A.1) is equivalent to (9.4A.2) (where we use $\int_{\infty}^{-\infty} = -\int_{-\infty}^{\infty}$).

The dependence on α in the integrand of (9.4A.1) can be determined by the additional transformation $z = \sqrt{\alpha} \, s$ ($dz = \sqrt{\alpha} \, ds$), in which case

$$g(x) = \frac{1}{\sqrt{\alpha}} e^{-x^2/4\alpha} \int_{-\infty}^{\infty} e^{-z^2} \, dz.$$

This yields the desired result, (9.4.9), since it is well known (but we will show) that

$$I = \int_{-\infty}^{\infty} e^{-z^2} \, dz = \sqrt{\pi}. \tag{9.4A.3}$$

Perhaps you have not seen (9.4A.3) derived. The $\int_{-\infty}^{\infty} e^{-z^2} \, dz$ can be evaluated by a remarkably unusual procedure. We do not know yet how to evaluate I, but we will show that I^2 is easy. Introducing different dummy integration variables for each I, we obtain

$$I^2 = \int_{-\infty}^{\infty} e^{-x^2} \, dx \int_{-\infty}^{\infty} e^{-y^2} \, dy = \int_{-\infty}^{\infty} \int_{-\infty}^{\infty} e^{-(x^2 + y^2)} \, dx \, dy.$$

We will evaluate this double integral, although each single integral is unknown. Polar coordinates are suggested:

$$\begin{array}{ccc} x = r \cos \theta & & \\ y = r \sin \theta & x^2 + y^2 = r^2 & dx \, dy = r \, dr \, d\theta. \end{array}$$

The region of integration is the entire two-dimensional plane. Thus,

$$I^2 = \int_{0}^{2\pi} \int_{0}^{\infty} e^{-r^2} r \, dr \, d\theta = \int_{0}^{2\pi} d\theta \int_{0}^{\infty} r e^{-r^2} \, dr.$$

Both of these integrals are easily evaluated; $I^2 = 2\pi \cdot \frac{1}{2} = \pi$, completing the proof.

EXERCISES 9.4

9.4.1. Show that the Fourier transform is a linear operator; that is, show that:
(a) $\mathcal{F}[c_1 f(x) + c_2 g(x)] = c_1 F(\omega) + c_2 G(\omega)$
(b) $\mathcal{F}[f(x)g(x)] \neq F(\omega)G(\omega)$

9.4.2. Show that the inverse Fourier transform is a linear operator; that is, show that:
(a) $\mathcal{F}^{-1}[c_1 F(\omega) + c_2 G(\omega)] = c_1 f(x) + c_2 g(x)$
(b) $\mathcal{F}^{-1}[F(\omega)G(\omega)] \neq f(x)g(x)$

9.4.3. Let $F(\omega)$ be the Fourier transform of $f(x)$. Show that if $f(x)$ is real, then $F^*(\omega) = F(-\omega)$, where * denotes the complex conjugate.

9.4.4. Show that

$$\mathcal{F}\left[\int f(x; \alpha) \, d\alpha \right] = \int F(\omega; \alpha) \, d\alpha$$

9.4.5. If $F(\omega)$ is the Fourier transform of $f(x)$, show that the inverse Fourier transform of $e^{i\omega\beta}F(\omega)$ is $f(x - \beta)$. This result is known as the **shift theorem** for Fourier transforms.

*9.4.6. If

$$f(x) = \begin{cases} 0 & |x| > a \\ 1 & |x| < a, \end{cases}$$

determine the Fourier transform of $f(x)$. [*Hint*: See page 357.]

*9.4.7. If $F(\omega) = e^{-|\omega|\alpha}$ ($\alpha > 0$), determine the inverse Fourier transform of $F(\omega)$. [*Hint*: See page 357.]

9.4.8. If $F(\omega)$ is the Fourier transform of $f(x)$, show that $-i\,dF/d\omega$ is the Fourier transform of $xf(x)$.

9.4.9. (a) Multiply (9.4.6) (assuming that $\gamma = 1$) by $e^{-i\omega x}$ and integrate from $-L$ to L to show that

$$\int_{-L}^{L} F(\omega)e^{-i\omega x}\,d\omega = \frac{1}{2\pi}\int_{-\infty}^{\infty} f(\bar{x})\frac{2\sin L(\bar{x}-x)}{\bar{x}-x}\,d\bar{x}. \tag{9.4.10}$$

(b) Derive (9.4.7). For simplicity, assume that $f(x)$ is continuous. [*Hints*: Let $f(\bar{x}) = f(x) + f(\bar{x}) - f(x)$. Use the sine integral, $\int_0^\infty \frac{\sin s}{s}\,ds = \frac{\pi}{2}$. Integrate (9.4.10) by parts and then take the limit as $L \to \infty$.]

*9.4.10. Consider the circularly symmetric heat equation on an infinite two-dimensional domain:

$$\frac{\partial u}{\partial t} = \frac{k}{r}\frac{\partial}{\partial r}\left(r\frac{\partial u}{\partial r}\right)$$

$$u(0, t)\text{ bounded}$$

$$u(r, 0) = f(r).$$

(a) Solve by separation. It is usual to let

$$u(r, t) = \int_0^\infty A(s)J_0(sr)e^{-s^2 kt}s\,ds$$

in which case the initial condition is satisfied if

$$f(r) = \int_0^\infty A(s)J_0(sr)s\,ds. \tag{9.4.11}$$

$A(s)$ is called the **Fourier-Bessel** or **Hankel transform** of $f(r)$.

(b) Use Green's formula to evaluate $\int_0^L J_0(sr)J_0(s_1 r)r\,dr$. Determine an approximate expression for large L using (6.8.3).

(c) Apply the answer of part (b) to part (a) to derive $A(s)$ from $f(r)$. (*Hint*: See Exercise 9.4.9.)

9.4.11. (a) If $f(x)$ is a function with unit area, show that the scaled and stretched function $(1/\alpha)f(x/\alpha)$ also has unit area.

(b) If $F(\omega)$ is the Fourier transform of $f(x)$, show that $F(\alpha\omega)$ is the Fourier transform of $(1/\alpha)f(x/\alpha)$.

(c) Show that part (b) implies that broadly spread functions have sharply peaked Fourier transforms near $\omega = 0$, and vice versa.

9.4.12. Show that $\lim\limits_{b\to\infty} \int_b^{b+ix/2\alpha} e^{-\alpha s^2}\,ds = 0$, where $s = b + iy$ $(0 < y < x/2\alpha)$.

9.4.13. Evaluate $I = \int_0^\infty e^{-k\omega^2 t}\cos \omega x\,d\omega$ in the following way. Determine $\partial I/\partial x$, and then integrate by parts.

9.4.14. The gamma function $\Gamma(x)$ is defined as follows:

$$\Gamma(x) = \int_0^\infty t^{x-1}e^{-t}\,dt.$$

Show that:

(a) $\Gamma(1) = 1$ (b) $\Gamma(x + 1) = x\Gamma(x)$

(c) $\Gamma(n + 1) = n!$ (d) $\Gamma(\tfrac{1}{2}) = 2\int_0^\infty e^{-t^2}\,dt = \sqrt{\pi}$

(e) What is $\Gamma(\tfrac{3}{2})$?

9.4.15. (a) Using the definition of the gamma function in Exercise 9.4.14, show that

$$\Gamma(x) = 2 \int_0^\infty u^{2x-1} e^{-u^2} \, du.$$

(b) Using double integrals in polar coordinates, show that

$$\Gamma(z)\Gamma(1 - z) = \frac{\pi}{\sin \pi z}$$

[*Hint*: It is known from complex variables that

$$2 \int_0^{\pi/2} (\tan \theta)^{2z-1} \, d\theta = \frac{\pi}{\sin \pi z}.]$$

*9.4.16. Evaluate

$$\int_0^\infty y^p e^{-ky^n} \, dy$$

in terms of the gamma function (see Exercise 9.4.14).

9.4.17. From complex variables, it is known that

$$\oint e^{-i\omega^3/3} \, d\omega = 0$$

for any closed contour. By considering the limit as $R \to \infty$ of the 30° pie-shaped wedge (of radius R) sketched in Fig. 9.4.4, show that

$$\int_0^\infty \cos\left(\frac{\omega^3}{3}\right) d\omega = \frac{\sqrt{3}}{2} 3^{-2/3} \Gamma\left(\frac{1}{3}\right)$$

$$\int_0^\infty \sin\left(\frac{\omega^3}{3}\right) d\omega = \frac{1}{2} 3^{-2/3} \Gamma\left(\frac{1}{3}\right)$$

Exercise 9.4.16 may be helpful.

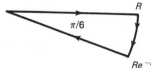

$Re^{-i\pi/6}$ **Figure 9.4.4**

9.4.18. (a) For what α does $\alpha e^{-\beta(x-x_0)^2}$ have unit area for $-\infty < x < \infty$?
(b) Show that the limit as $\beta \to \infty$ of the resulting function in part (a) satisfies the properties of the Dirac delta function $\delta(x - x_0)$.
(c) Obtain the Fourier transform of $\delta(x - x_0)$ in two ways:
    1. Take the transform of part (a) and take the limit as $\beta \to \infty$.
    2. Use integration properties of the Dirac delta function.
(d) Show that the transform of $\delta(x - x_0)$ is consistent with the following idea: "Transforms of sharply peaked functions are spread out (contain a significant 'amount' of many frequencies)."
(e) Show that the Fourier transform representation of the Dirac delta function is

$$\delta(x - x_0) = \frac{1}{2\pi} \int_{-\infty}^{\infty} e^{-i\omega(x-x_0)} \, d\omega. \tag{9.4.12}$$

Why is that not mathematically precise? However, what happens if $x = x_0$?

Similarly,

$$\delta(\omega - \omega_0) = \frac{1}{2\pi} \int_{-\infty}^{\infty} e^{-ix(\omega - \omega_0)} \, dx. \tag{9.4.13}$$

(f) Equation (9.4.13) may be interpreted as an orthogonality relation for the eigenfunctions $e^{-i\omega x}$. If

$$f(x) = \int_{-\infty}^{\infty} F(\omega) e^{-i\omega x} \, d\omega,$$

determine the "Fourier coefficient (transform)" using the orthogonality condition (9.4.13).

9.5 FOURIER TRANSFORM AND THE HEAT EQUATION

9.5.1 Heat Equation

In this subsection we begin to illustrate how to use Fourier transforms to solve the heat equation on an infinite interval. Earlier we showed $e^{-i\omega x} e^{-k\omega^2 t}$ for all ω solves the heat equation, $\partial u/\partial t = k \, \partial^2 u/\partial x^2$. A generalized principle of superposition showed that

$$u(x, t) = \int_{-\infty}^{\infty} c(\omega) e^{-i\omega x} e^{-k\omega^2 t} \, d\omega. \tag{9.5.1}$$

The initial condition $u(x, 0) = f(x)$ is satisfied if

$$f(x) = \int_{-\infty}^{\infty} c(\omega) e^{-i\omega x} \, d\omega. \tag{9.5.2}$$

From the definition of the Fourier transform ($\gamma = 1$), we observe that (9.5.2) is a Fourier integral representation of $f(x)$. Thus, $c(\omega)$ is the Fourier transform of the initial temperature distribution $f(x)$:

$$c(\omega) = \frac{1}{2\pi} \int_{-\infty}^{\infty} f(x) e^{i\omega x} \, dx. \tag{9.5.3}$$

Equations (9.5.1) and (9.5.3) describe the solution of our initial value problem for the heat equation.*

In this form, this solution is too complicated to be of frequent use. We therefore describe a simplification. We substitute $c(\omega)$ into the solution, recalling that the x in (9.5.3) is a dummy variable (and hence we introduce $\bar{x}$):

$$u(x, t) = \int_{-\infty}^{\infty} \left[\frac{1}{2\pi} \int_{-\infty}^{\infty} f(\bar{x}) e^{i\omega \bar{x}} \, d\bar{x} \right] e^{-i\omega x} e^{-k\omega^2 t} \, d\omega.$$

---

* In particular, in Exercise 9.5.2 we show that $u \to 0$ as $x \to \infty$, even though $e^{-i\omega x} \nrightarrow 0$ as $x \to \infty$.

Instead of doing the $\bar{x}$ integration first, we interchange the orders:

$$u(x, t) = \frac{1}{2\pi} \int_{-\infty}^{\infty} f(\bar{x}) \left[\int_{-\infty}^{\infty} e^{-k\omega^2 t} e^{-i\omega(x-\bar{x})} \, d\omega \right] d\bar{x}. \tag{9.5.4}$$

Equation (9.5.4) shows the importance of $g(x)$, the inverse Fourier transform of $e^{-k\omega^2 t}$:

$$g(x) = \int_{-\infty}^{\infty} e^{-k\omega^2 t} e^{-i\omega x} \, d\omega. \tag{9.5.5}$$

Thus, the integrand of (9.5.4) contains $g(x - \bar{x})$, not $g(x)$.

Influence function. We need to determine the function $g(x)$ whose Fourier transform is $e^{-k\omega^2 t}$ [and then make it a function of $x - \bar{x}$, $g(x - \bar{x})$]. $e^{-k\omega^2 t}$ is a Gaussian. From the previous section (or most tables of Fourier transforms; see Table 9.4.1), letting $\alpha = kt$, we obtain the Gaussian $g(x) = \sqrt{\pi/kt}\, e^{-x^2/4kt}$, and thus the solution of the heat equation is

$$\boxed{u(x, t) = \int_{-\infty}^{\infty} f(\bar{x}) \frac{1}{\sqrt{4\pi kt}} e^{-(x-\bar{x})^2/4kt} \, d\bar{x}.} \tag{9.5.6}$$

This form clearly shows the solution's dependence on the entire initial temperature distribution, $u(x, 0) = f(x)$. Each initial temperature "influences" the temperature at time t. We define

$$G(x, t; \bar{x}, 0) = \frac{1}{\sqrt{4\pi kt}} e^{-(x-\bar{x})^2/4kt} \tag{9.5.7}$$

and call it the **influence function**. Its relationship to an infinite space Green's function for the heat equation will be explained in Chapter 10. Equation (9.5.7) measures in some sense the effect of the initial ($\bar{t} = 0$) temperature at position $\bar{x}$ on the temperature at time t at location x. A sketch of the influence function (see Fig. 9.5.1) shows that the spread of influence is small when t is small, but large when t is large. At a fixed time, the largest influence occurs when $x = \bar{x}$; that is, the initial temperature at $\bar{x}$ has greatest influence at that same place, but also influences the temperature (to a lesser degree) at all other places. As $t \to 0$, the influence function becomes more and more concentrated. In fact, in Exercise

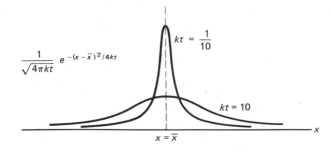

Figure 9.5.1 Influence function for the heat equation.

     Infinite Domain Problems   Chap. 9

9.4.18 it is shown that

$$\lim_{t \to 0+} \frac{1}{\sqrt{4\pi kt}} e^{-(x-\bar{x})^2/4kt} = \delta(x - \bar{x})$$

(the Dirac delta function of Chapter 8), thus verifying that (9.5.6) satisfies the initial conditions.

   The solution (9.5.6) of the heat equation on an infinite domain was derived in a complicated fashion using Fourier transforms. We required that $\int_{-\infty}^{\infty}|f(\bar{x})|d\bar{x} < \infty$, a restriction on the initial temperature distribution, in order for the Fourier transform to exist. However, the final form of the solution does not even refer to Fourier transforms. Thus, we never need to calculate any Fourier transforms to utilize (9.5.6). In fact, we claim that the restriction $\int_{-\infty}^{\infty}|f(x)|dx < \infty$ on (9.5.6) is not necessary. Equation (9.5.6) is valid (although the derivation we gave is not), roughly speaking, whenever the integral in (9.5.6) converges.

Example. Equation (9.5.6) may be used to solve the following interesting initial value problem:

$$u(x, 0) = f(x) = \begin{cases} 0 & x < 0 \\ 100 & x > 0. \end{cases}$$

We thus ask how the thermal energy, initially uniformly concentrated in the right half of a rod, diffuses into the entire rod. According to (9.5.6),

$$u(x, t) = \frac{100}{\sqrt{4\pi kt}} \int_0^\infty e^{-(x-\bar{x})^2/4kt} \, d\bar{x} = \frac{100}{\sqrt{\pi}} \int_{-x/\sqrt{4kt}}^\infty e^{-z^2} \, dz,$$

where the integral has been simplified by introducing the change of variables, $z = (\bar{x} - x)/\sqrt{4kt}(dz = d\bar{x}/\sqrt{4kt})$. The integrand no longer depends on any parameters. The integral represents the area under a Gaussian (or normal) curve as illustrated in Fig. 9.5.2. Due to the evenness of e^{-z^2},

$$\int_{-x/\sqrt{4kt}}^\infty = \int_0^\infty + \int_0^{x/\sqrt{4kt}}$$

Since, as shown earlier, $\int_0^\infty e^{-z^2} \, dz = \sqrt{\pi}/2$,

$$u(x, t) = 50 + \frac{100}{\sqrt{\pi}} \int_0^{x/\sqrt{4kt}} e^{-z^2} \, dz. \tag{9.5.8}$$

The temperature is constant whenever $x/\sqrt{4kt}$ is constant, the parabolas sketched in the $x-t$ plane in Fig. 9.5.3. $x/\sqrt{4kt}$ is called the **similarity variable**. For example, the distance between 60 and 75° increases proportional to $\sqrt{t}$. The

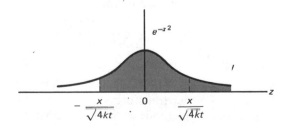

Figure 9.5.2 Area under a Gaussian.

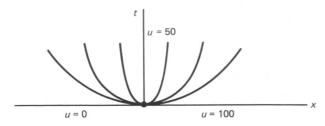

Figure 9.5.3 Constant temperatures.

temperature distribution spreads out, a phenomenon known as **diffusion**. The temperature distribution given by (9.5.8) is sketched in Fig. 9.5.4 for various fixed values of t.

We note that the temperature is nonzero at all x for any positive t ($t > 0$) even though $u = 0$ for $x < 0$ at $t = 0$. The **thermal energy spreads at an infinite propagation speed**. This is a fundamental property of the diffusion equation. It contrasts with the finite propagation speed of the wave equation, described in Chapter 11 (see also Sec. 9.7.1).

The area under the normal curve is well tabulated. We can express our solution in terms of the **error function**, erf $z = (2/\sqrt{\pi}) \int_0^z e^{-t^2} dt$, or the **complementary error function**, erfc $z = (2/\sqrt{\pi}) \int_z^\infty e^{-t^2} dt = 1 - $ erf z. Using these functions, the solution satisfying

$$u(x, 0) = \begin{cases} 0 & x < 0 \\ 100 & x > 0 \end{cases}$$

is

$$u(x, t) = 50 \left[1 + \mathrm{erf}\left(\frac{x}{\sqrt{4kt}}\right)\right] = 50 \left[2 - \mathrm{erfc}\left(\frac{x}{\sqrt{4kt}}\right)\right].$$

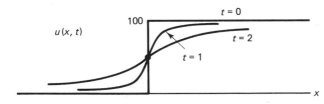

Figure 9.5.4 Temperature diffusion in a infinite rod.

9.5.2 Fourier Transforming the Heat Equation— Transforms of Derivatives

We have solved the heat equation on an infinite interval,

$$\frac{\partial u}{\partial t} = k \frac{\partial^2 u}{\partial x^2} \qquad -\infty < x < \infty$$

$$u(x, 0) = f(x).$$

$$(9.5.9)$$

Using separation of variables, we motivated the introduction of Fourier transforms. If we know that we should be using Fourier transforms, we can avoid separating variables. Here we describe this simpler method; we Fourier transform the entire problem. From the heat equation, (9.5.9), the Fourier transform of $\partial u/\partial t$ equals k times the Fourier transform of $\partial^2 u/\partial x^2$:

$$\mathcal{F}\left[\frac{\partial u}{\partial t}\right] = k\mathcal{F}\left[\frac{\partial^2 u}{\partial x^2}\right]. \tag{9.5.10}$$

Thus, we need to calculate Fourier transforms of derivatives of $u(x, t)$. We begin by defining the Fourier transform of $u(x, t)$,

$$\overline{U}(\omega, t) = \frac{1}{2\pi}\int_{-\infty}^{\infty} u(x, t)e^{i\omega x}\, dx. \tag{9.5.11}$$

Note that it is also a function of time; it is an ordinary Fourier transform with t fixed. To obtain a Fourier transform (in space), we multiply by $e^{i\omega x}$ and integrate. Spatial Fourier transforms of time derivatives are not difficult:

$$\mathcal{F}\left[\frac{\partial u}{\partial t}\right] = \frac{1}{2\pi}\int_{-\infty}^{\infty}\frac{\partial u}{\partial t}e^{i\omega x}\, dx = \frac{\partial}{\partial t}\left[\frac{1}{2\pi}\int_{-\infty}^{\infty} u(x, t)e^{i\omega x}\, dx\right] = \frac{\partial}{\partial t}\overline{U}(\omega, t). \tag{9.5.12}$$

The Fourier transform of a time derivative equals the time derivative of the Fourier transform.

More interesting (and useful) results occur for the Fourier transform of spatial derivatives:

$$\mathcal{F}\left[\frac{\partial u}{\partial x}\right] = \frac{1}{2\pi}\int_{-\infty}^{\infty}\frac{\partial u}{\partial x}e^{i\omega x}\, dx = \left.\frac{ue^{i\omega x}}{2\pi}\right|_{-\infty}^{\infty} - \frac{i\omega}{2\pi}\int_{-\infty}^{\infty} u(x, t)e^{i\omega x}\, dx,$$

which has been simplified using integration by parts:

$$df = \frac{\partial u}{\partial x}dx \qquad g = e^{i\omega x}$$

$$f = u \qquad dg = i\omega e^{i\omega x}\, dx.$$

If $u \to 0$ as $x \to \pm\infty$, then the end point contributions of integration by parts vanish. Furthermore, we note that

$$\boxed{\mathcal{F}\left[\frac{\partial u}{\partial x}\right] = i\omega\mathcal{F}[u] = -i\omega\overline{U}(\omega, t).} \tag{9.5.13}$$

In a similar manner, Fourier transforms of higher derivatives may be obtained:

$$\mathcal{F}\left[\frac{\partial^2 u}{\partial x^2}\right] = -i\omega\mathcal{F}\left[\frac{\partial u}{\partial x}\right] = (-i\omega)^2\overline{U}(\omega, t). \tag{9.5.14}$$

In general, **the Fourier transform of the nth derivative of a function** *with respect to x* **equals $(-i\omega)^n$ times the Fourier transform of the function,** assuming that $u(x, t) \to 0$ sufficiently fast as $x \to \pm\infty$.*

---

* We also need higher derivatives to vanish as $x \to \pm\infty$. Furthermore, each integration by parts is not valid unless the appropriate function is continuous.

By applying the Fourier transform to the heat equation, (9.5.9), we obtain (9.5.10). Because of the properties of the Fourier transform of derivatives, (9.5.10) becomes

$$\frac{\partial}{\partial t}\overline{U}(\omega, t) = k(-i\omega)^2\overline{U}(\omega, t) = -k\omega^2\overline{U}(\omega, t). \qquad (9.5.15)$$

The Fourier transform operation converts a linear partial differential equation with constant coefficients into an ordinary differential equation, since spatial derivatives are transformed into algebraic multiples of the transform.

Equation (9.5.15) is a first-order constant coefficient differential equation. Its general solution is

$$\overline{U}(\omega, t) = ce^{-k\omega^2 t}.$$

However, $\partial/\partial t$ is an ordinary derivative *keeping ω fixed*, and thus c is constant if ω is fixed. For other fixed values of ω, c may be a different constant; c depends on ω. c is actually *an arbitrary function of ω, $c(\omega)$*. Indeed, you may easily verify by substitution that

$$\boxed{\overline{U}(\omega, t) = c(\omega)e^{-k\omega^2 t}} \qquad (9.5.16)$$

solves (9.5.15). To determine $c(\omega)$ we note from (9.5.16) that $c(\omega)$ equals the initial value of the transform [obtained by transforming the initial condition, $f(x)$] $c(\omega) = 1/2\pi \int_{-\infty}^{\infty} f(x)e^{i\omega x}\, dx$. This is the same result as obtained by separation of variables. We could reproduce the entire solution obtained earlier. Instead, we will show a simpler way to obtain those results.

9.5.3 Convolution Theorem

We observe that $\overline{U}(\omega, t)$ is the product of two functions of ω, $c(\omega)$ and $e^{-k\omega^2 t}$, *both transforms of other functions*; $c(\omega)$ is the transform of the initial condition $f(x)$ and $e^{-k\omega^2 t}$ is the transform of some function (fortunately, since $e^{-k\omega^2 t}$ is a Gaussian, we know that it is the transform of another Gaussian, $\sqrt{\pi/kt}e^{-x^2/4kt}$). The mathematical problem of inverting a transform which is a product of transforms of known functions occurs very frequently (especially in using the Fourier transform to solve partial differential equations). Thus, we study this problem in some generality.

Suppose that $F(\omega)$ and $G(\omega)$ are the Fourier transforms of $f(x)$ and $g(x)$, respectively:

$$F(\omega) = \frac{1}{2\pi}\int_{-\infty}^{\infty} f(x)e^{i\omega x}\, dx \qquad G(\omega) = \frac{1}{2\pi}\int_{-\infty}^{\infty} g(x)e^{i\omega x}\, dx$$

$$\qquad (9.5.17)$$

$$f(x) = \int_{-\infty}^{\infty} F(\omega)e^{-i\omega x}\, d\omega \qquad g(x) = \int_{-\infty}^{\infty} G(\omega)e^{-i\omega x}\, d\omega.$$

We will determine the function $h(x)$ whose Fourier transform $H(\omega)$ equals the product of the two transforms:

$$H(\omega) = F(\omega)G(\omega) \qquad (9.5.18a)$$

$$h(x) = \int_{-\infty}^{\infty} H(\omega)e^{-i\omega x}\, d\omega = \int_{-\infty}^{\infty} F(\omega)G(\omega)e^{-i\omega x}\, d\omega. \qquad (9.5.18b)$$

We eliminate either $F(\omega)$ or $G(\omega)$ from (9.5.17); it does not matter which:

$$h(x) = \frac{1}{2\pi}\int_{-\infty}^{\infty} F(\omega)\left[\int_{-\infty}^{\infty} g(\bar{x})e^{i\omega\bar{x}}\, d\bar{x}\right]e^{-i\omega x}\, d\omega.$$

Assuming that we can interchange orders of integration, we obtain

$$h(x) = \frac{1}{2\pi}\int_{-\infty}^{\infty} g(\bar{x})\left[\int_{-\infty}^{\infty} F(\omega)e^{-i\omega(x-\bar{x})}\, d\omega\right] d\bar{x}.$$

We now recognize the inner integral as $f(x - \bar{x})$ [see (9.5.17)], and thus

$$h(x) = \frac{1}{2\pi}\int_{-\infty}^{\infty} g(\bar{x})f(x - \bar{x})\, d\bar{x}. \qquad (9.5.19)$$

The integral in (9.5.19) is called the **convolution** of $g(x)$ and $f(x)$; it is sometimes denoted $g * f$. **The inverse Fourier transform of the product of two Fourier transforms is $1/2\pi$ times the convolution of the two functions.** If we let $x - \bar{x} = w(d\bar{x} = -dw$ but $\int_{\infty}^{-\infty} = -\int_{-\infty}^{\infty})$, we obtain an alternative form,

$$h(x) = \frac{1}{2\pi}\int_{-\infty}^{\infty} f(w)g(x - w)\, dw, \qquad (9.5.20)$$

which would be denoted $f * g$. Thus, $g * f = f * g$.

Heat equation. We now apply the convolution theorem to our partial differential equation. The transform $\overline{U}(\omega, t)$ of the solution $u(x, t)$ is the product of $c(\omega)$ and $e^{-k\omega^2 t}$, where $c(\omega)$ is the transform of the initial temperature distribution and $e^{-k\omega^2 t}$ is the transform of $\sqrt{\pi/kt}\, e^{-x^2/4kt}$. Thus, according to the convolution theorem,

$$u(x, t) = \frac{1}{2\pi}\int_{-\infty}^{\infty} f(\bar{x})\sqrt{\frac{\pi}{kt}}\, e^{-(x-\bar{x})^2/4kt}\, d\bar{x}.$$

This is the same result as obtained (and discussed) earlier. In summary, the procedure that we follow is:

1. Fourier transform the partial differential equation.

2. Solve the ordinary differential equation.

3. Apply the initial conditions, determining the initial Fourier transform.

4. Use the convolution theorem.

By using the convolution theorem, we avoid for each problem substituting the inverse Fourier transform and interchanging the order of integration.

Parseval's identity. Since $h(x)$ is the inverse Fourier transform of $F(\omega)G(\omega)$, the convolution theorem can be stated in the following form:

$$\frac{1}{2\pi} \int_{-\infty}^{\infty} g(\bar{x})f(x - \bar{x})d\bar{x} = \int_{-\infty}^{\infty} F(\omega)G(\omega)e^{-i\omega x} \, d\omega. \tag{9.5.21}$$

Equation (9.5.21) is valid for all x. In particular, at $x = 0$,

$$\frac{1}{2\pi} \int_{-\infty}^{\infty} g(\bar{x})f(-\bar{x})d\bar{x} = \int_{-\infty}^{\infty} F(\omega)G(\omega)d\omega. \tag{9.5.22}$$

An interesting result occurs if we pick $g(x)$ such that

$$\boxed{g^*(x) = f(-x).} \tag{9.5.23}$$

Here * is the complex conjugate. [For real functions $g(x)$ is the reflection of $f(x)$ around $x = 0$.] In general, their Fourier transforms are related:

$$\begin{aligned} F(\omega) &= \frac{1}{2\pi} \int_{-\infty}^{\infty} f(x)e^{i\omega x} \, dx = \frac{1}{2\pi} \int_{-\infty}^{\infty} f(-\bar{x})e^{-i\omega\bar{x}} \, d\bar{x} \\ &= \frac{1}{2\pi} \int_{-\infty}^{\infty} g^*(x)e^{-i\omega x} \, dx = G^*(\omega), \end{aligned} \tag{9.5.24}$$

where we let $\bar{x} = -x$. Thus, (9.5.22) becomes **Parseval's identity**,

$$\boxed{\frac{1}{2\pi} \int_{-\infty}^{\infty} g(x)g^*(x) \, dx = \int_{-\infty}^{\infty} G(\omega)G^*(\omega) \, d\omega,} \tag{9.5.25}$$

where $g(x)g^*(x) = |g(x)|^2$ and $G(\omega)G^*(\omega) = |G(\omega)|^2$. We showed a similar relationship for all generalized Fourier series (see Sec. 5.10). This result, (9.5.25), is given the following interpretation. Often energy per unit distance is proportional to $|g(x)|^2$, and thus $1/2\pi \int_{-\infty}^{\infty} |g(x)|^2 \, dx$ represents the total energy. From (9.5.25), $|G(\omega)|^2$ may be defined as the energy per unit wave number (the **spectral energy density**). All the energy is contained within all the wave numbers. **The Fourier transform $G(\omega)$ of a function $g(x)$ is a complex quantity whose magnitude squared is the spectral energy density (the amount of energy per unit wave number).**

9.5.4 Summary of Properties of the Fourier Transform

Tables of Fourier transforms exist and can be very helpful. The results we have obtained are summarized in the short table on the next page.

$f(x) = \int_{-\infty}^{\infty} F(\omega)e^{-i\omega x}\, d\omega$	$F(\omega) = \dfrac{1}{2\pi}\int_{-\infty}^{\infty} f(x)e^{i\omega x}\, dx$	Reference
$e^{-\alpha x^2}$	$\dfrac{1}{\sqrt{4\pi\alpha}}e^{-\omega^2/4\alpha}$	Gaussian (Sec. 9.4.3)
$\sqrt{\dfrac{\pi}{\beta}}\, e^{-x^2/4\beta}$	$e^{-\beta\omega^2}$	
$\dfrac{\partial f}{\partial t}$	$\dfrac{\partial F}{\partial t}$	
$\dfrac{\partial f}{\partial x}$	$-i\omega F(\omega)$	Derivatives (Sec. 9.5.2)
$\dfrac{\partial^2 f}{\partial x^2}$	$(-i\omega)^2 F(\omega)$	
$\dfrac{1}{2\pi}\int_{-\infty}^{\infty} f(w)g(x-w)dw$	$F(\omega)G(\omega)$	Convolution (Sec. 9.5.3)
$\delta(x - x_0)$	$\dfrac{1}{2\pi}e^{i\omega x_0}$	Dirac delta function (Exercise 9.4.18)
$f(x - \beta)$	$e^{i\omega\beta}F(\omega)$	Shifting theorem (Exercise 9.4.5)
$xf(x)$	$-i\dfrac{dF}{d\omega}$	Multiplication by x (Exercise 9.4.8)
$\dfrac{2\alpha}{x^2 + \alpha^2}$	$e^{-|\omega|\alpha}$	Exercise 9.4.7
$f(x) = \begin{cases} 0 & |x| > a \\ 1 & |x| < a \end{cases}$	$\dfrac{1}{\pi}\dfrac{\sin a\omega}{\omega}$	Exercise 9.4.6

We list below some important and readily available tables of Fourier transforms. Beware of various different notations.

F. Oberhettinger, *Tabellen zur Fourier Transformation*, Springer-Verlag, New York, 1957.

R. V. Churchill, *Operational Mathematics*, 3rd ed., McGraw-Hill, New York, 1972.

G. A. Campbell and R. M. Foster, *Fourier Integrals for Practical Applications*, Van Nostrand, Princeton, N.J., 1948.

EXERCISES 9.5

9.5.1. Using Green's formula, show that:

$$\mathscr{F}\left[\frac{d^2f}{dx^2}\right] = -\omega^2 F(\omega) + \left.\frac{e^{i\omega x}}{2\pi}\left(\frac{df}{dx} - i\omega f\right)\right|_{-\infty}^{\infty}$$

9.5.2. For the heat equation, $u(x, t)$ is given by (9.5.1). Show that $u \to 0$ as $x \to \infty$ even though $\phi(x) = e^{-i\omega x}$ does not decay as $x \to \infty$. (*Hint:* Integrate by parts.)

9.5.3. *(a) Solve the diffusion equation with **convection**:

$$\frac{\partial u}{\partial t} = k\frac{\partial^2 u}{\partial x^2} + c\frac{\partial u}{\partial x} \qquad -\infty < x < \infty$$

$$u(x, 0) = f(x).$$

[*Hint*: Use the convolution theorem and the shift theorem (see Exercise 9.4.5).]

(b) Consider the initial condition to be $\delta(x)$. Sketch the corresponding $u(x, t)$ for various values of $t > 0$. Comment on the significance of the convection term $c\ \partial u/\partial x$.

9.5.4. (a) Solve

$$\frac{\partial u}{\partial t} = k\frac{\partial^2 u}{\partial x^2} - \gamma u \qquad -\infty < x < \infty$$

$$u(x, 0) = f(x).$$

(b) Does your solution suggest a simplifying transformation?

9.5.5. Consider

$$\frac{\partial u}{\partial t} = k\frac{\partial^2 u}{\partial x^2} + Q(x, t) \qquad -\infty < x < \infty$$

$$u(x, 0) = f(x).$$

(a) Show that a particular solution for the Fourier transform $\bar{u}$ is

$$\bar{u} = e^{-k\omega^2 t}\int_0^t \overline{Q}(\omega, \tau)e^{k\omega^2\tau}d\tau.$$

(b) Determine $\bar{u}$.

*(c) Solve for $u(x, t)$ (in the simplest form possible).

*9.5.6. The **Airy function** Ai (x) is the unique solution of

$$\frac{d^2 y}{dx^2} - xy = 0$$

which satisfies
(1) $\lim_{x \to \pm\infty} y = 0$
(2) $y(0) = 3^{-2/3}/\Gamma(\tfrac{2}{3}) = 3^{-2/3}\Gamma(\tfrac{1}{3})\sqrt{3}/2\pi = 1/\pi \int_0^\infty \cos(\omega^3/3)\ d\omega$ (it is not necessary to look at Exercises 9.4.15 and 9.4.17).

Determine a Fourier transform representation of the solution of this problem, Ai(x). (*Hint*: See Exercise 9.4.8.)

9.5.7. (a) Solve the **linearized Korteweg-deVries equation**

$$\frac{\partial u}{\partial t} = k\frac{\partial^3 u}{\partial x^3} \qquad -\infty < x < \infty$$

$$u(x, 0) = f(x).$$

(b) Use the convolution theorem to simplify.

*(c) See Exercise 9.5.6 for a further simplification.

(d) Specialize your result to the case in which

$$f(x) = \begin{cases} 0 & x < 0 \\ 1 & x > 0. \end{cases}$$

9.5.8. Solve

$$\frac{\partial^2 u}{\partial x^2} + \frac{\partial^2 u}{\partial y^2} = 0 \qquad \begin{array}{l} 0 < x < L \\ -\infty < y < \infty \end{array}$$

subject to

$$u(0, y) = g_1(y)$$
$$u(L, y) = g_2(y).$$

9.5.9. Solve

$$\frac{\partial^2 u}{\partial x^2} + \frac{\partial^2 u}{\partial y^2} = 0 \qquad \begin{matrix} y > 0 \\ -\infty < x < \infty \end{matrix}$$

subject to

$$u(x, 0) = f(x).$$

(*Hint:* If necessary, see Sec. 9.7.3.)

9.5.10. Solve

$$\frac{\partial^2 u}{\partial t^2} = c^2 \frac{\partial^2 u}{\partial x^2} \qquad -\infty < x < \infty$$

$$u(x, 0) = f(x)$$

$$\frac{\partial u}{\partial t}(x, 0) = 0.$$

(*Hint*: If necessary, see Sec. 9.7.1.)

9.5.11. Derive an expression for the Fourier transform of the product $f(x)g(x)$.

9.6 FOURIER SINE AND COSINE TRANSFORMS— THE HEAT EQUATION ON SEMI-INFINITE INTERVALS

9.6.1 Introduction

The Fourier series has been introduced to solve partial differential equations on the finite interval $-L < x < L$ with periodic boundary conditions. For problems defined on the interval $0 < x < L$, special cases of Fourier series, the sine and cosine series, were analyzed in order to satisfy the appropriate boundary conditions.

On an infinite interval, $-\infty < x < \infty$, instead we use the Fourier transform. In this section we show how to solve partial differential equations on a semi-infinite interval, $0 < x < \infty$. We will introduce special cases of the Fourier transform, known as the sine and cosine transforms. The modifications of the Fourier transform will be similar to the ideas we used for series on finite intervals.

9.6.2 Heat Equation on a Semi-Infinite Interval I

We will motivate the introduction of Fourier sine and cosine transforms by considering a simple physical problem. If the temperature is fixed at $0°$ at $x = 0$, then the mathematical problem for heat diffusion on a semi-infinite interval

$x > 0$ is

<table>
<tr><td>PDE:</td><td>$\dfrac{\partial u}{\partial t} = k \dfrac{\partial^2 u}{\partial x^2}, \qquad x > 0$</td><td>(9.6.1a)</td></tr>
<tr><td>BC:</td><td>$u(0, t) = 0$</td><td>(9.6.1b)</td></tr>
<tr><td>IC:</td><td>$u(x, 0) = f(x).$</td><td>(9.6.1c)</td></tr>
</table>

Here we have one boundary condition, which is homogeneous.

If we separate variables,

$$u(x, t) = \phi(x)h(t),$$

for the heat equation, we obtain as before

$$\frac{dh}{dt} = -\lambda k h \tag{9.6.2a}$$

$$\frac{d^2\phi}{dx^2} = -\lambda\phi. \tag{9.6.2b}$$

The boundary conditions to determine the allowable eigenvalues λ are

$$\phi(0) = 0 \tag{9.6.2c}$$

$$\lim_{x\to\infty} |\phi(x)| < \infty. \tag{9.6.2d}$$

The latter condition corresponds to $\lim_{x\to\infty} u(x, t) = 0$, since we usually assume that $\lim_{x\to\infty} f(x) = 0$.

There are nontrivial solutions of (9.6.2b)–(9.6.2d) only for all positive λ ($\lambda > 0$),

$$\phi(x) = c_1 \sin \sqrt{\lambda}\, x = c_1 \sin \omega x, \tag{9.6.3}$$

where, as with the Fourier transform, we prefer the variable $\omega = \sqrt{\lambda}$. Here $\omega > 0$ only. The corresponding time-dependent part is

$$h(t) = ce^{-\lambda kt} = ce^{-kt\omega^2}, \tag{9.6.4}$$

and thus product solutions are

$$u(x, t) = A \sin \omega x\, e^{-kt\omega^2}. \tag{9.6.5}$$

The generalized principle of superposition implies that we should seek a solution to the initial value problem in the form

$$u(x, t) = \int_0^\infty A(\omega) \sin \omega x\, e^{-k\omega^2 t}\, d\omega. \tag{9.6.6}$$

The initial condition $u(x, 0) = f(x)$ is satisfied if

$$f(x) = \int_0^\infty A(\omega) \sin \omega x \, d\omega. \tag{9.6.7}$$

In the next subsection, we will show that $A(\omega)$ is the Fourier sine transform of $f(x)$; we will show that $A(\omega)$ can be determined from (9.6.7):

$$A(\omega) = \frac{2}{\pi} \int_0^\infty f(x) \sin \omega x \, dx. \tag{9.6.8}$$

9.6.3 Fourier Sine and Cosine Transforms

In the preceding subsection we are asked to represent a function only using sine functions. We already know how to represent a function using complex exponentials, the Fourier transform:

$$f(x) = \frac{1}{\gamma} \int_{-\infty}^\infty F(\omega) e^{-i\omega x} \, d\omega \tag{9.6.9a}$$

$$F(\omega) = \frac{\gamma}{2\pi} \int_{-\infty}^\infty f(x) e^{i\omega x} \, dx. \tag{9.6.9b}$$

Recall that (9.6.9) is valid for any γ.

Fourier sine transform. Since we only want to use $\sin \omega x$ (for all ω), we consider cases in which $f(x)$ is an odd function. If our physical region is $x \geq 0$, then our functions do not have physical meaning for $x < 0$. In this case we can define these functions in any way we choose for $x < 0$; we introduce the odd extension of the given $f(x)$. If $f(x)$ is odd in this way, then its Fourier transform $F(\omega)$ can be simplified:

$$F(\omega) = \frac{\gamma}{2\pi} \int_{-\infty}^\infty f(x)(\cos \omega x + i \sin \omega x) dx = \frac{2i\gamma}{2\pi} \int_0^\infty f(x) \sin \omega x \, dx, \tag{9.6.10}$$

since $f(x) \cos \omega x$ is odd in x and $f(x) \sin \omega x$ is even in x. Note that $F(\omega)$ is an odd function of ω [when $f(x)$ is an odd function of x]. Thus, in a similar manner

$$f(x) = \frac{1}{\gamma} \int_{-\infty}^\infty F(\omega)(\cos \omega x - i \sin \omega x) d\omega = \frac{-2i}{\gamma} \int_0^\infty F(\omega) \sin \omega x \, d\omega. \tag{9.6.11}$$

We can choose γ in any way that we wish. Note that the product of the coefficients in front of both integrals must be $2i\gamma/2\pi \cdot -2i/\gamma = 2/\pi$ rather than $1/2\pi$ for Fourier transforms. For convenience, we let $-2i/\gamma = 1$ (i.e., $\gamma = -2i$), so that *if $f(x)$ is odd,*

$$f(x) = \int_0^\infty F(\omega) \sin \omega x \, d\omega \tag{9.6.12a}$$

$$F(\omega) = \frac{2}{\pi} \int_0^\infty f(x) \sin \omega x \, dx. \tag{9.6.12b}$$

Others may prefer a symmetric definition, $-2i/\gamma = \sqrt{2/\pi}$. These are called the **Fourier sine transform pair**. $F(\omega)$ is called the **sine transform** of $f(x)$, sometimes denoted $S[f(x)]$, while $f(x)$ is the **inverse sine transform** of $F(\omega)$, $S^{-1}[F(\omega)]$. Equations (9.6.12) are just related to the formulas for the Fourier transform of an odd function of x.

If the sine transform representation of $f(x)$ (9.6.12a) is utilized, then zero is always obtained at $x = 0$. This occurs even if $\lim_{x \to 0} f(x) \neq 0$. Equation (9.6.12a) as written is not always valid at $x = 0$. Instead, the odd extension of $f(x)$ has a jump discontinuity at $x = 0$ [if $\lim_{x \to 0} f(x) \neq 0$], from $-f(0)$ to $f(0)$. The Fourier sine transform representation of $f(x)$ converges to the average, which is zero (at $x = 0$).

Fourier cosine transform. Similarly, *if $f(x)$ is an even function*, we can derive the **Fourier cosine transform pair**:

$$f(x) = \int_0^\infty F(\omega) \cos \omega x \, d\omega \qquad \text{(9.6.13a)}$$

$$F(\omega) = \frac{2}{\pi} \int_0^\infty f(x) \cos \omega x \, dx. \qquad \text{(9.6.13b)}$$

Other forms are equivalent (as long as the product of the two numerical factors is again $2/\pi$). $F(\omega)$ is called the **cosine transform** of $f(x)$, sometimes denoted $C[f(x)]$, while $f(x)$ is the **inverse cosine transform** of $F(\omega)$, $C^{-1}[F(\omega)]$. Again if $f(x)$ is only defined for $x > 0$, then in order to use the Fourier cosine transform, we must introduce the even extension of $f(x)$.

A short table of both the Fourier sine and cosine transforms appears on pages 365 and 366.

9.6.4 Transforms of Derivatives

In Sec. 9.5.2 we derived important properties of the Fourier transform of derivatives. Here, similar results for the Fourier sine and cosine transform will be shown.

Our definitions of the Fourier cosine and sine transforms are:

$$C[f(x)] = \frac{2}{\pi} \int_0^\infty f(x) \cos \omega x \, dx \qquad \text{(9.6.14)}$$

$$S[f(x)] = \frac{2}{\pi} \int_0^\infty f(x) \sin \omega x \, dx. \qquad \text{(9.6.15)}$$

Integration by parts can be used to obtain formulas for the transforms of first derivatives:

$$C\left[\frac{df}{dx}\right] = \frac{2}{\pi}\int_0^\infty \frac{df}{dx}\cos \omega x\, dx = \frac{2}{\pi}f(x)\cos \omega x\,\Big|_0^\infty + \omega\frac{2}{\pi}\int_0^\infty f(x)\sin \omega x\, dx$$

$$S\left[\frac{df}{dx}\right] = \frac{2}{\pi}\int_0^\infty \frac{df}{dx}\sin \omega x\, dx = \frac{2}{\pi}f(x)\sin \omega x\,\Big|_0^\infty - \omega\frac{2}{\pi}\int_0^\infty f(x)\cos \omega x\, dx.$$

We have assumed that $f(x)$ is continuous. We obtain the following formulas:

$$C\left[\frac{df}{dx}\right] = -\frac{2}{\pi}f(0) + \omega S[f] \tag{9.6.16}$$

$$S\left[\frac{df}{dx}\right] = -\omega C[f], \tag{9.6.17}$$

assuming that $f(x) \to 0$ as $x \to \infty$. Sine or cosine transforms of first derivatives always involve the other type of semi-infinite transform. Thus, if a partial differential equation contains first derivatives with respect to a potential variable to be transformed, the Fourier sine or Fourier cosine transform will never work. Do not use Fourier sine or cosine transforms in this situation. Note that for the heat equation, $\partial u/\partial t = k\,\partial^2 u/\partial x^2$, the variable to be transformed is x. No first derivatives in x appear.

Transforms of second derivatives have simpler formulas. According to (9.6.16) and (9.6.17),

$$C\left[\frac{d^2f}{dx^2}\right] = -\frac{2}{\pi}\frac{df}{dx}(0) + \omega S\left[\frac{df}{dx}\right] = -\frac{2}{\pi}\frac{df}{dx}(0) - \omega^2 C[f] \tag{9.6.18}$$

$$S\left[\frac{d^2f}{dx^2}\right] = -\omega C\left[\frac{df}{dx}\right] = \frac{2}{\pi}\omega f(0) - \omega^2 S[f]. \tag{9.6.19}$$

We learn some important principles from (9.6.18) and (9.6.19). **In order to use the Fourier cosine transform** to solve a partial differential equation (containing a second derivative) defined on a semi-infinite interval ($x \geqslant 0$), df/dx **(0) must be known.** Similarly, **the Fourier sine transform may be used for semi-infinite problems if** $f(0)$ **is given.** Furthermore, problems are more readily solved if the boundary conditions are homogeneous. If $f(0) = 0$, then a Fourier sine transform will often yield a relatively simple solution. If df/dx (0) $= 0$, then a Fourier cosine transform will often be extremely convenient. These conditions are not surprising. If $f(0) = 0$, separation of variables motivates the use of sines only. Similarly df/dx (0) $= 0$ implies the use of cosines.

9.6.5 Heat Equation on a Semi-Infinite Interval II

Let us show how to utilize the formulas for the transforms of derivatives to solve partial differential equations. We consider a problem that is somewhat more general than the one presented earlier. Suppose that we are interested in the heat flow in a semi-infinite region with the temperature prescribed as a function of time at $x = 0$:

PDE:
$$\frac{\partial u}{\partial t} = k\frac{\partial^2 u}{\partial x^2} \qquad (9.6.20a)$$

BC:
$$u(0, t) = g(t) \qquad (9.6.20b)$$

IC:
$$u(x, 0) = f(x). \qquad (9.6.20c)$$

The boundary condition $u(0, t) = g(t)$ is nonhomogeneous. We cannot use the method of separation of variables. Since $0 < x < \infty$, we may wish to use a transform. Since u is specified at $x = 0$, we should try to use Fourier sine transforms (and *not* the Fourier cosine transform). Thus, we introduce $\overline{U}(\omega, t)$, the Fourier sine transform of $u(x, t)$:

$$\overline{U}(\omega, t) = \frac{2}{\pi}\int_0^\infty u(x, t) \sin \omega x \, dx. \qquad (9.6.21)$$

The partial differential equation (9.6.20a) becomes an ordinary differential equation,

$$\frac{\partial \overline{U}}{\partial t} = k\left(\frac{2}{\pi}\omega g(t) - \omega^2\overline{U}\right), \qquad (9.6.22a)$$

using (9.6.19). The initial condition yields the initial value of the Fourier sine transform:

$$\overline{U}(\omega, 0) = \frac{2}{\pi}\int_0^\infty f(x) \sin \omega x \, dx. \qquad (9.6.22b)$$

Solving (9.6.22a) is somewhat complicated in general (involving the integrating factor $e^{k\omega^2 t}$; see Sec. 7.3). We leave discussion of this to the exercises.

Example. *In the special case with homogeneous boundary conditions*, $g(t) = 0$, it follows from (9.6.22a) that
$$\overline{U}(\omega, t) = c(\omega)e^{-k\omega^2 t}, \qquad (9.6.23)$$

where from the initial condition

$$c(\omega) = \frac{2}{\pi}\int_0^\infty f(x) \sin \omega x \, dx. \qquad (9.6.24)$$

The solution is thus

$$u(x, t) = \int_0^\infty c(\omega)e^{-k\omega^2 t} \sin \omega x \, d\omega. \qquad (9.6.25)$$

This is the solution obtained earlier by separation of variables. To simplify this solution, we note that $c(\omega)$ is an odd function of ω. Thus,

$$u(x, t) = \frac{1}{2} \int_{-\infty}^{\infty} c(\omega) e^{-k\omega^2 t} \sin \omega x \, d\omega = \int_{-\infty}^{\infty} \frac{c(\omega)}{2i} e^{-k\omega^2 t} e^{i\omega x} \, d\omega. \quad (9.6.26)$$

If we introduce the odd extension of $f(x)$, then

$$\frac{c(\omega)}{2i} = \frac{2}{\pi} \int_0^{\infty} f(x) \frac{\sin \omega x}{2i} \, dx = \frac{1}{\pi} \int_{-\infty}^{\infty} f(x) \frac{\sin \omega x}{2i} \, dx = \frac{1}{2\pi} \int_{-\infty}^{\infty} f(x) e^{-i\omega x} \, dx. \quad (9.6.27)$$

We note that (9.6.26) and (9.6.27) are exactly the results for the heat equation on an infinite interval. Thus,

$$u(x, t) = \frac{1}{\sqrt{4\pi k t}} \int_{-\infty}^{\infty} f(\bar{x}) e^{-(x - \bar{x})^2/4kt} \, d\bar{x}.$$

Here, $f(x)$ has been extended to $-\infty < x < \infty$ as an odd function $[f(-x) = -f(x)]$. In order to only utilize $f(\bar{x})$ for $\bar{x} > 0$, we use the oddness property

$$u(x, t) = \frac{1}{\sqrt{4\pi k t}} \left[\int_{-\infty}^0 -f(-\bar{x}) e^{-(x - \bar{x})^2/4kt} \, d\bar{x} + \int_0^{\infty} f(\bar{x}) e^{-(x - \bar{x})^2/4kt} \, d\bar{x} \right].$$

In the first integral we let $\bar{x} = -\bar{\bar{x}}$ (and then we replace $\bar{\bar{x}}$ by $\bar{x}$). In this manner

$$u(x, t) = \frac{1}{\sqrt{4\pi k t}} \int_0^{\infty} f(\bar{x}) [e^{-(x - \bar{x})^2/4kt} - e^{-(x + \bar{x})^2/4kt}] \, d\bar{x}. \quad (9.6.28)$$

The influence function for the initial condition is in brackets above. We will discuss this solution further in Chapter 10. An equivalent (and simpler) method to obtain (9.6.28) from (9.6.23) and (9.6.24) is to use the convolution theorem for Fourier sine transforms (see Exercise 9.6.6).

FOURIER SINE TRANSFORM

$f(x) - \int_0^{\infty} F(\omega) \sin \omega x \, d\omega$	$S[f(x)] - F(\omega) = \frac{2}{\pi} \int_0^{\infty} f(x) \sin \omega x \, dx$	Reference
$\dfrac{df}{dx}$	$-\omega C[f(x)]$	Derivatives (Sec. 9.6.4)
$\dfrac{d^2 f}{dx^2}$	$\dfrac{2}{\pi} \omega f(0) - \omega^2 F(\omega)$	
$\dfrac{x}{x^2 + \beta^2}$	$e^{-\omega\beta}$	Exercise 9.6.1
$e^{-\varepsilon x}$	$\dfrac{2}{\pi} \cdot \dfrac{\omega}{\varepsilon^2 + \omega^2}$	Exercise 9.6.2
1	$\dfrac{2}{\pi} \cdot \dfrac{1}{\omega}$	Exercise 9.6.9
$\dfrac{1}{\pi} \int_0^{\infty} f(\bar{x}) [g(x - \bar{x}) - g(x + \bar{x})] \, d\bar{x}$ $= \dfrac{1}{\pi} \int_0^{\infty} g(\bar{x}) [f(x + \bar{x}) - f(\bar{x} - x)] d\bar{x}$	$S[f(x)] C[g(x)]$	Convolution (Exercise 9.6.6)

$f(x) = \int_0^\infty F(\omega) \cos \omega x \, d\omega$	$C[f(x)] = F(\omega) = \dfrac{2}{\pi} \int_0^\infty f(x) \cos \omega x \, dx$	Reference
$\dfrac{df}{dx}$	$-\dfrac{2}{\pi} f(0) + \omega S[f(x)]$	Derivatives
$\dfrac{d^2 f}{dx^2}$	$-\dfrac{2}{\pi} \dfrac{df}{dx}(0) - \omega^2 F(\omega)$	(Sec. 9.6.4)
$\dfrac{\beta}{x^2 + \beta^2}$	$e^{-\omega \beta}$	Exercise 9.6.1
$e^{-\varepsilon x}$	$\dfrac{2}{\pi} \cdot \dfrac{\varepsilon}{\varepsilon^2 + \omega^2}$	Exercise 9.6.2
$e^{-\alpha x^2}$	$2 \dfrac{1}{\sqrt{4\pi\alpha}} e^{-\omega^2/4\alpha}$	Exercise 9.6.3
$\dfrac{1}{\pi} \int_0^\infty g(\bar{x})[f(x - \bar{x}) + f(x + \bar{x})]d\bar{x}$	$F(\omega)G(\omega)$	Convolution (Exercise 9.6.7)

EXERCISES 9.6

9.6.1. Consider $F(\omega) = e^{-\omega \beta}$, $\beta > 0$ ($\omega \geq 0$).
  (a) Derive the inverse Fourier sine transform of $F(\omega)$.
  (b) Derive the inverse Fourier cosine transform of $F(\omega)$.

9.6.2. Consider $f(x) = e^{-\alpha x}$, $\alpha > 0$ ($x \geq 0$).
  (a) Derive the Fourier sine transform of $f(x)$.
  (b) Derive the Fourier cosine transform of $f(x)$.

*9.6.3. Derive *either* the Fourier cosine transform of $e^{-\alpha x^2}$ or the Fourier sine transform of $e^{-\alpha x^2}$

9.6.4. **(a)** Derive (9.6.18) using Green's formula.
  (b) Do the same for (9.6.19).

9.6.5. **(a)** Show that the Fourier sine transform of $f(x)$ is an odd function of ω (if defined for all ω).
  (b) Show that the Fourier cosine transform of $f(x)$ is an even function of ω (if defined for all ω).

9.6.6. There is an interesting convolution-type theorem for Fourier sine transforms. Suppose that we want $h(x)$, but know its sine transform $H(\omega)$ to be a product

$$H(\omega) = \overline{S}(\omega)\overline{C}(\omega),$$

where $\overline{S}(\omega)$ is the sine transform of $s(x)$ and $\overline{C}(\omega)$ is the *cosine* transform of $c(x)$. Assuming that $c(x)$ is even and $s(x)$ is odd, show that

$$h(x) = \frac{1}{\pi} \int_0^\infty s(\bar{x})[c(x - \bar{x}) - c(x + \bar{x})]d\bar{x} = \frac{1}{\pi} \int_0^\infty c(\bar{x})[s(x + \bar{x}) - s(\bar{x} - x)]d\bar{x}.$$

9.6.7. Derive the following. If a Fourier cosine transform in $x, H(\omega)$, is the product of two Fourier cosine transforms,

$$H(\omega) = F(\omega)G(\omega),$$

Then

$$h(x) = \frac{1}{\pi} \int_0^\infty g(\bar{x})[f(x - \bar{x}) + f(x + \bar{x})]d\bar{x}.$$

In this result f and g can be interchanged.

9.6.8. Solve (9.6.1) using the convolution theorem of Exercise 9.6.6. Exercise 9.6.3 may be of some help.

9.6.9. Let $S[f(x)]$ designate the Fourier sine transform.
  (a) Show that

$$S[e^{-\varepsilon x}] = \frac{2}{\pi} \frac{\omega}{\varepsilon^2 + \omega^2} \qquad \text{for } \varepsilon > 0.$$

  Show that $\lim_{\varepsilon \to 0+} S[e^{-\varepsilon x}] = 2/\pi\omega$. We will let $S[1] = 2/\pi\omega$. Why isn't $S[1]$ technically defined?
  (b) Show that

$$S^{-1}\left[\frac{2/\pi}{\omega}\right] = \frac{2}{\pi} \int_0^\infty \frac{\sin z}{z} dz,$$

  which is known to equal 1.

*9.6.10. Determine the inverse cosine transform of $\omega e^{-\omega \alpha}$. (*Hint:* Use differentiation with respect to a parameter.)

*9.6.11. Consider

$$\frac{\partial u}{\partial t} = k\frac{\partial^2 u}{\partial x^2} \qquad \begin{array}{l} x > 0 \\ t > 0 \end{array}$$

$$u(0, t) = 1$$

$$u(x, 0) = f(x).$$

  (a) Solve directly using sine transforms. (*Hint:* Use Exercise 9.6.8 and the convolution theorem, Exercise 9.6.6.)
  (b) If $f(x) \to 1$ as $x \to \infty$, let $v(x, t) = u(x, t) - 1$ and solve for $v(x, t)$.
  (c) Compare part (b) to part (a).

9.6.12. Solve

$$\frac{\partial u}{\partial t} = k\frac{\partial^2 u}{\partial x^2} \qquad x > 0$$

$$\frac{\partial u}{\partial x}(0, t) = 0$$

$$u(x, 0) = f(x).$$

9.6.13. Solve (9.6.20) by solving (9.6.22a).

9.6.14. Consider

$$\frac{\partial u}{\partial t} = k\frac{\partial^2 u}{\partial x^2} - v_0\frac{\partial u}{\partial x} \qquad (x > 0)$$

$$u(0, t) = 0$$

$$u(x, 0) = f(x).$$

  (a) Show that the Fourier sine transform does not yield an immediate solution.
  (b) Instead, introduce

$$u = e^{[x - (v_0/2)t]v_0/2k}w,$$

and show that

$$\frac{\partial w}{\partial t} = k\frac{\partial^2 w}{\partial x^2}$$

$$w(0, t) = 0$$

$$w(x, 0) = f(x)e^{-v_0 x/2k}.$$

(c) Use part (b) to solve for $u(x, t)$.

9.6.15. Solve

$$\frac{\partial^2 u}{\partial x^2} + \frac{\partial^2 u}{\partial y^2} = 0 \qquad \begin{array}{l} 0 < x < L \\ 0 < y < \infty \end{array}$$

$$u(x, 0) = 0$$

$$u(0, y) = g_1(y)$$

$$u(L, y) = g_2(y).$$

(*Hint*: If necessary, see Sec. 9.7.2.)

9.6.16. Solve

$$\frac{\partial^2 u}{\partial x^2} + \frac{\partial^2 u}{\partial y^2} = 0 \qquad \begin{array}{l} 0 < x < \infty \\ 0 < y < \infty \end{array}$$

$$u(0, y) = g(y)$$

$$\frac{\partial}{\partial y}u(x, 0) = 0.$$

(*Hint*: If necessary, see Sec. 9.7.4.)

9.6.17. The effect of periodic surface heating (either daily or seasonal) on the interior of the earth may be modeled by

$$\frac{\partial u}{\partial t} = k\frac{\partial^2 u}{\partial x^2} \qquad 0 < x < \infty$$

$$u(x, 0) = 0$$

$$u(0, t) = Ae^{i\sigma_0 t},$$

where the real part of $u(x, t)$ is the temperature (and x measures distance from the surface).
   (a) Determine $\overline{U}(\omega, t)$, the Fourier sine transform of $u(x, t)$.
   *(b) Approximate $\overline{U}(\omega, t)$ for large t.
   (c) Determine the inverse sine transform of part (b) in order to obtain an approximation for $u(x, t)$ valid for large t. [*Hint*: See Exercise 9.6.2(a) or tables.]
   (d) Sketch the approximate temperature (for fixed large t).
   (e) At what distance below the surface are temperature variations negligible?

9.6.18. Reconsider Exercise 9.6.17. Determine $u(x, t)$ exactly. (*Hint*: See Exercise 9.6.6.)

9.6.19. (a) Determine a particular solution of Exercise 9.6.17, satisfying the boundary condition but not the initial condition, of the form $u(x, t) = F(x)G(t)$.
   (b) Compare part (a) with either Exercise 9.6.17 or 9.6.18.

9.7 WORKED EXAMPLES USING TRANSFORMS

9.7.1 One-Dimensional Wave Equation on an Infinite Interval

Previously, we have analyzed vibrating strings on a finite interval, usually $0 \leq x \leq L$. Here we will study a vibrating string on an infinite interval. The best way to analyze vibrating strings on an infinite (or semi-infinite) interval is to use the method of characteristics, which we describe in Chapter 11. There concepts of wave propagation are more completely discussed. Here we will analyze only the following example, to show briefly how Fourier transforms can be used to solve the one-dimensional wave equation:

PDE:
$$\frac{\partial^2 u}{\partial t^2} = c^2 \frac{\partial^2 u}{\partial x^2} \qquad -\infty < x < \infty \qquad (9.7.1a)$$

IC:
$$u(x, 0) = f(x) \qquad (9.7.1b)$$

$$\frac{\partial u}{\partial t}(x, 0) = 0. \qquad (9.7.1c)$$

We give the initial position $f(x)$ of the string, but insist that the string is at rest, $\partial u / \partial t(x, 0) = 0$, in order to simplify the mathematics.

Product solutions can be obtained by separation of variables. Instead, we will introduce the Fourier transform of $u(x, t)$:

$$\overline{U}(\omega, t) = \frac{1}{2\pi} \int_{-\infty}^{\infty} u(x, t) e^{i\omega x} \, dx \qquad (9.7.2a)$$

$$u(x, t) = \int_{-\infty}^{\infty} \overline{U}(\omega, t) e^{-i\omega x} \, d\omega. \qquad (9.7.2b)$$

Taking the Fourier transform of the one-dimensional wave equation yields

$$\frac{\partial^2 \overline{U}}{\partial t^2} = -c^2 \omega^2 \overline{U}, \qquad (9.7.3a)$$

with the initial conditions becoming

$$\overline{U}(\omega, 0) = \frac{1}{2\pi} \int_{-\infty}^{\infty} f(x) e^{i\omega x} \, dx \qquad (9.7.3b)$$

$$\frac{\partial}{\partial t} \overline{U}(\omega, 0) = 0. \qquad (9.7.3c)$$

The general solution of (9.7.3a) is a linear combination of sines and cosines:
$$\overline{U}(\omega, t) = A(\omega) \cos c\omega t + B(\omega) \sin c\omega t. \qquad (9.7.4)$$

The initial conditions imply that
$$B(\omega) = 0 \qquad (9.7.5a)$$

$$A(\omega) = \overline{U}(\omega, 0) = \frac{1}{2\pi} \int_{-\infty}^{\infty} f(x) e^{i\omega x} \, dx. \qquad (9.7.5b)$$

Using the inverse Fourier transform, the solution of the one-dimensional wave equation is

$$u(x, t) = \int_{-\infty}^{\infty} \overline{U}(\omega, 0) \cos c\omega t \, e^{-i\omega x} \, d\omega, \qquad (9.7.6)$$

where $\overline{U}(\omega, 0)$ is the Fourier transform of the initial position.

The solution can be considerably simplified. Using Euler's formula,

$$u(x, t) = \frac{1}{2} \int_{-\infty}^{\infty} \overline{U}(\omega, 0)[e^{-i\omega(x - ct)} + e^{-i\omega(x + ct)}] d\omega. \qquad (9.7.7)$$

However, $\overline{U}(\omega, 0)$ is the Fourier transform of $f(x)$, and hence

$$f(x) = \int_{-\infty}^{\infty} \overline{U}(\omega, 0)e^{-i\omega x} \, d\omega. \qquad (9.7.8)$$

By comparing (9.7.7) and (9.7.8), we obtain

$$u(x, t) = \frac{1}{2}[f(x - ct) + f(x + ct)]. \qquad (9.7.9)$$

For an infinite string (*started at rest*), the solution is the sum of two terms, $\frac{1}{2}f(x - ct)$ and $\frac{1}{2}f(x + ct)$. $\frac{1}{2}f(x - ct)$ is a waveform of fixed shape. Its height stays fixed if $x - ct = $ constant, and thus $dx/dt = c$. For example, the origin corresponds to $x = ct$. Assuming that $c > 0$, this fixed shape moves to the right with velocity c. It is called a **traveling wave**. Similarly, $\frac{1}{2}f(x + ct)$ is a wave of fixed shape traveling to the left (at velocity $-c$). Our interpretation of this result is that the initial position of the string breaks in two, *if started at rest*, half moving to the left and half moving to the right at equal speeds c; the solution is the simple sum of these two traveling waves.

9.7.2 Laplace's Equation in a Semi-Infinite Strip

The mathematical problem for steady-state heat conduction in a semi-infinite strip $(0 < x < L, y > 0)$ is

$$\nabla^2 u = \frac{\partial^2 u}{\partial x^2} + \frac{\partial^2 u}{\partial y^2} = 0 \qquad (9.7.10a)$$

$$u(0, y) = g_1(y) \qquad (9.7.10b)$$
$$u(L, y) = g_2(y) \qquad (9.7.10c)$$
$$u(x, 0) = f(x). \qquad (9.7.10d)$$

We assume that $g_1(y)$ and $g_2(y)$ approach zero as $y \to \infty$. In Fig. 9.7.1 we illustrate the three nonhomogeneous boundary conditions and the useful

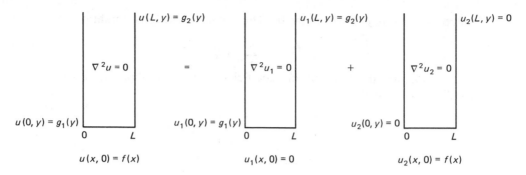

Figure 9.7.1 Laplace's equation in a semi-infinite strip.

simplification

$$u(x, y) = u_1(x, y) + u_2(x, y). \tag{9.7.11}$$

Here both u_1 and u_2 satisfy Laplace's equation. Separation of variables for Laplace's equation in the semi-infinite strip geometry provides motivation for our approach:

$$u(x, y) = \phi(x)\theta(y), \tag{9.7.12}$$

in which case

$$\frac{d^2\phi}{dx^2} = -\lambda\psi \tag{9.7.13a}$$

$$\frac{d^2\theta}{dy^2} = \lambda\theta. \tag{9.7.13b}$$

Two homogeneous boundary conditions are necessary for an eigenvalue problem. That is why we divided our problem into two parts.

Zero-temperature sides. For the u_2-problem, $u_2 = 0$ at both $x = 0$ and $x = L$. Thus, the differential equation in x is an eigenvalue problem, defined over a *finite* interval. The boundary conditions are exactly those of a Fourier sine series in x. The y-dependent solutions are exponentials. For u_2, separated solutions are

$$u_2(x, y) = \sin\frac{n\pi x}{L} e^{-n\pi y/L} \quad \text{and} \quad u_2(x, y) = \sin\frac{n\pi x}{L} e^{+n\pi y/L}.$$

The principle of superposition implies that

$$u_2(x, y) = \sum_{n=1}^{\infty} a_n \sin\frac{n\pi x}{L} e^{-n\pi y/L} + \sum_{n=1}^{\infty} b_n \sin\frac{n\pi x}{L} e^{n\pi y/L}. \tag{9.7.14}$$

There are two other conditions on u_2:

$$u_2(x, 0) = f(x) \tag{9.7.15a}$$

$$\lim_{y\to\infty} u_2(x, y) = 0. \tag{9.7.15b}$$

Since $u_2 \rightarrow 0$ as $y \rightarrow \infty$, $b_n = 0$. The nonhomogeneous condition is

$$f(x) = \sum_{n=1}^{\infty} a_n \sin \frac{n\pi x}{L},$$ (9.7.16)

and thus a_n are the Fourier sine coefficients of the nonhomogeneous boundary condition at $y = 0$:

$$a_n = \frac{2}{L} \int_0^L f(x) \sin \frac{n\pi x}{L} \, dx.$$ (9.7.17)

Using these coefficients, the solution is

$$u_2(x, y) = \sum_{n=1}^{\infty} a_n \sin \frac{n\pi x}{L} e^{-n\pi y/L}.$$ (9.7.18)

There is no need to use Fourier transforms for the u_2-problem. The partial differential equation for u_2 could have been analyzed much earlier in this text (e.g., in Chapter 2).

Zero-temperature bottom. For the u_1-problem, the second homogeneous boundary condition is less apparent:

$$u_1(x, 0) = 0$$ (9.7.19a)

$$\lim_{y \to \infty} u_1(x, y) = 0.$$ (9.7.19b)

The y-dependent part is the boundary value problem. As $y \rightarrow \infty$ the "separated" solution must remain bounded (they do not necessarily vanish). The appropriate solutions of (9.7.13b) are sines and cosines (corresponding to $\lambda < 0$). The homogeneous boundary condition at $y = 0$, (9.7.19a), implies that only sines should be used. Instead of continuing to discuss the method of separation of variables, we now introduce the Fourier sine transform in y:

$$u_1(x, y) = \int_0^{\infty} \overline{U}_1(x, \omega) \sin \omega y \, d\omega$$ (9.7.20a)

$$\overline{U}_1(x, \omega) = \frac{2}{\pi} \int_0^{\infty} u_1(x, y) \sin \omega y \, dy.$$ (9.7.20b)

We directly take the Fourier sine transform with respect to y of Laplace's equation, (9.7.10a). The properties of the transform of derivatives shows that Laplace's equation becomes an ordinary differential equation:

$$\frac{\partial^2}{\partial x^2} \overline{U}_1(x, \omega) - \omega^2 \overline{U}_1(x, \omega) = 0.$$ (9.7.21)

The boundary conditions at $y = 0$, $u_1(x, 0) = 0$, has been used to simplify our result. The solution of (9.7.21) is a linear combination of nonoscillatory (exponential) functions. It is most convenient (although not necessary) to utilize the following hyperbolic functions:

$$\overline{U}_1(x, \omega) = a(\omega) \sinh \omega x + b(\omega) \sinh \omega(L - x). \qquad (9.7.22)$$

The two nonhomogeneous conditions at $x = 0$ and $x = L$ yield

$$\overline{U}_1(0, \omega) = b(\omega) \sinh \omega L = \frac{2}{\pi} \int_0^\infty g_1(y) \sin \omega y \, dy \qquad (9.7.23a)$$

$$\overline{U}_1(L, \omega) = a(\omega) \sinh \omega L = \frac{2}{\pi} \int_0^\infty g_2(y) \sin \omega y \, dy. \qquad (9.7.23b)$$

$\overline{U}_1(x, \omega)$ the Fourier sine transform of $u_1(x, y)$ is given by (9.7.22), where $a(\omega)$ and $b(\omega)$ are obtained from (9.7.23).* This completes the somewhat complicated solution of Laplace's equation in a semi-infinite channel. It is the sum of a solution obtained using a Fourier sine series and one using a Fourier sinc transform.

Nonhomogeneous boundary conditions. If desired, Laplace's equation with three nonhomogeneous boundary conditions, (9.7.10), can be solved by directly applying a Fourier sine transform in y without decomposing the problem into two:

$$u(x, y) = \int_0^\infty \overline{U}(x, \omega) \sin \omega y \, d\omega. \qquad (9.7.24)$$

Since the boundary condition at $y = 0$, $u(x, 0) = f(x)$, is nonhomogeneous, an extra term is introduced into the Fourier sine transform of Laplace's equation (9.7.10a):

$$\frac{\partial^2 \overline{U}}{\partial x^2} - \omega^2 \overline{U} = -\frac{2}{\pi} \omega f(x). \qquad (9.7.25)$$

In this case, the Fourier sine transform satisfies a second-order linear constant-coefficient *nonhomogeneous* ordinary differential equation. This equation must be solved with two nonhomogeneous boundary conditions at $x = 0$ and $x = L$. Equation (9.7.25) can be solved by variation of parameters. This solution is probably more complicated than the one consisting of a sum of a series and a transform. Furthermore, this solution has a jump discontinuity at $y = 0$; the integral in (9.7.24) equals zero at $y = 0$, but converges to $f(x)$ as $y \to 0$. Usually, breaking the problem into two problems is preferable.

---

* Unfortunately, $\overline{U}_1(x, \omega)$ is not the product of the transforms of two simple functions. Hence, we do not use the convolution theorem.

9.7.3 Laplace's Equation in a Half-Plane

If the temperature is specified to equal $f(x)$ on an infinite wall, $y = 0$, then the steady-state temperature distribution for $y > 0$ satisfies Laplace's equation,

$$\nabla^2 u = \frac{\partial^2 u}{\partial x^2} + \frac{\partial^2 u}{\partial y^2} = 0, \qquad (9.7.26a)$$

subject to the boundary condition,

$$u(x, 0) = f(x). \qquad (9.7.26b)$$

If $f(x) \to 0$ as $x \to \pm\infty$, then there are three other implied boundary conditions,

$$\lim_{x \to +\infty} u(x, y) = 0 \qquad (9.7.26c)$$

$$\lim_{x \to -\infty} u(x, y) = 0 \qquad (9.7.26d)$$

$$\lim_{y \to +\infty} u(x, y) = 0; \qquad (9.7.26e)$$

the temperature approaches zero at large distances from the wall.

The methods of separation of variables suggests the use of a Fourier transform in x, since there are two homogeneous boundary conditions as $x \to \pm\infty$:

$$u(x, y) = \int_{-\infty}^{\infty} \overline{U}(\omega, y)e^{-i\omega x}\, d\omega \qquad (9.7.27a)$$

$$\overline{U}(\omega, y) = \frac{1}{2\pi} \int_{-\infty}^{\infty} u(x, y)e^{i\omega x}\, dx. \qquad (9.7.27b)$$

By taking the Fourier transform in x of (9.7.26a), we obtain the ordinary differential equation satisfied by the Fourier transform,

$$\frac{\partial^2 \overline{U}}{\partial y^2} - \omega^2 \overline{U} = 0. \qquad (9.7.28a)$$

Since $u(x, y) \to 0$ as $y \to \infty$, its Fourier transform in x also vanishes as $y \to \infty$,

$$\overline{U}(\omega, y) \to 0, \qquad \text{as } y \to \infty. \qquad (9.7.28b)$$

In addition, at $y = 0$, $\overline{U}(\omega, 0)$ is the Fourier transform of the boundary condition,

$$\overline{U}(\omega, 0) = \frac{1}{2\pi} \int_{-\infty}^{\infty} f(x)e^{i\omega x}\, dx. \qquad (9.7.28c)$$

We must be careful solving (9.7.28a). The general solution is

$$\overline{U}(\omega, y) = a(\omega)e^{\omega y} + b(\omega)e^{-\omega y}, \qquad (9.7.29)$$

which is of interest *for all* ω [see (9.7.27a)]. There are two boundary conditions to determine the two arbitrary functions $a(\omega)$ and $b(\omega)$. Equation (9.7.28b) states that $\overline{U}(\omega, y) \to 0$ as $y \to \infty$. At first you might think that this implies that $a(\omega) = 0$, but that is not correct. Instead, for $\overline{U}(\omega, y)$ to vanish as $y \to \infty$,

$a(\omega) = 0$ only for $\omega > 0$. If $\omega < 0$, then $b(\omega)e^{-\omega y}$ exponentially grows as $y \to \infty$. Thus, $b(\omega) = 0$ for $\omega < 0$. We have shown that

$$\overline{U}(\omega, y) = \begin{cases} a(\omega)e^{\omega y} & \text{for } \omega < 0 \\ b(\omega)e^{-\omega y} & \text{for } \omega > 0, \end{cases}$$

where $a(\omega)$ is arbitrary for $\omega < 0$ and $b(\omega)$ is arbitrary for $\omega > 0$. It is more convenient to note that this is equivalent to

$$\boxed{\overline{U}(\omega, y) = c(\omega)e^{-|\omega|y},} \tag{9.7.30}$$

for all ω. The nonhomogeneous boundary condition (9.7.26b) now shows that $c(\omega)$ is the Fourier transform of the temperature at the wall, $f(x)$. This completes our solution. However, we will determine a simpler representation of the solution.

Application of the convolution theorem. The easiest way to simplify the solution is to note that $\overline{U}(\omega, y)$ is the product of two Fourier transforms. $f(x)$ has the Fourier transform $c(\omega)$ and some function $g(x, y)$, as yet unknown, has the Fourier transform $e^{-|\omega|y}$. Using the convolution theorem, the solution of our problem is

$$u(x, y) = \frac{1}{2\pi} \int_{-\infty}^{\infty} f(\overline{x})g(x - \overline{x}, y)d\overline{x}. \tag{9.7.31}$$

We now need to determine what function $g(x, y)$ has the Fourier transform $e^{-|\omega|y}$. According to the inversion integral,

$$g(x, y) = \int_{-\infty}^{\infty} e^{-|\omega|y}e^{-i\omega x}\, d\omega,$$

which may be integrated directly:

$$g(x, y) = \int_{-\infty}^{0} e^{\omega y}e^{-i\omega x}\, d\omega + \int_{0}^{\infty} e^{-\omega y}e^{-i\omega x}\, d\omega$$

$$= \frac{e^{\omega(y - ix)}}{y - ix}\Big|_{-\infty}^{0} + \frac{e^{-\omega(y + ix)}}{-(y + ix)}\Big|_{0}^{\infty} = \frac{1}{y - ix} + \frac{1}{y + ix} = \frac{2y}{x^2 + y^2}.$$

Thus, the solution to Laplace's equation in semi-infinite space $(y > 0)$ subject to $u(x, 0) = f(x)$ is

$$\boxed{u(x, y) = \frac{1}{2\pi} \int_{-\infty}^{\infty} f(x)\frac{2y}{(x - \overline{x})^2 + y^2}d\overline{x}.} \tag{9.7.32}$$

This solution was derived assuming $f(x) \to 0$ as $x \to \pm\infty$. In fact, it is valid in other cases, roughly speaking as long as the integral is convergent. In Chapter 8 we obtained (9.7.32) through use of the Green's function. It was shown [see (8.5.37)] that the influence function for the nonhomogeneous boundary condition [see (9.7.32)] is the outward normal derivative (with respect to source points) of the Green's function:

$$-\frac{\partial}{\partial \bar{y}} G(x, y; \bar{x}, \bar{y}) \bigg|_{\bar{y}=0} = \frac{y}{\pi} \frac{1}{(x - \bar{x})^2 + y^2},$$

where $G(\mathbf{x}, \mathbf{x}_0)$ is the Green's function, the influence function for sources *in* the half-plane ($y > 0$).

Example. A simple, but interesting, solution arises if

$$f(x) = \begin{cases} 0 & x < 0 \\ 1 & x > 0. \end{cases} \tag{9.7.33}$$

This boundary condition corresponds to the wall uniformly heated to two different temperatures. We will determine the equilibrium temperature distribution for $y > 0$. From (9.7.32),

$$u(x, y) = \frac{1}{2\pi} \int_0^\infty \frac{2y}{y^2 + (x - \bar{x})^2} d\bar{x} = \frac{1}{\pi} \tan^{-1}\left(\frac{\bar{x} - x}{y}\right) \bigg|_0^\infty \tag{9.7.34}$$

$$= \frac{1}{\pi} \left[\tan^{-1}(\infty) - \tan^{-1}\left(\frac{-x}{y}\right) \right] = \frac{1}{\pi} \left[\frac{\pi}{2} + \tan^{-1}\left(\frac{x}{y}\right) \right].$$

Some care must be used in evaluating the inverse tangent function along a continuous branch. Figure 9.7.2, in which the tangent and inverse tangent functions are sketched, is helpful. If we introduce the angle θ *from the y-axis*,

$$\theta = \tan^{-1}\left(\frac{x}{y}\right),$$

then the temperature distribution becomes

$$\boxed{u(x, y) = \frac{1}{2} + \frac{\theta}{\pi}.} \tag{9.7.35}$$

We can check this answer independently. Reconsider this problem, but pose it in these rotated polar coordinates. Laplace's equation is

$$\frac{1}{r} \frac{\partial}{\partial r} \left(r \frac{\partial u}{\partial r} \right) + \frac{1}{r^2} \frac{\partial^2 u}{\partial \theta^2} = 0,$$

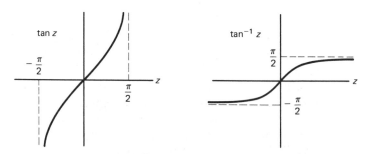

Figure 9.7.2 Tangent and inverse tangent functions.

Infinite Domain Problems   Chap. 9

and the boundary conditions are $u(r, \pi/2) = 1$ and $u(r, -\pi/2) = 0$. The solution depends only on the angle, $u(r, \theta) = u(\theta)$, in which case

$$\frac{d^2u}{d\theta^2} = 0,$$

subject to $u(\pi/2) = 1$ and $u(-\pi/2) = 0$, confirming (9.7.35). This result and more complicated ones for Laplace's equation can be obtained using conformal mappings in the complex plane.

9.7.4 Laplace's Equation in a Quarter-Plane

In this subsection we consider the steady-state temperature distribution within a quarter-plane ($x > 0$, $y > 0$) with the temperature given on one of the semi-infinite walls and the heat flow given on the other:

$$\nabla^2 u = \frac{\partial^2 u}{\partial x^2} + \frac{\partial^2 u}{\partial y^2} = 0 \qquad (9.7.36a)$$

$$u(0, y) = g(y) \qquad y > 0 \qquad (9.7.36b)$$
$$\frac{\partial u}{\partial y}(x, 0) = f(x) \qquad x > 0. \qquad (9.7.36c)$$

We assume that $g(y) \to 0$ as $y \to \infty$ and $f(x) \to 0$ as $x \to \infty$, such that $u(x, y) \to 0$ both as $x \to \infty$ and $y \to \infty$. There are two nonhomogeneous boundary conditions. Thus, it is convenient to decompose the problem into two, as illustrated in Fig. 9.7.3:

$$u = u_1(x, y) + u_2(x, y), \qquad (9.7.37)$$

where

$\nabla^2 u_1 = 0$	(9.7.38a)		$\nabla^2 u_2 = 0$	(9.7.39a)
$u_1(0, y) = g(y)$	(9.7.38b)		$u_2(0, y) = 0$	(9.7.39b)
$\dfrac{\partial}{\partial y} u_1(x, 0) = 0$	(9.7.38c)		$\dfrac{\partial}{\partial y} u_2(x, 0) = f(x).$	(9.7.39c)

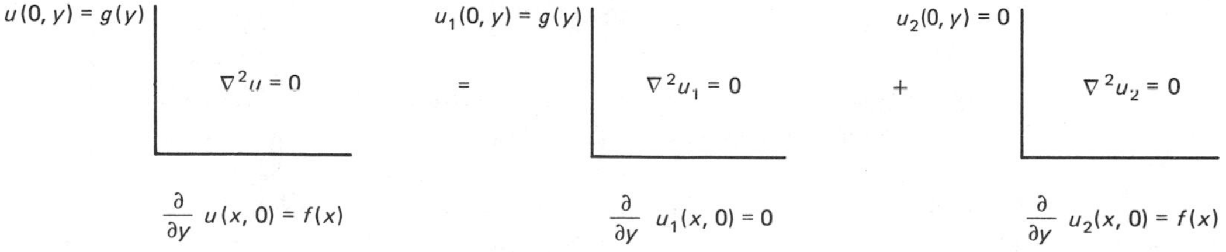

Figure 9.7.3 Laplace's equation in a quarter-plane.

Here, we will only analyze the problem for u_1, leaving the u_2-problem for an exercise.

Cosine transform in y. The u_1-problem can be analyzed in two different ways. The problem is semi-infinite in both x and y. Since u_1 is given at $x = 0$, a Fourier sine transform in x can be used. However, $\partial u_1/\partial y$ is given at $y = 0$ and hence a Fourier cosine transform in y can also be used. In fact, $\partial u_1/\partial y = 0$ at $y = 0$. Thus, we prefer to use a Fourier cosine transform in y since we expect the resulting ordinary differential equation to be *homogeneous*:

$$u_1(x, y) = \int_0^\infty \overline{U}_1(x, \omega) \cos \omega y \, d\omega \qquad (9.7.40a)$$

$$\overline{U}_1(x, \omega) = \frac{2}{\pi} \int_0^\infty u_1(x, y) \cos \omega y \, dy. \qquad (9.7.40b)$$

If the u_1-problem is Fourier cosine transformed in y, then we obtain

$$\frac{\partial^2 \overline{U}_1}{\partial x^2} - \omega^2 \overline{U}_1 = 0. \qquad (9.7.41a)$$

The variable x ranges from 0 to ∞. The two boundary conditions for this ordinary differential equation are

$$\overline{U}_1(0, \omega) = \frac{2}{\pi} \int_0^\infty g(y) \cos \omega y \, dy \qquad \text{and} \qquad \lim_{x \to \infty} \overline{U}_1(x, \omega) = 0. \quad (9.7.41b,c)$$

The general solution of (9.7.41a) is

$$\overline{U}_1(x, \omega) = a(\omega)e^{-\omega x} + b(\omega)e^{\omega x}, \qquad (9.7.42)$$

for $x > 0$ and $\omega > 0$ only. From the boundary conditions (9.7.41b,c), it follows that

$$b(\omega) = 0 \qquad \text{and} \qquad a(\omega) = \frac{2}{\pi} \int_0^\infty g(y) \cos \omega y \, dy. \qquad (9.7.43)$$

Convolution theorem. A simpler form of the solution can be obtained using a convolution theorem for Fourier cosine transforms. We derive the following in Exercise 9.6.7, where we assume $f(x)$ is even:

If a Fourier cosine transform in x, $H(\omega)$, is the product of two Fourier cosine transforms, $H(\omega) = F(\omega)G(\omega)$, then

$$h(x) = \frac{1}{\pi} \int_0^\infty g(\overline{x})[f(x - \overline{x}) + f(x + \overline{x})] \, d\overline{x}. \qquad (9.7.44)$$

In our problem, $\overline{U}_1(x, \omega)$, the Fourier cosine transform of $u_1(x, y)$, is the product of $a(\omega)$, the Fourier cosine transform of $g(y)$, and $e^{-\omega x}$:

$$\overline{U}_1(x, \omega) = a(\omega)e^{-\omega x}.$$

We use the cosine transform pair [see (9.7.40)] to obtain the function $Q(y)$ which has the Fourier cosine transform $e^{-\omega x}$:

$$Q(y) = \int_0^\infty e^{-\omega x} \cos \omega y \, d\omega = \int_0^\infty e^{-\omega x} \frac{e^{i\omega y} + e^{-i\omega y}}{2} \, d\omega$$

$$= \frac{1}{2}\left(\frac{1}{x - iy} + \frac{1}{x + iy}\right) = \frac{x}{x^2 + y^2}$$

Thus, according to the convolution theorem,

$$u_1(x, y) = \frac{x}{\pi} \int_0^\infty g(\bar{y}) \left[\frac{1}{x^2 + (y - \bar{y})^2} + \frac{1}{x^2 + (y + \bar{y})^2}\right] d\bar{y}. \qquad (9.7.45)$$

This result also could have been obtained using the Green's function method (see Chapter 8). In this case appropriate image sources may be introduced so as to utilize the infinite space Green's function for Laplace's equation.

Sine transform in x. An alternative method to solve for $u_1(x, y)$ is to use the Fourier sine transform in x:

$$u_1(x, y) = \int_0^\infty \overline{U}_1(\omega, y) \sin \omega x \, d\omega$$

$$\overline{U}_1(\omega, y) = \frac{2}{\pi} \int_0^\infty u_1(x, y) \sin \omega x \, dx.$$

The ordinary differential equation for $\overline{U}_1(\omega, y)$ is nonhomogeneous:

$$\frac{\partial^2 \overline{U}_1}{\partial y^2} - \omega^2 \overline{U}_1 = -\frac{2}{\pi}\omega g(y).$$

This equation must be solved with the following boundary conditions at $y = 0$ and $y - \infty$:

$$\frac{\partial \overline{U}_1}{\partial y}(\omega, 0) = 0 \qquad \text{and} \qquad \lim_{y \to \infty} \overline{U}_1(\omega, y) = 0.$$

This approach is further discussed in the exercises.

9.7.5. Heat Equation in a Plane (Two-Dimensional Fourier Transforms)

Transforms can be used to solve problems that are infinite in both x and y. Consider the heat equation in the x–y plane, $-\infty < x < \infty$, $-\infty < y < \infty$:

$$\frac{\partial u}{\partial t} = k\left(\frac{\partial^2 u}{\partial x^2} + \frac{\partial^2 u}{\partial y^2}\right), \qquad (9.7.46a)$$

subject to the initial condition

$$u(x, y, 0) = f(x, y). \qquad (9.7\text{:}46b)$$

If we separate variables, we obtain product solutions of the form
$$u(x, y, t) = e^{-i\omega_1 x} e^{-i\omega_2 y} e^{-k(\omega_1^2 + \omega_2^2)t}$$
for all ω_1 and ω_2. Corresponding to $u \to 0$ as $x \to \pm\infty$ and $y \to \pm\infty$ are the boundary conditions that the separated solutions remain bounded as $x \to \pm\infty$ and $y \to \pm\infty$. Thus, these types of solutions are valid for all real ω_1 and ω_2. A generalized principle of superposition implies that the form of the integral representation of the solution is

$$u(x, y, t) = \int_{-\infty}^{\infty} \int_{-\infty}^{\infty} A(\omega_1, \omega_2) e^{-i\omega_1 x} e^{-i\omega_2 y} e^{-k(\omega_1^2 + \omega_2^2)t} \, d\omega_1 \, d\omega_2. \qquad (9.7.47)$$

The initial condition is satisfied if
$$f(x, y) = \int_{-\infty}^{\infty} \int_{-\infty}^{\infty} A(\omega_1, \omega_2) e^{-i\omega_1 x} e^{-i\omega_2 y} \, d\omega_1 \, d\omega_2.$$

We will show that we can determine $A(\omega_1, \omega_2)$,

$$A(\omega_1, \omega_2) = \frac{1}{(2\pi)^2} \int_{-\infty}^{\infty} \int_{-\infty}^{\infty} f(x, y) e^{i\omega_1 x} e^{i\omega_2 y} \, dx \, dy, \qquad (9.7.48)$$

completing the solution. $A(\omega_1, \omega_2)$ is called the **double Fourier transform** of $f(x, y)$.

Double Fourier transforms. We have used separation of variables to motivate **two-dimensional Fourier transforms**. Suppose that we have a function of two variables $f(x, y)$ which decays sufficiently fast as x and $y \to \pm\infty$. The Fourier transform in x (keeping y fixed, with transform variable ω_1) is

$$F(\omega_1, y) = \frac{1}{2\pi} \int_{-\infty}^{\infty} f(x, y) e^{i\omega_1 x} \, dx;$$

its inverse is

$$f(x, y) = \int_{-\infty}^{\infty} F(\omega_1, y) e^{-i\omega_1 x} \, d\omega_1.$$

$F(\omega_1, y)$ is a function of y which also can be Fourier transformed (here with transform variable ω_2):

$$\tilde{F}(\omega_1, \omega_2) = \frac{1}{2\pi} \int_{-\infty}^{\infty} F(\omega_1, y) e^{i\omega_2 y} \, dy$$

$$F(\omega_1, y) = \int_{-\infty}^{\infty} \tilde{F}(\omega_1, \omega_2) e^{-i\omega_2 y} \, d\omega_2.$$

Combining these we obtain the **two-dimensional (or double) Fourier transform pair**:

$$\tilde{F}(\omega_1, \omega_2) = \frac{1}{(2\pi)^2} \int_{-\infty}^{\infty} \int_{-\infty}^{\infty} f(x, y) e^{i\omega_1 x} e^{i\omega_2 y} \, dx \, dy \qquad (9.7.49a)$$

$$f(x, y) = \int_{-\infty}^{\infty} \int_{-\infty}^{\infty} \tilde{F}(\omega_1, \omega_2) e^{-i\omega_1 x} e^{-i\omega_2 y} \, d\omega_1 \, d\omega_2. \qquad (9.7.49b)$$

Wave number vector. Let us motivate a more convenient notation. When using $e^{i\omega x}$ we refer to ω as the wave number (the number of waves in 2π distance). Here we note that

$$e^{i\omega_1 x} e^{i\omega_2 y} = e^{i(\omega_1 x + \omega_2 y)} = e^{i\boldsymbol{\omega} \cdot \mathbf{r}},$$

where $\mathbf{r}$ is a **position vector*** and $\boldsymbol{\omega}$ is a **wave number vector**:

$$\mathbf{r} = x\hat{\mathbf{i}} + y\hat{\mathbf{j}} \qquad (9.7.50)$$

$$\boldsymbol{\omega} = \omega_1 \hat{\mathbf{i}} + \omega_2 \hat{\mathbf{j}}. \qquad (9.7.51)$$

To interpret $e^{i\boldsymbol{\omega} \cdot \mathbf{r}}$ we discuss, for example, its real part, $\cos(\boldsymbol{\omega} \cdot \mathbf{r}) = \cos(\omega_1 x + \omega_2 y)$. The crests are located at $\omega_1 x + \omega_2 y = n(2\pi)$, sketched in Fig. 9.7.4. The direction perpendicular to the crests is $\nabla(\omega_1 x + \omega_2 y) = \omega_1 \hat{\mathbf{i}} + \omega_2 \hat{\mathbf{j}} = \boldsymbol{\omega}$. This is called the direction of the wave. Thus, *the wave number vector is in the direction of the wave.* We introduce the magnitude ω of the wave number vector

$$\omega^2 = \boldsymbol{\omega} \cdot \boldsymbol{\omega} = |\boldsymbol{\omega}|^2 = \omega_1^2 + \omega_2^2.$$

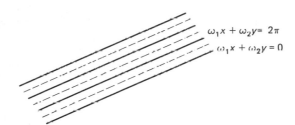

$$\omega_1 x + \omega_2 y = 2\pi$$
$$\omega_1 x + \omega_2 y = 0$$

Figure 9.7.4 Crests for a two-dimensional wave.

The unit vector in the wave direction is $\boldsymbol{\omega}/\omega$. If we move a distance s in the wave direction (from the origin), then $\mathbf{r} = s\boldsymbol{\omega}/\omega$. Thus,

$$\boldsymbol{\omega} \cdot \mathbf{r} = s\omega \qquad \text{and} \qquad \cos(\boldsymbol{\omega} \cdot \mathbf{r}) = \cos(\omega s).$$

Thus, ω is the number of waves in 2π distance (in the direction of the wave). We have justified the name wave number vector for $\boldsymbol{\omega}$; that is, the **wave number vector is in the direction of the wave and its magnitude is the number of waves in 2π distance (in the direction of the wave)**.

Using the position vector $\mathbf{r} = x\hat{\mathbf{i}} + y\hat{\mathbf{j}}$ and the wave number vector $\boldsymbol{\omega} =$

---

* In other contexts, we use the notation $\mathbf{x}$ for the position vector. Thus $\mathbf{r} = \mathbf{x}$.

$\omega_1\hat{\mathbf{i}} + \omega_2\hat{\mathbf{j}}$, the **double Fourier transform pair**, (9.7.49), becomes

$$F(\boldsymbol{\omega}) = \frac{1}{(2\pi)^2} \int_{-\infty}^{\infty} \int_{-\infty}^{\infty} f(\mathbf{r})e^{i\boldsymbol{\omega}\cdot\mathbf{r}}\, d^2r \qquad (9.7.52a)$$

$$f(\mathbf{r}) = \int_{-\infty}^{\infty} \int_{-\infty}^{\infty} F(\boldsymbol{\omega})e^{-i\boldsymbol{\omega}\cdot\mathbf{r}}\, d^2\omega, \qquad (9.7.52b)$$

where $f(\mathbf{r}) = f(x, y)$, $d^2r = dx\,dy$, $d^2\omega = d\omega_1\,d\omega_2$, and $F(\boldsymbol{\omega})$ is the double Fourier transform of $f(\mathbf{r})$.

Using the notation $\mathscr{F}[u(x, y)]$ for the double Fourier transform of $u(x, y)$, we have the following easily verified fundamental properties:

$$\mathscr{F}\left[\frac{\partial u}{\partial t}\right] = \frac{\partial}{\partial t}\mathscr{F}[u] \qquad (9.7.53a)$$

$$\mathscr{F}\left[\frac{\partial u}{\partial x}\right] = -i\omega_1\mathscr{F}[u] \qquad (9.7.53b)$$

$$\mathscr{F}\left[\frac{\partial u}{\partial y}\right] = -i\omega_2\mathscr{F}[u] \qquad (9.7.53c)$$

$$\mathscr{F}[\nabla^2 u] = -\omega^2\mathscr{F}[u], \qquad (9.7.53d)$$

where $\omega^2 = \boldsymbol{\omega} \cdot \boldsymbol{\omega} = \omega_1^2 + \omega_2^2$, as long as u decays sufficiently rapidly as x and $y \to \pm\infty$. A short table of the double Fourier transform appears on page 384.

Heat equation. Instead of using the method of separation of variables, the two-dimensional heat equation (9.7.46a) can be directly solved by double Fourier transforming it:

$$\frac{\partial \overline{U}}{\partial t} = -k\omega^2\overline{U}, \qquad (9.7.54)$$

where $\overline{U}$ is the double Fourier transform of $u(x, y, t)$:

$$\mathscr{F}[u] = \overline{U}(\boldsymbol{\omega}, t) = \frac{1}{(2\pi)^2} \int_{-\infty}^{\infty} \int_{-\infty}^{\infty} u(x, y, t)e^{i\boldsymbol{\omega}\cdot\mathbf{r}}\, dx\,dy. \qquad (9.7.55)$$

The elementary solution of (9.7.54) is

$$\overline{U}(\boldsymbol{\omega}, t) = A(\boldsymbol{\omega})e^{-k\omega^2 t}. \qquad (9.7.56)$$

Applying (9.7.46), $A(\boldsymbol{\omega})$ is the Fourier transform of the initial condition:

$$A(\boldsymbol{\omega}) = \overline{U}(\boldsymbol{\omega}, 0) = \frac{1}{(2\pi)^2} \int_{-\infty}^{\infty} \int_{-\infty}^{\infty} f(x, y)e^{i\boldsymbol{\omega}\cdot\mathbf{r}}\, dx\,dy. \qquad (9.7.57)$$

Thus, the solution of the two-dimensional heat equation is

$$
\begin{aligned}
u(x, y, t) &= \int_{-\infty}^{\infty} \int_{-\infty}^{\infty} \overline{U}(\boldsymbol{\omega}, t) e^{-i\boldsymbol{\omega}\cdot\mathbf{r}} \, d\omega_1 \, d\omega_2 \\
&= \int_{-\infty}^{\infty} \int_{-\infty}^{\infty} A(\boldsymbol{\omega}) e^{-k\omega^2 t} e^{-i\boldsymbol{\omega}\cdot\mathbf{r}} \, d\omega_1 \, d\omega_2 .
\end{aligned}
\tag{9.7.58}
$$

This verifies what was suggested earlier by separation of variables.

Application of convolution theorem.   In an exercise we show that a **convolution theorem** holds directly *for double Fourier transforms*: If $H(\boldsymbol{\omega}) = F(\boldsymbol{\omega})G(\boldsymbol{\omega})$, then

$$
h(\mathbf{r}) = \frac{1}{(2\pi)^2} \int_{-\infty}^{\infty} \int_{-\infty}^{\infty} f(\mathbf{r}_0) g(\mathbf{r} - \mathbf{r}_0) \, dx_0 \, dy_0 .
\tag{9.7.59}
$$

For the two-dimensional heat equation, we have shown that $\overline{U}(\boldsymbol{\omega}, t)$ is the product of $e^{-k\omega^2 t}$ and $A(\boldsymbol{\omega})$, the double Fourier transform of the initial condition. Thus, we need to determine the function whose double Fourier transform is $e^{-k\omega^2 t}$:

$$
\int_{-\infty}^{\infty} \int_{-\infty}^{\infty} e^{-k\omega^2 t} e^{-i\boldsymbol{\omega}\cdot\mathbf{r}} \, d\omega_1 \, d\omega_2 = \int_{-\infty}^{\infty} e^{-k\omega_1^2 t} e^{-i\omega_1 x} \, d\omega_1 \int_{-\infty}^{\infty} e^{-k\omega_2^2 t} e^{-i\omega_2 y} \, d\omega_2
$$

$$
= \sqrt{\frac{\pi}{kt}} e^{-x^2/4kt} \sqrt{\frac{\pi}{kt}} e^{-y^2/4kt} = \frac{\pi}{kt} e^{-r^2/4kt} .
$$

The inverse transform of $e^{-k\omega^2 t}$ is the product of the two one-dimensional inverse transforms; it is a two-dimensional Gaussian, $(\pi/kt)e^{-r^2/4kt}$, where $r^2 = x^2 + y^2$. In this manner, the solution of the initial value problem for the two-dimensional heat equation on an infinite plane is

$$
u(x, y, t) = \int_{-\infty}^{\infty} \int_{-\infty}^{\infty} f(x_0, y_0) \frac{1}{4\pi kt} \exp\left[-\frac{(x - x_0)^2 + (y - y_0)^2}{4kt} \right] dx_0 \, dy_0 .
$$

$$
\tag{9.7.60}
$$

The influence function for the initial condition is

$$
g(x, y, t; x_0, y_0, 0) = \frac{1}{4\pi kt} \exp\left[-\frac{(x - x_0)^2 + (y - y_0)^2}{4kt} \right] = \frac{1}{4\pi kt} e^{-|\mathbf{r} - \mathbf{r}_0|^2/4kt} .
$$

It expresses the effect at x, y (at time t) due to the initial heat energy at x_0, y_0. **The influence function for the two-dimensional heat equation is the product of the influence functions of two one-dimensional heat equations.**

$f(\mathbf{r}) = \int_{-\infty}^{\infty}\int_{-\infty}^{\infty} F(\boldsymbol{\omega})e^{-i\boldsymbol{\omega}\cdot\mathbf{r}}\,d\omega_1\,d\omega_2$	$F(\boldsymbol{\omega}) = \dfrac{1}{(2\pi)^2}\int_{-\infty}^{\infty}\int_{-\infty}^{\infty} f(\mathbf{r})e^{i\boldsymbol{\omega}\cdot\mathbf{r}}\,dx\,dy$	Reference
$\dfrac{\partial f}{\partial x},\ \dfrac{\partial f}{\partial y}$	$-i\omega_1 F(\boldsymbol{\omega}),\ -i\omega_2 F(\boldsymbol{\omega})$	Derivatives
$\nabla^2 f$	$-\omega^2 F(\boldsymbol{\omega})$	(Sec. 9.7.5)
$\dfrac{\pi}{\beta}e^{-r^2/4\beta}$	$e^{-\beta\omega^2}$	Gaussian (Sec. 9.7.5)
$f(\mathbf{r}-\boldsymbol{\beta})$	$e^{i\boldsymbol{\omega}\cdot\boldsymbol{\beta}}F(\boldsymbol{\omega})$	Exercise 9.7.8
$\dfrac{1}{(2\pi)^2}\int_{-\infty}^{\infty}\int_{-\infty}^{\infty} f(\mathbf{r}_0)g(\mathbf{r}-\mathbf{r}_0)\,dx_0\,dy_0$	$F(\boldsymbol{\omega})G(\boldsymbol{\omega})$	Convolution (Exercise 9.7.7)

EXERCISES 9.7

9.7.1. Solve

$$\frac{\partial^2 u}{\partial x^2} + \frac{\partial^2 u}{\partial y^2} = 0$$

for $0 < y < H$, $-\infty < x < \infty$ subject to:

*(a) $u(x, 0) = f_1(x)$ and $u(x, H) = f_2(x)$

(b) $\dfrac{\partial}{\partial y}\, u(x, 0) = f_1(x)$ and $u(x, H) = f_2(x)$

(c) $u(x, 0) = 0$ and $\dfrac{\partial u}{\partial y}(x, H) + hu(x, H) = f(x)$

9.7.2. Solve

$$\frac{\partial^2 u}{\partial x^2} + \frac{\partial^2 u}{\partial y^2} = 0 \qquad \text{for } 0 < x < L, y < 0$$

subject to the following boundary conditions. If there is a solvability condition, state it and explain it physically:

(a) $\dfrac{\partial u}{\partial x}(0, y) = g_1(y)$, $\dfrac{\partial u}{\partial x}(L, y) = g_2(y)$, $u(x, 0) = f(x)$

*(b) $u(0, y) = g_1(y)$, $\dfrac{\partial u}{\partial x}(L, y) = 0$, $\dfrac{\partial u}{\partial y}(x, 0) = 0$

(c) $\dfrac{\partial u}{\partial x}(0, y) = 0$, $\dfrac{\partial u}{\partial x}(L, y) = 0$, $\dfrac{\partial u}{\partial y}(x, 0) = f(x)$

(d) $\dfrac{\partial u}{\partial x}(0, y) = 0$, $\dfrac{\partial u}{\partial x}(L, y) = g(y)$, $\dfrac{\partial u}{\partial y}(x, 0) = 0$

9.7.3. (a) Solve

$$\frac{\partial^2 u}{\partial x^2} + \frac{\partial^2 u}{\partial y^2} = 0$$

for $x < 0$, $-\infty < y < \infty$ subject to $u(0, y) = g(y)$.

(b) Determine the simplest form of the solution if

$$g(y) = \begin{cases} 0 & |y| > 1 \\ 1 & |y| < 1. \end{cases}$$

9.7.4. Solve

$$\frac{\partial^2 u}{\partial x^2} + \frac{\partial^2 u}{\partial y^2} = 0$$

for $x > 0$, $y > 0$ subject to:

*(a) $u(0, y) = 0$ and $\frac{\partial u}{\partial y}(x, 0) = f(x)$

(b) $u(0, y) = 0$ and $u(x, 0) = f(x)$

9.7.5. Reconsider Exercise 9.7.4(a). Let $w = \partial u/\partial y$. Show that w satisfies Exercise 9.7.4(b). In this manner solve both Exercises 9.7.4(a) and (b).

9.7.6. Consider (9.7.38). In the text we introduce the Fourier cosine transform in y. Instead, here we introduce the Fourier sine transform in x (see page 379).
   (a) Solve for $\overline{U}_1(\omega, y)$ if

$$\frac{\partial^2 \overline{U}_1}{\partial y^2} - \omega^2 \overline{U}_1 = -\frac{2}{\pi}\omega g(y)$$

$$\frac{\partial \overline{U}_1}{\partial y}(\omega, 0) = 0$$

$$\lim_{y \to \infty} \overline{U}_1(\omega, y) = 0$$

[*Hint*: See (8.3.8) and (8.3.9) or (12.3.9), using exponentials.]
   (b) Derive $u_1(x, y)$. Show that (9.7.45) is valid.

9.7.7. Derive the two-dimensional convolution theorem, (9.7.59).

9.7.8. Derive the following shift theorem for two-dimensional Fourier transforms: The inverse transform of $e^{i\omega \cdot \beta} F(\omega)$ is $f(\mathbf{r} - \boldsymbol{\beta})$.

9.7.9. Solve

$$\frac{\partial u}{\partial t} + \mathbf{v}_0 \cdot \nabla u = k \nabla^2 u$$

subject to the initial condition

$$u(x, y, 0) = f(x, y).$$

(*Hint:* See Exercise 9.7.7.) Show how the influence function is altered by the convection term $\mathbf{v}_0 \cdot \nabla u$.

9.7.10. Solve

$$\frac{\partial u}{\partial t} = k_1 \frac{\partial^2 u}{\partial x^2} + k_2 \frac{\partial^2 u}{\partial y^2}$$

subject to the initial condition

$$u(x, y, 0) = f(x, y).$$

9.7.11. Consider

$$\frac{\partial u}{\partial t} = k\left(\frac{\partial^2 u}{\partial x^2} + \frac{\partial^2 u}{\partial y^2}\right) \qquad \begin{array}{c} x > 0 \\ y > 0 \end{array}$$

subject to the initial condition

$$u(x, y, 0) = f(x, y).$$

Solve with the following boundary conditions:
   *(a) $u(0, y, t) = 0$ and $u(x, 0, t) = 0$

(b) $\dfrac{\partial u}{\partial x}(0, y, t) = 0$ and $\dfrac{\partial u}{\partial y}(x, 0, t) = 0$

(c) $u(0, y, t) = 0$ and $\dfrac{\partial u}{\partial y}(x, 0, t) = 0$

9.7.12. Consider

$$\frac{\partial u}{\partial t} = k \left(\frac{\partial^2 u}{\partial x^2} + \frac{\partial^2 u}{\partial y^2} \right) \qquad \begin{array}{c} 0 < x < L \\ y > 0 \end{array}$$

subject to the initial condition

$$u(x, y, 0) = f(x, y).$$

Solve with the following boundary conditions:

*(a) $u(0, y, t) = 0,$ $u(L, y, t) = 0,$ $u(x, 0, t) = 0$

(b) $u(0, y, t) = 0,$ $u(L, y, t) = 0,$ $\dfrac{\partial u}{\partial y}(x, 0, t) = 0$

(c) $\dfrac{\partial u}{\partial x}(0, y, t) = 0,$ $\dfrac{\partial u}{\partial x}(L, y, t) = 0,$ $\dfrac{\partial u}{\partial y}(x, 0, t) = 0$

9.7.13. Solve

$$\frac{\partial u}{\partial t} = k \left(\frac{\partial^2 u}{\partial x^2} + \frac{\partial^2 u}{\partial y^2} \right) \qquad \begin{array}{c} 0 < y < H \\ -\infty < x < \infty \end{array}$$

subject to the initial condition

$$u(x, y, 0) = f(x, y)$$

and the boundary conditions

$$u(x, 0, t) = 0$$
$$u(x, H, t) = 0.$$

9.7.14. (a) Without deriving or applying a convolution theorem, solve for $u(x, y, z, t)$:

$$\left. \begin{array}{c} \dfrac{\partial u}{\partial t} = k \left(\dfrac{\partial^2 u}{\partial x^2} + \dfrac{\partial^2 u}{\partial y^2} + \dfrac{\partial^2 u}{\partial z^2} \right) \quad t > 0 \\[2mm] \text{such that} \\[2mm] u(x, y, z, 0) = f(x, y, z) \\[2mm] u(x, y, 0, t) = 0 \end{array} \right\} \quad \begin{array}{c} -\infty < x < \infty \\ -\infty < y < \infty \\ 0 < z < \infty. \end{array}$$

(b) Simplify your answer (developing convolution ideas, as needed.)

9.7.15. Consider

$$\frac{\partial^2 u}{\partial x^2} + \frac{\partial^2 u}{\partial y^2} + \frac{\partial^2 u}{\partial z^2} = 0 \qquad z > 0$$

$$u(x, y, 0) = f(x, y).$$

*(a) Determine the double Fourier transform of u.

*(b) Solve for $u(x, y, z)$ by calculating the inversion integral in polar coordinates. [*Hints:* It is easier to first solve for w, where $u = \partial w / \partial z$. Then the following integral may be useful:

$$\int_0^{2\pi} \frac{d\theta}{a^2 \sin^2 \theta + b^2 \cos^2 \theta} = \frac{2\pi}{ab}.$$

(This integral may be derived using the change of variables $z = e^{i\theta}$ and the theory of complex variables.)]

(c) Compare your result to the Green's function result.

9.7.16. Consider Laplace's equation

$$\nabla^2 u = 0$$

inside a quarter-circle (Fig. 9.7.5) (a finite region) subject to the boundary conditions:

$$u(a, \theta) = f(\theta), \qquad u(r, 0) = g_1(r), \qquad u\left(r, \frac{\pi}{2}\right) = g_2(r).$$

$u = f(\theta)$

$u = g_2(r)$

$u = g_1(r)$ **Figure 9.7.5**

(a) Divide into three problems, $u = u_1 + u_2 + u_3$ such that

$$u_1(a, \theta) = 0, \qquad u_2(a, \theta) = 0 \qquad u_3(a, \theta) = f(\theta)$$

$$u_1(r, 0) = g_1(r) \qquad u_2(r, 0) = 0 \qquad u_3(r, 0) = 0$$

$$u_1\left(r, \frac{\pi}{2}\right) = 0 \qquad u_2\left(r, \frac{\pi}{2}\right) = g_2(r) \qquad u_3\left(r, \frac{\pi}{2}\right) = 0.$$

Solve for $u_3(r, \theta)$.

*(b) Solve for $u_2(r, \theta)$. [*Hints:* Try to use the method of separation of variables, $u_2(r, \theta) = \phi(r)h(\theta)$. Show that $\phi(r) = \sin[\sqrt{\lambda} \ln (r/a)]$ for all $\lambda > 0$. It will be necessary to use a Fourier sine transform in the variable $\rho = -\ln (r/a)$.] [*Comments:* Here, a singular Sturm–Liouville problem on a *finite* interval occurs which has a continuous spectrum. For the wave equation on a quarter-circle, the corresponding singular Sturm–Liouville problem (involving Bessel functions) has a discrete spectrum.]

9.7.17. Reconsider the problem for $u_2(r, \theta)$ described in Exercise 9.7.16(b). Introduce the independent variable $\rho = -\ln (r/a)$ instead of r. [*Comment:* In complex variables this is the conformal transformation

$$w = -\ln\left(\frac{z}{a}\right), \qquad \text{where } z = x + iy$$

(i.e., $\ln (z/a) = \ln |z/a| + i\theta$).]

(a) Determine the partial differential equation for u_2 in the variables ρ and θ. Sketch the boundary in Cartesian coordinates, ρ and θ.

(b) Solve this problem. (*Hint:* Section 9.7.2 may be helpful.)

(c) Compare to Exercise 9.7.16.

*9.7.18. Solve

$$\frac{\partial^2 u}{\partial t^2} = c^2 \frac{\partial^2 u}{\partial x^2} \qquad -\infty < x < \infty$$

$$u(x, 0) = 0$$

$$\frac{\partial u}{\partial t}(x, 0) = g(x).$$

(*Hint:* Use the convolution theorem and see Exercise 9.4.6.)

9.7.19. For the problem in Sec. 9.7.1, we showed that

$$\overline{U}(\omega, t) = F(\omega) \cos c\omega t.$$

Obtain $u(x, t)$ using the convolution theorem. (*Hint:* $\cos c\omega t$ does not have an ordinary inverse Fourier transform. However, obtain the inverse Fourier transform of $\cos c\omega t$ using Dirac delta functions.)

9.7.20. Consider Exercise 9.7.17 for $u_3(r, \theta)$ rather than $u_2(r, \theta)$.

10

Green's functions
for Time-Dependent
Problems

10.1 INTRODUCTION

In Chapter 8 we had some success in obtaining Green's functions for time-independent problems. One particularly important idea was the use of infinite space Green's functions. However, to analyze infinite space problems, at times the Fourier transform is necessary, as discussed in Chapter 9. In this chapter we utilize the ideas of Chapter 8 on Green's functions and the ideas of Chapter 9 on Fourier transform solutions of partial differential equations. We analyze Green's functions for the heat and wave equations. Problems with one, two, and three spatial dimensions will be considered.

10.2 GREEN'S FUNCTIONS FOR THE WAVE EQUATION

10.2.1 Introduction

In this section we solve the wave equation with possibly time-dependent sources,

$$
\frac{\partial^2 u}{\partial t^2} = c^2 \, \nabla^2 u \, + \, Q(\mathbf{x}, t),
\tag{10.2.1a}
$$

subject to the two initial conditions,

$$u(\mathbf{x}, 0) = f(\mathbf{x}) \qquad (10.2.1b)$$

$$\frac{\partial u}{\partial t}(\mathbf{x}, 0) = g(\mathbf{x}). \qquad (10.2.1c)$$

If the problem is on a finite or semi-infinite region, then in general $u(\mathbf{x}, t)$ will satisfy nonhomogeneous conditions on the boundary. We will determine simultaneously how to solve this problem in one, two, and three dimensions. (In one dimension $\nabla^2 = \partial^2/\partial x^2$.)

We introduce the Green's function $G(\mathbf{x}, t; \mathbf{x}_0, t_0)$ as a solution due to a concentrated source at $\mathbf{x} = \mathbf{x}_0$ acting instantaneously only at $t = t_0$:

$$\frac{\partial^2 G}{\partial t^2} = c^2 \nabla^2 G + \delta(\mathbf{x} - \mathbf{x}_0)\, \delta(t - t_0), \qquad (10.2.2a)$$

where $\delta(\mathbf{x} - \mathbf{x}_0)$ is the Dirac delta function of the appropriate dimension. For finite or semi-infinite problems, G will satisfy the related homogeneous boundary conditions corresponding to the nonhomogeneous ones satisfied by $u(\mathbf{x}, t)$.

The Green's function is the response at $\mathbf{x}$ at time t due to a source located at $\mathbf{x}_0$ at time t_0. Since we desire the Green's function G to be the response only due to this source acting at $t = t_0$ (not due to some nonzero earlier conditions), we insist that the response G will be zero before the source acts ($t < t_0$):

$$G(\mathbf{x}, t; \mathbf{x}_0, t_0) = 0 \qquad \text{for } t < t_0, \qquad (10.2.2b)$$

known as the **causality principle** (see Sec. 8.2).

The Green's function $G(\mathbf{x}, t; \mathbf{x}_0, t_0)$ only depends on the time after the occurrence of the concentrated source. If we introduce the **elapsed** time, $T = t - t_0$,

$$\frac{\partial^2 G}{\partial T^2} = c^2 \nabla^2 G + \delta(\mathbf{x} - \mathbf{x}_0)\, \delta(T)$$

$$G = 0 \qquad \text{for } T < 0,$$

then G is also seen to be the response due to a concentrated source at $\mathbf{x} = \mathbf{x}_0$ at $T = 0$. We call this the **translation** property,

$$G(\mathbf{x}, t; \mathbf{x}_0, t_0) = G(\mathbf{x}, t - t_0; \mathbf{x}_0, 0). \qquad (10.2.2c)$$

10.2.2 Green's Formula

Before solving for the Green's function (in various dimensions), we will show how the solution of the nonhomogeneous wave equation (10.2.1a) (with nonhomogeneous initial and boundary conditions) is obtained using the Green's function. For time-independent problems (nonhomogeneous Sturm–Liouville type or the Poisson equation), the relationship between the nonhomogeneous solution and the Green's function was obtained using Green's formula:

Sturm–Liouville operator $[L = d/dx(p\, d/dx) + q]$:

$$\int_a^b (uL(v) - vL(u))\, dx = p\left(u\frac{dv}{dx} - v\frac{du}{dx}\right)\bigg|_a^b \qquad (10.2.3)$$

Three-dimensional Laplacian $(L = \nabla^2)$:

$$\iiint [uL(v) - vL(u)]d^3x = \oiint (u\,\nabla v - v\,\nabla u)\cdot \hat{\mathbf{n}}\, dS, \qquad (10.2.4)$$

where $d^3x = dV = dx\, dy\, dz$. There is a corresponding result for the two-dimensional Laplacian.

To extend these ideas to the nonhomogeneous wave equation, we introduce the appropriate linear differential operator

$$L = \frac{\partial^2}{\partial t^2} - c^2\, \nabla^2. \qquad (10.2.5)$$

Using this notation, the nonhomogeneous wave equation (10.2.1a) satisfies

$$L(u) - Q(\mathbf{x}, t), \qquad (10.2.6)$$

while the Green's function (10.2.2a) satisfies

$$L(G) = \delta(\mathbf{x} - \mathbf{x}_0)\, \delta(t - t_0). \qquad (10.2.7)$$

For the wave operator L [see (10.2.5)] we will derive a Green's formula analogous to (10.2.3) and (10.2.4). We will use a *notation* corresponding to three dimensions, but will make clear modifications (when necessary) for one and two dimensions. For time-dependent problems L has both space and time variables. Formulas analogous to (10.2.3) and (10.2.4) are expected to exist, but integration will occur over both space $\mathbf{x}$ and time t. Since for the wave operator

$$uL(v) - vL(u) = u\frac{\partial^2 v}{\partial t^2} - v\frac{\partial^2 u}{\partial t^2} - c^2(u\,\nabla^2 v - v\,\nabla^2 u),$$

the previous Green's formulas will yield the new **"Green's formula"**:

$$\int_{t_i}^{t_f} \iiint [uL(v) - vL(u)]d^3x\, dt$$
$$= \iiint \left(u\frac{\partial v}{\partial t} - v\frac{\partial u}{\partial t}\right)\bigg|_{t_i}^{t_f} d^3x - c^2 \int_{t_i}^{t_f}\left(\oiint (u\,\nabla v - v\,\nabla u)\cdot \hat{\mathbf{n}}\, dS\right) dt, \qquad (10.2.8)$$

where $\iiint$ indicates integration over the three-dimensional space ($\int_a^b$ for one-dimensional problems) and $\oiint$ indicates integration over its boundary ($\big|_a^b$ for one-dimensional problems). The terms on the right-hand side represent contributions from the boundaries—the spatial boundaries for all time and the temporal boundaries ($t = t_i$ and $t = t_f$) for all space. These space-time boundaries are illustrated (for a one-dimensional problem) in Fig. 10.2.1.

For example, *if* both u and v satisfy the usual type of *homogeneous* boundary conditions (in space, for all time), then $\oint (u \, \nabla v - v \, \nabla u) \cdot \hat{\mathbf{n}} \, dS = 0$, but

$$\iiint \left(u \frac{\partial v}{\partial t} - v \frac{\partial u}{\partial t} \right) \Bigg|_{t_i}^{t_f} d^3x$$

may not equal zero due to contributions from the "initial" time t_i and "final" time t_f.

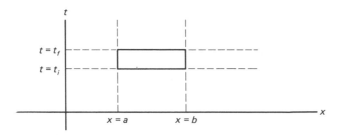

Figure 10.2.1 Space-time boundaries for a one-dimensional wave equation.

10.2.3 Reciprocity

For time-independent problems, we have shown that the Green's function is symmetric, $G(\mathbf{x}, \mathbf{x}_0) = G(\mathbf{x}_0, \mathbf{x})$. We proved this result using Green's formula for two different Green's functions [$G(\mathbf{x}, \mathbf{x}_1)$ and $G(\mathbf{x}, \mathbf{x}_2)$]. The result followed because the boundary terms in Green's formula vanished.

For the wave equation there is a somewhat analogous property. The Green's function $G(\mathbf{x}, t; \mathbf{x}_0, t_0)$ satisfies

$$\frac{\partial^2 G}{\partial t^2} - c^2 \, \nabla^2 G = \delta(\mathbf{x} - \mathbf{x}_0) \, \delta(t - t_0), \tag{10.2.9a}$$

subject to the causality principle,

$$G(\mathbf{x}, t; \mathbf{x}_0, t_0) = 0 \qquad \text{for } t < t_0. \tag{10.2.9b}$$

G will be nonzero for $t > t_0$. To utilize Green's formula (to prove reciprocity), we need a second Green's function. If we choose it to be $G(\mathbf{x}, t; \mathbf{x}_A, t_A)$, then the contribution $\int_{t_i}^{t_f} \oint (u \, \nabla v - v \, \nabla u) \cdot \hat{\mathbf{n}} \, dS \, dt$ on the spatial boundary (or infinity) vanishes, but the contribution

$$\iiint \left(u \frac{\partial v}{\partial t} - v \frac{\partial u}{\partial t} \right) \Bigg|_{t_i}^{t_f} d^3x$$

on the time boundary will not vanish at both $t = t_i$ and $t = t_f$. In time our problem is an initial value problem, not a boundary value problem. If we let $t_i \leq t_0$ in Green's formula, the "initial" contribution will vanish.

For a second Green's function we are interested in varying the source time t, $G(\mathbf{x}, t_1; \mathbf{x}_1, t)$, what we call the **source-varying Green's function**. From the translation property,

$$\boxed{G(\mathbf{x}, t_1; \mathbf{x}_1, t) = G(\mathbf{x}, -t; \mathbf{x}_1, -t_1),}$$ (10.2.10a)

since the elapsed times are the same $[-t - (-t_1) = t_1 - t]$. By causality, these are zero if $t_1 < t$ (or equivalently $-t < -t_1$):

$$G(\mathbf{x}, t_1; \mathbf{x}_1, t) = 0 \qquad t > t_1.$$ (10.2.10b)

We call this the **source-varying causality principle**. By introducing this Green's function, we will show that the "final" contribution from Green's formula may vanish.

To determine the differential equation satisfied by the source-varying Green's function, we let $t = -\tau$, in which case, from (10.2.10a),

$$G(\mathbf{x}, t_1; \mathbf{x}_1, t) = G(\mathbf{x}, \tau; \mathbf{x}_1, -t_1).$$

This is the ordinary (variable response position) Green's function with τ being the time variable. It has a concentrated source located at $\mathbf{x} = \mathbf{x}_1$ when $\tau - -t_1 (t - t_1)$:

$$\left(\frac{\partial^2}{\partial \tau^2} - c^2 \nabla^2\right) G(\mathbf{x}, t_1; \mathbf{x}_1, t) - \delta(\mathbf{x} - \mathbf{x}_1)\, \delta(t - t_1).$$

Since $\tau = -t$, from the chain rule $\partial/\partial\tau = -\partial/\partial t$, but $\partial^2/\partial\tau^2 = \partial^2/\partial t^2$. Thus, the wave operator is symmetric in time, and therefore

$$\boxed{\begin{aligned}\left(\frac{\partial^2}{\partial t^2} - c^2 \nabla^2\right) G(\mathbf{x}, t_1; \mathbf{x}_1, t) &= L[G(\mathbf{x}, t_1; \mathbf{x}_1, t)] \\ &= \delta(\mathbf{x} - \mathbf{x}_1)\, \delta(t - t_1).\end{aligned}}$$ (10.2.10c)

A reciprocity formula results from Green's formula (10.2.8) using two Green's functions, one with varying response time,

$$u = G(\mathbf{x}, t; \mathbf{x}_0, t_0),$$ (10.2.11)

and one with varying source time,

$$v = G(\mathbf{x}, t_1; \mathbf{x}_1, t).$$ (10.2.12)

Both satisfy partial differential equations involving the *same* wave operator, $L = \partial^2/\partial t^2 - c^2 \nabla^2$. We integrate from $t = -\infty$ to $t = +\infty$ in Green's formula (10.2.8) (i.e., $t_i = -\infty$ and $t_f = +\infty$). Since both Green's functions satisfy the same homogeneous boundary conditions, Green's formula (10.2.8) yields

$$\int_{-\infty}^{\infty} \iiint [u\, \delta(\mathbf{x} - \mathbf{x}_1)\, \delta(t - t_1) - v\, \delta(\mathbf{x} - \mathbf{x}_0)\, \delta(t - t_0)] d^3x\, dt$$

$$= \iiint \left(u \frac{\partial v}{\partial t} - v \frac{\partial u}{\partial t}\right)\Bigg|_{-\infty}^{+\infty} d^3x.$$ (10.2.13)

From the causality principles, u and $\partial u/\partial t$ vanish for $t < t_0$ and v and $\partial v/\partial t$ vanish for $t > t_1$. Thus, the r.h.s. of (10.2.13) vanishes. Consequently, using the properties of the Dirac delta function, u at $\mathbf{x} = \mathbf{x}_1$, $t = t_1$ equals v at $\mathbf{x} = \mathbf{x}_0$, $t = t_0$:

$$G(\mathbf{x}_1, t_1; \mathbf{x}_0, t_0) = G(\mathbf{x}_0, t_1; \mathbf{x}_1, t_0), \qquad (10.2.14)$$

the **reciprocity formula** for the Green's function for the wave equation. Assuming that $t_1 > t_0$, the response at $\mathbf{x}_1$ (at time t_1) due to a source at $\mathbf{x}_0$ (at time t_0) is the same as the response at $\mathbf{x}_0$ (at time t_1) due to a source at $\mathbf{x}_1$, *as long as the elapsed times from the sources are the same*. In this case it is seen that interchanging the source and location points has no effect, what we called Maxwell reciprocity for time-independent Green's functions.

10.2.4 Using the Green's Function

As with our earlier work, the relationship between the Green's function and the solution of the nonhomogeneous problem is established using the appropriate Green's formula, (10.2.8). We let

$$u = u(\mathbf{x}, t) \qquad (10.2.15a)$$

$$v = G(\mathbf{x}, t_0; \mathbf{x}_0, t) = G(\mathbf{x}_0, t_0; \mathbf{x}, t), \qquad (10.2.15b)$$

where $u(\mathbf{x}, t)$ is the solution of the nonhomogeneous wave equation satisfying

$$L(u) = Q(\mathbf{x}, t)$$

subject to the given initial conditions for $u(\mathbf{x}, 0)$ and $\partial u/\partial t(\mathbf{x}, 0)$, and where $G(\mathbf{x}, t_0; \mathbf{x}_0, t)$ is the source-varying Green's function satisfying (10.2.10c):

$$L[G(\mathbf{x}, t_0; \mathbf{x}_0, t)] = \delta(\mathbf{x} - \mathbf{x}_0)\,\delta(t - t_0)$$

subject to the source-varying causality principle

$$G(\mathbf{x}, t_0; \mathbf{x}_0, t) = 0 \qquad \text{for } t > t_0.$$

G satisfies homogeneous boundary conditions, but u may not. We use Green's formula (10.2.8) with $t_i = 0$ and $t_f = t_{0^+}$; we integrate just beyond the appearance of a concentrated source at $t = t_0$:

$$\int_0^{t_0^+} \iiint [u(\mathbf{x}, t)\,\delta(\mathbf{x} - \mathbf{x}_0)\,\delta(t - t_0) - G(\mathbf{x}, t_0; \mathbf{x}_0, t)Q(\mathbf{x}, t)]\,d^3x\,dt$$

$$= \iiint \left(u\frac{\partial v}{\partial t} - v\frac{\partial u}{\partial t} \right)\bigg|_0^{t_0^+} d^3x - c^2 \int_0^{t_0^+} \left[\oiint (u\,\nabla v - v\,\nabla u)\cdot\hat{\mathbf{n}}\,dS \right] dt.$$

At $t = t_{0^+}$, $v = 0$ and $\partial v/\partial t = 0$, since we are using the source-varying Green's function. We obtain, using the reciprocity formula (10.2.14),

$$u(\mathbf{x}_0, t_0) = \int_0^{t_0^+} \iiint G(\mathbf{x}_0, t_0; \mathbf{x}, t)Q(\mathbf{x}, t)\,d^3x\,dt$$

$$+ \iiint \left[\frac{\partial u}{\partial t}(\mathbf{x}, 0)G(\mathbf{x}_0, t_0; \mathbf{x}, 0) - u(\mathbf{x}, 0)\frac{\partial}{\partial t}G(\mathbf{x}_0, t_0; \mathbf{x}, 0) \right] d^3x$$

$$- c^2 \int_0^{t_0^+} \left[\oiint \left(u(\mathbf{x}, t)\,\nabla G(\mathbf{x}_0, t_0; \mathbf{x}, t) - G(\mathbf{x}_0, t_0; \mathbf{x}, t)\,\nabla u(\mathbf{x}, t) \right)\cdot\hat{\mathbf{n}}\,dS \right] dt.$$

It can be shown that t_{0+} may be replaced by t_0 in these limits. If the roles of $\mathbf{x}$ and $\mathbf{x}_0$ are interchanged (as well as t and t_0), we obtain a representation formula for $u(\mathbf{x}, t)$ in terms of the Green's function $G(\mathbf{x}, t; \mathbf{x}_0, t_0)$:

$$
\begin{aligned}
u(\mathbf{x}, t) = {} & \int_0^t \iiint G(\mathbf{x}, t; \mathbf{x}_0, t_0) Q(\mathbf{x}_0, t_0)\, d^3 x_0\, dt_0 \\
& + \iiint \left[\frac{\partial u}{\partial t_0}(\mathbf{x}_0, 0) G(\mathbf{x}, t; \mathbf{x}_0, 0) - u(\mathbf{x}_0, 0) \frac{\partial}{\partial t_0} G(\mathbf{x}, t; x_0, 0) \right] d^3 x_0 \\
& - c^2 \int_0^t \left[\oint \left(u(\mathbf{x}_0, t_0)\, \nabla_{\mathbf{x}_0} G(\mathbf{x}, t; \mathbf{x}_0, t_0) - G(\mathbf{x}, t; \mathbf{x}_0, t_0)\, \nabla_{\mathbf{x}_0} u(\mathbf{x}_0, t_0) \right) \cdot \hat{\mathbf{n}}\, dS_0 \right] dt_0.
\end{aligned}
$$

$$(10.2.16)$$

Note that $\nabla_{\mathbf{x}_0}$ means a derivative with respect to the source position. Equation (10.2.16) expresses the response due to the three kinds of nonhomogeneous terms: source terms, initial conditions, and nonhomogeneous boundary conditions. In particular, the initial position $u(\mathbf{x}_0, 0)$ has an influence function

$$
-\frac{\partial}{\partial t_0} G(\mathbf{x}, t; \mathbf{x}_0, 0)
$$

(meaning the source time derivative evaluated initially), while the influence function for the initial velocity is $G(\mathbf{x}, t; \mathbf{x}_0, 0)$.

Furthermore, for example, if u is given on the boundary, then G satisfies the related homogeneous boundary condition; that is, $G = 0$ on the boundary. In this case the boundary term in (10.2.16) simplifies to

$$
-c^2 \int_0^t \left[\oint u(\mathbf{x}_0, t_0)\, \nabla_{\mathbf{x}_0} G(\mathbf{x}, t; \mathbf{x}_0, t_0) \cdot \hat{\mathbf{n}}\, dS_0 \right] dt_0.
$$

The influence function for this nonhomogeneous boundary condition is $-c^2\, \nabla_{\mathbf{x}_0} G(\mathbf{x}, t; \mathbf{x}_0, t_0) \cdot \hat{\mathbf{n}}$. This is $-c^2$ times the source outward normal derivative of the Green's function.

10.2.5 Infinite Space Green's Functions

The Green's function for the wave equation satisfies

$$
\frac{\partial^2 G}{\partial t^2} - c^2 \nabla^2 G = \delta(\mathbf{x} - \mathbf{x}_0)\, \delta(t - t_0) \qquad (10.2.17a)
$$

$$
G(\mathbf{x}, t; \mathbf{x}_0, t_0) = 0 \qquad \text{for } t < t_0. \qquad (10.2.17b)
$$

For infinite space problems, there are no boundary conditions. We solve infinite space problems using Fourier transforms.

Fourier transform of Dirac delta function. In n-dimensional space the Fourier transform relationship is

$$f(\mathbf{x}) = \int F(\boldsymbol{\omega})e^{-i\boldsymbol{\omega}\cdot\mathbf{x}}\, d^n\omega \qquad (10.2.18a)$$

$$F(\boldsymbol{\omega}) = \frac{1}{(2\pi)^n}\int f(\mathbf{x})e^{i\boldsymbol{\omega}\cdot\mathbf{x}}\, d^n x. \qquad (10.2.18b)$$

The results for a Dirac delta function are easy and needed:

$$\mathscr{F}[\delta(\mathbf{x} - \mathbf{x}_0)] = \frac{1}{(2\pi)^n}\int_{-\infty}^{\infty}\delta(\mathbf{x} - \mathbf{x}_0)e^{i\boldsymbol{\omega}\cdot\mathbf{x}}\, d^n x = \frac{e^{i\boldsymbol{\omega}\cdot\mathbf{x}_0}}{(2\pi)^n} \qquad (10.2.19a)$$

$$\delta(\mathbf{x} - \mathbf{x}_0) = \int_{-\infty}^{\infty}\frac{e^{i\boldsymbol{\omega}\cdot\mathbf{x}_0}}{(2\pi)^n}e^{-i\boldsymbol{\omega}\cdot\mathbf{x}}\, d^n\omega = \frac{1}{(2\pi)^n}\int_{-\infty}^{\infty}e^{-i\boldsymbol{\omega}\cdot(\mathbf{x}-\mathbf{x}_0)}\, d^n\omega. \qquad (10.2.19b)$$

The complex amplitude of the Fourier transform of the Dirac delta function, $|e^{i\boldsymbol{\omega}\cdot\mathbf{x}_0}/(2\pi)^n| = 1/(2\pi)^n$, is a constant. Thus, the Dirac delta function is composed of all wave numbers (spatial frequencies) with equal weighting (in an absolute value sense). Equation (10.2.19b) is the Fourier transform representation of the Dirac delta function. Although (10.2.19b) is not valid, strictly speaking, since the integral in (10.2.19b) does not converge to zero for $\mathbf{x} \neq \mathbf{x}_0$, (10.2.19b) may be used with some care. To understand (10.2.19b), let us briefly consider the one-dimensional result

$$\delta(x - x_0) = \frac{1}{2\pi}\int_{-\infty}^{\infty}e^{-i\omega(x - x_0)}\, d\omega.$$

If $x = x_0$, the integrand is 1, yielding ∞ when integrated from $-\infty$ to $+\infty$. This agrees with the Dirac delta function at $x = x_0$. If $x \neq x_0$, then

$$\int_{-\infty}^{\infty}e^{-i\omega(x-x_0)}\, d\omega$$

represents the area under a complex exponential (real part a pure cosine and imaginary part a pure sine). Although the integral does not converge as the limits approach $\pm\infty$, we can think of each positive and negative contribution exactly canceling each other. In that way (10.2.19b) is not inaccurate for $\mathbf{x} \neq \mathbf{x}_0$.

Fourier transform of the Green's function. We introduce the Fourier transform of the Green's function, $G(\mathbf{x}, t; \mathbf{x}_0, t_0)$, and denote it $\overline{G}(\boldsymbol{\omega}, t; \mathbf{x}_0, t_0)$. Instead of solving for the Green's function, we will solve for its Fourier transform. The ordinary differential equation for the transform is obtained by transforming the defining partial differential equation for the Green's function (10.2.17a):

$$\frac{\partial^2\overline{G}}{\partial t^2} + c^2\omega^2\overline{G} = \frac{e^{i\boldsymbol{\omega}\cdot\mathbf{x}_0}}{(2\pi)^n}\delta(t - t_0), \qquad (10.2.20)$$

where

$$\omega^2 = \boldsymbol{\omega}\cdot\boldsymbol{\omega}.$$

The Fourier transform of (10.2.17b) yields a causality principle for the Fourier transform:

$$\overline{G}(\boldsymbol{\omega}, t; \mathbf{x}_0, t_0) = 0 \qquad \text{for } t < t_0. \qquad (10.2.21)$$

Equation (10.2.20) is not difficult to solve. For $t > t_0$,

$$\frac{\partial^2 \overline{G}}{\partial t^2} + c^2 \omega^2 \overline{G} = 0,$$

and hence

$$\overline{G} = \begin{cases} 0 & t < t_0 \\ A \cos c\omega(t - t_0) + B \sin c\omega(t - t_0) & t > t_0, \end{cases}$$

where A and B are functions of $\mathbf{x}_0$, t_0, and $\boldsymbol{\omega}$. The jump conditions at $t = t_0$ determine A and B. $\overline{G}$ is continuous at $t = t_0$:

$$\overline{G}(\boldsymbol{\omega}, t_0; \mathbf{x}_0, t_0) = 0 \quad \Rightarrow \quad 0 = A;$$

and the jump in the derivative is derived by integrating (10.2.20) from $t = t_{0-}$ to $t = t_{0+}$:

$$\frac{\partial \overline{G}}{\partial t}(\boldsymbol{\omega}, t_{0+}; \mathbf{x}_0, t_0) = \frac{e^{i\boldsymbol{\omega} \cdot \mathbf{x}_0}}{(2\pi)^n} \quad \Rightarrow \quad \frac{e^{i\boldsymbol{\omega} \cdot \mathbf{x}_0}}{(2\pi)^n} = c\omega B,$$

where we have used the fact that $\partial \overline{G}/\partial t = 0$ at $t = t_{0-}$. In this way, we obtain the Fourier transform of the Green's function:

$$\overline{G}(\boldsymbol{\omega}, t; \mathbf{x}_0, t_0) = \frac{e^{i\boldsymbol{\omega} \cdot \mathbf{x}_0}}{(2\pi)^n c\omega} \sin c\omega(t - t_0). \tag{10.2.22a}$$

We thus have the following Fourier integral representation of the Green's function:

$$G(\mathbf{x}, t; \mathbf{x}_0, t_0) = \frac{1}{(2\pi)^n} \int \frac{e^{-i[\boldsymbol{\omega} \cdot (\mathbf{x} - \mathbf{x}_0)]} \sin c\omega(t - t_0)}{c\omega} d^n\omega, \tag{10.2.22b}$$

for $t > t_0$. The evaluation of the integral in (10.2.22) depends on the dimension n.

10.2.6 One-Dimensional Infinite Space Green's Function (d'Alembert's Solution)

From (10.2.22a) we obtain the Fourier transform of the Green's function for the one-dimensional wave equation:

$$\overline{G}(\omega, t; x_0, t_0) = \frac{1}{2c} e^{i\omega x_0} \frac{\sin c\omega(t - t_0)}{\pi\omega}. \tag{10.2.23}$$

This Fourier transform can be readily inverted. From our tables (and Exercise 9.4.6), the inverse transform of

$$\frac{\sin c\omega(t - t_0)}{\pi\omega} \quad \text{is} \quad \begin{cases} 0 & |x| > c(t - t_0) \\ 1 & |x| < c(t - t_0). \end{cases}$$

Multiplication by $e^{i\omega x_0}$ corresponds to shifting x to $x - x_0$. Thus,

$$G(x, t; x_0, t_0) = \begin{cases} 0 & |x - x_0| > c(t - t_0) \\ \dfrac{1}{2c} & |x - x_0| < c(t - t_0). \end{cases}$$

The Green's function is a single rectangular-shaped expanding pulse. Initially (at $t = t_0$) it is located at one point, $x = x_0$. Each end spreads out at velocity c. We illustrate this Green's function for the one-dimensional wave equation in

Fig. 10.2.2. It can be represented in terms of the Heaviside unit step function,

$$H(z) = \begin{cases} 0 & z < 0 \\ 1 & z > 0: \end{cases}$$

$$G(x, t; x_0, t_0) = \frac{1}{2c}\{-H[(x - x_0) - c(t - t_0)]$$
$$+ H[(x - x_0) + c(t - t_0)]\}.$$

(10.2.24)

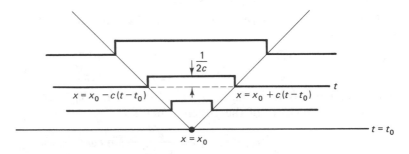

Figure 10.2.2 Green's function for the one-dimensional wave equation.

Example. To illustrate the use of this Green's function, consider the initial value problem for the wave equation without sources on an infinite domain $-\infty < x < \infty$ (see Sec. 9.7.1):

$$\frac{\partial^2 u}{\partial t^2} = c^2 \frac{\partial^2 u}{\partial x^2}$$

$$u(x, 0) = f(x)$$

$$\frac{\partial u}{\partial t}(x, 0) = g(x).$$

In the formula (10.2.16), the boundary contribution* vanishes since $G = 0$ for x sufficiently large (positive or negative); see Fig. 10.2.2. Since there are also no sources, $u(x, t)$ is caused only by the initial conditions:

$$u(x, t) = \int_{-\infty}^{\infty} \left[g(x_0)G(x, t; x_0, 0) - f(x_0)\frac{\partial}{\partial t_0}G(x, t; x_0, 0) \right] dx_0.$$

We need to calculate $\partial/\partial t_0\, G(x, t; x_0, 0)$ from (10.2.24). Using properties of the derivative of a step [see (8.3.28a)], it follows that

$$\frac{\partial}{\partial t_0}G(x, t; x_0, t_0) = \tfrac{1}{2}[-\delta(x - x_0 + c(t - t_0)) - \delta(x - x_0 - c(t - t_0))]$$

* The boundary contribution for an infinite problem is the limit as $L \to \infty$ of the boundaries of a finite region, $-L < x < L$.

and thus,

$$\frac{\partial}{\partial t_0} G(x, t; x_0, 0) = \tfrac{1}{2}[-\delta(x - x_0 + ct) - \delta(x - x_0 - ct)].$$

Finally, we obtain the solution of the initial value problem:

$$u(x, t) = \frac{f(x + ct) + f(x - ct)}{2} + \frac{1}{2c} \int_{x-ct}^{x+ct} g(x_0) \, dx_0. \qquad (10.2.25)$$

This is known as **d'Alembert's solution** of the wave equation. It can be obtained more simply by the method of characteristics (see Chapter 11). There we will discuss the physical interpretation of the one-dimensional wave equation.

Related problems. Semi-infinite or finite problems for the one-dimensional wave equation can be solved by obtaining the Green's function by the method of images. In some cases transform or series techniques may be used. Of greatest usefulness is the method of characteristics.

10.2.7 Three-Dimensional Infinite Space Green's Function (Huygens' Principle)

In Sec. 10.2.5 we obtained the Fourier integral representation of the Green's function for the three-dimensional wave equation:

$$G(\mathbf{x}, t; \mathbf{x}_0, t_0) = \frac{1}{(2\pi)^3} \iiint \frac{e^{-i[\boldsymbol{\omega}\cdot(\mathbf{x}-\mathbf{x}_0)]} \sin c\omega(t - t_0)}{c\omega} \, d^3\omega. \qquad (10.2.26)$$

Spherical coordinates. The integral may be evaluated if we integrate in spherical coordinates (centered at the origin $\omega = 0$ with "North Pole" $\phi = 0$, in the direction $\mathbf{x} - \mathbf{x}_0$, from source point $\mathbf{x}_0$ to physical location $\mathbf{x}$) as illustrated in Fig. 10.2.3. This is a convenient choice because

$$\boldsymbol{\omega} \cdot (\mathbf{x} - \mathbf{x}_0) = |\boldsymbol{\omega}||\mathbf{x} - \mathbf{x}_0| \cos \phi.$$

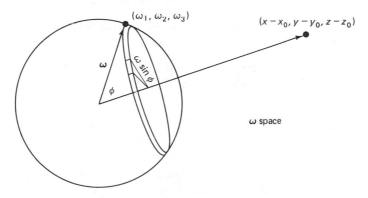

Figure 10.2.3 Spherical coordinates.

We introduce the length ω of the transform vector,

$$\omega = |\boldsymbol{\omega}|, \tag{10.2.27a}$$

and the distance ρ from source to space position,

$$\rho = |\mathbf{x} - \mathbf{x}_0|. \tag{10.2.27b}$$

Thus,

$$\boldsymbol{\omega} \cdot (\mathbf{x} - \mathbf{x}_0) = \omega\rho \cos \phi. \tag{10.2.28}$$

Some facts (as illustrated in Fig. 10.2.3) concerning spherical coordinates are needed. Consider a sphere of fixed radius ω. A constant angle ϕ to the North Pole corresponds to a cone that intersects the sphere along a circle of radius $\omega \sin \phi$. A point in $\boldsymbol{\omega}$-space is uniquely determined by giving its distance ω from the origin, its angle from the North Pole, ϕ, and its polar or *azimuthal* angle θ on the circle (of radius $\omega \sin \phi$) previously defined.

We need an expression for the differential volume in spherical coordinates. In other applications we need the differential surface area on the surface of a sphere (ω constant). As ϕ varies from ϕ to $\phi + \Delta\phi$, a thin "ribbon" of thickness $\omega \Delta\phi$ results. As θ varies from θ to $\theta + \Delta\theta$, a small "patch" on the surface of the sphere results. Its length is the radius of the circle $\omega \sin \phi$ times the angle subtended $\Delta\theta$. If $\Delta\theta$ and $\Delta\phi$ are very small, the area of the patch is approximately the product of its length and width; the surface area is approximately $(\omega \sin \phi \, \Delta\theta)(\omega \, \Delta\phi)$. The resulting small volume, formed between spheres of radius ω and $\omega + \Delta\omega$, has an approximate volume equal to the product of its small surface area $(\omega \sin \phi \, \Delta\theta)(\omega \, \Delta\phi)$ and its small thickness $\Delta\omega$. The correct expression for differential volume in spherical coordinates is thus

$$\boxed{d^3\omega = \omega^2 \sin \phi \, d\phi \, d\theta \, d\omega,} \tag{10.2.29}$$

while the differential surface area (on the sphere) is

$$dS = \omega^2 \sin \phi \, d\phi \, d\theta.$$

Such expressions in spherical coordinates can be found in numerous reference books.

Green's function.   Using (10.2.28) and (10.2.29), the Fourier integral representation of the Green's function (10.2.26) of the three-dimensional wave equation is

$$G(\mathbf{x}, t; \mathbf{x}_0, t_0) = \frac{1}{(2\pi)^3} \iiint \frac{e^{-i\omega\rho \cos \phi} \sin c\omega(t - t_0)}{c\omega} \omega^2 \sin \phi \, d\phi \, d\omega \, d\theta,$$

where $\rho = |\mathbf{x} - \mathbf{x}_0|$. We integrate over the entire space, corresponding to $0 < \phi < \pi$, $0 < \theta < 2\pi$, and $0 < \omega < \infty$, as indicated in Fig. 10.2.3. The integrand does not depend on the polar angle ($\int_0^{2\pi} d\theta = 2\pi$), and the ϕ integration can be performed:

$$\left(\int_0^\pi e^{-i\omega\rho\cos\phi} \omega \sin\phi \, d\phi = \frac{e^{-i\omega\rho\cos\phi}}{i\rho}\bigg|_0^\pi = \frac{e^{+i\omega\rho} - e^{-i\omega\rho}}{i\rho} = \frac{2}{\rho}\sin\omega\rho \right)$$

$$G(\mathbf{x}, t; \mathbf{x}_0, t_0) = \frac{2}{(2\pi)^2 c\rho} \int_0^\infty \sin\omega\rho \sin c\omega(t - t_0) \, d\omega$$

$$= \frac{1}{(2\pi)^2 c\rho} \int_0^\infty \{\cos\omega \, [\rho - c(t - t_0)] - \cos\omega[\rho + c(t - t_0)]\} \, d\omega.$$

We have shown that [see (10.2.19)] $1/2\pi \int_{-\infty}^\infty e^{-i\omega z} \, d\omega = \delta(z)$. The real part of this expression yields $1/\pi \int_0^\infty \cos\omega z \, d\omega = \delta(z)$, where the evenness of cosine was used. Therefore,

$$G(\mathbf{x}, t; \mathbf{x}_0, t_0) = \frac{1}{4\pi c\rho} \{\delta[\rho - c(t - t_0)] - \delta[\rho + c(t - t_0)]\}.$$

However, since $\rho > 0$ and $t > t_0$, the latter Dirac delta function is always zero. Consequently, we derive

$$\boxed{G(\mathbf{x}, t; \mathbf{x}_0, t_0) = \frac{1}{4\pi c\rho} \delta[\rho - c(t - t_0)],} \qquad (10.2.30)$$

where $\rho = |\mathbf{x} - \mathbf{x}_0|$. The Green's function for the three-dimensional wave equation is a spherical shell impulse spreading out from $\rho = 0$ (i.e., $\mathbf{x} = \mathbf{x}_0$) at radial velocity c with "amplitude" diminishing proportional to $1/\rho$.

Huygens' principle. We have shown that a concentrated source at $\mathbf{x}_0$ (at time t_0) only influences the position $\mathbf{x}$ (at time t) if $|\mathbf{x} - \mathbf{x}_0| = c(t - t_0)$. The distance from source to location equals c times the time. The point source emits a wave moving in all directions at velocity c. At time $t - t_0$ later, the source's effect is located on a spherical shell a distance $c(t - t_0)$ away. This is part of what is known as **Huygens' principle.**

Example. To be more specific, let us analyze the effect of sources, $Q(x, t)$. Consider the wave equation with sources in infinite three-dimensional space with zero initial conditions. According to Green's formula (10.2.16),

$$u(\mathbf{x}, t) = \int_0^t \iiint G(\mathbf{x}, t; \mathbf{x}_0, t_0) Q(\mathbf{x}_0, t_0) \, d^3 x_0 \, dt_0, \qquad (10.2.31)$$

since the "boundary" contribution vanishes. Using the infinite three-dimensional space Green's function,

$$u(\mathbf{x}, t) = \frac{1}{4\pi c} \int_0^t \iiint \frac{1}{\rho} \delta[\rho - c(t - t_0)] Q(\mathbf{x}_0, t_0) \, d^3 x_0 \, dt_0,$$

where $\rho = |\mathbf{x} - \mathbf{x}_0|$. The only sources that contribute satisfy $|\mathbf{x} - \mathbf{x}_0| = c(t - t_0)$. The effect at $\mathbf{x}$ at time t is caused by all received sources; the velocity of propagation of each source is c.

10.2.8 Summary

The Green's functions for the wave equation in one and three dimensions are similar. In both, information propagates at velocity c. However, in three dimensions the influence of a concentrated source is only felt on the surface of the expanding sphere. In one dimension, the influence is felt uniformly inside the expanding pulse.

The Green's function for the two-dimensional wave equation is more difficult to analyze. From Exercise 10.2.12 it can be shown that the influence of a concentrated source is felt within a circle expanding also at radial velocity c. The largest effect occurs on the circumference, but the effect diminishes behind the pulse. Huygens' principle is valid only in three dimensions.

EXERCISES 10.2

10.2.1. (a) Show that for $G(\mathbf{x}, t; \mathbf{x}_0, t_0)$, $\partial G/\partial t = -\partial G/\partial t_0$.
  (b) Use part (a) to show that the response due to $u(\mathbf{x}, 0) = f(\mathbf{x})$ is the time derivative of the response due to $\partial u/\partial t\,(\mathbf{x}, 0) = f(\mathbf{x})$.

10.2.2. Express (10.2.16) for a one-dimensional problem.

10.2.3. If $G(\mathbf{x}, t; \mathbf{x}_0, t_0) = 0$ for $\mathbf{x}$ on the boundary, explain why the corresponding term in (10.2.16) vanishes (for any $\mathbf{x}$).

10.2.4. For the one-dimensional wave equation, sketch $G(x, t; x_0, t_0)$ as a function of:
  (a) x with t fixed (x_0, t_0 fixed)
  (b) t with x fixed (x_0, t_0 fixed)

10.2.5. (a) For the one-dimensional wave equation, for what values of x_0 (x, t, t_0 fixed) is $G(x, t; x_0, t_0) \neq 0$?
  (b) Determine the answer to part (a) using the reciprocity property.

10.2.6. (a) Solve

$$\frac{\partial^2 u}{\partial t^2} = c^2 \frac{\partial^2 u}{\partial x^2} + Q(x, t) \qquad -\infty < x < \infty$$

  with $u(x, 0) = 0$ and $\dfrac{\partial u}{\partial t}(x, 0) = 0$.

  *(b) What space-time locations of the source $Q(x, t)$ influence u at position x_1 and time t_1?

10.2.7. Reconsider Exercise 10.2.6 if $Q(x, t) = g(x)e^{-i\omega t}$.
  *(a) Solve for $u(x, t)$. Show that the influence function for $g(x)$ is an outward-propagating wave.
  (b) Instead, determine a particular solution of the form $u(x, t) = \psi(x)e^{-i\omega t}$. (See Exercise 8.3.13.)
  (c) Compare parts (a) and (b).

10.2.8. *(a) In three-dimensional infinite space, solve

$$\frac{\partial^2 u}{\partial t^2} = c^2 \nabla^2 u + g(\mathbf{x})e^{-i\omega t}$$

  with zero initial conditions, $u(\mathbf{x}, 0) = 0$ and $\partial u/\partial t\,(\mathbf{x}, 0) = 0$. From your solution, show that the influence function for $g(x)$ is an outward-propagating wave.

(b) Compare with Exercise 8.5.10.

10.2.9. Consider the Green's function $G(\mathbf{x}, t; \mathbf{x}_0, t_0)$ for the wave equation. From (10.2.16) we easily obtain the influence functions for $Q(\mathbf{x}_0, t_0)$, $u(\mathbf{x}_0, 0)$, and $\partial u/\partial t_0 (\mathbf{x}_0, 0)$. These results may be obtained in the following alternative way:

(a) For $t > t_{0+}$, show that

$$\frac{\partial^2 G}{\partial t^2} = c^2 \nabla^2 G, \tag{10.2.32a}$$

where (by integrating from t_{0-} to t_{0+})

$$G(\mathbf{x}, t_{0+}; \mathbf{x}_0, t_0) = 0 \tag{10.2.32b}$$

$$\frac{\partial G}{\partial t}(\mathbf{x}, t_{0+}; \mathbf{x}_0, t_0) = \delta(\mathbf{x} - \mathbf{x}_0). \tag{10.2.32c}$$

From (10.2.32), briefly explain why $G(\mathbf{x}, t; \mathbf{x}_0, 0)$ is the influence function for $\partial u/\partial t_0(\mathbf{x}_0, 0)$.

(b) Let $\phi = \partial G/\partial t$. Show that for $t > t_{0+}$,

$$\frac{\partial^2 \phi}{\partial t^2} = c^2 \nabla^2 \phi \tag{10.2.33a}$$

$$\phi(\mathbf{x}, t_{0+}; \mathbf{x}_0, t_0) = \delta(\mathbf{x} - \mathbf{x}_0) \tag{10.2.33b}$$

$$\frac{\partial \phi}{\partial t}(\mathbf{x}, t_{0+}; \mathbf{x}_0, t_0) = 0. \tag{10.2.33c}$$

From (10.2.33), briefly explain why $-\partial G/\partial t_0(\mathbf{x}, t; \mathbf{x}_0, 0)$ is the influence function for $u(\mathbf{x}_0, 0)$.

10.2.10. Consider

$$\frac{\partial^2 u}{\partial t^2} = c^2 \frac{\partial^2 u}{\partial x^2} + Q(x, t) \qquad x > 0$$

$$u(x, 0) = f(x)$$

$$\frac{\partial u}{\partial t}(x, 0) = g(x)$$

$$u(0, t) = h(t).$$

(a) Determine the appropriate Green's function using the method of images.

***(b)** Solve for $u(x, t)$ if $Q(x, t) = 0$, $f(x) = 0$, and $g(x) = 0$.

(c) For what values of t does $h(t)$ influence $u(x_1, t_1)$? Briefly interpret physically.

10.2.11. Reconsider Exercise 10.2.10:

(a) if $Q(x, t) \neq 0$, but $f(x) = 0$, $g(x) = 0$ and $h(t) = 0$

(b) if $f(x) \neq 0$, but $Q(x, t) = 0$, $g(x) = 0$, and $h(t) = 0$

(c) if $g(x) \neq 0$, but $Q(x, t) = 0$, $f(x) = 0$, and $h(t) = 0$

10.2.12. Consider the Green's function $G(\mathbf{x}, t; \mathbf{x}_1, t_1)$ for the *two*-dimensional wave equation as the solution of the following *three*-dimensional wave equation:

$$\frac{\partial^2 u}{\partial t^2} = c^2 \nabla^2 u + Q(\mathbf{x}, t)$$

$$u(\mathbf{x}, 0) = 0$$

$$\frac{\partial u}{\partial t}(\mathbf{x}, 0) = 0$$

$$Q(\mathbf{x}, t) = \delta(x - x_1)\,\delta(y - y_1)\,\delta(t - t_1).$$

We will solve for the two-dimensional Green's function by this **method of descent** (descending from three dimensions to two dimensions).

*(a) Solve for $G(\mathbf{x}, t; \mathbf{x}_1, t_1)$ using the general solution of the three-dimensional wave equation. Here, the source $Q(\mathbf{x}, t)$ may be interpreted either as a point source in two dimensions or a line source in three dimensions. [*Hint:* $\int_{-\infty}^{\infty} \cdots dz_0$ may be evaluated by introducing the three-dimensional distance ρ from the point source,

$$\rho^2 = (x - x_1)^2 + (y - y_1)^2 + (z - z_0)^2.]$$

(b) Show that G is only a function of the elapsed time $t - t_1$ and the two-dimensional distance r from the line source,

$$r^2 = (x - x_1)^2 + (y - y_1)^2.$$

(c) Where is the effect of an impulse felt after a time τ has elapsed? Compare to the one- and three-dimensional problems.

(d) Sketch G for $t - t_1$ fixed.

(e) Sketch G for r fixed.

10.2.13. Consider the three-dimensional wave equation. Determine the response to a unit point source moving at the constant velocity $\mathbf{v}$:

$$Q(\mathbf{x}, t) = \delta(\mathbf{x} - \mathbf{v}t).$$

10.2.14. Solve the wave equation in infinite three-dimensional space without sources, subject to the initial conditions

(a) $u(\mathbf{x}, 0) = 0$ and $\dfrac{\partial u}{\partial t}(\mathbf{x}, 0) = g(\mathbf{x})$.

The answer is called **Kirchoff's formula**, although it is due to Poisson (according to Weinberger [1965]).

(b) $u(\mathbf{x}, 0) = f(\mathbf{x})$ and $\dfrac{\partial u}{\partial t}(\mathbf{x}, 0) = 0$ [*Hint:* Use (10.2.16).]

(c) Solve part (b) in the following manner. Let $v(\mathbf{x}, t) = \dfrac{\partial}{\partial t} u(\mathbf{x}, t)$, where $u(\mathbf{x}, t)$ satisfies part (a). [*Hint:* Show that $v(\mathbf{x}, t)$ satisfies the wave equation with $v(\mathbf{x}, 0) = g(\mathbf{x})$ and $\dfrac{\partial v}{\partial t}(\mathbf{x}, 0) = 0$.]

10.2.15. Derive the one-dimensional Green's function for the wave equation by considering a three-dimensional problem with $Q(\mathbf{x}, t) = \delta(x - x_1)\, \delta(t - t_1)$. [*Hint:* Use polar coordinates for the y_0, z_0 integration centered at $y_0 = y, z_0 = z$.]

10.3 GREEN'S FUNCTIONS FOR THE HEAT EQUATION

10.3.1 Introduction

We are interested in solving the heat equation with possibly time-dependent sources,

$$\frac{\partial u}{\partial t} = k\,\nabla^2 u + Q(\mathbf{x}, t), \tag{10.3.1}$$

subject to the initial condition $u(\mathbf{x}, 0) = g(\mathbf{x})$. We will analyze this problem in one, two, and three spatial dimensions. In this subsection we do not specify the geometric region or the possibly nonhomogeneous boundary conditions. There can be three nonhomogeneous terms: the source $Q(\mathbf{x}, t)$, the initial condition, and the boundary conditions.

We define the **Green's function** $G(\mathbf{x}, t; \mathbf{x}_0, t_0)$ as the solution of

$$\frac{\partial G}{\partial t} = k \nabla^2 G + \delta(\mathbf{x} - \mathbf{x}_0)\, \delta(t - t_0) \qquad (10.3.2a)$$

on the same region with the related homogeneous boundary conditions. Since the Green's function represents the temperature response at $\mathbf{x}$ (at time t) due to a concentrated thermal source at $\mathbf{x}_0$ (at time t_0), we will insist that this Green's function is zero before the source acts:

$$G(\mathbf{x}, t; \mathbf{x}_0, t_0) = 0 \qquad \text{for } t < t_0, \qquad (10.3.2b)$$

the **causality principle**.

Furthermore, we show that only the elapsed time $t - t_0$ (from the initiation time $t = t_0$) is needed:

$$G(\mathbf{x}, t; \mathbf{x}_0, t_0) = G(\mathbf{x}, t - t_0; \mathbf{x}_0, 0), \qquad (10.3.3)$$

the **translation property**. Equation (10.3.3) is shown by letting $T = t - t_0$, in which case the Green's function $G(\mathbf{x}, t; \mathbf{x}_0, t_0)$ satisfies

$$\frac{\partial G}{\partial T} = k \nabla^2 G + \delta(\mathbf{x} - \mathbf{x}_0)\, \delta(T)$$

$$G = 0 \qquad \text{for } T < 0.$$

This is precisely the response due to a concentrated source at $\mathbf{x} = \mathbf{x}_0$ at $T = 0$, implying (10.3.3).

We postpone until later subsections the actual calculation of the Green's function. For now, we will assume that the Green's function is known and ask how to represent the temperature $u(\mathbf{x}, t)$ in terms of the Green's function.

10.3.2 Non-Self-Adjoint Nature of the Heat Equation

To show how this problem relates to others discussed in this book, we introduce the linear operator notation,

$$L = \frac{\partial}{\partial t} - k\nabla^2, \qquad (10.3.4)$$

called the **heat** or **diffusion operator**. In previous problems the relation between the solution of the nonhomogeneous problem and its Green's function was based on Green's formulas. We have solved problems in which L is the Sturm-Liouville operator, the Laplacian, and most recently the wave operator.

The heat operator L is composed of two parts. ∇^2 is easily analyzed by Green's formula for the Laplacian [see (10.2.4)]. However, as innocuous as $\partial/\partial t$ appears, it is *much* harder to analyze than any of the other previous operators. To illustrate the difficulty presented by first derivatives, consider

$$L = \frac{\partial}{\partial t}.$$

For second-order Sturm-Liouville operators, elementary integrations yielded Green's formula. The same idea for $L = \partial/\partial t$ will not work. In particular,

$$\int [uL(v) - vL(u)] \, dt = \int \left(u \frac{\partial v}{\partial t} - v \frac{\partial u}{\partial t} \right) dt$$

cannot be simplified. There is no formula to evaluate $\int [uL(v) - vL(u)] \, dt$, the operator $L = \partial/\partial t$ is *not* self-adjoint. Instead, by standard integration by parts,

$$\int_a^b uL(v) \, dt = \int_a^b u \frac{\partial v}{\partial t} \, dt = uv \Big|_a^b - \int_a^b v \frac{\partial u}{\partial t} \, dt,$$

and thus

$$\int_a^b \left(u \frac{\partial v}{\partial t} + v \frac{\partial u}{\partial t} \right) dt = uv \Big|_a^b. \tag{10.3.5}$$

For the operator $L = \partial/\partial t$ we introduce the **adjoint operator**,

$$L^* = -\frac{\partial}{\partial t}.$$

From (10.3.5),

$$\int_a^b [uL^*(v) - vL(u)] \, dt = -uv \Big|_a^b.$$

This is analogous to Green's formula.†

10.3.3 Green's Formula

We now return to the nonhomogeneous heat problem:

$$L(u) = Q(\mathbf{x}, t) \tag{10.3.6}$$

$$L(G) = \delta(\mathbf{x} - \mathbf{x}_0) \, \delta(t - t_0), \tag{10.3.7}$$

---

† For a first-order operator, typically there is only one "boundary condition," $u(a) = 0$. For the integrated-by-parts term to vanish, we must introduce an *adjoint boundary condition*, $v(b) = 0$.

where

$$L = \frac{\partial}{\partial t} - k \nabla^2.$$

For the nonhomogeneous heat equation, our results are more complicated since we must introduce the **adjoint heat operator**,

$$L^* = -\frac{\partial}{\partial t} - k \nabla^2.$$

By a direct calculation

$$uL^*(v) - vL(u) = -u\frac{\partial v}{\partial t} - v\frac{\partial u}{\partial t} + k(v \nabla^2 u - u \nabla^2 v),$$

and thus

$$\int_{t_i}^{t_f} \iiint [uL^*(v) - vL(u)] \, d^3x \, dt$$
$$= -\iiint uv \Big|_{t_i}^{t_f} d^3x + k\int_{t_i}^{t_f} \oiint (v \nabla u - u\nabla v) \cdot \hat{\mathbf{n}} \, dS \, dt. \qquad (10.3.8)$$

We have integrated over all space and from some time $t = t_i$ to another time $t = t_f$. We have used (10.3.5) for the $\partial/\partial t$ terms and Green's formula (10.2.4) for the ∇^2 operator. The "boundary contributions" are of two types, the spatial part (over $\oiint$†) and a temporal part (at the initial t_i and final t_f times). *If* both u and v satisfy the same homogeneous boundary condition (of the usual types), then the spatial contributions vanish,

$$\oiint (v \nabla u - u \nabla v) \cdot \hat{\mathbf{n}} \, dS = 0.$$

Equation (10.3.8) will involve initial contributions (at $t = t_i$) and final contributions $(t = t_f)$.

10.3.4 Adjoint Green's Function

In order to eventually derive a representation formula for $u(\mathbf{x}, t)$ in terms of the Green's function $G(\mathbf{x}, t; \mathbf{x}_0, t_0)$, we must consider summing up various source times. Thus, we consider the **source-varying Green's function**,

$$G(\mathbf{x}, t_1; \mathbf{x}_1, t) = G(\mathbf{x}, -t; \mathbf{x}_1, -t_1),$$

where the translation property has been used. This is precisely the procedure we employed when analyzing the wave equation [see (10.2.10a)]. By causality,

---

† For infinite or semi-infinite geometries, we consider finite regions in some appropriate limit. The boundary terms at infinity will vanish if u and v decay sufficiently fast.

these are zero if $t > t_1$:

$$G(\mathbf{x}, t_1; \mathbf{x}_1, t) = 0 \qquad \text{for } t > t_1. \qquad (10.3.9)$$

Letting $\tau = -t$, we see that the source-varying Green's function $G(\mathbf{x}, t_1; \mathbf{x}_1, t)$ satisfies

$$\left(-\frac{\partial}{\partial t} - k \nabla^2 \right) G(\mathbf{x}, t_1; \mathbf{x}_1, t) = \delta(\mathbf{x} - \mathbf{x}_1)\, \delta(t - t_1), \qquad (10.3.10)$$

as well as the source-varying causality principle (10.3.9). The heat operator L does not occur. Instead, the adjoint heat operator L^* appears:

$$\boxed{L^*[G(\mathbf{x}, t_1; \mathbf{x}_1, t)] = \delta(\mathbf{x} - \mathbf{x}_1)\, \delta(t - t_1).} \qquad (10.3.11)$$

We see that $G(\mathbf{x}, t_1; \mathbf{x}_1, t)$ is the Green's function for the adjoint heat operator (with the source-varying causality principle). Sometimes it is called the **adjoint Green's function**, $G^*(\mathbf{x}, t; \mathbf{x}_1, t_1)$. However, it is unnecessary to ever calculate or use it since

$$G^*(\mathbf{x}, t; \mathbf{x}_1, t_1) = G(\mathbf{x}, t_1; \mathbf{x}_1, t) \qquad (10.3.12)$$

and both are zero for $t > t_1$.

10.3.5 Reciprocity

As with the wave equation, we derive a reciprocity formula. Here, there are some small differences because of the occurrence of the adjoint operator in Green's formula, (10.3.8). In (10.3.8) we introduce

$$u = G(\mathbf{x}, t; \mathbf{x}_0, t_0) \qquad (10.3.13a)$$

$$v = G(\mathbf{x}, t_1; \mathbf{x}_1, t), \qquad (10.3.13b)$$

the latter having been shown to be the source-varying or adjoint Green's function. Thus, the defining properties for u and v are:

$$L(u) = \delta(\mathbf{x} - \mathbf{x}_0)\, \delta(t - t_0) \qquad L^*(v) = \delta(\mathbf{x} - \mathbf{x}_1)\, \delta(t - t_1)$$

$$u = 0 \quad \text{for } t < t_0 \qquad\qquad v = 0 \quad \text{for } t > t_1.$$

We integrate from $t = -\infty$ to $t = +\infty$ [i.e., $t_i = -\infty$ and $t_f = +\infty$ in (10.3.8)], obtaining

$$\int_{-\infty}^{\infty} \iiint [G(\mathbf{x}, t; \mathbf{x}_0, t_0)\, \delta(\mathbf{x} - \mathbf{x}_1)\, \delta(t - t_1) - G(\mathbf{x}, t_1; \mathbf{x}_1, t)\, \delta(\mathbf{x} - \mathbf{x}_0)\, \delta(t - t_0)]\, d^3x\, dt$$

$$= -\iiint G(\mathbf{x}, t; \mathbf{x}_0, t_0) G(\mathbf{x}, t_1; \mathbf{x}_1, t) \Big|_{t=-\infty}^{t=\infty} d^3x,$$

since u and v both satisfy the same homogeneous boundary conditions, so that

$$\oiint (v\, \nabla u - u\, \nabla v) \cdot \hat{\mathbf{n}}\, dS$$

vanishes. The contributions also vanish at $t = \pm\infty$ due to causality. Using the properties of the Dirac delta function, we obtain **reciprocity**:

$$\boxed{G(\mathbf{x}_1, t_1; \mathbf{x}_0, t_0) = G(\mathbf{x}_0, t_1; \mathbf{x}_1, t_0).}$$

(10.3.14)

As we have shown for the wave equation [see (10.2.14)], interchanging the source and location positions does not alter the responses if the elapsed times from the sources are the same. In this sense the Green's function for the heat (diffusion) equation is symmetric.

10.3.6 Representation of the Solution Using Green's Functions

To obtain the relationship between the solution of the nonhomogeneous problem and the Green's function, we apply Green's formula (10.3.8) with u satisfying (10.3.1) subject to nonhomogeneous boundary and initial conditions. We let v equal the source-varying or adjoint Green's function, $v = G(\mathbf{x}, t_0; \mathbf{x}_0, t)$. Using the defining differential equations (10.3.6) and (10.3.7), Green's formula (10.3.8) becomes

$$\int_0^{t_0^+} \iiint [u\,\delta(\mathbf{x} - \mathbf{x}_0)\,\delta(t - t_0) - G(\mathbf{x}, t_0; \mathbf{x}_0, t)Q(\mathbf{x}, t)]\, d^3x\, dt$$

$$= \iiint u(\mathbf{x}, 0)G(\mathbf{x}, t_0; \mathbf{x}_0, 0)\, d^3x$$

$$+ k \int_0^{t_0^+} \oiint [G(\mathbf{x}, t_0; \mathbf{x}_0, t)\,\nabla u - u\,\nabla G(\mathbf{x}, t_0; \mathbf{x}_0, t)] \cdot \hat{\mathbf{n}}\, dS\, dt,$$

since $G = 0$ for $t > t_0$. Solving for u, we obtain

$$u(\mathbf{x}_0, t_0) = \int_0^{t_0} \iiint G(\mathbf{x}, t_0; \mathbf{x}_0, t)Q(\mathbf{x}, t)\, d^3x\, dt$$

$$+ \iiint u(\mathbf{x}, 0)G(\mathbf{x}, t_0; \mathbf{x}_0, 0)\, d^3x$$

$$+ k \int_0^{t_0} \oiint [G(\mathbf{x}, t_0; \mathbf{x}_0, t)\,\nabla u - u\,\nabla G(\mathbf{x}, t_0; \mathbf{x}_0, t)] \cdot \hat{\mathbf{n}}\, dS\, dt.$$

It can be shown that the limits t_{0+} may be replaced by t_0. We now (as before) interchange $\mathbf{x}$ with $\mathbf{x}_0$ and t with t_0. In addition, we use reciprocity and derive

$$\boxed{\begin{aligned} u(\mathbf{x}, t) &= \int_0^t \iiint G(\mathbf{x}, t; \mathbf{x}_0, t_0)Q(\mathbf{x}_0, t_0)\, d^3x_0\, dt_0 \\ &\quad + \iiint G(\mathbf{x}, t; \mathbf{x}_0, 0)u(\mathbf{x}_0, 0)\, d^3x_0 \\ &\quad + k \int_0^t \oiint [G(\mathbf{x}, t; \mathbf{x}_0, t_0)\,\nabla_{\mathbf{x}_0} u \\ &\quad - u(\mathbf{x}_0, t_0)\,\nabla_{\mathbf{x}_0} G(\mathbf{x}, t; \mathbf{x}_0, t_0)] \cdot \hat{\mathbf{n}}\, dS_0\, dt_0. \end{aligned}}$$

(10.3.15)

Equation (10.3.15) illustrates how the temperature $u(\mathbf{x}, t)$ is affected by the three nonhomogeneous terms. The Green's function $G(\mathbf{x}, t; \mathbf{x}_0, t_0)$ is the influence function for the source term $Q(\mathbf{x}_0, t_0)$ as well as for the initial temperature distribution $u(\mathbf{x}_0, 0)$ (if we evaluate the Green's function at $t_0 = 0$, as is quite reasonable). Furthermore, nonhomogeneous boundary conditions are accounted for by the term $k \int_0^t \oiint (G \nabla_{\mathbf{x}_0} u - u \nabla_{\mathbf{x}_0} G) \cdot \hat{\mathbf{n}} \, dS_0 \, dt_0$. Equation (10.3.15) illustrates the causality principle; at time t, the sources and boundary conditions have an effect only for $t_0 < t$. Equation (10.3.15) generalizes the results obtained by the method of eigenfunction expansion in Sec. 8.2 for the one-dimensional heat equation on a finite interval with zero boundary conditions.

Example. Both u and its normal derivative seem to be needed on the boundary. To clarify the effect of the nonhomogeneous boundary conditions, we consider an example in which the temperature is specified along the entire boundary:

$$u(\mathbf{x}, t) = u_B(\mathbf{x}, t) \qquad \text{along the boundary.}$$

The Green's function satisfies the related homogeneous boundary conditions, in this case

$$G(\mathbf{x}, t; \mathbf{x}_0, t_0) = 0 \qquad \text{for all } \mathbf{x} \text{ along the boundary.}$$

Thus, the effect of this imposed temperature distribution is

$$-k \int_0^t \oiint u_B(\mathbf{x}_0, t_0) \, \nabla_{\mathbf{x}_0} G(\mathbf{x}, t; \mathbf{x}_0, t_0) \cdot \hat{\mathbf{n}} \, dS_0 \, dt_0.$$

The influence function for the nonhomogeneous boundary conditions is minus k times the normal derivative of the Green's function (a dipole distribution).

One-dimensional case. It may be helpful to illustrate the modifications necessary for one-dimensional problems. Volume integrals $\iiint d^3x_0$ become one-dimensional integrals $\int_a^b dx_0$. Boundary contributions on the closed surface $\oiint dS_0$ become contributions at the two ends $x = a$ and $x = b$. For example, if the temperature is prescribed at both ends, $u(a, t) = A(t)$ and $u(b, t) = B(t)$, then these nonhomogeneous boundary conditions influence the temperature $u(\mathbf{x}, t)$:

$$-k \int_0^t \oiint u_B(\mathbf{x}_0, t_0) \, \nabla_{\mathbf{x}_0} G(\mathbf{x}, t; \mathbf{x}_0, t_0) \cdot \hat{\mathbf{n}} \, dS_0 \, dt_0 \text{ becomes}$$

$$-k \int_0^t \left[B(t_0) \frac{\partial G}{\partial x_0}(x, t; b, t_0) - A(t_0) \frac{\partial G}{\partial x_0}(x, t; a, t_0) \right] dt_0.$$

This agrees with results that could be obtained by the method of eigenfunction expansions (Chapter 7) for nonhomogeneous boundary conditions.

10.3.7 Green's Function for the Heat Equation on an Infinite Domain

The Green's function $G(\mathbf{x}, t; \mathbf{x}_0, t_0)$ for the heat equation without boundaries satisfies

$$\frac{\partial G}{\partial t} = k \nabla^2 G + \delta(\mathbf{x} - \mathbf{x}_0) \, \delta(t - t_0) \qquad (10.3.16a)$$

subject to the causality principle

$$G(\mathbf{x}, t; \mathbf{x}_0, t_0) = 0 \qquad \text{for } t < t_0. \tag{10.3.16b}$$

We will obtain the solution in one, two, and three dimensions. To solve this problem we introduce the Fourier transform of the Green's function

$$\overline{G}(\boldsymbol{\omega}, t; \mathbf{x}_0, t_0) = \frac{1}{(2\pi)^n} \int_{-\infty}^{\infty} G(\mathbf{x}, t; \mathbf{x}_0, t_0) e^{i\boldsymbol{\omega}\cdot\mathbf{x}} \, d^n x \tag{10.3.17a}$$

$$G(\mathbf{x}, t; \mathbf{x}_0, t_0) = \int_{-\infty}^{\infty} \overline{G}(\boldsymbol{\omega}, t; \mathbf{x}_0, t_0) e^{-i\boldsymbol{\omega}\cdot\mathbf{x}} \, d^n \omega. \tag{10.3.17b}$$

If we Fourier transform (10.3.16a) using techniques of Chapter 9, we obtain an ordinary differential equation for the transform of the Green's function:

$$\frac{\partial \overline{G}}{\partial t} = -k\omega^2 \overline{G} + \frac{e^{i\boldsymbol{\omega}\cdot\mathbf{x}_0}}{(2\pi)^n} \delta(t - t_0), \tag{10.3.18a}$$

where $\omega^2 = \boldsymbol{\omega} \cdot \boldsymbol{\omega}$ and where the Fourier transform of the n-dimensional Dirac delta function (10.2.19a) has been utilized. In addition, by transforming (10.3.16b), we get

$$\overline{G}(\boldsymbol{\omega}, t; \mathbf{x}_0, t_0) = 0 \qquad \text{for } t < t_0; \tag{10.3.18b}$$

the Fourier transform of the Green's function also solves a causality principle. There are many ways to solve (10.3.18a). The transform of the Green's function solves a homogeneous differential equation for $t > t_0$:

$$\frac{\partial \overline{G}}{\partial t} = -k\omega^2 G. \tag{10.3.19}$$

The general solution of (10.3.19) is

$$\overline{G}(\boldsymbol{\omega}, t; \mathbf{x}_0, t_0) = C(\boldsymbol{\omega}) e^{-k\omega^2(t - t_0)}.$$

$C(\boldsymbol{\omega})$ is usually determined by initial conditions at $t = t_0$. Since $\overline{G} = 0$ for $t < t_0$ and the Dirac delta function acts at $t = t_0$, here $C(\boldsymbol{\omega})$ is determined by calculating the jump condition at $t = t_0$. By integrating (10.3.18a) from $t = t_{0-}$ to $t = t_{0+}$, we obtain

$$\overline{G}(t_{0+}) - \overline{G}(t_{0-}) = \frac{e^{i\boldsymbol{\omega}\cdot\mathbf{x}_0}}{(2\pi)^n}.$$

However, the causality principle implies that $G(t_{0-}) = 0$. Thus, $C(\boldsymbol{\omega})$ is determined using $\overline{G}(t_{0+}) = e^{i\boldsymbol{\omega}\cdot\mathbf{x}_0}/(2\pi)^n$. It follows that the Fourier transform of the Green's function is given by

$$\overline{G}(\boldsymbol{\omega}, t; \mathbf{x}_0, t_0) = \frac{e^{i\boldsymbol{\omega}\cdot\mathbf{x}_0}}{(2\pi)^n} e^{-k\omega^2(t - t_0)}.$$

Using the inverse Fourier transform, we obtain the Fourier transform representation of the Green's function:

$$G(\mathbf{x}, t; \mathbf{x}_0, t_0) = \int_{-\infty}^{\infty} \frac{e^{-k\omega^2(t - t_0)}}{(2\pi)^n} e^{-i\boldsymbol{\omega}\cdot(\mathbf{x} - \mathbf{x}_0)} \, d^n \omega.$$

The Green's function depends on $\mathbf{x} - \mathbf{x}_0$; its transform is a Gaussian. Thus,

$G(\mathbf{x}, t; \mathbf{x}_0, t_0)$ is a Gaussian centered at $\mathbf{x} = \mathbf{x}_0$:

$$G(\mathbf{x}, t; \mathbf{x}_0, t_0) = \frac{1}{(2\pi)^n} \left[\frac{\pi}{k(t - t_0)} \right]^{n/2} e^{-(\mathbf{x} - \mathbf{x}_0)^2/4k(t - t_0)}. \tag{10.3.20}$$

where n is the dimension ($n = 1, 2, 3$). this Green's function shows the symmetry of the response and source positions as long as the elapsed time is the same. As with one-dimensional problems discussed in Sec. 9.4, the influence of a concentrated heat source diminishes exponentially as one moves away from the source. For small times (t near t_0) the decay is especially strong.

In this manner we can obtain the solution of the heat equation with sources on an infinite domain:

$$\frac{\partial u}{\partial t} = k \nabla^2 u + Q(\mathbf{x}, t)$$

$$u(\mathbf{x}, 0) = f(\mathbf{x}).$$

According to (10.3.15) and (10.3.20), the solution is

$$\begin{aligned} u(\mathbf{x}, t) &= \int_0^t \int_{-\infty}^{\infty} \left[\frac{1}{4\pi k(t - t_0)} \right]^{n/2} e^{-(\mathbf{x} - \mathbf{x}_0)^2/4k(t - t_0)} Q(\mathbf{x}_0, t_0) \, d^n x_0 \, dt_0 \\ &\quad + \int_{-\infty}^{\infty} \left(\frac{1}{4\pi kt} \right)^{n/2} e^{-(\mathbf{x} - \mathbf{x}_0)^2/4kt} f(\mathbf{x}_0) \, d^n x_0. \end{aligned} \tag{10.3.21}$$

If $Q(\mathbf{x}, t) = 0$, this simplifies to the solution obtained in Chapter 8 using Fourier transforms directly without using the Green's function.

10.3.8 Green's Function for the Heat Equation (Semi-Infinite Domain)

In this subsection we obtain the Green's function needed to solve the nonhomogeneous heat equation on the semi-infinite interval in one dimension ($x > 0$), subject to a nonhomogeneous boundary condition at $x = 0$:

$$\text{PDE:} \quad \frac{\partial u}{\partial t} = k \frac{\partial^2 u}{\partial x^2} + Q(x, t) \qquad x > 0, \tag{10.3.22a}$$

$$\text{BC:} \quad u(0, t) = A(t) \tag{10.3.22b}$$

$$\text{IC:} \quad u(x, 0) = f(x). \tag{10.3.22.c}$$

Equation (10.3.15) can be used to determine $u(x, t)$ if we can obtain the Green's function. The Green's function $G(x, t; x_0, t_0)$ is the response due to a concentrated source:

$$\frac{\partial G}{\partial t} = k \frac{\partial^2 G}{\partial x^2} + \delta(x - x_0)\,\delta(t - t_0).$$

The Green's function satisfies the corresponding *homogeneous* boundary condition,

$$G(0, t; x_0, t_0) = 0,$$

and the causality principle,

$$G(x, t; x_0, t_0) = 0 \qquad \text{for } t < t_0.$$

The Green's function is determined by the method of images (see Sec. 8.5.8). Instead of a semi-infinite interval with one concentrated positive source at $x = x_0$, we consider an infinite interval with an additional negative source (the image source) located at $x = -x_0$. By symmetry the temperature G will be zero at $x = 0$ for all t. The Green's function is thus the sum of two infinite space Green's functions,

$$G(x, t; x_0, t_0) = \frac{1}{\sqrt{4\pi k(t - t_0)}} \left\{ \exp\left[-\frac{(x - x_0)^2}{4k(t - t_0)} \right] \right. \qquad (10.3.23)$$
$$\left. - \exp\left[-\frac{(x + x_0)^2}{4k(t - t_0)} \right] \right\}.$$

We note that the boundary condition at $x = 0$ is automatically satisfied.

10.3.9 Green's Function for the Heat Equation (on a Finite Region)

For a one-dimensional rod, $0 < x < L$, we have already determined in Chapter 8 the Green's function for the heat equation by the method of eigenfunction expansions. With zero boundary conditions at both ends,

$$G(x, t; x_0, t_0) = \sum_{n=1}^{\infty} \frac{2}{L} \sin \frac{n\pi x}{L} \sin \frac{n\pi x_0}{L} e^{-k(n\pi/L)^2(t - t_0)}. \qquad (10.3.24)$$

We can obtain an alternative representation for this Green's function by utilizing the method of images. By symmetry (see Fig. 10.3.1) the boundary conditions at $x = 0$ and at $x = L$ are satisfied if positive concentrated sources are located at $x = x_0 + 2Ln$ and negative concentrated sources are located at $x = -x_0 +$

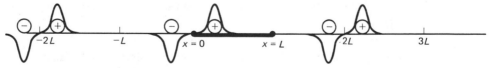

Figure 10.3.1 Multiple image sources for the Green's function for the heat equation for a finite one-dimensional rod.

$2Ln$ (for all integers n, $-\infty < n < \infty$). Using the infinite space Green's function, we have an alternative representation of the Green's function for a one-dimensional rod:

$$G(x, t; x_0, t_0) = \frac{1}{\sqrt{4\pi k(t - t_0)}} \sum_{n=-\infty}^{\infty} \left\{ \exp \left[-\frac{(x - x_0 - 2Ln)^2}{4k(t - t_0)} \right] \right.$$
$$\left. - \exp \left[-\frac{(x + x_0 - 2Ln)^2}{4k(t - t_0)} \right] \right\}. \tag{10.3.25}$$

Each form has its own advantage. The eigenfunction expansion, (10.3.24), is an infinite series which converges rapidly if $(t - t_0)k/L^2$ is large. It is thus most useful for $t \gg t_0$. In fact, if $t \gg t_0$,

$$G(x, t; x_0, t_0) \approx \frac{2}{L} \sin \frac{\pi x_0}{L} \sin \frac{\pi x}{L} e^{-k(\pi/L)^2(t - t_0)}.$$

However, if the elapsed time $t - t_0$ is small, then many terms of the infinite series are needed.

Using the method of images, the Green's function is also represented by an infinite series, (10.3.25). The infinite space Green's function (at fixed t) exponentially decays away from the source position,

$$\frac{1}{\sqrt{4\pi k(t - t_0)}} e^{-(x - x_0)^2/4k(t - t_0)}.$$

It decays in space very sharply if t is near t_0. If t is near t_0, then only sources near the response location x are important; sources far away will not be important (if t is near t_0); see Fig. 10.3.1. Thus, the image sources can be neglected if t is near t_0 (and if x or x_0 is neither near the boundaries 0 or L, as is explained in Exercise 10.3.8). As an approximation,

$$G(x, t; x_0, t_0) \approx \frac{1}{\sqrt{4\pi k(t - t_0)}} e^{-(x - x_0)^2/4k(t - t_0)};$$

if t is near t_0, the Green's function with boundaries can be approximated (in regions away from the boundaries) **by the infinite space Green's function.** This means that for small times the boundary can be neglected (away from the boundary).

To be more precise, the effect of every image source is much smaller than the actual source if $L^2/k(t - t_0)$ is large. This yields a better understanding of a "small time" approximation. The Green's function may be approximated by the infinite space Green's function if $t - t_0$ is small (i.e., if $t - t_0 \ll L^2/k$, where L^2/k is a ratio of physically measurable quantities). Alternatively, this approximation is valid for a "long rod" (in the sense that $L \gg \sqrt{k(t - t_0)}$).

In summary, the image method yields a rapidly convergent infinite series for the Green's function if $L^2/k(t - t_0) \gg 1$, while the eigenfunction expansion yields a rapidly convergent infinite series representation of the Green's function if $L^2/k(t - t_0) \ll 1$. If $L^2/k(t - t_0)$ is neither small nor large, then the two expansions are competitive, but both require at least a moderate number of terms.

EXERCISES 10.3

10.3.1. Show that for (10.3.2)
$$G^*(\mathbf{x}, t; \mathbf{x}_0, t_0) = G(\mathbf{x}_0, t_0; \mathbf{x}, t).$$

10.3.2.*(a) Suppose that
$$L = p(x) \frac{d^2}{dx^2} + r(x) \frac{d}{dx} + q(x).$$

Consider
$$\int_a^b v L(u) \, dx.$$

By repeated integration by parts, determine the adjoint operator L^* such that
$$\int_a^b [u L^*(v) - v L(u)] \, dx = H(x) \Big|_a^b.$$

What is $H(x)$? Under what conditions does $L = L^*$, the **self-adjoint** case? [*Hint:* Show that
$$L^* = p \frac{d^2}{dx^2} + \left(2 \frac{dp}{dx} - r \right) \frac{d}{dx} + \left(\frac{d^2 p}{dx^2} - \frac{dr}{dx} + q \right).]$$

(b) If
$$u(0) = 0 \qquad \text{and} \qquad \frac{du}{dx}(L) + u(L) = 0,$$

what boundary conditions should $v(x)$ satisfy for $H(x)|_0^L = 0$, called the adjoint boundary conditions?

10.3.3. Using the result of Exercise 10.3.2, show that eigenfunctions of $L(\phi) + \lambda \sigma \phi = 0$ are orthogonal to eigenfunctions of $L^*(\psi) + \lambda \sigma \psi = 0$ if the eigenvalues are different. Assume that ψ satisfies adjoint boundary conditions.

10.3.4. Using the result of Exercise 10.3.2, prove the following part of the **Fredholm alternative** (for operators that are not necessarily self-adjoint): A solution of $L(\phi) = f(x)$ subject to homogeneous boundary conditions may exist only if $f(x)$ is orthogonal to all solutions of the homogeneous adjoint problem.

10.3.5. If L is the following first-order linear differential operator
$$L = p(x) \frac{d}{dx},$$

then determine the adjoint operator L^* such that
$$\int_a^b [u L^*(v) - v L(u)] \, dx = B(x) \Big|_a^b.$$

What is $B(x)$? [*Hint:* Consider $\int_a^b v L(u) \, dx$ and integrate by parts.]

10.3.6. Consider

$$\frac{\partial u}{\partial t} = k\frac{\partial^2 u}{\partial x^2} + Q(x, t) \qquad x > 0$$

$$u(0, t) = A(t)$$

$$u(x, 0) = f(x).$$

(a) Solve if $A(t) = 0$ and $f(x) = 0$. Simplify this result if $Q(x, t) = 1$.
(b) Solve if $Q(x, t) = 0$ and $A(t) = 0$. Simplify this result if $f(x) = 1$.
*(c) Solve if $Q(x, t) = 0$ and $f(x) = 0$. Simplify this result if $A(t) = 1$.

*10.3.7. Determine the Green's function for

$$\frac{\partial u}{\partial t} = k\frac{\partial^2 u}{\partial x^2} + Q(x, t) \qquad x > 0$$

$$\frac{\partial u}{\partial x}(0, t) = A(t)$$

$$u(x, 0) = f(x).$$

10.3.8. Consider (10.3.23), the Green's function for (10.3.22). Show that the Green's function for this semi-infinite problem may be approximated by the Green's function for the infinite problem if

$$\frac{xx_0}{k(t - t_0)} \gg 1 \qquad (\text{i.e., } t - t_0 \text{ small}).$$

Explain physically why this approximation fails if x or x_0 is near the boundary.

10.3.9. Consider

$$\frac{\partial u}{\partial t} = k\frac{\partial^2 u}{\partial x^2} + Q(x, t)$$

$$u(x, 0) = f(x)$$

$$\frac{\partial u}{\partial x}(0, t) = A(t)$$

$$\frac{\partial u}{\partial x}(L, t) = B(t).$$

(a) Solve for the appropriate Green's function using the method of eigenfunction expansion.
(b) Approximate the Green's function of part (a). Under what conditions is your approximation valid?
(c) Solve for the appropriate Green's function using the infinite space Green's function.
(d) Approximate the Green's function of part (c). Under what conditions is your approximation valid?
(e) Solve for $u(x, t)$ in terms of the Green's function.

10.3.10. Determine the Green's function for the heat equation subject to zero boundary conditions at $x = 0$ and $x = L$ by applying the method of eigenfunction expansions directly to the defining differential equation. [*Hint:* The answer is given by (10.3.24).]

The Method of Characteristics for Linear and Quasi-Linear Wave Equations

11.1 INTRODUCTION

In previous chapters we obtained certain results concerning the one-dimensional wave equation,

$$\frac{\partial^2 u}{\partial t^2} = c^2 \frac{\partial^2 u}{\partial x^2}, \tag{11.1.1a}$$

subject to the initial conditions

$$u(x, 0) = f(x) \tag{11.1.1b}$$

$$\frac{\partial u}{\partial t}(x, 0) = g(x). \tag{11.1.1c}$$

For a vibrating string with zero displacement at $x = 0$ and $x = L$, we obtained a somewhat complicated Fourier sine series solution by the method of separation of variables in Chapter 4:

$$u(x, t) = \sum_{n=1}^{\infty} \sin \frac{n\pi x}{L} \left(a_n \cos \frac{n\pi c t}{L} + b_n \sin \frac{n\pi c t}{L} \right). \tag{11.1.2}$$

Further analysis of this solution [see (4.4.14) and Exercises 4.4.7 and 4.4.8] shows that the solution can be represented as the sum of a forward and backward moving wave. In particular,

$$u(x, t) = \frac{f(x - ct) + f(x + ct)}{2} + \frac{1}{2c} \int_{x-ct}^{x+ct} g(x_0) \, dx_0, \tag{11.1.3}$$

where $f(x)$ and $g(x)$ and the odd periodic extensions of the functions given in

(11.1.1b) and (11.1.1c). We also obtained (11.1.3) in Chapter 10 for the one-dimensional wave equation without boundaries, using the infinite space Green's function.

In this chapter we introduce the more powerful method of characteristics to solve the one-dimensional wave equation. We will show in general that $u(x, t) = F(x - ct) + G(x + ct)$, where F and G are arbitrary functions. We will show that (11.1.3) follows for infinite space problems. Then we will discuss modifications needed to solve semi-infinite and finite domain problems. In Section 11.6, the method of characteristics will be applied to quasi-linear partial differential equations. There shock waves will be introduced when characteristics intersect.

11.2 CHARACTERISTICS FOR FIRST-ORDER WAVE EQUATIONS

11.2.1 Introduction

The one-dimensional wave equation can be rewritten as

$$\frac{\partial^2 u}{\partial t^2} - c^2 \frac{\partial^2 u}{\partial x^2} = 0.$$

A short calculation shows that it can be "factored" in two ways:

$$\left(\frac{\partial}{\partial t} + c \frac{\partial}{\partial x}\right)\left(\frac{\partial u}{\partial t} - c \frac{\partial u}{\partial x}\right) = 0$$

$$\left(\frac{\partial}{\partial t} - c \frac{\partial}{\partial x}\right)\left(\frac{\partial u}{\partial t} + c \frac{\partial u}{\partial x}\right) = 0,$$

since the mixed second-derivative terms vanish in both. If we let

$$w = \frac{\partial u}{\partial t} - c \frac{\partial u}{\partial x} \qquad (11.2.1a)$$

$$v = \frac{\partial u}{\partial t} + c \frac{\partial u}{\partial x}, \qquad (11.2.1b)$$

we see that the one-dimensional wave equation (involving second derivatives) yields two **first-order wave equations**:

$$\frac{\partial w}{\partial t} + c \frac{\partial w}{\partial x} = 0 \qquad (11.2.2a)$$

$$\frac{\partial v}{\partial t} - c \frac{\partial v}{\partial x} = 0. \qquad (11.2.2b)$$

11.2.2 *Method of Characteristics for First-Order Partial Differential Equations*

We begin by discussing one of these simple first-order partial differential equations:

$$\frac{\partial w}{\partial t} + c \frac{\partial w}{\partial x} = 0. \tag{11.2.3}$$

The methods we will develop will be helpful in analyzing the one-dimensional wave equation. We consider the rate of change of $w(x(t), t)$ as measured by a moving observer, $x = x(t)$. The chain rule* implies that

$$\frac{d}{dt} w(x(t), t) = \frac{\partial w}{\partial t} + \frac{dx}{dt} \frac{\partial w}{\partial x}. \tag{11.2.4}$$

The first term $\partial w/\partial t$ represents the change in w at the fixed position, while $(dx/dt)(\partial w/\partial x)$ represents the change due to the fact that the observer moves into a region of possibly different w. Compare (11.2.4) with the partial differential equation for w, equation (11.2.3). It is apparent that *if* the observer moves with velocity c, that is, if

$$\frac{dx}{dt} = c, \tag{11.2.5a}$$

then

$$\frac{dw}{dt} = 0. \tag{11.2.5b}$$

Thus, w is constant. An observer moving with this special speed c would measure no changes in w.

Characteristics. In this way, the partial differential equation (11.2.3) has been replaced by two ordinary differential equations, (11.2.5a) and (11.2.5b). Integrating (11.2.5a) yields

$$x = ct + x_0, \tag{11.2.6}$$

the equation for the family of parallel **characteristics**† of (11.2.3), sketched in Fig. 11.2.1. Note that at $t = 0$, $x = x_0$. $w(x, t)$ is constant *along this line* (not

---

* Here d/dt as measured by a moving observer is sometimes called the substantial derivative.

† A characteristic is a curve along which a PDE reduces to an ODE.

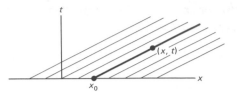

Figure 11.2.1 Characteristics for the first-order wave equation.

necessarily constant everywhere). w **propagates** as a **wave** with **wave speed** c [see (11.2.5a)].

General solution. If $w(x, t)$ is given initially at $t = 0$,

$$w(x, 0) = P(x), \tag{11.2.7}$$

then let us determine w at the point (x, t). Since w is constant along the characteristic,

$$w(x, t) = w(x_0, 0) = P(x_0).$$

Given x and t, the parameter is known from the characteristic, $x_0 = x - ct$, and thus

$$\boxed{w(x, t) = P(x - ct),} \tag{11.2.8}$$

which we call the general solution of (11.2.3).

We can think of $P(x)$ as being an arbitrary function. To verify this, we substitute (11.2.8) back into the partial differential equation (11.2.3). Using the chain rule

$$\frac{\partial w}{\partial x} = \frac{dP}{d(x - ct)} \frac{\partial (x - ct)}{\partial x} = \frac{dP}{d(x - ct)}$$

and

$$\frac{\partial w}{\partial t} = \frac{dP}{d(x - ct)} \frac{\partial (x - ct)}{\partial t} = -c \frac{dP}{d(x - ct)}.$$

Thus, it is verified that (11.2.3) is satisfied by (11.2.8). The general solution of a first-order partial differential equation contains an arbitrary function, while the general solution to ordinary differential equations contains arbitrary constants.

Example. Consider

$$\frac{\partial w}{\partial t} + 2 \frac{\partial w}{\partial x} = 0,$$

subject to the initial condition

$$w(x, 0) = \begin{cases} 4x & 0 < x < 1 \\ 0 & x < 0 \text{ and } x > 1. \end{cases}$$

We have shown that w is constant along the characteristics $x - 2t = \text{constant}$, keeping its same shape moving at velocity 2 (to the right). The important characteristics, $x = 2t + 0$ and $x = 2t + 1$, as well as a sketch of the solution at various times, appear in Fig. 11.2.2. $w(x, t) = 0$ if $x > 2t + 1$ or if $x < 2t$.

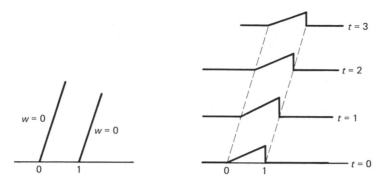

Figure 11.2.2 Propagation for the first-order wave equation.

Otherwise, by shifting

$$w(x, t) = 4(x - 2t) \qquad \text{if } 2t < x < 2t + 1.$$

To derive this analytic solution, we use the characteristic which starts at $x = x_0$:

$$x = 2t + x_0.$$

Along this characteristic, $w(x, t)$ is constant. If $0 < x_0 < 1$, then

$$w(x, t) = w(x_0, 0) = 4x_0 = 4(x - 2t),$$

as before. This is valid if $0 < x_0 < 1$ or equivalently $0 < x - 2t < 1$.

Same shape. In general $w(x, t) = P(x - ct)$. *At fixed t, the solution of the first-order wave equation is the same shape shifted a distance ct* (distance = velocity times time). We illustrate this in Fig. 11.2.3.

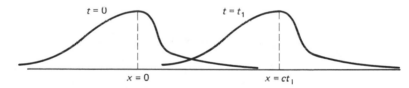

Figure 11.2.3 Shape invariance for the first-order wave equation.

Summary. The method of characteristics solves the first-order wave equation (11.2.3). In Sections 11.3–11.5, this method is applied to solve the wave equation (11.1.1a). The reader may proceed directly to Section 11.6 where the method of characteristics is described for quasi-linear partial differential equations.

EXERCISES 11.2

11.2.1. Show that the wave equation can be considered as the following system of two coupled first-order partial differential equations:

$$\frac{\partial u}{\partial t} - c \frac{\partial u}{\partial x} = w$$

$$\frac{\partial w}{\partial t} + c \frac{\partial w}{\partial x} = 0.$$

*11.2.2. Solve

$$\frac{\partial w}{\partial t} - 3 \frac{\partial w}{\partial x} = 0 \qquad \text{with } w(x, 0) = \cos x.$$

11.2.3. Solve

$$\frac{\partial w}{\partial t} + 4 \frac{\partial w}{\partial x} = 0 \qquad \text{with } w(0, t) = \sin 3t.$$

11.2.4. Solve

$$\frac{\partial w}{\partial t} + c \frac{\partial w}{\partial x} = 0 \qquad (c > 0)$$

for $x > 0$ and $t > 0$ if

$$w(x, 0) = f(x) \qquad x > 0$$
$$w(0, t) = h(t) \qquad t > 0.$$

11.2.5. Solve using the method of characteristics (if necessary, see Section 11.6):

(a) $\dfrac{\partial w}{\partial t} + c \dfrac{\partial w}{\partial x} = e^{2x}$ with $w(x, 0) = f(x)$

*(b) $\dfrac{\partial w}{\partial t} + x \dfrac{\partial w}{\partial x} = 1$ with $w(x, 0) = f(x)$

(c) $\dfrac{\partial w}{\partial t} + t \dfrac{\partial w}{\partial x} = 1$ with $w(x, 0) = f(x)$

*(d) $\dfrac{\partial w}{\partial t} + 3t \dfrac{\partial w}{\partial x} = w$ with $w(x, 0) = f(x)$

*11.2.6. Consider (if necessary, see Section 11.6):

$$\frac{\partial u}{\partial t} + 2u \frac{\partial u}{\partial x} = 0$$

$$u(x, 0) = f(x).$$

Show that the characteristics are straight lines.

11.2.7. Consider Exercise 11.2.6 with

$$u(x, 0) = f(x) = \begin{cases} 1 & x < 0 \\ 1 + x/L & 0 < x < L \\ 2 & x > L. \end{cases}$$

(a) Determine equations for the characteristics. Sketch the characteristics.
(b) Determine the solution $u(x, t)$. Sketch $u(x, t)$ for t fixed.

*11.2.8. Consider Exercise 11.2.6 with

$$u(x, 0) = f(x) = \begin{cases} 1 & x < 0 \\ 2 & x > 0. \end{cases}$$

Obtain the solution $u(x, t)$ by considering the limit as $L \to 0$ of the characteristics obtained in Exercise 11.2.7. Sketch characteristics and $u(x, t)$ for t fixed.

11.2.9. As motivated by the analysis of a moving observer, make a change of independent variables from (x, t) to a coordinate system moving with velocity c, (ξ, t'), where $\xi = x - ct$ and $t' = t$, in order to solve (11.2.3).

11.2.10. For the first-order "quasi-linear" partial differential equation

$$a \frac{\partial u}{\partial x} + b \frac{\partial u}{\partial y} = c,$$

where a, b, and c are functions of x, y and u, show that the method of characteristics (if necessary, see Section 11.6) yields

$$\frac{dx}{a} = \frac{dy}{b} = \frac{du}{c}.$$

11.3 METHOD OF CHARACTERISTICS FOR THE ONE-DIMENSIONAL WAVE EQUATION

11.3.1 Introduction

From the one-dimensional wave equation,

$$\boxed{\frac{\partial^2 u}{\partial t^2} - c^2 \frac{\partial^2 u}{\partial x^2} = 0,} \tag{11.3.1}$$

we derived two first-order partial differential equations, $\partial w/\partial t + c\, \partial w/\partial x = 0$ and $\partial v/\partial t - c\, \partial v/\partial x = 0$, where $w = \partial u/\partial t - c\, \partial u/\partial x$ and $v = \partial u/\partial t + c\, \partial u/\partial x$. We have shown that w remains the same shape moving at velocity c:

$$w = \frac{\partial u}{\partial t} - c \frac{\partial u}{\partial x} = P(x - ct). \tag{11.3.2}$$

The problem for v is identical (replace c by $-c$). Thus, we could have shown that v is translated unchanged at velocity $-c$:

$$v = \frac{\partial u}{\partial t} + c \frac{\partial u}{\partial x} = Q(x + ct). \tag{11.3.3}$$

By combining (11.3.2) and (11.3.3) we obtain, for example,

$$\frac{\partial u}{\partial t} = \frac{1}{2}[P(x - ct) + Q(x + ct)],$$

and thus

$$\boxed{u(x, t) = F(x - ct) + G(x + ct),} \tag{11.3.4}$$

where F and G are arbitrary functions ($-cF' = \frac{1}{2}P$ and $cG' = \frac{1}{2}Q$). This result was obtained by d'Alembert in 1747.

The general solution is the sum of $F(x - ct)$, a wave of fixed shape moving to the right with velocity c, and $G(x + ct)$, a wave of fixed shape moving to the left with velocity $-c$. The solution may be sketched if $F(x)$ and $G(x)$ are known. We shift $F(x)$ to the right a distance ct and shift $G(x)$ to the left a distance ct and add the two. Although each shape is unchanged, the sum will in general be a shape that is changing in time. In Sec. 11.3.2 we will show how to determine $F(x)$ and $G(x)$ from initial conditions.

Characteristics. Part of the solution is constant along the family of characteristics $x - ct = $ constant, while a different part of the solution is constant along $x + ct = $ constant. For the one-dimensional wave equation, (11.3.1), there are two families of characteristic curves, as sketched in Fig. 11.3.1.

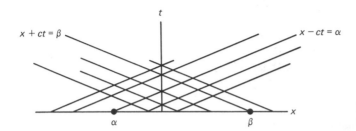

Figure 11.3.1 Characteristics for the one-dimensional wave equation.

11.3.2 Initial Value Problem (Infinite Domain)

In Sec. 11.3.1 we showed that the general solution of the one-dimensional wave equation is

$$u(x, t) = F(x - ct) + G(x + ct). \qquad (11.3.5)$$

Here we will determine the arbitrary functions in order to satisfy the initial conditions:

$$u(x, 0) = f(x) \qquad -\infty < x < \infty \qquad (11.3.6a)$$

$$\frac{\partial u}{\partial t}(x, 0) = g(x) \qquad -\infty < x < \infty. \qquad (11.3.6b)$$

These initial conditions imply that

$$f(x) = F(x) + G(x) \qquad (11.3.7a)$$

$$\frac{g(x)}{c} = -\frac{dF}{dx} + \frac{dG}{dx}. \qquad (11.3.7b)$$

We solve for $G(x)$ by eliminating $F(x)$; for example, adding the derivative of (11.3.7a) to (11.3.7b) yields

$$\frac{dG}{dx} = \frac{1}{2}\left(\frac{df}{dx} + \frac{g(x)}{c}\right).$$

By integrating this, we obtain

$$G(x) = \frac{1}{2}f(x) + \frac{1}{2c}\int_0^x g(\bar{x})\,d\bar{x} \qquad + k \qquad\qquad (11.3.8a)$$

$$F(x) = \frac{1}{2}f(x) - \frac{1}{2c}\int_0^x g(\bar{x})\,d\bar{x} \qquad - k, \qquad\qquad (11.3.8b)$$

where the latter equation was obtained from (11.3.7a). k can be neglected since $u(x, t)$ is obtained from (11.3.5) by adding (11.3.8a) and (11.3.8b) (with appropriate shifts).

Sketching technique. The solution $u(x, t)$ can be graphed based on (11.3.5) in the following straightforward manner:

1. Given $f(x)$ and $g(x)$. Obtain the graphs of

$$\frac{1}{2}f(x) \qquad \text{and} \qquad \frac{1}{2c}\int_0^x g(\bar{x})\,d\bar{x},$$

   the latter by integrating first.
2. By addition and subtraction, form $F(x)$ and $G(x)$; see (11.3.8).
3. Translate (shift) $F(x)$ to the right a distance ct and $G(x)$ to the left ct.
4. Add the two shifted functions, thus satisfying (11.3.5).

Initially at rest. If a vibrating string is initially at rest $[\partial u/\partial t(x, 0) = g(x) = 0]$, then from (11.3.8) $F(x) = G(x) = \frac{1}{2}f(x)$. Thus,

$$u(x, t) = \frac{1}{2}[f(x - ct) + f(x + ct)]. \qquad\qquad (11.3.9)$$

The initial condition $u(x, 0) = f(x)$ splits into two parts; half moves to the left and half to the right.

Example. Suppose that an infinite vibrating string is initially stretched into the shape of a single rectangular pulse and is let go from rest. The corresponding initial conditions are

$$u(x, 0) = f(x) = \begin{cases} 1 & |x| < h \\ 0 & |x| > h \end{cases}$$

and

$$\frac{\partial u}{\partial t}(x, 0) = g(x) = 0.$$

The solution is given by (11.3.9). By adding together these two rectangular pulses, we obtain Fig. 11.3.2. The pulses overlap until the left end of the right-moving one passes the right end of the other. Since each is traveling at speed c, they are moving apart at velocity $2c$. The ends are initially a distance $2h$ apart, and hence the time at which the two pulses separate is

$$t = \frac{\text{distance}}{\text{velocity}} = \frac{2h}{2c} = \frac{h}{c}.$$

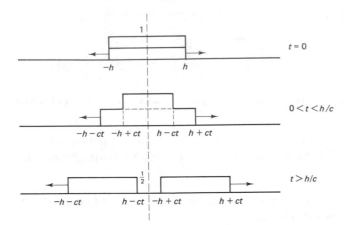

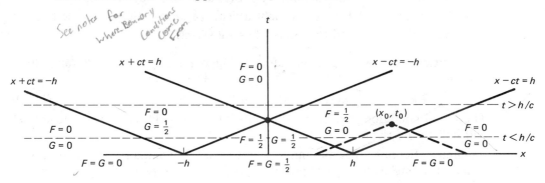

Figure 11.3.2 Initial value problem for the one-dimensional equation.

Important characteristics are sketched in Fig. 11.3.3. F stays constant moving to the right at velocity c, while G stays constant moving to the left. From (11.3.8)

$$F(x) = G(x) = \begin{cases} \frac{1}{2} & |x| < h \\ 0 & |x| > h. \end{cases}$$

This information also appears in Fig. 11.3.3.

Figure 11.3.3 Method of characteristics for the one-dimensional wave equation.

Example. Suppose that an infinite string is initially horizontally stretched with prescribed initial velocity as follows:

$$u(x, 0) = f(x) = 0$$

$$\frac{\partial u}{\partial t}(x, 0) = g(x) = \begin{cases} 1 & |x| < h \\ 0 & |x| > h. \end{cases}$$

In Exercise 11.3.2 it is shown that this corresponds to instantaneously applying a constant impulsive force to the entire region $|x| < h$, as though the string is being struck by a broad ($|x| < h$) hammer. The calculation of the solution of the wave equation with these initial conditions is more involved than in the preceding example. From (11.3.8), we need $\int_0^x g(\bar{x})\, d\bar{x}$, representing the area under $g(x)$ from 0 to x:

$$2cG(x) = -2cF(x) = \int_0^x g(\bar{x})\, d\bar{x} = \begin{cases} -h & x < -h \\ x & -h < x < h \\ h & x > h. \end{cases}$$

The solution $u(x, t)$ is the sum of $F(x)$ shifted to the right (at velocity c) and $G(x)$ shifted to the left (at velocity c). $F(x)$ and $G(x)$ are sketched in Fig. 11.3.4, as is their shifted sum. The striking of the broad hammer causes the displacement of the string to gradually increase near where the hammer hit and to have this disturbance spread out to the left and right as time increases. Eventually, the string reaches an elevated rest position. Alternatively, the solution can be obtained in an algebraic way (see Exercise 11.3.5). The characteristics sketched in Fig. 11.3.3 are helpful.

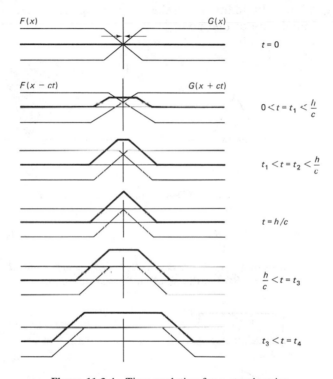

Figure 11.3.4 Time evolution for a struck string.

11.3.3 d'Alembert's Solution

The general solution of the one-dimensional wave equation can be simplified somewhat. Substituting (11.3.8) into the general solution (11.3.5) yields

$$u(x, t) = \frac{f(x + ct) + f(x - ct)}{2} + \frac{1}{2c}\left[\int_0^{x+ct} g(\bar{x})\, d\bar{x} - \int_0^{x-ct} g(\bar{x})\, d\bar{x}\right]$$

or

$$u(x, t) = \frac{f(x - ct) + f(x + ct)}{2} + \frac{1}{2c} \int_{x-ct}^{x+ct} g(\bar{x}) \, d\bar{x}, \qquad (11.3.10)$$

known as **d'Alembert's solution** (previously obtained by Fourier transform methods). It is a very elegant result. However, for sketching solutions often it is easier to work directly with (11.3.8), where these are shifted according to (11.3.5).

Domain of dependence and range of influence. The importance of the characteristics $x - ct = $ constant and $x + ct = $ constant is clear. At position x at time t the initial position data are needed at $x \pm ct$, while all the initial velocity data between $x - ct$ and $x + ct$ is needed. The region between $x - ct$ and $x + ct$ is called the **domain of dependence** of the solution at (x, t) as sketched in Fig. 11.3.5. In addition, we sketch the **range of influence**, the region affected by the initial data at one point.

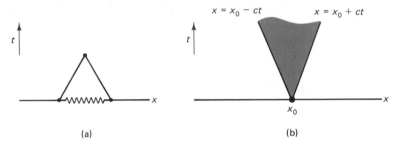

(a) (b)

Figure 11.3.5 (a) Domain of dependence; (b) range of influence.

EXERCISES 11.3

11.3.1. Suppose that $u(x, t) = F(x - ct) + G(x + ct)$, where F and G are sketched in Fig. 11.3.6. Sketch the solution for various times.

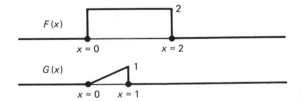

Figure 11.3.6

11.3.2. Suppose that a stretched string is unperturbed (horizontal, $u = 0$) and at rest ($\partial u / \partial t = 0$). If an impulsive force is applied at $t = 0$, the initial value problem is

$$\frac{\partial^2 u}{\partial t^2} = c^2 \frac{\partial^2 u}{\partial x^2} + \alpha(x)\delta(t)$$

$$u(x, t) = 0 \qquad t < 0.$$

(a) Without using explicit solutions, show that this is equivalent to

$$\frac{\partial^2 u}{\partial t^2} = c^2 \frac{\partial^2 u}{\partial x^2} \qquad t > 0$$

subject to $\quad u(x, 0) = 0 \quad$ and $\quad \dfrac{\partial u}{\partial t}(x, 0) = \alpha(x)$.

Thus, the initial velocity $\alpha(x)$ is equivalent to an impulsive force.

(b) Do part (a) using the explicit solution of both problems.

11.3.3. An alternative way to solve the one-dimensional wave equation (11.3.1) is based on (11.3.2) and (11.3.3). Solve the wave equation by introducing a change of variables from (x, t) to two moving coordinates (ξ, η), one moving to the left (with velocity $-c$) and one moving to the right (with velocity c):

$$\xi = x - ct$$
$$\eta = x + ct.$$

*11.3.4. Suppose that $u(x, t) = F(x - ct)$. Evaluate:

(a) $\dfrac{\partial u}{\partial t}(x, 0)$

(b) $\dfrac{\partial u}{\partial x}(0, t)$

11.3.5. Determine analytic formulas for $u(x, t)$ if

$$u(x, 0) = f(x) = 0$$

$$\frac{\partial u}{\partial t}(x, 0) = g(x) = \begin{cases} 1 & |x| < h \\ 0 & |x| > h. \end{cases}$$

(*Hint:* Using characteristics as sketched in Fig. 11.3.3, show there are two distinct regions $t < h/c$ and $t > h/c$. In each, show that the solution has five different forms, depending on x.)

11.3.6. Consider the three-dimensional wave equation

$$\frac{\partial^2 u}{\partial t^2} = c^2 \, \nabla^2 u.$$

Assume that the solution is **spherically symmetric**, so that

$$\nabla^2 u = (1/\rho^2) \, (\partial/\partial\rho) \, (\rho^2 \partial u / \partial\rho).$$

(a) Make the transformation $u = (1/\rho) \, w(\rho, t)$ and verify that

$$\frac{\partial^2 w}{\partial t^2} = c^2 \frac{\partial^2 w}{\partial\rho^2}.$$

(b) Show that the most general spherically symmetric solution of the wave equation consists of the sum of two spherically symmetric waves, one moving outward at speed c and the other inward at speed c. Note the decay of the amplitude.

11.4 SEMI-INFINITE STRINGS AND REFLECTIONS

We will solve the one-dimensional wave equation on a semi-infinite interval, $x > 0$:

PDE:

$$\frac{\partial^2 u}{\partial t^2} = c^2 \frac{\partial^2 u}{\partial x^2}$$

(11.4.1a)

IC:

$$u(x, 0) = f(x)$$

(11.4.1b)

$$\frac{\partial u}{\partial t}(x, 0) = g(x).$$

(11.4.1c)

A condition is necessary at the boundary $x = 0$. We suppose that the string is fixed at $x = 0$:

BC: $u(0, t) = 0.$

(11.4.1d)

Although a Fourier sine transform can be used, we prefer to indicate how to use the general solution and the method of characteristics:

$$u(x, t) = F(x - ct) + G(x + ct).$$

(11.4.2)

As in Sec. 11.3, the initial conditions are satisfied if

$$G(x) = \frac{1}{2}f(x) + \frac{1}{2c}\int_0^x g(\bar{x})\, d\bar{x} \qquad x > 0$$

(11.4.3a)

$$F(x) = \frac{1}{2}f(x) - \frac{1}{2c}\int_0^x g(\bar{x})\, d\bar{x} \qquad x > 0.$$

(11.4.3b)

However, it is very important to note that (unlike the case of the infinite string) (11.4.3) is valid only for $x > 0$; the arbitrary functions are only determined for positive arguments. In the general solution, $G(x + ct)$ requires only positive arguments of G (since $x > 0$ and $t > 0$). On the other hand, $F(x - ct)$ requires positive arguments if $x > ct$, but requires negative arguments if $x < ct$. As indicated by a space–time diagram, Fig. 11.4.1, the information that there is a fixed end at $x = 0$ travels at a finite velocity c. Thus, if $x > ct$, the string does not know that there is any boundary. In this case ($x > ct$), the solution is obtained as before [using (11.4.3)],

$$u(x, t) = \frac{f(x - ct) + f(x + ct)}{2} + \frac{1}{2c}\int_{x-ct}^{x+ct} g(\bar{x})\, d\bar{x}, \qquad x > ct, \qquad (11.4.4)$$

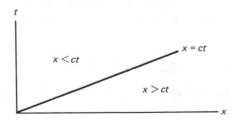

Figure 11.4.1 Characteristic eminating from the boundary.

The Method of Characteristics   Chap. 11

d'Alembert's solution. However, here this is not valid if $x < ct$. Since $x + ct > 0$,

$$G(x + ct) = \frac{1}{2}f(x + ct) + \frac{1}{2c}\int_0^{x+ct} g(\bar{x})\,d\bar{x},$$

as determined earlier. To obtain F for negative arguments, we cannot use the initial conditions. Instead, the boundary condition must be utilized. $u(0, t) = 0$ implies that [from (11.4.2)]

$$0 = F(-ct) + G(ct) \qquad \text{for } t > 0. \tag{11.4.5}$$

Thus, F for negative arguments is $-G$ of the corresponding positive argument:

$$F(z) = -G(-z) \qquad \text{for } z < 0. \tag{11.4.6}$$

Thus, the solution for $x - ct < 0$ is

$$u(x, t) = F(x - ct) + G(x + ct) = G(x + ct) - G(ct - x)$$
$$= \frac{1}{2}[f(x + ct) - f(ct - x)] + \frac{1}{2c}\left[\int_0^{x+ct} g(\bar{x})\,d\bar{x} - \int_0^{ct-x} g(\bar{x})\,d\bar{x}\right]$$
$$= \frac{1}{2}[f(x + ct) - f(ct - x)] + \frac{1}{2c}\int_{ct-x}^{x+ct} g(\bar{x})\,d\bar{x}.$$

To interpret this solution, the method of characteristics is helpful. Recall that for infinite problems $u(x, t)$ is the sum of F (moving to the right) and G (moving to the left). For semi-infinite problems with $x > ct$, the boundary does *not* affect the characteristics (see Fig. 11.4.2). If $x < ct$, then Fig. 11.4.3 shows the left-moving characteristic (G constant) not affected by the boundary, but the right-moving characteristic emanates from the boundary. F is constant moving to the right. Due to the boundary condition, $F + G = 0$ at $x = 0$, the right-moving wave is minus the left-moving wave. The wave inverts as it "bounces off" the boundary. The resulting right-moving wave $-G(ct - x)$ is called the

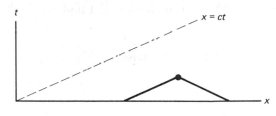

Figure 11.4.2 Characteristics.

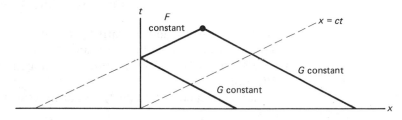

Figure 11.4.3 Reflected characteristics.

reflected wave. For $x < ct$, the total solution is the reflected wave plus the as yet unreflected left-moving wave:

$$u(x, t) = G(x + ct) - G(-(x - ct)).$$

The negatively reflected wave $-G(-(x - ct))$ moves to the right. It behaves *as if* initially at $t = 0$ it were $-G(-x)$. If there were no boundary, the right-moving wave $F(x - ct)$ would be initially $F(x)$. Thus, the reflected wave is exactly the wave that would have occurred if

$$F(x) = -G(-x) \qquad \text{for } x < 0,$$

or equivalently

$$\frac{1}{2}f(x) - \frac{1}{2c}\int_0^x g(\bar{x}) \, d\bar{x} = -\frac{1}{2}f(-x) - \frac{1}{2c}\int_0^{-x} g(\bar{x}) \, d\bar{x}.$$

One way to obtain this is to extend the initial position $f(x)$ for $x > 0$ as an odd function [such that $f(-x) = -f(x)$] and also extend the initial velocity $g(x)$ for $x > 0$ as an odd function [then its integral, $\int_0^x g(\bar{x}) \, d\bar{x}$, will be an even function]. *In summary*, **the solution of the semi-infinite problem with $u = 0$ at $x = 0$ is the same as an infinite problem with the initial positions and velocities extended as odd functions.**

As further explanation, suppose that $u(x, t)$ is any solution of the wave equation. Since the wave equation is unchanged when x is replaced by $-x$, $u(-x, t)$ (and any multiple of it) is also a solution of the wave equation. If the initial conditions satisfied by $u(x, t)$ are odd functions of x, then both $u(x, t)$ and $-u(-x, t)$ solve these initial conditions and the wave equation. Since the initial value problem has a unique solution, $u(x, t) = -u(-x, t)$; that is, $u(x, t)$, which is odd initially, will remain odd for all time. Thus, odd initial conditions yield a solution that will satisfy a zero boundary condition at $x = 0$.

Example. Consider a semi-infinite string $x > 0$ with a fixed end $u(0, t) = 0$, which is initially at rest, $\partial u / \partial t \, (x, 0) = 0$, with an initial unit rectangular pulse,

$$f(x) = \begin{cases} 1 & 4 < x < 5 \\ 0 & \text{otherwise}. \end{cases}$$

Since $g(x) = 0$, it follows that

$$F(x) = G(x) = \frac{1}{2}f(x) = \begin{cases} \frac{1}{2} & 4 < x < 5 \\ 0 & \text{otherwise (with } x > 0). \end{cases}$$

F moves to the right; G moves to the left, negatively reflecting off $x = 0$. This can also be interpreted as an initial condition (on an infinite domain) with $f(x)$ and $g(x)$ extended as odd functions. The solution is sketched in Fig. 11.4.4. Note the *negative* reflection.

Problems with nonhomogeneous boundary conditions at $x = 0$ can be analyzed in a similar way.

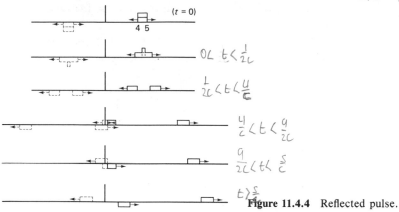

Draw Characteristic Equations.

$(t = 0)$

4 5

$0 < t < \frac{1}{2c}$

$\frac{1}{2c} < t < \frac{4}{c}$

$\frac{4}{c} < t < \frac{9}{2c}$

$\frac{9}{2c} < t < \frac{5}{c}$

$t > \frac{5}{c}$

Figure 11.4.4 Reflected pulse.

EXERCISES 11.4

*11.4.1. Solve by the method of characteristics:

$$\frac{\partial^2 u}{\partial t^2} = c^2 \frac{\partial^2 u}{\partial x^2} \qquad x > 0$$

subject to $u(x, 0) = 0,\ \dfrac{\partial u}{\partial t}(x, 0) = 0,$ and $u(0, t) = h(t)$.

*11.4.2. Determine $u(x, t)$ if

$$\frac{\partial^2 u}{\partial t^2} - c^2 \frac{\partial^2 u}{\partial x^2} \qquad \text{for } x < 0 \text{ only,}$$

where

$$u(x, 0) = \cos x \qquad x < 0$$

$$\frac{\partial u}{\partial t}(x, 0) = 0 \qquad x < 0$$

$$u(0, t) = e^{-t} \qquad t > 0.$$

Do _not_ sketch the solution. However, draw a space–time diagram, including all important characteristics.

11.4.3. Consider the wave equation on a semi-infinite interval

$$\frac{\partial^2 u}{\partial t^2} = c^2 \frac{\partial^2 u}{\partial x^2} \qquad \text{for } 0 < x < \infty$$

with the free boundary condition

$$\frac{\partial u}{\partial x}(0, t) = 0$$

and the initial conditions

$$u(x, 0) = \begin{cases} 0 & 0 < x < 2 \\ 1 & 2 < x < 3 \\ 0 & x > 3 \end{cases} \qquad \frac{\partial u}{\partial t}(x, 0) = 0.$$

Determine the solution. Sketch the solution for various times. (Assume that u is continuous at $x = 0$, $t = 0$.)

11.4.4. (a) Solve for $x > 0$, $t > 0$ (using the method of characteristics)

$$\frac{\partial^2 u}{\partial t^2} = c^2 \frac{\partial^2 u}{\partial x^2}$$

$$\left.\begin{array}{l} u(x, 0) = f(x) \\[2mm] \dfrac{\partial u}{\partial t}(x, 0) = g(x) \end{array}\right\} \quad x > 0$$

$$\frac{\partial u}{\partial x}(0, t) = 0 \qquad t > 0.$$

(Assume that u is continuous at $x = 0$, $t = 0$.)

(b) Show that the solution of part (a) may be obtained by extending the initial position and velocity as even functions (around $x = 0$).

(c) Sketch the solution if $g(x) = 0$ and

$$f(x) = \begin{cases} 1 & 4 < x < 5 \\ 0 & \text{otherwise.} \end{cases}$$

11.4.5. (a) Show that if $u(x, t)$ and $\partial u / \partial t$ are initially even around $x = x_0$, $u(x, t)$ will remain even for all time.

(b) Show that this type of even initial condition yields a solution that will satisfy a zero derivative boundary condition at $x = x_0$.

*11.4.6. Solve ($x > 0$, $t > 0$)

$$\frac{\partial^2 u}{\partial t^2} = c^2 \frac{\partial^2 u}{\partial x^2}$$

subject to the conditions $u(x, 0) = 0$, $\dfrac{\partial u}{\partial t}(x, 0) = 0$, and $\dfrac{\partial u}{\partial x}(0, t) = h(t)$.

*11.4.7. Solve

$$\frac{\partial^2 u}{\partial t^2} = c^2 \frac{\partial^2 u}{\partial x^2} \qquad \begin{array}{l} x > 0 \\ t > 0 \end{array}$$

subject to $u(x, 0) = f(x)$, $\dfrac{\partial u}{\partial t}(x, 0) = 0$, and $\dfrac{\partial u}{\partial x}(0, t) = h(t)$.

(Assume that u is continuous at $x = 0$, $t = 0$.)

11.4.8. Solve

$$\frac{\partial^2 u}{\partial t^2} = c^2 \frac{\partial^2 u}{\partial x^2} \quad \text{with} \quad u(x, 0) = 0 \quad \text{and} \quad \frac{\partial u}{\partial t}(x, 0) = 0,$$

subject to $u(x, t) = g(t)$ along $x = \dfrac{c}{2}t$ ($c > 0$).

11.5 METHOD OF CHARACTERISTICS FOR A VIBRATING STRING OF FIXED LENGTH

In Chapter 2 we solved for the vibration of a finite string satisfying

$$\frac{\partial^2 u}{\partial t^2} = c^2 \frac{\partial^2 u}{\partial x^2} \tag{11.5.1a}$$

BC:

$$u(0, t) = 0 \tag{11.5.1b}$$
$$u(L, t) = 0 \tag{11.5.1c}$$

IC:

$$u(x, 0) = f(x) \tag{11.5.1d}$$
$$\frac{\partial u}{\partial t}(x, 0) = g(x), \tag{11.5.1e}$$

using Fourier series methods. We can obtain an equivalent, but in some ways more useful, result by using the general solution of the one-dimensional wave equation:

$$u(x, t) = F(x - ct) + G(x + ct). \tag{11.5.2}$$

The initial conditions are prescribed only for $0 < x < L$, and hence the formulas for $F(x)$ and $G(x)$ previously obtained are valid only for $0 < x < L$:

$$F(x) = \frac{1}{2}f(x) - \frac{1}{2c}\int_0^x g(\bar{x}) \, d\bar{x} \tag{11.5.3a}$$

$$G(x) = \frac{1}{2}f(x) + \frac{1}{2c}\int_0^x g(\bar{x}) \, d\bar{x}. \tag{11.5.3b}$$

If $0 < x - ct < L$ and $0 < x + ct < L$ as shown shaded in Fig. 11.5.1, then d'Alembert's solution is valid:

$$u(x, t) = \frac{f(x - ct) + f(x + ct)}{2} + \frac{1}{2c}\int_{x-ct}^{x+ct} g(\bar{x}) \, d\bar{x}. \tag{11.5.4}$$

In this region the string does not know that either boundary exists; the information that there is a boundary propagates at velocity c from $x = 0$ and $x = L$.

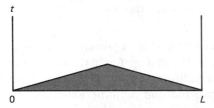

$-x \rightarrow$ **Figure 11.5.1** Characteristics.

If one's position and time is such that signals from the boundary have already arrived, then modifications in (11.5.4) must be made. The boundary condition at $x = 0$ implies that

$$0 = F(-ct) + G(ct) \qquad \text{for } t > 0, \tag{11.5.5a}$$

while at $x = L$ we have

$$0 = F(L - ct) + G(L + ct) \qquad \text{for } t > 0. \tag{11.5.5b}$$

These in turn imply reflections and multiple reflections, as illustrated in Fig. 11.5.2.

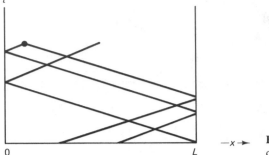

Figure 11.5.2 Multiply reflected characteristics.

Alternatively, a solution on an infinite domain without boundaries can be considered which is odd around $x = 0$ and odd around $x = L$, as sketched in Fig. 11.5.3. In this way, the zero condition at both $x = 0$ and $x = L$ will be satisfied. We note that $u(x, t)$ is periodic with period $2L$. In fact, we ignore the oddness around $x = L$, since periodic functions that are odd around $x = 0$ are automatically odd around $x = L$. Thus, the simplest way to obtain the solution is to **extend the initial conditions as odd functions (around $x = 0$) which are periodic (with period $2L$).** With these odd periodic initial conditions, the method of characteristics can be utilized as well as d'Alembert's solution (11.5.4).

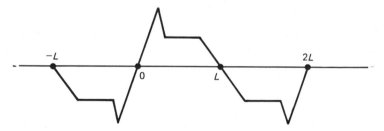

Figure 11.5.3 Odd periodic extension.

Example. Suppose that a string is initially at rest with prescribed initial conditions $u(x, 0) = f(x)$. The string is fixed at $x = 0$ and $x = L$. Instead of using Fourier series methods, we extend the initial conditions as odd functions around $x = 0$ and $x = L$. Equivalently, we introduce the *odd periodic extension*. (The odd periodic extension is also used in the Fourier series solution.) Since the string is initially at rest, $g(x) = 0$; the odd periodic extension is $g(x) = 0$ for all x. Thus, the solution of the one-dimensional wave equation is the sum of two simple waves:

$$u(x, t) = \frac{1}{2}[f(x - ct) + f(x + ct)],$$

where $f(x)$ is the odd periodic extension of the given initial position. This solution is much simpler than the summation of the first 100 terms of its Fourier sine series.

EXERCISES 11.5

11.5.1. Consider

$$\frac{\partial^2 u}{\partial t^2} = c^2 \frac{\partial^2 u}{\partial x^2}$$

$$\left.\begin{array}{l} u(x, 0) = f(x) \\ \dfrac{\partial u}{\partial t}(x, 0) = g(x) \end{array}\right\} \qquad \boxed{0 < x < L}$$

$$u(0, t) = 0$$
$$u(L, t) = 0.$$

  (a) Obtain the solution by Fourier series techniques.
  *(b) If $g(x) = 0$, show that part (a) is equivalent to the results of Chapter 11.
  (c) If $f(x) = 0$, show that part (a) is equivalent to the results of Chapter 11.

11.5.2. Solve using the method of characteristics:

$$\frac{\partial^2 u}{\partial t^2} - c^2 \frac{\partial^2 u}{\partial x^2}$$

$$u(x, 0) - 0 \qquad u(0, t) - h(t)$$

$$\frac{\partial u}{\partial t}(x, 0) = 0 \qquad u(L, t) = 0.$$

11.5.3. Consider

$$\frac{\partial^2 u}{\partial t^2} = c^2 \frac{\partial^2 u}{\partial x^2} \qquad 0 < x < 10$$

$$u(x, 0) = f(x) = \begin{cases} 1 & 4 < x < 5 \\ 0 & \text{otherwise} \end{cases} \qquad u(0, t) = 0$$

$$\frac{\partial u}{\partial t}(x, 0) = g(x) = 0 \qquad \frac{\partial u}{\partial x}(L, t) = 0.$$

  (a) Sketch the solution using the method of characteristics.
  (b) Obtain the solution using Fourier-series-type techniques.
  (c) Obtain the solution by converting to an equivalent problem on an infinite domain.

11.5.4. How should initial conditions be extended if $\partial u/\partial x\,(0, t) = 0$ and $u(L, t) = 0$?

11.6 THE METHOD OF CHARACTERISTICS FOR QUASI-LINEAR PARTIAL DIFFERENTIAL EQUATIONS

11.6.1 Method of Characteristics

Most of this text describes methods for solving linear partial differential equations (separation of variables, eigenfunction expansions, Fourier and Laplace transforms, Green's functions) that cannot be extended to nonlinear problems. However, the method of characteristics, used to solve the wave equation, can be applied to partial differential equations of the form

$$\frac{\partial \rho}{\partial t} + c \frac{\partial \rho}{\partial x} = Q, \qquad (11.6.1)$$

where c and Q may be functions of x, t, and ρ. When the coefficient c depends on the unknown solution ρ, (11.6.1) is not linear. Superposition is not valid. Nonetheless (11.6.1) is called a **quasi-linear** partial differential equation, since it is linear in the first partial derivatives, $\partial \rho / \partial t$ and $\partial \rho / \partial x$. To solve (11.6.1), we again consider an observer moving in some prescribed way $x(t)$. By comparing (11.2.4) and (11.6.1), we obtain

$$\boxed{\frac{d\rho}{dt} = Q(\rho, x, t),} \qquad (11.6.2a)$$

if

$$\boxed{\frac{dx}{dt} = c(\rho, x, t).} \qquad (11.6.2b)$$

The partial differential equation (11.6.1) reduces to two coupled ordinary differential equations along the special trajectory or direction defined by (11.6.2b), known as a **characteristic curve**, or simply a **characteristic** for short. The velocity defined by (11.6.2b) is called the **characteristic velocity**, or **local wave velocity**. A characteristic starting from $x = x_0$, as illustrated in Fig. 11.6.1, is determined from the coupled differential equations (11.6.2) using the initial conditions $\rho(x, 0) = f(x)$. Along the characteristic, the solution ρ changes according to (11.6.2a). Other initial positions yield other characteristics, generating a family of characteristics.

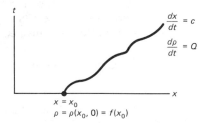

Figure 11.6.1 Characteristic starting from $x = x_0$ at time $t = 0$.

Example. If the local wave velocity c is a constant c_0 and $Q = 0$, then the quasi-linear partial differential equation (11.6.1) becomes the linear one, (11.2.3), which arises in the analysis of the wave equation. In this example, the characteristics may be obtained by directly integrating (11.6.2b) without using (11.6.2a). Each characteristic has the same constant velocity, c_0. The family of characteristics are parallel straight lines, as sketched in Fig. 11.2.1.

11.6.2 Traffic Flow

Traffic density and flow.   As an approximation it is possible to model a congested one-directional highway by a quasi-linear partial differential equation. We introduce the **traffic density** $\rho(x, t)$, the number of cars per mile at time t located at position x. An easily observed and measured quantity is the **traffic flow** $q(x, t)$, the number of cars per hour passing a fixed place x (at time t).

Conservation of cars.   We consider an arbitrary section of roadway, between $x = a$ and $x = b$. If there are neither entrances nor exits on this segment of the road, then the number of cars between $x = a$ and $x = b$ ($N = \int_a^b \rho(x, t)\, dx$, the definite integral of the density) might still change in time. The rate of change of the number of cars, dN/dt, equals the number per unit time entering at $x = a$ (the traffic flow $q(a, t)$ there) minus the number of cars per unit time leaving at $x = b$ (the traffic flow $q(b, t)$ there):

$$\frac{d}{dt} \int_a^b \rho(x, t)\, dx = q(a, t) \quad q(b, t). \tag{11.6.3}$$

Equation (11.6.3) is called the integral form of **conservation of cars.** As with heat flow, a partial differential equation may be derived from (11.6.3) in several equivalent ways. One way is to note that the boundary contribution may be expressed as an integral over the region:

$$q(a, t) - q(b, t) - -\int_a^b \frac{\partial}{\partial x} q(x, t)\, dx. \tag{11.6.4}$$

Thus, by taking the time-derivative inside the integral (making it a partial derivative) and using (11.6.4), it follows that

$$\frac{\partial \rho}{\partial t} + \frac{\partial q}{\partial x} - 0, \tag{11.6.5}$$

since a and b are arbitrary (see Section 1.2). We call (11.6.5) **conservation of cars.**

Car velocity.   The number of cars per hour passing a place equals the density of cars times the velocity of cars. By introducing $u(x, t)$ as the **car velocity,** we have

$$q = \rho u. \tag{11.6.6}$$

In the mid-1950s, Lighthill and Whitham* and, independently, Richards† made

---

* Lighthill, M. J. and Whitham, G. B., "On kinematic waves II. A theory of traffic flow on long crowded roads," *Proc. Roy. Soc. A*, 229, 317–345 (1955).

† Richards, P. I., "Shock waves on the highway," *Operations Research* 4, 42–51 (1956).

a simplifying assumption, namely, that the car velocity depends only on the density, $u = u(\rho)$, with cars slowing down as the traffic density increases—i.e., $du/d\rho \leq 0$. For further discussion, the interested reader is referred to Whitham [1974] and Haberman [1977]. Under this assumption, the traffic flow is only a function of the traffic density, $q = q(\rho)$. In this case, conservation of cars (11.6.5) becomes

$$\frac{\partial \rho}{\partial t} + c(\rho) \frac{\partial \rho}{\partial x} = 0, \qquad (11.6.7)$$

where $c(\rho) = q'(\rho)$, a quasi-linear partial differential equation with $Q = 0$ (see (11.6.1)). Here $c(\rho)$ is considered to be a known function of the unknown solution ρ. In any physical problem in which a density ρ is conserved and the flow q is a function of density, ρ satisfies (11.6.7).

11.6.3 Method of Characteristics (Q = 0)

The equations for the characteristics for (11.6.7) are

$$\frac{d\rho}{dt} = 0 \qquad (11.6.8a)$$

along

$$\frac{dx}{dt} = c(\rho). \qquad (11.6.8b)$$

The characteristic velocity c is not constant but depends on the density ρ. It is known as the **density wave velocity**. From (11.6.8a), it follows that the density ρ remains constant along each as yet undetermined characteristic. The velocity of each characteristic, $c(\rho)$, will be constant, since ρ is constant. Each characteristic is thus a straight line (as in the case in which $c(\rho)$ is a constant c_0). However, different characteristics will move at different constant velocities because they may start with different densities. The characteristics, though each is straight, are not parallel to one another. Consider the characteristic that is initially at the position $x = x_0$, as shown in Fig. 11.6.2. Along the curve $dx/dt = c(\rho)$, $d\rho/dt = 0$ or ρ is constant. Initially ρ equals the value at $x = x_0$ (i.e., at $t = 0$). Thus, along this one characteristic,

$$\rho(x, t) = \rho(x_0, 0) = f(x_0), \qquad (11.6.9)$$

which is a known constant. The local wave velocity that determines the char-

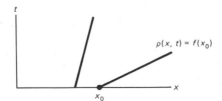

$\rho(x, t) = f(x_0)$

x_0

Figure 11.6.2 Possibly nonparallel straight-line characteristics.

The Method of Characteristics   Chap. 11

acteristic is a constant, $dx/dt = c(f(x_0))$. Consequently, this characteristic is a straight line,

$$x = c(f(x_0))t + x_0, \qquad (11.6.10)$$

since $x = x_0$ at $t = 0$. Different values of x_0 yield different straight-line characteristics, perhaps as illustrated in Fig. 11.6.2. Along each characteristic, the traffic density ρ is a constant; see (11.6.9). To determine the density at some later time, the characteristic with parameter x_0 that goes through that space-time point must be obtained from (11.6.10).

Graphical solution. In practice, it is often difficult and not particularly interesting actually to determine x_0 from (11.6.10) as an explicit function of x and t. Instead, a graphical procedure may be used to determine $\rho(x, t)$. Suppose the initial density is as sketched in Fig. 11.6.3. We know that each density ρ_0 stays the same moving at its own constant density wave velocity $c(\rho_0)$. At time t, the density ρ_0 will have moved a distance $c(\rho_0)t$, as illustrated by the arrow in Fig. 11.6.3. This process must be carried out for a large number of points (as is elementary to do on any computer). In this way, we could obtain the density at time t.

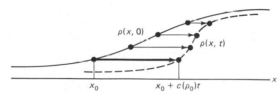

Figure 11.6.3 Graphical solution.

Fan-like characteristics. As an example of the method of characteristics, we consider the following initial value problem:

$$\frac{\partial \rho}{\partial t} + 2\rho \frac{\partial \rho}{\partial x} = 0$$

$$\rho(x, 0) = \begin{cases} 3 & x < 0 \\ 4 & x > 0. \end{cases}$$

The density $\rho(x, t)$ is constant moving with the characteristic velocity 2ρ:

$$\frac{dx}{dt} = 2\rho.$$

Thus, the characteristics are given by

$$x = 2\rho(x_0, 0)t + x_0. \qquad (11.6.11)$$

If $x_0 > 0$, then $\rho(x_0, 0) = 4$, while if $x_0 < 0$, then $\rho(x_0, 0) = 3$. The characteristics, sketched in Fig. 11.6.4, show that

$$\rho(x, t) = \begin{cases} 4 & x > 8t \\ 3 & x < 6t, \end{cases}$$

as illustrated in Fig. 11.6.5. The distance between $\rho = 3$ and $\rho = 4$ is increasing; we refer to the solution as an **expansion wave**. But, what happens for $6t < x < 8t$?

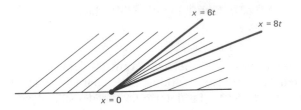

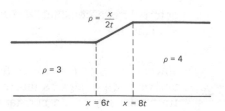

Figure 11.6.4 Characteristics (including the fan-like ones).

Figure 11.6.5 Expansion wave.

The difficulty is caused by the initial density having a discontinuity at $x = 0$.
We imagine that all values of ρ between 3 and 4 are present initially at $x = 0$.
There will be a straight line characteristic along which ρ equals each value
between 3 and 4. Since these characteristics start from $x = 0$ at $t = 0$, it follows
from (11.6.11) that the equation for these characteristics is

$$x = 2\rho t, \qquad \text{for } 3 < \rho < 4,$$

also sketched in Fig. 11.6.4. In this way, we obtain the density in the wedge-
shaped region

$$\rho = \frac{x}{2t} \qquad \text{for} \qquad 6t < x < 8t,$$

which is linear in x (for fixed t). We note that the characteristics fan out from
$x = 6t$ to $x = 8t$ and hence are called **fan-like characteristics**. The resulting
density is sketched in Fig. 11.6.5. It could also be obtained by the graphical
procedure.

11.6.4 Shock Waves

Intersecting characteristics. The method of characteristics will not always
work as we have previously described. For quasi-linear partial differential equations, it is quite usual for characteristics to intersect. The resolution will require
the introduction of moving discontinuities called *shock waves*. In order to make
the mathematical presentation relatively simple, we restrict our attention to quasi-
linear partial differential equations with $Q = 0$, in which case

$$\frac{\partial \rho}{\partial t} + c(\rho) \frac{\partial \rho}{\partial x} = 0. \tag{11.6.12}$$

In Fig. 11.6.6 two characteristics are sketched, one starting at $x = x_1$, with $\rho = f(x_1, 0) \equiv \rho_1$ and the other starting at $x = x_2$ with $\rho = f(x_2, 0) \equiv \rho_2$. These
characteristics intersect if $c(\rho_1) > c(\rho_2)$, the faster catching up to the slower.

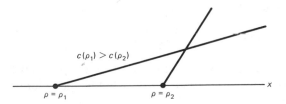

Figure 11.6.6 Intersecting
characteristics.

The density is constant along characteristics. As time increases, the distance between the densities ρ_1 and ρ_2 decreases. Thus, this is called a **compression wave.** We sketch the initial condition in Fig. 11.6.7(a). The density distribution becomes steeper as time increases (Fig. 11.6.7(b) and (c)). Eventually characteristics intersect; the theory predicts the density is simultaneously ρ_1 and ρ_2. If we continue to apply the method of characteristics, the faster-moving characteristic passes the slower. Then we obtain Fig. 11.6.7(d). The method of characteristics predicts that the density becomes a "multivalued" function of position; that is, at some later time our mathematics predicts there will be three densities at some positions [as illustrated in Fig. 11.6.7(d)]. We say the density wave *breaks*. However, in many physical problems (such as traffic flow) it makes no sense to have three values of density at one place.* The density must be a single-valued function of position.

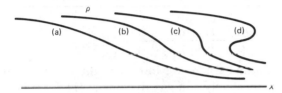

Figure 11.6.7 Density wave steepens (density becomes triple-valued).

Discontinuous solutions. On the basis of the quasi-linear partial differential equation (11.6.12), we predicted the physically impossible phenomenon that the density becomes multivalued. Since the method of characteristics is mathematically justified, it is the partial differential equation itself which must not be entirely valid. Some approximation or assumption that we used must at times be invalid. We will assume that the density (as illustrated in Fig. 11.6.8) and velocity have a jump-discontinuity, which we call a **shock wave**, or simply a **shock**.† The shock

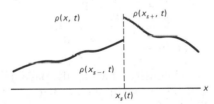

Figure 11.6.8 Density discontinuity at $x = x_s(t)$.

---

* The partial differential equations describing the height of water waves near the shore (i.e., in shallow water) are similar to the equations for traffic density waves. In this situation the prediction of breaking is then quite significant!

† The terminology *shock wave* is introduced because of the analogous behavior that occurs in gas dynamics. There, changes in pressure and density of air, for example, propagate and are heard (due to the sensitivity of the human ear). They are called *sound waves*. When fluctuations of pressure and density are small, the equations describing sound waves can be linearized. Then sound is propagated at a constant speed known as the *sound speed*. However, if the amplitudes of the fluctuations of pressure and density are not small, then the partial differential equations are quasi-linear. Characteristics may intersect. In this case, the pressure and density can be mathematically modeled as being discontinuous, the result being called a *shock wave*. Examples are the sound emitted from an explosion or the thunder resulting from lightning. If a shock wave results from exceeding the sound barrier, it is known as a *sonic boom*.

occurs at some unknown position x_s and propagates in time, so that $x_s(t)$. We introduce the notation x_{s-} and x_{s+} for the position of the shock on the two sides of the discontinuity. The shock velocity, dx_s/dt, is as yet unknown.

Shock velocity. On either side of the shock, the quasi-linear partial differential equation applies, $\partial\rho/\partial t + c(\rho)\partial\rho/\partial x = 0$, where $c(\rho) = dq(\rho)/d\rho$. We need to determine how the discontinuity propagates. If ρ is conserved even at a discontinuity, then the flow relative to the moving shock on one side of the shock must equal the flow relative to the moving shock on the other side. This statement of relative inflow equaling relative outflow becomes

$$\rho(x_{s-}, t)\left[u(x_{s-}, t) - \frac{dx_s}{dt}\right] = \rho(x_{s+}, t)\left[u(x_{s+}, t) - \frac{dx_s}{dt}\right], \quad (11.6.13)$$

since flow equals density times velocity (here relative velocity). Solving for the shock velocity from (11.6.13) yields

$$\frac{dx_s}{dt} = \frac{q(x_{s+}, t) - q(x_{s-}, t)}{\rho(x_{s+}, t) - \rho(x_{s-}, t)} = \frac{[q]}{[\rho]}, \quad (11.6.14)$$

where we recall that $q = \rho u$ and where we introduce the notation $[q]$ and $[\rho]$ for the jumps in q and ρ, respectively. In gas dynamics, (11.6.14) is called the Rankine-Hugoniot condition. *In summary, for the conservation law $\partial\rho/\partial t + \partial q/\partial x = 0$ (if the quantity $\int \rho\, dx$ is actually conserved), the shock velocity equals the jump in the flow divided by the jump in the density of the conserved quantity.* At points of discontinuity, this shock condition replaces the use of the partial differential equation, which is valid elsewhere. However, we have not yet explained where shocks occur and how to determine $\rho(x_{s+}, t)$ and $\rho(x_{s-}, t)$.

Example. We consider the initial value problem

$$\frac{\partial\rho}{\partial t} + 2\rho\frac{\partial\rho}{\partial x} = 0$$

$$\rho(x, 0) = \begin{cases} 4 & x < 0 \\ 3 & x > 0. \end{cases}$$

We assume that ρ is a conserved density. Putting the partial differential equation in conservation form ($\partial\rho/\partial t + \partial q/\partial x = 0$) shows that the flow $q = \rho^2$. Thus, if there is a discontinuity, the shock velocity satisfies $dx/dt = [q]/[\rho] = [\rho^2]/[\rho]$. The density $\rho(x, t)$ is constant moving at the characteristic velocity 2ρ:

$$\frac{dx}{dt} = 2\rho.$$

Therefore, the equation for the characteristics is

$$x = 2\rho(x_0, 0)\, t + x_0.$$

If $x_0 < 0$, then $\rho(x_0, 0) = 4$. This parallel group of characteristics intersects those starting from $x_0 > 0$ (with $\rho(x_0, 0) = 3$) in the cross-hatched region in Fig. 11.6.9a. The method of characteristics yields a multi-valued solution of the partial differential equation. This difficulty is remedied by introducing a shock wave (Fig. 11.6.9b), a propagating wave indicating the path at which densities and

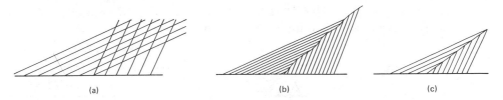

Figure 11.6.9 Shock caused by intersecting characteristics.

velocities abruptly change (i.e., are discontinuous). On one side of the shock, the method of characteristics suggests the density is constant $\rho = 4$, and on the other side $\rho = 3$. We do not know as yet the path of the shock. The theory for such a discontinuous solution implies that the path for any shock must satisfy the shock condition, (11.6.14). Substituting the jumps in flow and density yields the following equation for the shock velocity:

$$\frac{dx_s}{dt} = \frac{q(4) - q(3)}{4 - 3} = \frac{4^2 - 3^2}{4 - 3} = 7,$$

since in this case $q = \rho^2$. Thus, the shock moves at a constant velocity. The initial position of the shock is known, giving a condition for this first-order ordinary differential equation. In this case, the shock must initiate at $x_s = 0$ at $t = 0$. Consequently, applying the initial condition results in the position of the shock, $x_s = 7t$. The resulting space-time diagram is sketched in Fig. 11.6.9(c). For any time $t > 0$, the traffic density is discontinuous, as shown in Fig. 11.6.10.

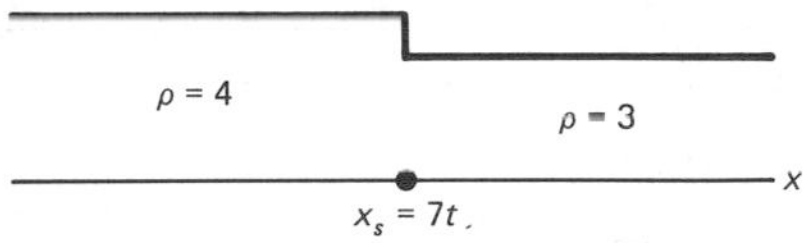

$\rho = 4$ $\rho = 3$

$x_s = 7t$.

Figure 11.6.10 Density shock wave.

Initiation of a shock. We have described the propagation of shock waves. In the example considered, the density was initially discontinuous; thus, the shock wave formed immediately. However, we will now show that shock waves take a finite time to form if the initial density is continuous. Suppose that the first shock occurs at $t = \tau$ due to the intersection of two characteristics initially a distance Δx (not necessarily small) apart. However, any characteristic starting between the two at $t = 0$ will almost always intersect one of the other two characteristics before $t = \tau$. Thus shocks cannot first occur due to characteristics that are a finite distance Δx apart. Instead, the first shock actually occurs due to the intersection of neighboring characteristics (the limit as $\Delta x \rightarrow 0$). We will show that even though $\Delta x \rightarrow 0$, the first intersection occurs at a finite positive time, the time of the earliest shock. The density ρ is constant along characteristics, satisfying $dx/dt = c(\rho)$. We will analyze neighboring characteristics. Consider the characteristic emanating from $x = x_0$ at $t = 0$, where $\rho(x, 0) = f(x)$,

$$x = c[f(x_0)]t + x_0 \tag{11.6.15}$$

and the characteristic starting from $x = x_0 + \Delta x$ at $t = 0$,

$$x = c[f(x_0 + \Delta x)]t + x_0 + \Delta x.$$

Only if $c[f(x_0)] > c[f(x_0 + \Delta x)]$ will these *characteristics intersect* (in a positive time). Solving for the intersection point by eliminating x yields
$$c[f(x_0)]t + x_0 = c[f(x_0 + \Delta x)]t + x_0 + \Delta x.$$
Therefore, the time at which nearly neighboring curves intersect is
$$t = \frac{\Delta x}{c[f(x_0)] - c[f(x_0 + \Delta x)]} = \frac{1}{\{c[f(x_0)] - c[f(x_0 + \Delta x)]\}/\Delta x}.$$
The characteristics are paths of observers following constant density. Then this equation states that the time of intersection of the two characteristics is the initial distance between the characteristics divided by the relative velocity of the two characteristics. Although the distance in between is small, the relative velocity is also small. To consider neighboring characteristics, the limit as $\Delta x \to 0$ must be calculated:
$$t = \frac{-1}{dc/dx_0}. \tag{11.6.16}$$
Characteristics will intersect ($t > 0$) only if $(d/dx_0)\,c[f(x_0)] < 0$. Thus, we conclude that all neighboring characteristics that emanate from regions where the characteristic velocity is *locally* decreasing will always intersect. To determine the first time at which an intersection (shock) occurs, we must minimize the intersection time over all possible neighboring characteristics, i.e., find the *absolute* minimum of t given by equation (11.6.16). This can be calculated by determining where $d^2/dx_0^2\,c[f(x_0)] = 0$.

Shock dynamics. We will show that the slope of the solution is infinite where neighboring characteristics intersect. Since $\rho(x, t) = \rho(x_0, 0)$, we have
$$\frac{\partial \rho}{\partial x} = \frac{d\rho}{dx_0}\frac{\partial x_0}{\partial x} = \frac{d\rho}{dx_0}\bigg/\left[1 + \frac{dc}{dx_0}t\right].$$
This has also used the result of partial differentiation of (11.6.15) with respect to x:
$$1 = \frac{\partial x_0}{\partial x}\left[1 + \frac{dc}{dx_0}t\right],$$
The slope is infinite at those places satisfying (11.6.16). This shows that the turning points of the triple-valued solution correspond to the intersection of neighboring characteristics (the envelope of the characteristics). Within the envelope of characteristics, the solution is triple-valued. It is known that the envelope of the characteristics is cusp-shaped, as indicated in Fig. 11.6.11. How-

Shock

Figure 11.6.11 Envelope of characteristics, locus of intersections of neighboring characteristics.

x

The Method of Characteristics   Chap. 11

ever, the triple-valued solution (within the cusp region) makes no sense. Instead, as discussed earlier, a shock wave exists satisfying (11.6.14), initiating at the cusp point. The shock is located within the envelope. In fact, the triple-valued solution, obtained by the method of characteristics, may be used to determine the location of the shock. Whitham [1974] has shown that the correct location of the shock may be determined by cutting the lobes off to form equal areas (Fig. 11.6.12). The reason for this is that the method of characteristics conserves cars and that, when a shock is introduced, the number of cars (represented by the area $\int \rho \, dx$) must also be the same as it is initially.

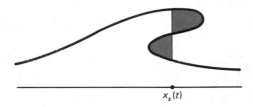

$x_s(t)$

Figure 11.6.12 Whitham's equal-area principle.

EXERCISES 11.6

11.6.1. Determine the solution $\rho(x, t)$ satisfying the initial condition $\rho(x, 0) = f(x)$ if

   *(a) $\dfrac{\partial \rho}{\partial t} = 0$ **(b)** $\dfrac{\partial \rho}{\partial t} = -3\rho + 4e^{7t}$

   *(c) $\dfrac{\partial \rho}{\partial t} = -3x\rho$ **(d)** $\dfrac{\partial \rho}{\partial t} = x^2 t \rho$

*11.6.2. Determine the solution of $\partial \rho / \partial t = \rho$, which satisfies $\rho(x, t) = 1 + \sin x$ along $x = -2t$.

11.6.3. Suppose $\dfrac{\partial \rho}{\partial t} + c_0 \dfrac{\partial \rho}{\partial x} = 0$ with c_0 constant.

   *(a) Determine $\rho(x, t)$ if $\rho(x, 0) = \sin x$.
   *(b) If $c_0 > 0$, determine $\rho(x, t)$ for $x > 0$ and $t > 0$, where
        $\rho(x, 0) = f(x)$ for $x > 0$ and $\rho(0, t) = g(t)$ for $t > 0$.
   (c) Show that part (b) cannot be solved if $c_0 < 0$.

*11.6.4. If $u(\rho) = \alpha + \beta\rho$, determine α and β such that $u(0) = u_{max}$ and $u(\rho_{max}) = 0$.
   (a) What is the flow as a function of density? Graph the flow as a function of the density.
   (b) At what density is the flow maximum? What is the corresponding velocity? What is the maximum flow (called the *capacity*)?

11.6.5. Redo Exercise 11.6.4 if $u(\rho) = u_{max}(1 - \rho^3/\rho_{max}^3)$.

11.6.6. Consider the traffic flow problem

$$\frac{\partial \rho}{\partial t} + c(\rho) \frac{\partial \rho}{\partial x} = 0.$$

Assume $u(\rho) = u_{max}(1 - \rho/\rho_{max})$. Solve for $\rho(x, t)$ if the initial conditions are
   (a) $\rho(x, 0) = \rho_{max}$ for $x < 0$ and $\rho(x, 0) = 0$ for $x > 0$. This corresponds to the traffic density that results after an infinite line of stopped traffic is started by a red light turning green.

(b) $\rho(x, 0) = \begin{cases} \rho_{max} & x < 0 \\ \dfrac{\rho_{max}}{2} & 0 < x < a \\ 0 & x > a \end{cases}$
  (c) $\rho(x, 0) = \begin{cases} \dfrac{3\rho_{max}}{5} & x < 0 \\ \dfrac{\rho_{max}}{5} & x > 0 \end{cases}$

11.6.7. Solve the following problems:

(a) $\dfrac{\partial \rho}{\partial t} + \rho^2 \dfrac{\partial \rho}{\partial x} = 0$

$$\rho(x, 0) = \begin{cases} 3 & x < 0 \\ 4 & x > 0 \end{cases}$$

(b) $\dfrac{\partial \rho}{\partial t} + 4\rho \dfrac{\partial \rho}{\partial x} = 0$

$$\rho(x, 0) = \begin{cases} 2 & x < 1 \\ 3 & x > 1 \end{cases}$$

(c) $\dfrac{\partial \rho}{\partial t} + 3\rho \dfrac{\partial \rho}{\partial x} = 0$

$$\rho(x, 0) = \begin{cases} 1 & x < 0 \\ 2 & 0 < x < 1 \\ 4 & x > 1 \end{cases}$$

(d) $\dfrac{\partial \rho}{\partial t} + 6\rho \dfrac{\partial \rho}{\partial x} = 0$ for $x > 0$ only.

$$\rho(x, 0) = 5 \quad x > 0$$
$$\rho(0, t) = 2 \quad t > 0$$

11.6.8. Solve subject to the initial condition $\rho(x, 0) = f(x)$:

*(a) $\dfrac{\partial \rho}{\partial t} + c \dfrac{\partial \rho}{\partial x} = e^{-3x}$
  (b) $\dfrac{\partial \rho}{\partial t} + 3x \dfrac{\partial \rho}{\partial x} = 4$

*(c) $\dfrac{\partial \rho}{\partial t} + t \dfrac{\partial \rho}{\partial x} = 5$
  (d) $\dfrac{\partial \rho}{\partial t} + 5t \dfrac{\partial \rho}{\partial x} = 3\rho$

*(e) $\dfrac{\partial \rho}{\partial t} - t^2 \dfrac{\partial \rho}{\partial x} = -\rho$
  (f) $\dfrac{\partial \rho}{\partial t} + t^2 \dfrac{\partial \rho}{\partial x} - 0$

*(g) $\dfrac{\partial \rho}{\partial t} + x \dfrac{\partial \rho}{\partial x} = t.$

11.6.9. Determine a parametric representation of the solution satisfying $\rho(x, 0) = f(x)$:

*(a) $\dfrac{\partial \rho}{\partial t} - \rho^2 \dfrac{\partial \rho}{\partial x} = 3\rho$
  (b) $\dfrac{\partial \rho}{\partial t} + \rho \dfrac{\partial \rho}{\partial x} = t$

*(c) $\dfrac{\partial \rho}{\partial t} + t^2 \rho \dfrac{\partial \rho}{\partial x} = -\rho$
  (d) $\dfrac{\partial \rho}{\partial t} + \rho \dfrac{\partial \rho}{\partial x} = -x\rho$

11.6.10. Solve $\dfrac{\partial \rho}{\partial t} + t^2 \dfrac{\partial \rho}{\partial x} = 4\rho$ for $x > 0$ and $t > 0$ with $\rho(0, t) = h(t)$ and $\rho(x, 0) = 0$.

11.6.11. Solve $\dfrac{\partial \rho}{\partial t} + (1 + t) \dfrac{\partial \rho}{\partial x} = 3\rho$ for $t > 0$ and $x > -t/2$ with $\rho(x, 0) = f(x)$ for $x > 0$ and $\rho(x, t) = g(t)$ along $x = -t/2$.

11.6.12. Consider (11.6.3) if there is a moving shock x_s such that $a < x_s(t) < b$. By differentiating the integral [with a discontinuous integrand at $x_s(t)$], *derive* (11.6.14).

11.6.13. Suppose that, instead of $u = U(\rho)$, a car's velocity u is

$$u = U(\rho) - \dfrac{\nu}{\rho} \dfrac{\partial \rho}{\partial x},$$

where ν is a constant.
(a) What sign should ν have for this expression to be physically reasonable?
(b) What equation now describes conservation of cars?
(c) Assume that $U(\rho) = u_{max}(1 - \rho/\rho_{max})$. Derive **Burgers' equation**:

$$\dfrac{\partial \rho}{\partial t} + u_{max}\left[1 - \dfrac{2\rho}{\rho_{max}}\right] \dfrac{\partial \rho}{\partial x} = \nu \dfrac{\partial^2 \rho}{\partial x^2}.$$

11.6.14. Consider Burgers' equation as derived in Exercise 11.6.13. Suppose that a solution exists as a density wave moving without change of shape at velocity V, $\rho(x, t) = f(x - Vt)$.
  *(a) What ordinary differential equation is satisfied by f?
  (b) Integrate this differential equation once. By graphical techniques show that a solution exists such that $f \to \rho_2$ as $x \to +\infty$ and $f \to \rho_1$ as $x \to -\infty$ only if $\rho_2 > \rho_1$. Roughly sketch this solution. Give a physical interpretation of this result.
  *(c) Show that the velocity of wave propagation, V, is the same as the shock velocity separating $\rho = \rho_1$ from $\rho = \rho_2$ (occurring if $\nu = 0$).

11.6.15. Consider Burgers' equation as derived in Exercise 11.6.13. Show that the change of dependent variables

$$\rho = \frac{\nu \rho_{\max}}{u_{\max}} \frac{\phi_x}{\phi},$$

introduced independently by E. Hopf and J. D. Cole, transforms Burgers' equation into a diffusion equation, $\dfrac{\partial \phi}{\partial t} + u_{\max} \dfrac{\partial \phi}{\partial x} = \nu \dfrac{\partial^2 \phi}{\partial x^2}$. Use this to solve the initial value problem $\rho(x, 0) = f(x)$ for $-\infty < x < \infty$. [In Whitham [1974] it is shown that this exact solution can be asymptotically analyzed as $\nu \to 0$ using Laplace's method for exponential integrals to show that $\rho(x, t)$ approaches the solution obtained for $\nu = 0$ using the method of characteristics with shock dynamics.]

11.6.16. Suppose that the initial traffic density is $\rho(x, 0) = \rho_0$ for $x < 0$ and $\rho(x, 0) = \rho_1$ for $x > 0$. Consider the two cases, $\rho_0 < \rho_1$ and $\rho_1 < \rho_0$. For which of the preceding cases is a density shock necessary? Briefly explain.

11.6.17. Consider a traffic problem, with $u(\rho) = u_{\max}(1 - \rho/\rho_{\max})$. Determine $\rho(x, t)$ if

$$\text{*(a)} \quad \rho(x, 0) = \begin{cases} \dfrac{\rho_{\max}}{5} & x < 0 \\ \dfrac{3\rho_{\max}}{5} & x > 0 \end{cases} \qquad \text{(b)} \quad \rho(x, 0) = \begin{cases} \dfrac{\rho_{\max}}{3} & x < 0 \\ \dfrac{2\rho_{\max}}{3} & x > 0 \end{cases}$$

11.6.18. Assume that $u(\rho) = u_{\max}(1 - \rho^2/\rho_{\max}^2)$. Determine the traffic density ρ (for $t > 0$) if $\rho(x, 0) = \rho_1$ for $x < 0$ and $\rho(x, 0) = \rho_2$ for $x > 0$.
  (a) Assume that $\rho_2 > \rho_1$.   ***(b)** Assume that $\rho_2 < \rho_1$.

11.6.19. Solve the following problems:

(a) $\dfrac{\partial \rho}{\partial t} + \rho^2 \dfrac{\partial \rho}{\partial x} = 0$

$$\rho(x, 0) = \begin{cases} 4 & x < 0 \\ 3 & x > 0 \end{cases}$$

(b) $\dfrac{\partial \rho}{\partial t} + 4\rho \dfrac{\partial \rho}{\partial x} = 0$

$$\rho(x, 0) = \begin{cases} 3 & x < 1 \\ 2 & x > 1 \end{cases}$$

(c) $\dfrac{\partial \rho}{\partial t} + 3\rho \dfrac{\partial \rho}{\partial x} = 0$

$$\rho(x, 0) = \begin{cases} 4 & x < 0 \\ 2 & 0 < x < 1 \\ 1 & x > 1 \end{cases}$$

(d) $\dfrac{\partial \rho}{\partial t} + 6\rho \dfrac{\partial \rho}{\partial x} = 0$ for $x > 0$ only.

$$\rho(x, 0) = 2 \qquad x > 0$$
$$\rho(0, t) = 5 \qquad t > 0$$

12

A Brief Introduction
to Laplace Transform Solution
of Partial Differential Equations

12.1 INTRODUCTION

We have introduced some techniques to solve linear partial differential equations. For problems with a simple geometry, the method of separation of variables motivates using Fourier series, its various generalizations, or variants of the Fourier transform. Of most importance is the type of boundary condition, including whether the domain is finite, infinite or semi-infinite. In some problems a Green's function can be utilized, while for the one-dimensional wave equation the method of characteristics exists. Whether or not any of these methods may be appropriate, numerical methods (as will be discussed in Chapter 13) are often most efficient.

Another technique, to be elaborated on in this chapter, relies on the use of Laplace transforms. *Most problems in partial differential equations that can be analyzed by Laplace transforms also can be analyzed by one of our earlier techniques, and substantially equivalent answers can be obtained.* The use of Laplace transforms is advocated by those who feel more comfortable with them than with our other methods. Instead of taking sides, we will present the elementary aspects of Laplace transforms in order to enable the reader to become somewhat familiar with them. However, whole books* have been written concerning their use in partial differential equations. Consequently, in this chapter we only briefly discuss Laplace transforms and describe their application to partial differential equations with only a few examples.

---

* For example, Churchill [1972].

12.2 ELEMENTARY PROPERTIES OF THE LAPLACE TRANSFORM

12.2.1 Introduction

Definition. One technique for solving ordinary differential equations (mostly with constant coefficients) is to introduce the **Laplace transform of** $f(t)$ as follows:

$$\mathscr{L}[f(t)] = F(s) = \int_0^\infty f(t)e^{-st}\,dt. \qquad (12.2.1)$$

For the Laplace transform to be defined the integral in (12.2.1) must converge. For many functions, $f(t)$, s is restricted. For example, if $f(t)$ approaches a nonzero constant as $t \to \infty$, then the integral converges if $s > 0$. If s is complex, $s = \text{Re}(s) + i\,\text{Im}(s)$ and $e^{-st} = e^{-\text{Re}(s)t}[\cos(\text{Im}(s)t) - i\sin(\text{Im}(s)t)]$, then it follows in this case that $\text{Re}(s) > 0$ for convergence.

We will assume that the reader has studied (at least briefly) Laplace transforms. We will review quickly the important properties of Laplace transforms. Tables exist,* and we include a short one here. The Laplace transform of some elementary functions can be obtained by direct integration. Some fundamental properties can be derived from the definition; these and others are summarized in Table 12.2.1 (see next page).

From the definition of the Laplace transform, $f(t)$ is only needed for $t > 0$. So there is no confusion we usually define $f(t)$ to be zero for $t < 0$. One formula (12.2.2l) requires the Heaviside unit step function:

$$H(t - b) = \begin{cases} 0 & t < b \\ 1 & t > b. \end{cases} \qquad (12.2.3)$$

Inverse Laplace transforms. If instead we are given $F(s)$ and want to calculate $f(t)$, then we can also use the same tables. $f(t)$ is called the **inverse Laplace transform of** $F(s)$. The notation $f(t) = \mathscr{L}^{-1}[F(s)]$ is also used. For example, from the tables the inverse Laplace transform of $1/(s - 3)$ is e^{3t}, $\mathscr{L}^{-1}[1/(s - 3)] = e^{3t}$.

Not all functions of s have inverse Laplace transforms. From (12.2.1) we notice that if $f(t)$ is any type of ordinary function, then $F(s) \to 0$ as $s \to \infty$. All functions in our table have this property.

12.2.2 Singularities of the Laplace Transform

We note that when $f(t)$ is a simple exponential, $f(t) = e^{at}$, the growth rate a is also the point at which its Laplace transform $F(s) = 1/(s - a)$ has a singularity. As $s \to a$, the Laplace transform approaches ∞. We claim *in general* that as a check in *any* calculation **the singularities of a Laplace transform $F(s)$** (*the zeros*

---

* Some better ones are in Churchill [1972]; *CRC Standard Mathematical Tables* [1981] (from Churchill's table); Abramowitz and Stegun [1965] (also has Churchill's table); and Roberts and Kaufman [1966].

TABLE 12.2.1 LAPLACE TRANSFORMS (short table of formulas and properties)

$f(t)$	$F(s) \equiv \mathcal{L}[f(t)] \equiv \int_0^\infty f(t)e^{-st}\, dt$	
1	$\dfrac{1}{s}$	(12.2.2a)
$t^n\,(n > -1)$	$n!\, s^{-(n+1)}$	(12.2.2b)
e^{at}	$\dfrac{1}{s-a}$	(12.2.2c)
$\sin \omega t$	$\dfrac{\omega}{s^2 + \omega^2}$	(12.2.2d)
$\cos \omega t$	$\dfrac{s}{s^2 + \omega^2}$	(12.2.2e)
$\sinh at = \dfrac{1}{2}(e^{at} - e^{-at})$	$\dfrac{1}{2}\left(\dfrac{1}{s-a} - \dfrac{1}{s+a}\right) = \dfrac{a}{s^2 - a^2}$	(12.2.2f)
$\cosh at = \dfrac{1}{2}(e^{at} + e^{-at})$	$\dfrac{1}{2}\left(\dfrac{1}{s-a} + \dfrac{1}{s+a}\right) = \dfrac{s}{s^2 - a^2}$	(12.2.2g)
$\dfrac{df}{dt}$	$sF(s) - f(0)$	(12.2.2h)
$\dfrac{d^2f}{dt^2}$	$s^2F(s) - sf(0) - \dfrac{df}{dt}(0)$	(12.2.2i)
$-tf(t)$	$\dfrac{dF}{ds}$	(12.2.2j)
$e^{at}f(t)$	$F(s - a)$	(12.2.2k)
$H(t - b)f(t - b)$	$e^{-bs}F(s)\quad (b > 0)$	(12.2.2l)
$\displaystyle\int_0^t f(t - \bar{t})g(\bar{t})\, d\bar{t}$	$F(s)G(s)$	(12.2.2m)
$\delta(t - b)$	$e^{-bs}\quad (b \geq 0)$	(12.2.2n)
$\dfrac{1}{2\pi i}\displaystyle\int_{\gamma - i\infty}^{\gamma + i\infty} F(s)e^{st}\, ds$	$F(s)$	(12.2.2o)
$t^{-1/2}e^{-a^2/4t}$	$\sqrt{\dfrac{\pi}{s}}\, e^{-a\sqrt{s}}\quad (a \geq 0)$	(12.2.2p)
$t^{-3/2}e^{-a^2/4t}$	$\dfrac{2\sqrt{\pi}}{a}\, e^{-a\sqrt{s}}\quad (a > 0)$	(12.2.2q)

Elementary functions (Exercises 12.2.1 and 12.2.2)

Fundamental properties (Sec. 12.2.3 and Exercise 12.2.3)

Convolution (Sec. 12.2.4)

Dirac delta function (Sec. 12.2.4)

Inverse transform (Sec. 12.7)

Miscellaneous (Exercise 12.2.9)

of its denominator) **correspond** (*in some way*) **to the exponential growth rates of** $f(t)$. We refer to this as the **singularity property** of Laplace transforms. Later we will show this using complex variables. Throughout this chapter we illustrate this correspondence.

Examples. For now we briefly discuss some examples. Both the Laplace transforms $\omega/(s^2 + \omega^2)$ and $s/(s^2 + \omega^2)$ have singularities at $s^2 + \omega^2 = 0$ or

$s = \pm i\omega$. Thus, their inverse Laplace transforms will involve exponentials e^{st}, where $s = \pm i\omega$. According to the tables their inverse Laplace transforms are respectively $\sin \omega t$ and $\cos \omega t$, which we know from Euler's formulas can be represented as linear combinations of $e^{\pm i\omega t}$.

As another example, consider the Laplace transform $F(s) = 3/[s(s^2 + 4)]$. One method to determine $f(t)$ is to use partial fractions (with real factors):

$$\frac{3}{s(s^2 + 4)} = \frac{a}{s} + \frac{bs + c}{s^2 + 4} = \frac{\frac{3}{4}}{s} + \frac{-\frac{3}{4}s}{s^2 + 4}.$$

Now the inverse transform is easy to obtain using tables:

$$f(t) = \tfrac{3}{4} - \tfrac{3}{4} \cos 2t.$$

As a check we note that $3/[s(s^2 + 4)]$ has singularities at $s = 0$ and $s = \pm 2i$. The singularity property of Laplace transforms then implies that its inverse Laplace transform must be a linear combination of e^{0t} and $e^{\pm 2it}$, as we have already seen.

Partial fractions. In doing inverse Laplace transforms, we are frequently faced with the ratio of two polynomials $q(s)/p(s)$. To be a Laplace transform, it must approach 0 as $s \to \infty$. Thus, we can assume that the degree of p is greater than the degree of q. A partial-fraction expansion will yield immediately the desired inverse Laplace transform. We only describe this technique in the case in which the roots of the denominator are simple; there are no repeated or multiple roots. First we factor the denominator

$$p(s) = \alpha(s - s_1)(s - s_2) \cdots (s - s_n),$$

where $s_1, \ldots, s_n$ are the n *distinct* roots of $p(s)$, also called the **simple poles** of $q(s)/p(s)$. The partial-fraction expansion of $q(s)/p(s)$ is

$$\frac{q(s)}{p(s)} = \frac{c_1}{s - s_1} + \frac{c_2}{s - s_2} + \cdots + \frac{c_n}{s - s_n}. \tag{12.2.4}$$

The coefficients c_i of the partial-fraction expansion can be obtained by cumbersome algebraic manipulations using a common denominator. A more elegant and sometimes quicker method utilizes the singularities s_i of $p(s)$. To determine c_i, we multiply (12.2.4) by $s - s_i$ and then take the limit as $s \to s_i$. All the terms except c_i vanish on the right:

$$c_i = \lim_{s \to s_i} \frac{(s - s_i)q(s)}{p(s)}. \tag{12.2.5}$$

Often, this limit is easy to evaluate. Since $s - s_i$ is a factor of $p(s)$ we cancel it in (12.2.5), and then evaluate the limit.

Example. Using complex roots,

$$\frac{3}{s(s^2 + 4)} = \frac{c_1}{s} + \frac{c_2}{s + 2i} + \frac{c_3}{s - 2i},$$

where

$$c_1 = \lim_{s \to 0} s \frac{3}{s(s^2 + 4)} = \frac{3}{4}$$

$$c_2 = \lim_{s \to -2i} (s + 2i) \frac{3}{s(s^2 + 4)} = \lim_{s \to -2i} \cancel{(s + 2i)} \frac{3}{s\cancel{(s + 2i)}(s - 2i)} = -\frac{3}{8}$$

$$c_3 = \lim_{s \to 2i} (s - 2i) \frac{3}{s(s^2 + 4)} = \lim_{s \to 2i} \cancel{(s - 2i)} \frac{3}{s(s + 2i)\cancel{(s - 2i)}} = -\frac{3}{8}.$$

Simple poles. In some problems, we can make the algebra even easier. The limit in (12.2.5) is 0/0 since $p(s_i) = 0$ [$s = s_i$ is a root of $p(s)$]. L'Hôpital's rule for evaluating 0/0 yields

$$c_i = \lim_{s \to s_i} \frac{d/ds\,[(s - s_i)q(s)]}{d/ds\,p(s)} = \frac{q(s_i)}{p'(s_i)}. \qquad (12.2.6)$$

Equation (12.2.6) is valid only for simple poles.

Once we have a partial-fraction expansion of a Laplace transform, its inverse transform may be easily obtained. In summary, if

$$F(s) = \frac{q(s)}{p(s)}, \qquad (12.2.7a)$$

then by inverting (12.2.4),

$$f(t) = \sum_{i=1}^{n} \frac{q(s_i)}{p'(s_i)} e^{s_i t}, \qquad (12.2.7b)$$

where we assumed that $p(s)$ has only simple poles at $s = s_i$.

Example. To apply this formula for $q(s)/p(s) = 3/[s(s^2 + 4)]$, we let $q(s) = 3$ and $p(s) = s(s^2 + 4) = s^3 + 4s$. We need $p'(s) = 3s^2 + 4$. Thus, if

$$F(s) = \frac{3}{s(s^2 + 4)} = \frac{c_1}{s} + \frac{c_2}{s + 2i} + \frac{c_3}{s - 2i},$$

then

$$c_1 = \frac{q(0)}{p'(0)} = \frac{3}{4}, \qquad c_2 = \frac{q(-2i)}{p'(-2i)} = \frac{3}{-8}, \qquad \text{and} \qquad c_3 = \frac{q(2i)}{p'(2i)} = \frac{3}{-8},$$

as before. For this example,

$$f(t) = \tfrac{3}{4} - \tfrac{3}{8}e^{-2it} - \tfrac{3}{8}e^{2it} = \tfrac{3}{4} - \tfrac{3}{4}\cos 2t.$$

Quadratic expressions (completing the square). Inverse Laplace transforms for quadratic expressions

$$F(s) = \frac{\alpha s + \beta}{as^2 + bs + c}$$

can be obtained by partial fractions if the roots are real or complex. However, if the roots are complex, it is often easier to complete the square. For example, consider

$$F(s) = \frac{1}{s^2 + 2s + 8} = \frac{1}{(s + 1)^2 + 7},$$

whose roots are $s = -1 \pm i\sqrt{7}$. Since a function of $s + 1$ appears, we use the shift theorem:

$$F(s) = G(s + 1) \qquad \text{where} \qquad G(s) = \frac{1}{s^2 + 7}.$$

According to the shift theorem the inverse transform of $G(s + 1)$ is (using $a = -1$) $f(t) = e^{-t}g(t)$, where $g(t)$ is the inverse transform of $1/(s^2 + 7)$. From the tables $g(t) = (1/\sqrt{7}) \sin \sqrt{7}\, t$, and thus

$$f(t) = \frac{1}{\sqrt{7}} e^{-t} \sin \sqrt{7}\, t.$$

This result is consistent with the singularity property; the solution is a linear combination of e^{st}, where $s = -1 \pm i\sqrt{7}$.

12.2.3 Transforms of Derivatives

One of the most useful properties of the Laplace transform is the way in which it operates on derivatives. For example, by elementary integration by parts:

$$\boxed{\mathscr{L}\left[\frac{df}{dt}\right]} = \int_0^\infty \frac{df}{dt} e^{-st}\, dt = fe^{-st}\Big|_0^\infty + s\int_0^\infty fe^{-st}\, dt \boxed{= sF(s) - f(0).}$$

$$(12.2.8)$$

Similarly,

$$\boxed{\mathscr{L}\left[\frac{d^2f}{dt^2}\right]} = s\mathscr{L}\left[\frac{df}{dt}\right] - \frac{df}{dt}(0) = s(sF(s) - f(0)) - \frac{df}{dt}(0)$$

$$(12.2.9)$$

$$\boxed{= s^2F(s) - sf(0) - \frac{df}{dt}(0).}$$

This property shows that the transform of derivatives can be evaluated in terms of the transform of the function. Certain "initial" conditions are needed. For the transform of the first derivative df/dt, $f(0)$ is needed. For the transform of the second derivative d^2f/dt^2, $f(0)$ and df/dt (0) are needed. These are just the types of information that are known if the variable t is time. Usually, *if a Laplace transform is used, the transformed variable t is time*. Furthermore, **the Laplace transform method will often simplify if the initial conditions are all zero.**

Application to ordinary differential equations. For ordinary differential equations, the use of the Laplace transform reduces the problem to an algebraic equation. For example, consider

$$\frac{d^2y}{dt^2} + 4y = 3$$

$$\text{with} \qquad y(0) = 1$$

$$\frac{dy}{dt}(0) = 5.$$

Taking the Laplace transform of the differential equation yields

$$s^2 Y(s) - s - 5 + 4Y(s) = \frac{3}{s},$$

where $Y(s)$ is the Laplace transform of $y(t)$. Thus,

$$Y(s) = \frac{1}{s^2 + 4}\left(\frac{3}{s} + s + 5\right) = \frac{3}{s(s^2 + 4)} + \frac{s}{s^2 + 4} + \frac{5}{s^2 + 4}.$$

The inverse transforms of $s/(s^2 + 4)$ and $5/(s^2 + 4)$ are easily found in tables. The function whose transform is $3/[s(s^2 + 4)]$ has been obtained in different ways. Thus,

$$y(t) = \tfrac{3}{4} - \tfrac{3}{4}\cos 2t + \cos 2t + \tfrac{5}{2}\sin 2t.$$

12.2.4 Convolution Theorem

Another method to obtain the inverse Laplace transform of $3/[s(s^2 + 4)]$ is to use the convolution theorem. We begin by stating and deriving the convolution theorem. Often, as in this example, we need to obtain the function whose Laplace transform is the product of two transforms, $F(s)G(s)$. The **convolution theorem** states that

$$\mathscr{L}^{-1}[F(s)G(s)] = g * f = \int_0^t g(\bar{t})f(t - \bar{t})\,d\bar{t}, \qquad (12.2.10a)$$

where $g * f$ is called the **convolution** of g and f. Equivalently,

$$\mathscr{L}\left[\int_0^t g(\bar{t})f(t - \bar{t})\,d\bar{t}\right] = F(s)G(s). \qquad (12.2.10b)$$

Earlier, when studying Fourier transforms (see Sec. 9.5) we also introduced the convolution of g and f in a slightly different way,

$$g * f \equiv \int_{-\infty}^{\infty} f(t - \bar{t})g(\bar{t})\,d\bar{t}.$$

However, in the context of Laplace transforms, both $f(t)$ and $g(t)$ are zero for $t < 0$, and thus (12.2.10a) follows since $f(t - \bar{t}) = 0$ for $\bar{t} > t$ and $g(\bar{t}) = 0$ for $\bar{t} < 0$.

Laplace transform of Dirac delta functions. One derivation of the convolution theorem uses the Laplace transform of a Dirac delta function:

$$\mathscr{L}\{\delta(t - b)\} = \int_0^\infty \delta(t - b)e^{-st} \, dt = e^{-sb}, \tag{12.2.11}$$

if $b > 0$. Thus, the inverse Laplace transform of an exponential is a Dirac delta function:

$$\mathscr{L}^{-1}\{e^{-sb}\} = \delta(t - b). \tag{12.2.12}$$

In the limit as $b \to 0$, we also obtain

$$\mathscr{L}\{\delta(t - 0+)\} = 1 \quad \text{and} \quad \mathscr{L}^{-1}\{1\} = \delta(t - 0+). \tag{12.2.13}$$

Derivation of convolution theorem. To derive the convolution theorem, we introduce two transforms $F(s)$ and $G(s)$ and their product $F(s)G(s)$:

$$F(s) = \int_0^\infty f(t)e^{-st} \, dt \tag{12.2.14a}$$

$$G(s) = \int_0^\infty g(t)e^{-st} \, dt \tag{12.2.14b}$$

$$F(s)G(s) = \int_0^\infty \int_0^\infty f(\bar{t})g(\bar{\bar{t}})e^{-s(\bar{t}+\bar{\bar{t}})} \, d\bar{\bar{t}} \, d\bar{t}. \tag{12.2.14c}$$

$h(t)$ is the inverse Laplace transform of $F(s)G(s)$:

$$h(t) = \mathscr{L}^{-1}\{F(s)G(s)\} = \int_0^\infty \int_0^\infty f(\bar{t})g(\bar{\bar{t}})\mathscr{L}^{-1}\{e^{-s(\bar{t}+\bar{\bar{t}})}\} \, d\bar{t} \, d\bar{\bar{t}},$$

where the linearity of the inverse Laplace transform has been utilized. However, the inverse Laplace transform of an exponential is a Dirac delta function (12.2.12), and thus

$$h(t) = \int_0^\infty \int_0^\infty f(\bar{t})g(\bar{\bar{t}}) \, \delta[t - (\bar{t} + \bar{\bar{t}})] \, d\bar{t} \, d\bar{\bar{t}}.$$

Performing the $\bar{\bar{t}}$ integration first, we obtain a contribution only at $\bar{\bar{t}} = t - \bar{t}$. Therefore, the fundamental property of Dirac delta functions implies that

$$h(t) = \int_0^t f(\bar{t})g(t - \bar{t}) \, d\bar{t},$$

the convolution theorem for Laplace transforms. By letting $t - \bar{t} = w$, we also derive that $g * f = f * g$; the order is of no importance.

Example. Determine the function whose Laplace transform is $3/[s(s^2 + 4)]$, using the convolution theorem. We introduce

$$F(s) = \frac{3}{s} \quad \text{(so that } f(t) = 3\text{)} \quad \text{and} \quad G(s) = \frac{1}{s^2 + 4} \quad \text{(so that } g(t) = \tfrac{1}{2} \sin 2t\text{)}.$$

It follows from the convolution theorem that the inverse transform of $(3/s)[1/(s^2 + 4)]$ is $\int_0^t f(t - \bar{t})g(\bar{t}) \, d\bar{t}$:

$$\int_0^t 3 \cdot \tfrac{1}{2} \sin 2\bar{t} \, d\bar{t} = -\tfrac{3}{4} \cos 2\bar{t} \Big|_0^t = \tfrac{3}{4}(1 - \cos 2t),$$

as we obtained earlier using the partial-fraction expansion.

EXERCISES 12.2

12.2.1. From the definition of the Laplace transform (i.e., using explicit integration), determine the Laplace transform of $f(t) = $:
  (a) 1 (b) e^{at}
  (c) $\sin \omega t$ [*Hint:* $\sin \omega t = \text{Im } (e^{i\omega t})$.]   (d) $\cos \omega t$ [*Hint:* $\cos \omega t = \text{Re } (e^{i\omega t})$.]
  (e) $\sinh at$ (f) $\cosh at$
  (g) $H(t - t_0), \; t_0 > 0$

12.2.2. The gamma function $\Gamma(x)$ was defined in Exercise 9.4.14. Derive that $\mathcal{L}[t^n] = \Gamma(n + 1)/s^{n+1}$ for $n > -1$. Why is this not valid for $n \leqslant -1$?

12.2.3. Derive the following fundamental properties of Laplace transforms:
  (a) $\mathcal{L}[-tf(t)] = dF/ds$
  (b) $\mathcal{L}[e^{at}f(t)] = F(s - a)$
  (c) $\mathcal{L}[H(t - b)f(t - b)] = e^{-bs}F(s) \quad (b > 0)$

*12.2.4. Using the tables, determine the Laplace transform of

$$\int_0^t f(\bar{t}) \, d\bar{t}$$

in terms of $F(s)$.

12.2.5. Using the tables of Laplace transforms, determine the Laplace transform of $f(t) = $:
  (a) $t^3 e^{-2t}$ ***(b)** $t \sin 4t$
  (c) $H(t - 3)$ ***(d)** $e^{3t} \sin 4t$
  (e) $te^{-4t} \cos 6t$ ***(f)** $f(t) = \begin{cases} 0 & t < 5 \\ t^2 & 5 < t < 8 \\ 0 & 8 < t \end{cases}$
  (g) $t^2 H(t - 1)$ ***(h)** $(t - 1)^4 H(t - 1)$

12.2.6. Using the tables for Laplace transforms, determine the inverse Laplace transform of $F(s) = $:
  (a) $\dfrac{1}{s^2 + 4}$ (b) $\dfrac{e^{-3s}}{s^2 - 4}$
  (c) s^{-3} (d) $(s - 4)^{-7}$
  *(e) $\dfrac{s}{s^2 + 8s + 7}$ (f) $\dfrac{2s + 1}{s^2 - 4s + 9}$
  (g) $\dfrac{s}{(s^2 + 1)(s^2 + 4)}$ (h) $\dfrac{s}{s^2 - 4s - 5}$

(i) $\dfrac{s}{s^2 - 4s - 5}(1 - 4e^{-7s})$

*(j) $\dfrac{s + 2}{s(s^2 + 9)}(1 - 5e^{-4s})$

(k) $\dfrac{1}{(s + 1)^2}$

(l) $\dfrac{1}{(s^2 + 1)^2}$

12.2.7. Solve the following ordinary differential equations using Laplace transforms:

(a) $\dfrac{d^2y}{dt^2} + 3\dfrac{dy}{dt} + y = t^3$ with $y(0) = 7$ and $\dfrac{dy}{dt}(0) = 5$

*(b) $\dfrac{dy}{dt} + y = 1$ with $y(0) = 2$

(c) $\dfrac{dy}{dt} + 3y = \begin{cases} 4e^{-t} & t < 8 \\ 2 & t > 8 \end{cases}$ with $y(0) = 1$

*(d) $\dfrac{d^2y}{dt^2} + 5\dfrac{dy}{dt} - 6y = \begin{cases} 0 & 0 < t < 3 \\ e^{-t} & t > 3 \end{cases}$ with $y(0) = 3$ and $\dfrac{dy}{dt}(0) = 7$

(e) $\dfrac{d^2y}{dt^2} + y = \cos t$ with $y(0) = 0$ and $\dfrac{dy}{dt}(0) = 0$

*(f) $\dfrac{d^2y}{dt^2} + 4y = \sin t$ with $y(0) = 0$ and $\dfrac{dy}{dt}(0) = 0$

12.2.8. Derive the convolution theorem for Laplace transforms without using the Dirac delta function. [*Hint:* Introduce the variable $z = \bar{t} + \bar{\bar{t}}$ in order to evaluate the double integral (12.2.14c).]

12.2.9. In this problem we will determine

$$I = \mathcal{L}\{t^{-3/2}e^{-a^2/4t}\} \quad \text{and} \quad J = \mathcal{L}\{t^{-1/2}e^{-a^2/4t}\}.$$

(a) Determine a relationship between I and J by substituting $u = s^{1/2}t^{1/2} - (a/2)t^{-1/2}$ into $\int_{-\infty}^{\infty} e^{-u^2} du = \sqrt{\pi}$.

(b) Determine a relationship between I and J by introducing the change of variables $sw = (a^2/4)t^{-1}$ into the definition of I.

(c) Derive that $I = (2\sqrt{\pi}/a)e^{-a\sqrt{s}}$ and $J = (\sqrt{\pi/s})e^{-a\sqrt{s}}$ using parts (a) and (b).

12.3 GREEN'S FUNCTIONS FOR INITIAL VALUE PROBLEMS FOR ORDINARY DIFFERENTIAL EQUATIONS

The convolution theorem is very useful in solving nonhomogeneous ordinary differential equations. For example, consider

$$\alpha\frac{d^2y}{dt^2} + \beta\frac{dy}{dt} + \gamma y = f(t), \tag{12.3.1a}$$

subject to zero* initial conditions

$$y(0) = 0 \tag{12.3.1b}$$

$$\frac{dy}{dt}(0) = 0. \tag{12.3.1c}$$

* Nonzero initial conditions can be analyzed by adding appropriate homogeneous solutions of the differential equation.

Taking the Laplace transform of the differential equation (12.3.1) yields

$$(\alpha s^2 + \beta s + \gamma)Y(s) = F(s) \qquad \text{or} \qquad Y(s) = F(s) \cdot \frac{1}{\alpha s^2 + \beta s + \gamma}, \qquad (12.3.2)$$

where $Y(s)$ and $F(s)$ are the Laplace transforms of $y(t)$ and $f(t)$, respectively. The solution $y(t)$ can be obtained using the convolution theorem

$$y(t) = \int_0^t f(t_0)q(t - t_0)\, dt_0, \qquad (12.3.3)$$

where $q(t)$ is the inverse Laplace transform of $1/(\alpha s^2 + \beta s + \gamma)$. We can determine $q(t)$ using tables and/or partial fractions. This result, (12.3.3), will be equivalent to the solution of nonhomogeneous problems as is usually obtained by the method of variation of parameters in most elementary texts on ordinary differential equations.

There is an important alternative interpretation of this result. $q(t)$ is the solution of (12.3.1) if $F(s) = 1$. The inverse Laplace transform of $F(s) = 1$ is $f(t) = \delta(t - 0+)$ [see (12.2.13)]. Thus, $q(t)$ is the response due to an impulse at $t = 0+$:

$$\alpha \frac{d^2q}{dt^2} + \beta \frac{dq}{dt} + \gamma q = \delta(t - 0+)$$

$$q(0) = 0 \qquad (12.3.4)$$

$$\frac{dq}{dt}(0) = 0.$$

We can introduce the terminology of Chapters 8 and 9. We call $q(t)$ the **Green's function for the initial value problem**, the response at t due to a concentrated source at $t = 0$:

$$q(t) = G(t, 0).$$

The convolution theorem shows that we are interested in $q(t - t_0)$:

$$q(t - t_0) = G(t - t_0, 0).$$

However, due to the constant coefficients present in (12.3.4), **the response at t due to an impulse at t_0, $G(t, t_0)$, is the same as the response due to an impulse at 0, if the elapsed time is the same,

$$G(t, t_0) = G(t - t_0, 0), \qquad (12.3.5)$$

the **translation property** of the Green's function. Thus,

$$q(t - t_0) = G(t, t_0).$$

Therefore, from (12.3.3), through the use of Laplace transforms, we have obtained a representation of the solution of the nonhomogeneous initial value problem (12.3.1) involving the Green's function

$$y(t) = \int_0^t f(t_0)G(t, t_0)\, dt_0. \qquad (12.3.6)$$

The solution is the generalized superposition of all sources acting before the time t, an example of the **causality principle** for initial value problems for ordinary differential equations. In this form the result appears quite similar to our results concerning Green's functions for boundary value problems for ordinary and partial differential equations. Here the Green's function $h(t) = G(t, 0)$ is simply the inverse Laplace transform of $1/(\alpha s^2 + \beta s + \gamma)$.

Example. Consider the differential equation

$$\alpha^2 \frac{d^2 y}{dt^2} - \gamma^2 y = f(t). \tag{12.3.7}$$

The solution that satisfies zero initial conditions is

$$y(t) = \int_0^t f(t_0) G(t, t_0)\, dt_0,$$

where the Green's function $G(t, t_0)$ satisfies $q(t - t_0) = G(t, t_0)$. We introduce the Laplace transform of $q(t)$:

$$\mathscr{L}[q(t)] = \frac{1}{\alpha^2 s^2 - \gamma^2} = \frac{1}{\alpha^2(s^2 - \gamma^2/\alpha^2)}.$$

Directly from tables, we may obtain the Green's function, $G(t, 0)$,

$$G(t, 0) = q(t) = \frac{1}{\alpha^2} \frac{\alpha}{\gamma} \sinh \frac{\gamma}{\alpha} t.$$

Thus, the solution of (12.3.7) is

$$y(t) = \frac{1}{\alpha \gamma} \int_0^t f(t_0) \sinh \frac{\gamma}{\alpha}(t - t_0)\, dt_0, \tag{12.3.8}$$

where $y(0) = 0$ and $dy/dt\,(0) = 0$.

EXERCISES 12.3

12.3.1. By using Laplace transforms determine the effect of the initial conditions in terms of the Green's function for the initial value problem,

$$\alpha \frac{d^2 y}{dt^2} + \beta \frac{dy}{dt} + \gamma y = 0$$

subject to $y(0) = y_0$ and $\dfrac{dy}{dt}(0) = v_0$.

*12.3.2. What is the Green's function for

$$\frac{d^2 y}{dt^2} + y = f(t)$$

with $y(0) = 0$ and $\dfrac{dy}{dt}(0) = 0$? Solve for $y(t)$.

12.3.3. (a) Do Exercise 8.3.25(a).
(b) Do Exercise 8.3.25(b).

12.3.4. Show that for $t > t_0$, $G(t, t_0)$ for (12.3.1) satisfies

$$\alpha \frac{d^2G}{dt^2} + \beta \frac{dG}{dt} + \gamma G = 0$$

with $G(t_0, t_0) = 0$ and $\dfrac{dG}{dt}(t_0, t_0) = \dfrac{1}{\alpha}$.

12.3.5. Solve Exercise 12.3.2 using Exercise 12.3.4.

12.4 AN ELEMENTARY SIGNAL PROBLEM FOR THE WAVE EQUATION

Using Laplace transforms to solve partial differential equations often requires great skill in the use of Laplace transforms. We only pursue some relatively simple examples.* Consider a semi-infinite string ($x > 0$), whose motion is caused only by a time-dependent boundary condition at $x = 0$:

PDE:
$$\frac{\partial^2 u}{\partial t^2} = c^2 \frac{\partial^2 u}{\partial x^2} \qquad (12.4.1a)$$

BC:
$$u(0, t) = f(t) \qquad (12.4.1b)$$

IC:
$$u(x, 0) = 0 \qquad (12.4.1c)$$
$$\frac{\partial u}{\partial t}(x, 0) = 0. \qquad (12.4.1d)$$

The string is initially horizontal and at rest. The left end is being moved vertically (while maintaining the large tension). Since the problem is defined for all $x > 0$, we need a boundary condition as $x \to \infty$,

$$\lim_{x \to \infty} u(x, t) = 0. \qquad (12.4.1e)$$

All the initial conditions of this problem are zero. Consequently, the use of Laplace transforms is expected to yield a simple solution:

$$\mathscr{L}[u(x, t)] = \overline{U}(x, s) = \int_0^\infty u(x, t)e^{-st}\, dt.$$

As with ordinary differential equations, we take the Laplace transform of (12.4.1a),

$$\mathscr{L}\left[\frac{\partial^2 u}{\partial t^2}\right] = s^2 \mathscr{L}[u] - su(x, 0) - \frac{\partial u}{\partial t}(x, 0) = s^2 \mathscr{L}[u]. \qquad (12.4.2)$$

Here, we also need the Laplace transforms of partial derivatives with respect to x. We obtain

$$\mathscr{L}\left[\frac{\partial^2 u}{\partial x^2}\right] = \int_0^\infty \frac{\partial^2 u}{\partial x^2}e^{-st}\, dt = \frac{\partial^2}{\partial x^2}\int_0^\infty u(x, t)e^{-st}\, dt = \frac{\partial^2}{\partial x^2}\mathscr{L}[u]. \qquad (12.4.3)$$

In this manner, the Laplace transform of a partial differential equation yields an

* For more difficult examples, see Churchill [1972] and Weinberger [1965].

"ordinary" differential equation

$$s^2 \overline{U}(x, s) = c^2 \frac{\partial^2 \overline{U}}{\partial x^2}, \qquad (12.4.4a)$$

defined for $0 < x < \infty$. At $x = 0$, $u(x, t)$ is given for all t, and thus its Laplace transform is known:

$$\overline{U}(0, s) = \int_0^\infty u(0, t)e^{-st}\, dt = \int_0^\infty f(t)e^{-st}\, dt = F(s), \qquad (12.4.4b)$$

where $F(s)$ is the Laplace transform of the boundary condition. Also since $u(x, t) \to 0$ as $x \to \infty$ (for all fixed t), we have the same result for its Laplace transform,

$$\lim_{x \to \infty} \overline{U}(x, s) = 0. \qquad (12.4.4c)$$

The general solution of (12.4.4a) is

$$\overline{U}(x, s) = A(s)e^{-(s/c)x} + B(s)e^{(s/c)x},$$

where $A(s)$ and $B(s)$ are arbitrary functions of the transform variable s. For $s > 0$ [more precisely Re $(s) > 0$], $B(s) = 0$ to satisfy the decay as $x \to \infty$, (12.4.4c). In addition, the boundary condition at $x = 0$, (12.4.4b), implies that $A(s) = F(s)$, and thus

$$\overline{U}(x, s) = F(s)e^{-(s/c)x}. \qquad (12.4.5)$$

To invert this transform, we could use the convolution theorem. Instead, a quick glance at our table shows that an exponential multiple in the transform yields a time shift in the solution. Consequently,

$$\boxed{u(x, t) = H\left(t - \frac{x}{c}\right)f\left(t - \frac{x}{c}\right),} \qquad (12.4.6)$$

where H is the Heaviside unit step function. The solution is zero for $x > ct$. In fact, the solution is constant whenever $x - ct$ is constant. The solution travels as a wave of fixed shape at velocity c. We obtained similar results in Chapter 11, using the method of characteristics. We illustrate this in a space-time diagram in Fig. 12.4.1. The signal propagates with velocity c, and thus at time t it has traveled only a distance ct. If $x > ct$, the "wiggling" of the string at $x = 0$ has not been noticed.

However, it is educational to obtain the same result using the convolution theorem. Since $\overline{U}(x, s) = F(s)e^{-(s/c)x}$,

$$u(x, t) = \int_0^t f(t_0)g(t - t_0)\, dt_0,$$

where $e^{-(s/c)x}$ is the Laplace transform of $g(t)$. From a table of Laplace transforms (as can be easily verified), $g(t) = \delta(t - x/c)$. Thus,

$$u(x, t) = \int_0^t f(t_0)\, \delta\left(t - t_0 - \frac{x}{c}\right) dt_0 = \begin{cases} 0 & t < x/c \\ f(t - x/c) & t > x/c, \end{cases}$$

which is equivalent to (12.4.6).

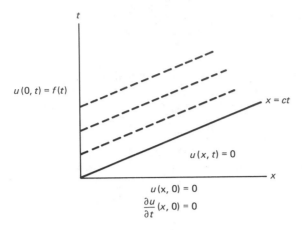

$u(0, t) = f(t)$

$x = ct$

$u(x, t) = 0$

$u(x, 0) = 0$

$\dfrac{\partial u}{\partial t}(x, 0) = 0$

Figure 12.4.1 Signal problem for the one-dimensional wave equation.

EXERCISES 12.4

12.4.1. Solve

$$\frac{\partial^2 u}{\partial t^2} = c^2 \frac{\partial^2 u}{\partial x^2}$$

subject to $\dfrac{\partial u}{\partial x}(0, t) = f(t)$, $u(x, 0) = 0$ and $\dfrac{\partial u}{\partial t}(x, 0) = 0$.

12.4.2. Solve

$$\frac{\partial w}{\partial t} + c \frac{\partial w}{\partial x} = 0 \qquad c > 0, x > 0, t > 0$$

$$w(0, t) = f(t)$$

$$w(x, 0) = 0.$$

*12.4.3. Solve

$$\frac{\partial^2 u}{\partial t^2} = c^2 \frac{\partial^2 u}{\partial x^2} \qquad -\infty < x < \infty$$

$$u(x, 0) = \sin x$$

$$\frac{\partial u}{\partial t}(x, 0) = 0.$$

*12.4.4. Consider

$$\frac{\partial u}{\partial t} = k \frac{\partial^2 u}{\partial x^2} \qquad x > 0$$

$$u(x, 0) = 0$$

$$u(0, t) = f(t).$$

Determine the Laplace transform of $u(x, t)$. Invert to obtain $u(x, t)$. (*Hint:* See a table of Laplace transforms.)

12.4.5. Reconsider Exercise 12.4.4 if instead the boundary and initial conditions are

$$u(x, 0) = 0 \qquad \text{and} \qquad \frac{\partial u}{\partial x}(0, t) = f(t).$$

12.4.6. Reconsider Exercise 12.4.4 if $f(t) = Ae^{i\sigma_0 t}$ (see Exercise 9.6.17).

    (a) Determine an expression for $u(x, t)$ using Laplace transforms.

    (b) Simplify part (a) with the change of variables $w = x/2\sqrt{t - \bar{t}}$, where $\bar{t}$ is the variable of integration in part (a).

    (c) Approximate $u(x, t)$ if t is large.

12.5 A SIGNAL PROBLEM FOR A VIBRATING STRING OF FINITE LENGTH

In partial differential equations the Laplace inversion integrals that are needed are often not as simple as in Sec. 12.4. We illustrate this for a vibrating string of length L initially at rest in the horizontal equilibrium position subject to the following time-dependent boundary condition at one end $x = L$:

PDE:
$$\frac{\partial^2 u}{\partial t^2} = c^2 \frac{\partial^2 u}{\partial x^2} \tag{12.5.1a}$$

BC:
$$u(0, t) = 0 \tag{12.5.1b}$$
$$u(L, t) = b(t) \tag{12.5.1c}$$

IC:
$$u(x, 0) = 0 \tag{12.5.1d}$$
$$\frac{\partial u}{\partial t}(x, 0) = 0. \tag{12.5.1e}$$

The zero initial conditions facilitate the use of the Laplace transform in t of $u(x, t)$:

$$\overline{U}(x, s) = \int_0^\infty e^{-st} u(x, t)\, dt. \tag{12.5.2}$$

By transforming (12.5.1), $\overline{U}(x, s)$ satisfies the ordinary differential equation

$$s^2 \overline{U} = c^2 \frac{\partial^2 \overline{U}}{\partial x^2}, \tag{12.5.3a}$$

subject to the boundary conditions

$$\overline{U}(0, s) = 0 \tag{12.5.3b}$$

$$\overline{U}(L, s) = B(s), \tag{12.5.3c}$$

where $B(s)$ is the Laplace transform of $b(t)$. We can easily determine $\overline{U}(x, s)$:

$$\overline{U}(x, s) = B(s) \frac{\sinh (s/c)x}{\sinh (s/c)L}. \tag{12.5.4}$$

The convolution theorem implies that

$$u(x, t) = \int_0^t b(t_0) f(t - t_0)\, dt_0, \tag{12.5.5}$$

where $f(t)$ is the inverse Laplace transform of

$$F(s) = \frac{\sinh (s/c)x}{\sinh (s/c)L}. \tag{12.5.6}$$

To obtain this inverse Laplace transform is not straightforward. One method to obtain the inverse transform of (12.5.6) is to attempt to use our elementary tables, in which primarily exponentials appear. We note that

$$\frac{\sinh (s/c)x}{\sinh (s/c)L} = \frac{e^{(s/c)x} - e^{-(s/c)x}}{e^{(s/c)L} - e^{-(s/c)L}} = \frac{e^{(s/c)x} - e^{-(s/c)x}}{e^{(s/c)L}(1 - e^{-(2s/c)L})}.$$

However, due to the denominator, this cannot be analyzed in a simple way. Instead, we can introduce an infinite series of exponentials based on the geometric series $[1/(1 - x) = 1 + x + x^2 + \cdots,$ if $|x| < 1]$:

$$\begin{aligned}
F(s) &= \frac{e^{(s/c)x} - e^{-(s/c)x}}{e^{(s/c)L}(1 - e^{-(2s/c)L})} \\
&= e^{-(s/c)L}(e^{(s/c)x} - e^{-(s/c)x})(1 + e^{-(2L/c)s} + e^{-(4L/c)s} + \cdots) \\
&= \sum_{n=0}^{\infty} \left\{ \exp\left[-s\left(\frac{2nL - x + L}{c} \right) \right] - \exp\left[-s\left(\frac{2nL + x + L}{c} \right) \right] \right\}.
\end{aligned}$$

(12.5.7)

These are all decaying exponentials since $x < L$, and hence each can be inverted using (12.2.2n). The Laplace transform is a linear combination of

$$\exp\left[-s\left(\frac{2nL \pm x + L}{c} \right) \right] \qquad (n \geq 0).$$

The inverse Laplace transform of $F(s)$ is thus a linear combination of Dirac delta functions, $\delta[t - (2nL \pm x + L)/c]$:

$$f(t) = \sum_{n=0}^{\infty} \left[\delta\left(t - \frac{2nL - x + L}{c} \right) - \delta\left(t - \frac{2nL + x + L}{c} \right) \right].$$

Since $f(t - t_0)$ in (12.5.5) is the influence function for the boundary condition, these represent signals whose travel times are $(2nL \pm x + L)/c$ [elapsed from the signal time t_0]. These can be interpreted as direct signals and their reflections off the boundaries $x_0 = 0$ and $x_0 = L$. Since the nonhomogeneous boundary condition is at $x_0 = L$, we imagine these signals are initiated there (at $t = t_0$). The signal can travel to x in different ways as illustrated in Fig. 12.5.1. The direct signal must travel a distance $L - x$ at velocity c, yielding the retarded time $(L - x)/c$ (corresponding to $n = 0$). A signal can also arrive at x by additionally making an integral number of complete circuits, an added travel distance of $2Ln$. The other terms correspond to waves first reflecting off the wall $x_0 = 0$ before impinging on x. In this case the total travel distance is $L + x + 2nL$ $(n \geq 0)$. Further details of the solution are left as exercises.

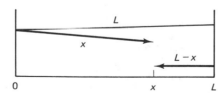

0   x L **Figure 12.5.1** Space-time signal paths.

Using Laplace transforms (and inverting in the way described in this section) yields a representation of the solution as an infinite sequence of reflecting waves. Similar results can be obtained by the method of characteristics (see Chapter 11) or (in some cases) by using the method of images (sequences of infinite space Green's functions).

Alternatively, in subsequent sections, we will describe the use of contour integrals in the complex plane to invert Laplace transforms. This technique will yield a significantly different representation of the same solution.

EXERCISES 12.5

12.5.1. Consider

$$\frac{\partial^2 u}{\partial t^2} - c^2 \frac{\partial^2 u}{\partial x^2} \qquad -\infty < x < \infty$$

$$u(x, 0) = f(x)$$

$$\frac{\partial u}{\partial t}(x, 0) = g(x).$$

Solve using Laplace transforms:
(a) if $g(x) = 0$
(b) if $f(x) = 0$

12.5.2. (a) Using the results of this section, invert (12.5.4) based on the convolution theorem. Solve for $u(x, t)$.
(b) Without using the convolution theorem, from (12.5.4) and (12.5.7), determine $u(x, t)$.

*12.5.3. Solve for $u(x, t)$ using Laplace transforms:

$$\frac{\partial^2 u}{\partial t^2} = c^2 \frac{\partial^2 u}{\partial x^2}$$

$$u(0, t) = 0 \qquad u(x, 0) = 0$$

$$\frac{\partial u}{\partial x}(L, t) - b(t) \qquad \frac{\partial u}{\partial t}(x, 0) - 0.$$

12.5.4. Reconsider Exercise 12.5.3 if instead

(a) $\dfrac{\partial u}{\partial x}(0, t) = 0$ and $u(L, t) = b(t)$

(b) $\dfrac{\partial u}{\partial x}(0, t) = 0$ and $\dfrac{\partial u}{\partial x}(L, t) = b(t)$

12.5.5. Solve for $u(x, t)$ using Laplace transforms

$$\frac{\partial u}{\partial t} = k \frac{\partial^2 u}{\partial x^2} \qquad -\infty < x < \infty$$

$$u(x, 0) = f(x).$$

(*Hint:* See a table of Laplace transforms.)

12.5.6. Solve for $u(x, t)$ using Laplace transforms:

$$\frac{\partial u}{\partial t} = k \frac{\partial^2 u}{\partial x^2}$$

subject to $u(x, 0) = f(x)$, $u(0, t) = 0$, and $u(L, t) = 0$. By what other method(s) can this representation of the solution be obtained?

12.6 *THE WAVE EQUATION AND ITS GREEN'S FUNCTION*

The Laplace transform can be used to determine the relationship between solutions of nonhomogeneous partial differential equations and its corresponding Green's function. Consider the wave equation on a finite interval $(0 < x < L)$ with sources and time-dependent boundary conditions

PDE:
$$\frac{\partial^2 u}{\partial t^2} = c^2 \frac{\partial^2 u}{\partial x^2} + q(x, t) \tag{12.6.1a}$$

BC:
$$u(0, t) = a(t) \tag{12.6.1b}$$
$$u(L, t) = b(t) \tag{12.6.1c}$$

IC:
$$u(x, 0) = f(x) \tag{12.6.1d}$$
$$\frac{\partial u}{\partial t}(x, 0) = g(x). \tag{12.6.1e}$$

Green's function. The Green's function $G(x, t; x_0, t_0)$ satisfies

$$\frac{\partial^2 G}{\partial t^2} = c^2 \frac{\partial^2 G}{\partial x^2} + \delta(x - x_0)\, \delta(t - t_0) \tag{12.6.2a}$$

$$G(0, t; x_0, t_0) = 0 \tag{12.6.2b}$$

$$G(L, t; x_0, t_0) = 0 \tag{12.6.2c}$$

subject to the causality principle

$$G(x, t; x_0, t_0) = 0 \quad \text{for } t < t_0. \tag{12.6.2d}$$

Laplace transform of Green's function. In this section we determine the Green's function using the Laplace transform. The transform of (12.6.2) yields

$$s^2 \overline{G} = c^2 \frac{\partial^2 \overline{G}}{\partial x^2} + \delta(x - x_0)e^{-st_0} \tag{12.6.3a}$$

$$\overline{G}(0, s; x_0, t_0) = 0 \tag{12.6.3b}$$

$$\overline{G}(L, s; x_0, t_0) = 0, \tag{12.6.3c}$$

where $\overline{G}(x, s; x_0, t_0)$ is the Laplace transform in time of $G(x, t; x_0, t_0)$. The transform of (12.6.2a) simplifies because the causality principle implies that G satisfies zero initial conditions (if $t_0 > 0$).

The Laplace transform of the Green's function satisfies (12.6.3a), an ordinary differential equation of the Green's function type. To satisfy the boundary conditions, $\overline{G}(x, s; x_0, t_0)$ must be proportional to $\sinh(s/c)x$ for $x < x_0$ and proportional to $\sinh(s/c)(L - x)$ for $x > x_0$. Since it will be symmetric, we know that

$$\overline{G}(x, s; x_0, t_0) = \begin{cases} \gamma \sinh \dfrac{s}{c}(L - x_0) \sinh \dfrac{s}{c}x & x < x_0 \\ \gamma \sinh \dfrac{s}{c}x_0 \sinh \dfrac{s}{c}(L - x) & x > x_0. \end{cases} \tag{12.6.4a}$$

where γ is a constant (independent of x and x_0). In this manner the continuity of $\overline{G}$ at $x = x_0$ is automatically satisfied. The additional jump condition,

$$0 = c^2 \frac{d\overline{G}}{dx}\bigg|_{x_0-}^{x_0+} + e^{-st_0},$$

determines γ:

$$0 = -c^2 \gamma \frac{s}{c} \left[\sinh \frac{s}{c}x_0 \cosh \frac{s}{c}(L - x_0) + \sinh \frac{s}{c}(L - x_0) \cosh \frac{s}{c}x_0 \right] + e^{-st_0}.$$

By using an addition formula for hyperbolic functions [$\sinh (a + b) = \sinh a \cosh b + \cosh a \sinh b$], we obtain

$$\gamma = \frac{e^{-st_0}}{cs \sinh (s/c)L}. \tag{12.6.4b}$$

Representation of solution in terms of the Green's function. In Exercise 12.6.2 the Green's function itself is obtained by determining the inverse Laplace transform of (12.6.4). Instead, we will investigate further using Laplace transforms the relationships between $u(x, t)$ and its Green's function. To simplify some of our work, we consider the special case of Section 12.5, $q(x, t) = 0$, $a(t) = 0$, $f(x) = 0$, and $g(x) = 0$; the only nonhomogeneous term is the boundary condition at $x = L$. In the preceding section we showed (12.5.4):

$$\overline{U}(x, s) = B(s) \frac{\sinh (s/c)x}{\sinh (s/c)L}. \tag{12.6.5}$$

Here we will relate this to the Laplace transform of the Green's function (12.6.4). Since the source satisfies $x_0 = L$, we need $\overline{G}$ for $x < x_0$:

$$\overline{G}(x, s; x_0, t_0) = \frac{e^{-st_0} \sinh (s/c)(L - x_0) \sinh (s/c)x}{cs \sinh (s/c)L}. \tag{12.6.6}$$

To compare this with (12.6.5), we take the derivative with respect to x_0:

$$\frac{\partial \overline{G}}{\partial x_0}(x, s; x_0, t_0) = -\frac{e^{-st_0} \cosh (s/c)(L - x_0) \sinh (s/c)x}{c^2 \sinh (s/c)L}.$$

We note that at $x_0 = L$ and $t_0 = 0$,

$$\frac{\partial \overline{G}}{\partial x_0}(x, s; L, 0) = -\frac{\sinh (s/c)x}{c^2 \sinh (s/c)L},$$

similar to the term appearing in (12.6.5). Thus,

$$\overline{U}(x, s) = -c^2 B(s) \frac{\partial \overline{G}}{\partial x_0}(x, s; L, 0).$$

Using the convolution theorem, we obtain

$$u(x, t) = -c^2 \int_0^t b(t_0) \frac{\partial G}{\partial x_0}(x, t - t_0; L, 0) \, dt_0,$$

which may be replaced by the more usual expression,

$$u(x, t) = -c^2 \int_0^t b(t_0) \frac{\partial G}{\partial x_0}(x, t; L, t_0) \, dt_0, \qquad (12.6.7)$$

due to the time-translation invariance of the Green's function. Equation (12.6.7), obtained using Laplace transforms, is equivalent to the representation formula (10.2.16) obtained using Green's formula for the special case $q(x, t) = 0$, $a(t) = 0$, $f(x) = 0$, and $g(x) = 0$. The general case (10.2.16) may be derived in the same way.

EXERCISES 12.6

12.6.1. (a) Determine the Laplace transform of $u(x, t)$ satisfying (12.6.1).
  (b) Represent $u(x, t)$ in terms of its Green's function using Laplace transforms [i.e., use (12.6.4)].

12.6.2. Determine the Green's function by inverting (12.6.4). Show that signals are appropriately reflected.

12.6.3. Determine the Laplace transform of the Green's function for the wave equation if the boundary conditions are

  (a) $u(0, t) = a(t)$ and $\frac{\partial u}{\partial x}(L, t) = b(t)$

  (b) $\frac{\partial u}{\partial x}(0, t) = a(t)$ and $\frac{\partial u}{\partial x}(L, t) = b(t)$

12.6.4. Consider

$$\frac{\partial u}{\partial t} = k \frac{\partial^2 u}{\partial x^2} + q(x, t) \qquad \begin{matrix} x > 0 \\ t > 0 \end{matrix}$$

$$u(0, t) = h(t)$$

$$u(x, 0) = f(x).$$

  *(a) Determine the Laplace transform of the Green's function for this problem.
  (b) Determine $\overline{U}(x, s)$ if $f(x) = 0$ and $q(x, t) = 0$ (see Exercise 12.4.4).
  (c) By comparing parts (a) and (b), derive a representation of $u(x, t)$ in terms of the Green's function [if $f(x) = 0$ and $q(x, t) = 0$]. Compare to (10.3.15).
  (d) From part (a), determine the Green's function. (*Hint:* See a table of Laplace transforms.)

12.6.5. Reconsider Exercise 12.6.4 if:
  (a) $h(t) = 0$ and $q(x, t) = 0$
  (b) $h(t) = 0$ and $f(x) = 0$

12.6.6. Reconsider Exercise 12.6.4 if, instead, the boundary condition were

$$\frac{\partial u}{\partial x}(0, t) = h(t).$$

[Restrict attention to $f(x) = 0$ and $q(x, t) = 0$.]

12.7 INVERSION OF LAPLACE TRANSFORMS USING CONTOUR INTEGRALS IN THE COMPLEX PLANE

Laplace transforms sometimes can be inverted by using tables. However, one of the most important properties of Laplace transforms is that they can be inverted by a contour integral in the complex plane. Furthermore, we will show how to evaluate this integral using results from the theory of functions of a complex variable.

Fourier and Laplace transforms. First we show that the Laplace transform can be considered a special case of the Fourier transform. As a review we introduce $g(x)$ and its Fourier transform $G(\omega)$:

$$G(\omega) = \frac{1}{2\pi} \int_{-\infty}^{\infty} g(x)e^{i\omega x}\,dx \qquad (12.7.1a)$$

$$g(x) = \int_{-\infty}^{\infty} G(\omega)e^{-i\omega x}\,d\omega. \qquad (12.7.1b)$$

Suppose, as is usual for functions that will be Laplace transformed, we discuss functions $g(x)$ which are zero for $x < 0$:

$$g(x) = \begin{cases} 0 & x < 0 \\ 2\pi f(x)e^{-\gamma x} & x > 0. \end{cases} \qquad (12.7.2)$$

The $e^{-\gamma x}$ is introduced (and γ chosen) so that $g(x)$ automatically decays sufficiently rapidly as $x \to \infty$ for certain $f(x)$. For this function, the Fourier transform pair (12.7.1) becomes

$$G(\omega) = \int_{0}^{\infty} f(x)e^{-(-i\omega + \gamma)x}\,dx$$

$$2\pi f(x)e^{-\gamma x} = \int_{-\infty}^{\infty} G(\omega)e^{-i\omega x}\,d\omega \qquad (x > 0).$$

If we introduce

$$s = \gamma - i\omega \qquad (ds = -i\,d\omega)$$

and

$$F(s) \equiv G(\omega) = G\left(\frac{s - \gamma}{-i}\right), \qquad (12.7.3)$$

then using t instead of x ($x = t$) yields

$$F(s) = \int_{0}^{\infty} f(t)e^{-st}\,dt \qquad (12.7.4a)$$

$$f(t) = \frac{1}{2\pi i} \int_{\gamma - i\infty}^{\gamma + i\infty} F(s)e^{st}\,ds \qquad (t > 0). \qquad (12.7.4b)$$

Equation (12.7.4a) shows that $F(s)$ *is the Laplace transform of* $f(t)$. (We usually use t instead of x when discussing Laplace transforms.) $F(s)$ *is also the Fourier transform of*

$$g(t) = \begin{cases} 0 & t < 0 \\ 2\pi f(t)e^{-\gamma t} & t > 0. \end{cases}$$

More importantly, **given the Laplace transform $F(s)$, (12.7.4b) shows how to compute its inverse Laplace transform**. It involves a line integral in the complex s-plane as illustrated in Fig. 12.7.1. From the theory of complex variables, it can be shown that the line integral is to the right of all singularities of $F(s)$. Other than that, the evaluation of the integral is independent of the value of γ. All singularities are in the "left-half plane."

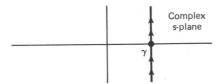

Figure 12.7.1 Line integral in complex s-plane for inverse Laplace transforms.

Cauchy's theorem and residues. We give only an extremely brief discussion of evaluating integrals using the theory of complex variables. The fundamental tool is **Cauchy's theorem**, which states that if $g(s)$ is analytic (no singularities) at all points inside and on a closed contour C, then the closed line integral is zero:

$$\oint_c g(s)\,ds = 0. \tag{12.7.5}$$

Closed line integrals are nonzero only due to singularities of $g(s)$. The **residue theorem** states that the closed line integral (counterclockwise) can be evaluated in terms of contributions (called **residues**) of the singularities s_n *inside the contour* (if there are no branch points, which usually are square-root or logarithmic type singularities):

$$\oint_c g(s)\,ds = 2\pi i \sum_n \text{res}\,(s_n). \tag{12.7.6}$$

The evaluation of residues is often straightforward. *If $g(s) = R(s)/Q(s)$ has simple poles at simple zeros s_n of $Q(s)$ inside the contour*, then in complex variables it is shown that

$$\text{res}\,(s_n) = \frac{R(s_n)}{Q'(s_n)}, \tag{12.7.7}$$

and thus

$$\oint_c g(s)\,ds = 2\pi i \sum_n \frac{R(s_n)}{Q'(s_n)}. \tag{12.7.8}$$

Inversion integral. The inversion integral for Laplace transforms is not a closed line integral but instead an infinite straight line (with constant real part γ) to the right of all singularities. In order to utilize a finite closed line integral, we can consider either of the two large semicircles illustrated in Fig. 12.7.2. We

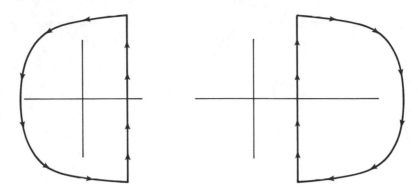

Figure 12.7.2 Closing the line integrals.

will allow the radius to approach infinity so that the straight part of the closed contour approaches the desired infinite straight line. We want the line integral along the arc of the circle to vanish as the radius approaches infinity. The integrand $F(s)e^{st}$ in the inversion integral (12.7.4b) must be sufficiently small. Since $F(s) \to 0$ as $s \to \infty$ [see Sec. 12.2 or (12.7.4a)], we will need e^{st} to vanish as the radius approaches infinity. If $t < 0$, e^{st} exponentially decays as the radius increases only on the right-facing semicircle (real part of $s > 0$). Thus, if $t < 0$, we "close the contour" to the right. Since there are no singularities to the right and the contribution of the large semicircle vanishes, we conclude that

$$f(t) = \frac{1}{2\pi i} \int_{\gamma - i\infty}^{\gamma + i\infty} F(s)e^{st}\,ds = 0$$

if $t < 0$; when we use Laplace transforms, we insist that $f(t) = 0$ for $t < 0$. Of more immediate importance is our analysis of the inversion integral (12.7.4b) for $t > 0$. If $t > 0$, e^{st} exponentially decays in the left-half plane (real part of $s < 0$). Thus, if $t > 0$, we close the contour to the left. There is a contribution to the integral from all the singularities. For $t > 0$, the inverse Laplace transform of $F(s)$ is

$$f(t) = \frac{1}{2\pi i} \int_{\gamma - i\infty}^{\gamma + i\infty} F(s)e^{st}\,ds = \frac{1}{2\pi i} \oint F(s)e^{st}\,ds = \sum_{n} \text{res}\,(s_n). \quad (12.7.9)$$

This is valid if $F(s)$ has no branch points. The summation includes all singularities (since the path is to the right of all singularities).

Simple poles. If $F(s) = p(s)/q(s)$ [so that $g(s) = p(s)e^{st}/q(s)$] and if all the singularities of the Laplace transform $F(s)$ are simple poles [at simple zeros of $q(s)$], then res $(s_n) = p(s_n)\,e^{s_n t}/q'(s_n)$ and thus

$$f(t) = \sum_{n} \frac{p(s_n)}{q'(s_n)}e^{s_n t}. \quad (12.7.10)$$

Equation (12.7.10) is the same result we derived earlier by partial fractions if $F(s)$ is a rational function [i.e., if $p(s)$ and $q(s)$ are polynomials].

Example of the calculation of the inverse Laplace transform. Consider $F(s) = (s^2 + 2s + 4)/[s(s^2 + 1)]$. The inverse transform yields

$$f(t) = \sum_n \text{res }(s_n).$$

The poles of $F(s)$ are the zeros of $s(s^2 + 1)$, namely $s = 0, \pm i$. The residues at these simple poles are

$$\text{res }(0) = 4e^{0t} = 4$$

$$\text{res }(s_n = \pm i) = \frac{s_n^2 + 2s_n + 4}{3s_n^2 + 1}e^{s_n t} = \frac{3 + 2s_n}{-2}e^{s_n t}.$$

Thus,

$$f(t) = 4 + (-\tfrac{3}{2} - i)e^{it} + (-\tfrac{3}{2} + i)e^{-it} = 4 - \tfrac{3}{2} \cdot 2 \cos t - i \cdot 2i \sin t$$
$$= 4 - 3 \cos t + 2 \sin t,$$

using Euler's formulas. In the next section we apply these ideas to solve for the inverse Laplace transform that arises in a partial differential equation's problem.

EXERCISES 12.7

12.7.1. Use the inverse theory for Laplace transforms to determine $f(t)$ if $F(s) = :$
   (a) $1/(s - a)$ *(b) $1/(s^2 + 9)$ (c) $(s + 3)/(s^2 + 16)$

12.7.2. The residue b_{-1} at a singularity s_0 of $f(s)$ is the coefficient of $1/(s - s_0)$ in an expansion of $f(s)$ valid near $s = s_0$. In general,

$$f(s) = \sum_{m=-\infty}^{\infty} b_m(s - s_0)^m,$$

called a **Laurent series** or **expansion**.
   (a) For a simple pole, the most negative power is $m = -1$ ($b_m = 0$ for $m < -1$). In this case, show that

$$\text{res }(s_0) = \lim_{s \to s_0} (s - s_0)f(s).$$

   (b) If s_0 is a simple pole and $f(s) = R(s)/Q(s)$ [with $Q(s_0) = 0$, $R(s_0) \neq 0$, and $dQ/ds\,(s_0) \neq 0$], show that

$$\text{res }(s_0) = \frac{R(s_0)}{Q'(s_0)},$$

   assuming that both $R(s)$ and $Q(s)$ have Taylor series around s_0.
   (c) For an Mth-order pole, the most negative power is $m = -M$ ($b_m = 0$ for $m < -M$). In this case, show that

$$\text{res }(s_0) = \frac{1}{(M-1)!} \frac{d^{M-1}}{ds^{M-1}}[(s - s_0)^M f(s)]\bigg|_{s=s_0}. \qquad (12.7.11)$$

   (d) In part (c), show that the M appearing in (12.7.11) may be replaced by any integer greater than M.

12.7.3. Using Exercise 12.7.2, determine $f(t)$ if $F(s) =$
   (a) $1/s^3$
   (b) $1/(s^2 + 4)^2$

12.7.4. If $|F(s)| < \alpha/r^2$ for large $r \equiv |s|$, prove that

$$\int_{C_R} F(s)e^{st}\,ds \longrightarrow 0 \qquad \text{as } r \longrightarrow \infty \text{ for } t > 0,$$

where C_R is any arc of a circle in the left-half plane (Re $s \le 0$). [If $|F(s)| < \alpha/r$ instead, it is more difficult to prove the same conclusion. The latter case is equivalent to **Jordan's lemma** in complex variables.]

12.8 SOLVING THE WAVE EQUATION USING LAPLACE TRANSFORMS (WITH COMPLEX VARIABLES)

In Sec. 12.6 we showed that

PDE:	$\dfrac{\partial^2 u}{\partial t^2} = c^2\,\dfrac{\partial^2 u}{\partial x^2}$	(12.8.1a)
BC:	$u(0,\,t) = 0$	(12.8.1b)
	$u(L,\,t) = b(t)$	(12.8.1c)
IC:	$u(x,\,0) = 0$	(12.8.1d)
	$\dfrac{\partial u}{\partial t}(x,\,0) = 0$	(12.8.1e)

could be analyzed by introducing $\overline{U}(x,\,s)$, the Laplace transform of $u(x,\,t)$. We obtained $u(x,\,t)$,

$$u(x,\,t) = -c^2 \int_0^t b(t_0)\,\frac{\partial G}{\partial x_0}(x,\,t;\,L,\,t_0)\,dt_0, \tag{12.8.2}$$

in terms of the Green's function. The Laplace transform of this influence function was known [see below (12.6.6)]:

$$\frac{\partial \overline{G}}{\partial x_0}(x,\,s;\,L,\,0) - F(s) - -\frac{\sinh (x/c)s}{c^2 \sinh (L/c)s}. \tag{12.8.3}$$

In this section we use the complex inversion integral for the Laplace transform:

$$\frac{\partial G}{\partial x_0}(x,\,t;\,L,\,0) = \frac{1}{2\pi i}\int_{\gamma - i\infty}^{\gamma + i\infty} -\frac{\sinh (x/c)s}{c^2 \sinh (L/c)s}\,e^{st}\,ds. \tag{12.8.4}$$

The singularities of $F(s)$ only are simple poles s_n, located at the zeros of the denominator:

$$\sinh \frac{L}{c}\,s_n = 0. \tag{12.8.5}$$

However, $s = 0$ is not a pole since, near $s = 0$, $F(s) \approx -(x/c)s/[c^2(L/c)s] \ne \infty$. There are an infinite number of these poles located on the imaginary axis:

$$\frac{L}{c}\,s_n = in\pi, \qquad n = \pm 1,\ \pm 2,\ \pm 3, \ldots \tag{12.8.6}$$

The location of the poles, $s_n = ic(n\pi/L)$, corresponds to the eigenvalues, $\lambda = (n\pi/L)^2$. This follows from the singularity property of Laplace transforms (see Sec. 12.2). The singularity of the Laplace transform $[s = ic(n\pi/L)]$ corresponds to a complex exponential solution $[e^{ic(n\pi/L)t}]$.

The residue at each pole may be evaluated:

$$\operatorname{res}(s_n) = \frac{R(s_n)}{Q'(s_n)} = \frac{-\sinh(x/c)s_n e^{s_n t}}{cL \cosh(L/c)s_n} = \frac{-i \sin n\pi x/L \ e^{i(n\pi ct/L)}}{cL \cos n\pi},$$

since $\sinh ix = i \sin x$ and $\cosh ix = \cos x$. Thus, the influence function for this problem is

$$
\begin{aligned}
\frac{\partial}{\partial x_0} G(x, t; L, 0) &= \sum_{\substack{n=-\infty \\ (n \neq 0)}}^{\infty} \frac{-i \sin n\pi x/L \ e^{i(n\pi ct/L)}}{cL \cos n\pi} \\
&= \frac{2}{cL} \sum_{n=1}^{\infty} (-1)^n \sin \frac{n\pi x}{L} \sin \frac{n\pi ct}{L},
\end{aligned}
\tag{12.8.7}
$$

where the positive and negative n contributions have been combined into one term. Finally, using (12.8.2), we obtain

$$u(x, t) = \sum_{n=1}^{\infty} A_n(t) \sin \frac{n\pi x}{L}, \quad \text{where } A_n = -(-1)^n \frac{2c}{L} \int_0^t b(t_0) \sin \frac{n\pi c}{L}(t - t_0)\, dt_0,$$

the same result as would be obtained by the method of eigenfunction expansion.

The influence function is an infinite series of the eigenfunctions. For homogeneous problems with homogeneous boundary conditions, inverting the Laplace transform using an infinite sequence of poles also will yield a series of eigenfunctions, the same result as obtained by separation of variables. In fact, it is the Laplace transform method that is often used to *prove* the validity of the method of separation of variables.

EXERCISES 12.8

*12.8.1. Solve for $u(x, t)$ using Laplace transforms

$$\frac{\partial^2 u}{\partial t^2} = c^2 \frac{\partial^2 u}{\partial x^2}$$

$$u(x, 0) = f(x) \qquad u(0, t) = 0$$

$$\frac{\partial u}{\partial t}(x, 0) = 0 \qquad u(L, t) = 0.$$

Invert the Laplace transform of $u(x, t)$ using the residue theorem for contour integrals in the s-plane. Show that this yields the same result as derivable by separation of variables.

12.8.2. Modify Exercise 12.8.1 if instead

(a) $u(x, 0) = 0$ and $\dfrac{\partial u}{\partial t}(x, 0) = g(x)$

(b) $u(0, t) = 0$ and $\dfrac{\partial u}{\partial x}(L, t) = 0$

(c) $\dfrac{\partial u}{\partial x}(0, t) = 0$ and $\dfrac{\partial u}{\partial x}(L, t) = 0$

12.8.3. Solve for $u(x, t)$ using Laplace transforms

$$\frac{\partial u}{\partial t} = k\frac{\partial^2 u}{\partial x^2}$$

subject to $u(x, 0) = f(x)$, $u(0\ t) = 0$, and $u(L, t) = 0$.

Invert the Laplace transform of $u(x, t)$ using the residue theorem for contour integrals in the complex s-plane. By what other method can this representation of the solution be obtained? (Compare to Exercise 12.5.6).

12.8.4. Consider

$$\frac{\partial^2 u}{\partial t^2} = c^2\frac{\partial^2 u}{\partial x^2} + \sin \sigma_0 t$$

$$u(x, 0) = 0 \qquad u(0, t) = 0$$

$$\frac{\partial u}{\partial t}(x, 0) = 0 \qquad u(L, t) = 0.$$

(a) Solve using Laplace transforms (with contour inversion) if $\sigma_0 \neq c(m\pi/L)$.
(b) Solve if $\sigma_0 = c(3\pi/L)$. Show that resonance occurs (see Sec. 7.5).

An Elementary Discussion of Finite Difference Numerical Methods for Partial Differential Equations

13.1 INTRODUCTION

Partial differential equations are often classified. Equations with the same classification have qualitatively similar mathematical and physical properties. We have studied primarily the simplest prototypes. The heat equation ($\partial u/\partial t = k\ \partial^2 u/\partial x^2$) is an example of a **parabolic** partial differential equation. Solutions usually exponentially decay in time and approach an equilibrium solution. Information and discontinuities propagate at an infinite velocity. The wave equation ($\partial^2 u/\partial t^2 = c^2\ \partial^2 u/\partial x^2$) typifies **hyperbolic** partial differential equations. There are modes of vibration. Information propagates at a finite velocity and thus discontinuities persist. Laplace's equation ($\partial^2 u/\partial x^2 + \partial^2 u/\partial y^2 = 0$) is an example of an **elliptic** partial differential equation. Solutions usually satisfy maximum principles. The terminology *parabolic, hyperbolic*, and *elliptic* result from transformation properties of the conic sections (e.g., see Weinberger [1965]).

In the previous chapters, we have studied various methods to obtain explicit solutions of some partial differential equations of physical interest. Except for the one-dimensional wave equation, the solutions were rather complicated, involving an infinite series or an integral representation. In many current situations, detailed numerical calculations of solutions of partial differential equations are needed. Our previous analyses suggest computational methods (e.g., the first 100 terms of a Fourier series). However, usually there are more efficient methods to obtain numerical results, especially if a computer is to be utilized. In this chapter we develop finite difference methods to numerically approximate solutions of the different types of partial differential equations (i.e., parabolic, hyperbolic, and elliptic). We will describe only simple cases, the heat, wave, and Laplace's equations, but algorithms for more complicated problems (including *nonlinear* ones) will become apparent.

13.2 FINITE DIFFERENCES AND TRUNCATED TAYLOR SERIES

Polynomial approximations. The fundamental technique for finite difference numerical calculations is based on polynomial approximations to $f(x)$ near $x = x_0$. We let $x = x_0 + \Delta x$, so that $\Delta x = x - x_0$. If we approximate $f(x)$ by a constant near $x = x_0$, we choose $f(x_0)$. A better approximation (see Fig. 13.2.1) to $f(x)$ is its tangent line at $x = x_0$:

$$f(x) \approx f(x_0) + \underbrace{(x - x_0)}_{\Delta x}\frac{df}{dx}(x_0),\tag{13.2.1}$$

a linear approximation (a first order polynomial). We can also consider a quadratic approximation to $f(x)$, $f(x_0) + \Delta x f'(x_0) + (\Delta x)^2 f''(x_0)/2!$, whose value, first, and second derivatives at $x = x_0$ equal that for $f(x)$. Each such succeeding higher degree polynomial approximation to $f(x)$ is more and more accurate if x is near enough to x_0 (i.e., if Δx is small).

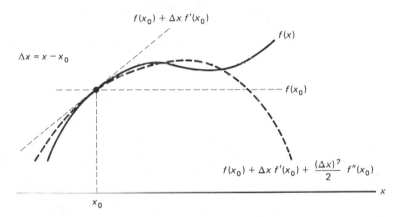

Figure 13.2.1 Taylor polynomials.

Truncation error. A formula for the error in these polynomial approximations is obtained directly from

$$f(x) = f(x_0) + \Delta x f'(x_0) + \cdots + \frac{(\Delta x)^n}{n!}f^{(n)}(x_0) + R_n,\tag{13.2.2}$$

known as **the Taylor series with remainder.** The remainder R_n (also called the **truncation error**) is known to be in the form of the next term of the series, but evaluated at a usually unknown intermediate point:

$$R_n = \frac{(\Delta x)^{n+1}}{(n+1)!}f^{(n+1)}(\xi_{n+1}), \quad \text{where} \quad x_0 < \xi_{n+1} < x = x_0 + \Delta x.\tag{13.2.3}$$

For this to be valid, $f(x)$ must have $n + 1$ continuous derivatives.

Example. The error in the tangent line approximation is given by (13.2.3) with $n = 1$:

$$f(x_0 + \Delta x) = f(x_0) + \Delta x \frac{df}{dx}(x_0) + \frac{(\Delta x)^2}{2!} \frac{d^2 f}{dx^2}(\xi_2), \qquad (13.2.4)$$

called the **extended mean value theorem**. If Δx is small, then ξ_2 is contained in a small interval, and the truncation error is almost determined (provided that $d^2 f / dx^2$ is continuous),

$$R \approx \frac{(\Delta x)^2}{2} \frac{d^2 f}{dx^2}(x_0).$$

We say that the truncation error is $O(\Delta x)^2$, "order delta-x squared," meaning that

$$|R| \leq C(\Delta x)^2,$$

since we usually assume that $d^2 f / dx^2$ is bounded ($|d^2 f / dx^2| < M$). Thus, $C = M/2$.

First derivative approximations. Through the use of Taylor series, we are able to approximate derivatives in various ways. For example, from (13.2.4)

$$\frac{df}{dx}(x_0) = \frac{f(x_0 + \Delta x) - f(x_0)}{\Delta x} - \frac{\Delta x}{2} \frac{d^2 f}{dx^2}(\xi_2). \qquad (13.2.5a)$$

We thus introduce a finite difference approximation, the **forward difference** approximation to df/dx:

$$\boxed{\frac{df}{dx}(x_0) \approx \frac{f(x_0 + \Delta x) - f(x_0)}{\Delta x}.} \qquad (13.2.5b)$$

This is nearly the definition of the derivative. Here we use a forward difference (but do not take the limit as $\Delta x \to 0$). Since (13.2.5a) is valid for all Δx, we can let Δx be replaced by $-\Delta x$ and derive the **backward difference** approximation to df/dx:

$$\frac{df}{dx}(x_0) = \frac{f(x - \Delta x) - f(x_0)}{-\Delta x} + \frac{\Delta x}{2} \frac{d^2 f}{dx^2}(\bar{\xi}_2) \qquad (13.2.6a)$$

$$\boxed{\frac{df}{dx}(x_0) \approx \frac{f(x_0 - \Delta x) - f(x_0)}{-\Delta x} = \frac{f(x_0) - f(x_0 - \Delta x)}{\Delta x}.} \qquad (13.2.6b)$$

By comparing (13.2.5a) to (13.2.5b) and (13.2.6a) to (13.2.6b), we observe that the truncation error is $O(\Delta x)$ and nearly identical for both forward and backward difference approximations of the first derivative.

To obtain a more accurate approximation for df/dx (x_0), we can average the forward and backward approximations. By adding (13.2.5a) and (13.2.6a):

$$2\frac{df}{dx}(x_0) = \frac{f(x_0 + \Delta x) - f(x_0 - \Delta x)}{\Delta x} + \frac{\Delta x}{2}\left[\frac{d^2 f}{dx^2}(\bar{\xi}_2) - \frac{d^2 f}{dx^2}(\xi_2)\right]. \qquad (13.2.6c)$$

Since $\bar{\xi}_2$ is near ξ_2, we expect the error nearly to cancel and thus be much less than $O(\Delta x)$. To derive the error in this approximation, we return to the Taylor series and subtract $f(x_0 - \Delta x)$ from $f(x_0 + \Delta x)$:

$$f(x_0 + \Delta x) = f(x_0) + \Delta x f'(x_0) + \frac{(\Delta x)^2}{2!} f''(x_0) + \frac{(\Delta x)^3}{3!} f'''(x_0) + \cdots \qquad (13.2.7a)$$

$$f(x_0 - \Delta x) = f(x_0) - \Delta x f'(x_0) + \frac{(\Delta x)^2}{2!} f''(x_0) - \frac{(\Delta x)^3}{3!} f'''(x_0) + \cdots \qquad (13.2.7b)$$

$$f(x_0 + \Delta x) - f(x_0 - \Delta x) = 2\Delta x f'(x_0) + \frac{2}{3!}(\Delta x)^3 f'''(x_0) + \cdots$$

We thus expect that

$$f'(x_0) = \frac{f(x_0 + \Delta x) - f(x_0 - \Delta x)}{2\Delta x} - \frac{(\Delta x)^2}{6} f'''(\xi_3), \qquad (13.2.8a)$$

which is proved in an exercise. This leads to the **centered difference** approximation to $df/dx(x_0)$:

$$\boxed{f'(x_0) \approx \frac{f(x_0 + \Delta x) - f(x_0 - \Delta x)}{2\Delta x}.} \qquad (13.2.8b)$$

Equation (13.2.8b) is usually preferable since it is more accurate [the truncation error is $O(\Delta x)^2$] and involves the same number (2) of function evaluations as both the forward and backward difference formulas. However, as we will show later, it is *not* always better to use the centered difference formula.

These finite difference approximations to df/dx are **consistent**, meaning that the truncation error vanishes as $\Delta x \to 0$. More accurate finite difference formulas exist, but they are used less frequently.

Example. Consider the numerical approximation of $df/dx(1)$ for $f(x) = \log x$ using $\Delta x = 0.1$. Unlike practical problems, here we know the exact answer, $df/dx(1) = 1$. Using a hand calculator [$x_0 = 1$, $\Delta x = 0.1$, $f(x_0 + \Delta x) = f(1.1) = \log(1.1) = 0.0953102$ and $f(x_0 - \Delta x) = \log(0.9) = -0.1053605$] yields the numerical results reported in Table 13.2.1. Theoretically, the error should be an order of magnitude Δx smaller for the centered difference. We observe this phenomenon. To understand the error further, we calculate the expected error E using an estimate of the remainder. For forward or backward differences,

$$E \approx \left| \frac{\Delta x}{2} \frac{d^2 f}{dx^2}(1) \right| = \frac{0.1}{2} = +0.05,$$

TABLE 13.2.1

	Forward	Backward	Centered
Difference formula	0.953102	1.053605	1.00335
|Error|	4.6898%	5.3605%	0.335%

whereas for a centered difference,

$$E \approx \frac{(\Delta x)^2}{6} \frac{d^3 f}{dx^3}(1) = \frac{(0.1)^2}{6} 2 = 0.00333 \dots.$$

These agree quite well with our actual tabulated errors. Estimating the errors this way is rarely appropriate since usually the second and third derivatives are not known to begin with.

Second derivatives. By adding (13.2.7a) and (13.2.7b), we obtain

$$f(x_0 + \Delta x) + f(x_0 - \Delta x) = 2f(x_0) + (\Delta x)^2 f''(x_0) + \frac{2(\Delta x)^4}{4!} f^{(iv)}(x_0) + \cdots$$

We thus expect that

$$f''(x_0) = \frac{f(x_0 + \Delta x) - 2f(x_0) + f(x_0 - \Delta x)}{(\Delta x)^2} - \frac{(\Delta x)^2}{12} f^{(iv)}(\xi). \qquad (13.2.9a)$$

This yields a finite difference approximation for the second derivative with an $O(\Delta x)^2$ truncation error:

$$\boxed{\frac{d^2 f}{dx^2}(x_0) \approx \frac{f(x_0 + \Delta x) - 2f(x_0) + f(x_0 - \Delta x)}{(\Delta x)^2}.} \qquad (13.2.9b)$$

Equation (13.2.9b) is called the **centered difference** approximation for the second derivative since it also can be obtained by repeated application of the centered difference formulas for first derivatives (see Exercise 13.2.2). The centered difference approximation for the second derivative involves three function evaluations $f(x_0 - \Delta x)$, $f(x_0)$, and $f(x_0 + \Delta x)$. The respective "weights," $1/(\Delta x)^2$, $-2/(\Delta x)^2$, and $1/(\Delta x)^2$, are illustrated in Fig. 13.2.2. In fact, in general, *the weights must sum to zero for any finite difference approximation to any derivative.*

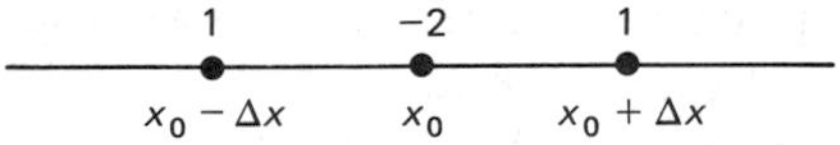

Figure 13.2.2 Weights for centered difference approximation of second derivative.

Partial derivatives. In solving partial differential equations, we analyze functions of two or more variables, for example $u(x, y)$, $u(x, t)$, and $u(x, y, t)$. Numerical methods often use finite difference approximations. Some (but not all) partial derivatives may be obtained using our earlier results for functions of one variable. For example if $u(x, y)$, then $\partial u/\partial x$ is an ordinary derivative du/dx, keeping y fixed. We may use the forward, backward, or centered difference formulas. Using the centered difference formula,

$$\frac{\partial u}{\partial x}(x_0, y_0) \approx \frac{u(x_0 + \Delta x, y_0) - u(x_0 - \Delta x, y_0)}{2\Delta x}.$$

For $\partial u/\partial y$ we keep x fixed and thus obtain

$$\frac{\partial u}{\partial y}(x_0, y_0) \approx \frac{u(x_0, y_0 + \Delta y) - u(x_0, y_0 - \Delta y)}{2\Delta y}$$

using the centered difference formula. These are both two-point formulas, which we illustrate in Fig. 13.2.3.

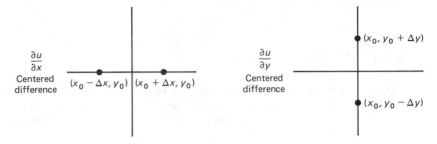

$$\frac{\partial u}{\partial x}$$
Centered difference

$$\frac{\partial u}{\partial y}$$
Centered difference

Figure 13.2.3 Points for first partial derivatives.

In physical problems we often need the Laplacian, $\nabla^2 u = \partial^2 u/\partial x^2 + \partial^2 u/\partial y^2$. We use the centered difference formula for second derivatives (13.2.9b), adding the formula for x fixed to the one for y fixed:

$$\nabla^2 u(x_0, y_0) \approx \frac{u(x_0 + \Delta x, y_0) - 2u(x_0, y_0) + u(x_0 - \Delta x, y_0)}{(\Delta x)^2}$$

$$+ \frac{u(x_0, y_0 + \Delta y) - 2u(x_0, y_0) + u(x_0, y_0 - \Delta y)}{(\Delta y)^2}. \qquad (13.2.10)$$

Here the error is the largest of $O(\Delta x)^2$ and $O(\Delta y)^2$. We often let $\Delta x = \Delta y$, obtaining the standard five-point finite difference approximation to the Laplacian ∇^2,

$$\nabla^2 u (x_0, y_0) \approx \frac{\begin{aligned} u(x_0 + \Delta x, y_0) + u(x_0 - \Delta x, y_0) + u(x_0, y_0 + \Delta y) \\ + u(x_0, y_0 - \Delta y) - 4u(x_0, y_0) \end{aligned}}{(\Delta x)^2},$$

$$(13.2.11)$$

as illustrated in Fig. 13.2.4, where $\Delta x = \Delta y$. Note that the relative weights again sum to zero.

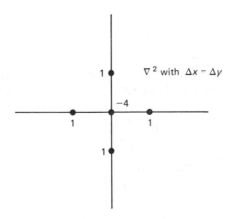

∇^2 with $\Delta x = \Delta y$

Figure 13.2.4 Weights for the Laplacian ($\Delta x = \Delta y$).

Other formulas for derivatives may be found in "Numerical Interpolation, Differentiation, and Integration" by P. J. Davis and I. Polonsky (Chapter 25 of Abramowitz and Stegun [1965]).

EXERCISES 13.2

13.2.1. (a) Show that the truncation error for the centered difference approximation of the first derivative (13.2.8b) is $-(\Delta x)^2 f'''(\xi_3)/6$. [*Hint:* Consider the Taylor series of $g(\Delta x) = f(x + \Delta x) - f(x - \Delta x)$ as a function of Δx around $\Delta x = 0$.]

(b) Explicitly show that (13.2.8b) is exact for any quadratic polynomial.

13.2.2. Derive (13.2.9b) by twice using the centered difference approximation for first derivatives.

13.2.3. Derive the truncation error for the centered difference approximation of the second derivative.

13.2.4. Suppose that we did not know (13.2.9), but thought it possible to approximate $d^2f/dx^2 (x_0)$ by an unknown linear combination of the three function values, $f(x_0 - \Delta x)$, $f(x_0)$, and $f(x_0 + \Delta x)$:

$$\frac{d^2f}{dx^2} \approx af(x_0 - \Delta x) + bf(x_0) + cf(x_0 + \Delta x).$$

Determine a, b, and c, by expanding the right-hand side in a Taylor series around x_0 using (13.2.7) and equating coefficients through $d^2f/dx^2(x_0)$.

13.2.5. Derive the most accurate five-point approximation for $f'(x_0)$ involving $f(x_0)$, $f(x_0 \pm \Delta x)$, and $f(x_0 \pm 2\Delta x)$. What is the order of magnitude of the truncation error?

*13.2.6. Derive an approximation for $\partial^2 u/\partial x \partial y$ whose truncation error is $O(\Delta x)^2$. (*Hint:* Twice apply the centered difference approximations for first-order partial derivatives.)

13.2.7. How well does $\frac{1}{2}[f(x) + f(x + \Delta x)]$ approximate $f(x + \Delta x/2)$ (i.e., what is the truncation error)?

13.3 HEAT EQUATION

13.3.1 Introduction

In this subsection we introduce a numerical finite difference method to solve the one-dimensional heat equation without sources on a finite interval $0 < x < L$:

$$\frac{\partial u}{\partial t} = k \frac{\partial^2 u}{\partial x^2} \tag{13.3.1a}$$

$$u(0, t) = 0 \tag{13.3.1b}$$

$$u(L, t) = 0 \tag{13.3.1c}$$

$$u(x, 0) = f(x). \tag{13.3.1d}$$

13.3.2 A Partial Difference Equation

We will begin by replacing the partial differential equation at the point $x = x_0$, $t = t_0$ by an approximation based on our finite difference formulas for the derivatives. We can do this in many ways. Eventually, we will learn why some ways are good and others bad. Somewhat arbitrarily, we choose a forward difference in time for $\partial u/\partial t$:

$$\frac{\partial u}{\partial t}(x_0, t_0) = \frac{u(x_0, t_0 + \Delta t) - u(x_0, t_0)}{\Delta t} - \frac{\Delta t}{2}\frac{\partial^2 u}{\partial t^2}(x_0, \eta_1)$$

where $t_0 < \eta_1 < t_0 + \Delta t$. For spatial derivatives we introduce our centered difference scheme

$$\frac{\partial^2 u}{\partial x^2}(x_0, t_0) = \frac{u(x_0 + \Delta x, t_0) - 2u(x_0, t_0) + u(x_0 - \Delta x, t_0)}{(\Delta x)^2} - \frac{(\Delta x)^2}{12}\frac{\partial^4 u}{\partial x^4}(\xi_1, t_0),$$

where $x_0 < \xi_1 < x_0 + \Delta x$. The heat equation at any point $x = x_0$, $t = t_0$, becomes

$$\frac{u(x_0, t_0 + \Delta t) - u(x_0, t_0)}{\Delta t} = k\frac{u(x_0 + \Delta x, t_0) - 2u(x_0, t_0) + u(x_0 - \Delta x, t_0)}{(\Delta x)^2} + E,$$

$$(13.3.2)$$

exactly, where the discretization (or truncation) error is

$$E = \frac{\Delta t}{2}\frac{\partial^2 u}{\partial t^2}(x_0, \eta_1) - \frac{k(\Delta x)^2}{12}\frac{\partial^4 u}{\partial x^4}(\xi_1, t_0). \qquad (13.3.3)$$

Since E is unknown, we cannot solve (13.3.2). Instead, we introduce the *approximation* that results by ignoring the truncation error:

$$\frac{u(x_0, t_0 + \Delta t) - u(x_0, t_0)}{\Delta t} \approx k\frac{u(x_0 + \Delta x, t_0) - 2u(x_0, t_0) + u(x_0 - \Delta x, t_0)}{(\Delta x)^2}.$$

$$(13.3.4)$$

To be more precise, we introduce $\bar{u}(x_0, t_0)$, an approximation at the point $x = x_0$, $t = t_0$ of the exact solution $u(x_0, t_0)$. We let the approximation $\bar{u}(x_0, t_0)$ solve (13.3.4) exactly,

$$\frac{\bar{u}(x_0, t_0 + \Delta t) - \bar{u}(x_0, t_0)}{\Delta t} = k\frac{\bar{u}(x_0 + \Delta x, t_0) - 2\bar{u}(x_0, t_0) + \bar{u}(x_0 - \Delta x, t_0)}{(\Delta x)^2}.$$

$$(13.3.5)$$

$\bar{u}(x_0, t_0)$ is the exact solution of an equation that is only approximately correct. We hope that the desired solution $u(x_0, t_0)$ is accurately approximated by $\bar{u}(x_0, t_0)$.

Equation (13.3.5) involves points separated a distance Δx in space and Δt in time. We thus introduce a uniform mesh Δx and a constant discretization time Δt. A space-time diagram (Fig. 13.3.1) illustrates our mesh and time discretization on the domain of our initial boundary value problem. We divide the rod of length L into N equal intervals, $\Delta x = L/N$. We have $x_0 = 0$, $x_1 = \Delta x$,

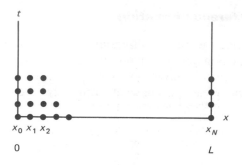

Figure 13.3.1 Space-time discretization.

$x_2 = 2\Delta x, \ldots, x_N = N \Delta x = L$. In general,

$$x_j = j \, \Delta x. \tag{13.3.6}$$

Similarly, we introduce time step sizes Δt such that

$$t_m = m \, \Delta t. \tag{13.3.7}$$

The exact temperature at the mesh point $u(x_j, t_m)$ is approximately $\bar{u}(x_j, t_m)$, which satisfies (13.3.5). We introduce the following *notation*:

$$\bar{u}(x_j, t_m) \equiv u_j^{(m)}, \tag{13.3.8}$$

indicating the exact solution of (13.3.5) at the jth mesh point at time t_m. Equation (13.3.5) will be satisfied at each mesh point $x_0 = x_j$ at each time $t_0 = t_m$ (excluding the space-time boundaries). Note that $x_0 + \Delta x$ becomes $x_j + \Delta x = x_{j+1}$ and $t_0 + \Delta t$ becomes $t_m + \Delta t = t_{m+1}$. Thus,

$$\frac{u_j^{(m+1)} - u_j^{(m)}}{\Delta t} = k \frac{u_{j+1}^{(m)} - 2u_j^{(m)} + u_{j-1}^{(m)}}{(\Delta x)^2}, \tag{13.3.9}$$

for $j = 1, \ldots, N - 1$ and m starting from 1. We call (13.3.9) a **partial difference equation**. The local truncation error is given by (13.3.3); it is the larger of $O(\Delta t)$ and $O(\Delta x)^2$. Since $E \to 0$ as $\Delta x \to 0$ and $\Delta t \to 0$, the approximation (13.3.9) is said to be **consistent** with the partial differential equation (13.3.1).

In addition, we insist that $u_j^{(m)}$ satisfies the initial conditions (at the mesh points)

$$u_j^{(0)} = u(x, 0) = f(x) = f(x_j), \tag{13.3.10}$$

where $x_j = j \, \Delta x$ for $j = 0, \ldots, N$. Similarly, $u_j^{(m)}$ satisfies the boundary conditions (at each time step)

$$u_0^{(m)} = u(0, t) = 0 \tag{13.3.11a}$$

$$u_N^{(m)} = u(L, t) = 0. \tag{13.3.11b}$$

If there is a physical (and thus mathematical) discontinuity at the initial time at any boundary point, then we can analyze $u_0^{(0)}$ or $u_N^{(0)}$ in different numerical ways.

13.3.3 Computations

Our finite difference scheme (13.3.9) involves four points, three at the time t_m, one at the advanced time $t_{m+1} = t_m + \Delta t$, as illustrated by Fig. 13.3.2. We can "march forward in time" by solving for $u_j^{(m+1)}$, starred in Fig. 13.3.2:

$$u_j^{(m+1)} = u_j^{(m)} + s(u_{j+1}^{(m)} - 2u_j^{(m)} + u_{j+1}^{(m)}), \qquad (13.3.12)$$

where s is a dimensionless parameter,

$$s = k\frac{\Delta t}{(\Delta x)^2}. \qquad (13.3.13)$$

$u_j^{(m+1)}$ is a linear combination of the specified three earlier values. We begin our computation using the initial condition $u_j^{(0)} = f(x_j)$, for $j = 1, \ldots, N - 1$. Then (13.3.12) specifies the solution $u_j^{(1)}$ at time Δt, and we continue the calculation. For mesh points adjacent to the boundary (i.e., $j = 1$ or $j = N - 1$), (13.3.12) requires the solution on the boundary points ($j = 0$ or $j = N$). We obtain these values from the boundary conditions. In this way we can easily solve our discrete problem numerically. Our proposed scheme is easily programmed for a computer.

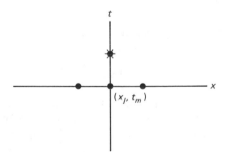

Figure 13.3.2 Marching forward in time.

Propagation speed of disturbances. As a simple example, suppose that the initial conditions at the mesh points are zero except for 1 at some interior mesh point far from the boundary. At the first time step, (13.3.12) will imply that the solution is zero everywhere except at the original nonzero mesh point and its two immediate neighbors. This process continues as illustrated in Fig. 13.3.3. Stars represent nonzero values. The isolated initial nonzero value spreads out at a constant speed (until the boundary has been reached). This disturbance propagates at velocity $\Delta x / \Delta t$. However, for the heat equation, disturbances move at an infinite speed (see Chapter 9). In some sense our numerical scheme poorly approximates this property of the heat equation. However, if the parameter

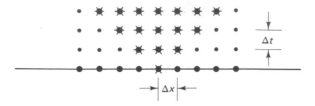

Figure 13.3.3 Propagation speed of disturbances.

s is fixed, then the numerical propagation speed is

$$\frac{\Delta x}{\Delta t} = \frac{k\,\Delta x}{s(\Delta x)^2} = \frac{k}{s\,\Delta x}.$$

As $\Delta x \to 0$ (with s fixed) this speed approaches ∞ as is desired.

Computed example. To compute with (13.3.12), we must specify Δx and Δt. Presumably, our solution will be more accurate with smaller Δx and Δt. Certainly, the local truncation error will be reduced. An obvious disadvantage of decreasing Δx and Δt will be the resulting increased time (and money) necessary for any numerical computation. This trade-off usually occurs in numerical calculations. However, there is a more severe difficulty which we will need to analyze. To indicate the problem, we will compute using (13.3.12). First, we must choose Δx and Δt. In our calculations we fix $\Delta x = L/10$ (nine interior points and two boundary points). Since our partial difference equation (13.3.12) depends primarily on $s = k\,\Delta t/(\Delta x)^2$, we pick Δt so that, as examples, $s = 1/4$ and $s = 1$. In both cases we assume initially that $u(x, 0) = f(x)$, as sketched in Fig. 13.3.4, and the zero boundary conditions (13.3.11). The exact solution of the partial differential equation is

$$u(x, t) = \sum_{n=1}^{\infty} a_n \sin \frac{n\pi x}{L} e^{-k(n\pi/L)^2 t}$$

$$a_n = \frac{2}{L} \int_0^L f(x) \sin \frac{n\pi x}{L} \, dx,$$

(13.3.14)

as described in Chapter 2. It shows that the solution decays exponentially in time and approaches a simple sinusoidal ($\sin \pi x/L$) shape in space for large t. Elementary computer calculations of our numerical scheme, (13.3.12), for $s = \frac{1}{4}$ and $s = 1$, are sketched in Fig. 13.3.5 (with smooth curves sketched through the nine interior points at fixed values of t). For $s = \frac{1}{4}$ these results seem quite reasonable, agreeing with our qualitative understanding of the exact solution. On the other hand, the solution of (13.3.12) for $s = 1$ is absurd. Its most obvious difficulty is the negative temperatures. The solution then grows wildly with rapid oscillations in space and time. None of these phenomena are associated with

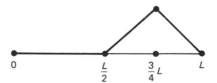

Figure 13.3.4 Initial condition.

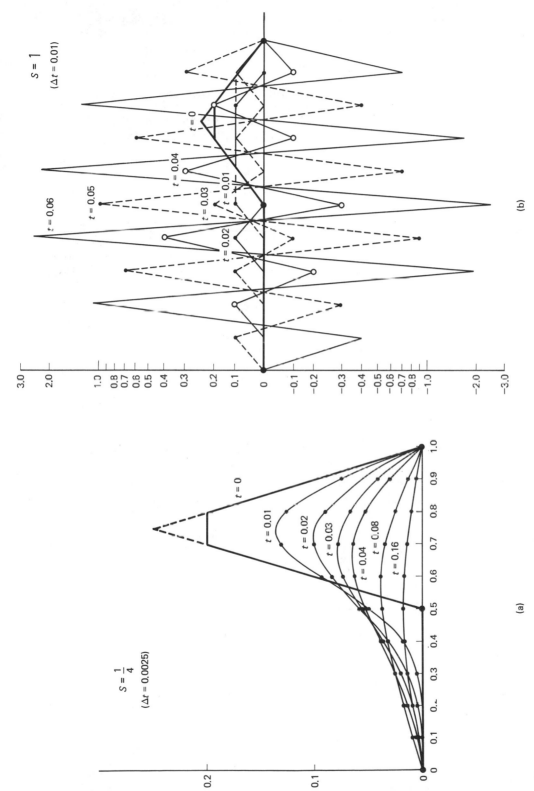

Figure 13.3.5 Computations for the heat equation, $s = k(\Delta t)/(\Delta x)^2$: (a) $s = \frac{1}{4}$ stable and (b) $s = 1$ unstable.

(a)

(b)

$s = \frac{1}{4}$

$(\Delta t = 0.0025)$

$S = 1$

$(\Delta t = 0.01)$

the heat equation. The finite difference approximation yields unusable results if $s = 1$. In the next subsection we explain these results. We must understand how to pick $s = k\,(\Delta t)/(\Delta x)^2$ so that we are able to obtain reasonable numerical solutions.

13.3.4 Fourier–von Neumann Stability Analysis

Introduction. In this subsection we analyze the finite difference scheme for the heat equation obtained by using a forward difference in time and a centered difference in space:

pde*:

$$u_j^{(m+1)} = u_j^{(m)} + s(u_{j+1}^{(m)} - 2u_j^{(m)} + u_{j-1}^{(m)}) \qquad (13.3.15a)$$

IC:

$$u_j^{(0)} = f(x_j) = f_j \qquad (13.3.15b)$$

BC:

$$u_0^{(m)} = 0 \qquad (13.3.15c)$$

$$u_N^{(m)} = 0, \qquad (13.3.15d)$$

where $s = k\,\Delta t/(\Delta x)^2$, $x_j = j\,\Delta x$, $t = m\,\Delta t$ and hopefully $u(x_j, t) \approx u_j^{(m)}$. We will develop von Neumann's ideas of the 1940s based on Fourier-type analysis.

Eigenfunctions and product solutions. In Sec. 13.3.5 we show that the method of separation of variables can be applied to the partial difference equation. There are special product solutions with wave number α of the form

$$u_j^{(m)} = e^{i\alpha x}Q^{t/\Delta t} = e^{i\alpha j \Delta x}Q^m. \qquad (13.3.16)$$

By substituting (13.3.16) into (13.3.15) and canceling $e^{i\alpha x}Q^m$, we obtain

$$Q = 1 + s(e^{i\alpha\Delta x} - 2 + e^{-i\alpha\Delta x}) = 1 - 2s[1 - \cos(\alpha\,\Delta x)]. \qquad (13.3.17)$$

Q is the same for positive and negative α. Thus a linear combination of $e^{\pm i\alpha x}$ may be used. The boundary condition $u_0^{(m)} = 0$ implies that $\sin \alpha x$ is appropriate, while $u_N^{(m)} = 0$ implies that $\alpha = n\pi/L$. Thus, there are solutions of (13.3.15) of the form

$$u_j^{(m)} = \sin\frac{n\pi x}{L}\,Q^{t/\Delta t}, \qquad (13.3.18)$$

where Q is determined from (13.3.17),

$$Q = 1 - 2s\left[1 - \cos\left(\frac{n\pi\,\Delta x}{L}\right)\right], \qquad (13.3.19)$$

and $n = 1, 2, 3, \ldots, N - 1$, as will be explained. For partial *differential*

---

* pde here means partial *difference* equation.

equations there are an infinite number of eigenfunctions ($\sin n\pi x/L$, $n = 1, 2, 3, \ldots$). However, we will show that for our partial *difference* equation there are only $N - 1$ independent eigenfunctions ($\sin n\pi x/L$, $n = 1, 2, 3, \ldots, N - 1$):

$$\phi_j = \sin \frac{n\pi x}{L} = \sin \frac{n\pi j \, \Delta x}{L} = \sin \frac{n\pi j}{N}, \tag{13.3.20}$$

the same eigenfunctions as for the partial differential equation (in this case). For example, for $n = N$, $\phi_j = \sin \pi j = 0$ (for all j). Furthermore, ϕ_j for $n = N + 1$ is equivalent to ϕ_j for $n = N - 1$ since

$$\sin \frac{(N + 1)\pi j}{N} = \sin \left(\frac{\pi j}{N} + j\pi \right) = \sin \left(\frac{\pi j}{N} - j\pi \right) = -\sin \frac{(N - 1)\pi j}{N}.$$

In Fig. 13.3.6 we sketch some of these "eigenfunctions" (for $N = 10$) For the partial difference equation due to the discretization, the solution is composed of only $N - 1$ waves. This number of waves equals the number of independent mesh points (excluding the end points). The wave with the smallest wavelength is

$$\sin \frac{(N - 1)\pi x}{L} = \sin \frac{(N - 1)\pi j}{N} = (-1)^{j+1} \sin \frac{\pi j}{N},$$

which alternates signs at every point. The general solution is obtained by the principle of superposition, introducing $N - 1$ constants β_n:

$$u_j^{(m)} = \sum_{n=1}^{N-1} \beta_n \sin \frac{n\pi x}{L} \left[1 - 2s \left(1 - \cos \frac{n\pi}{N} \right) \right]^{t/\Delta t}, \tag{13.3.21}$$

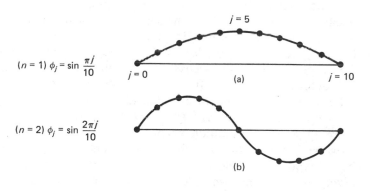

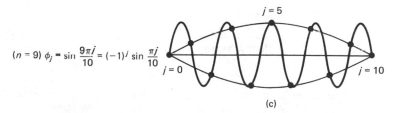

Figure 13.3.6 Eigenfunctions for the discrete problem.

where

$$s = \frac{k\,\Delta t}{(\Delta x)^2}.$$

These coefficients can be determined from the $N - 1$ initial conditions, using the discrete orthogonality of the eigenfunctions $\sin n\pi j/N$. The analysis of this discrete Fourier series is described in Exercises 13.3.3 and 13.3.4.

Comparison to partial differential equation. The product solutions $u_j^{(m)}$ of the partial difference equation may be compared with the product solutions $u(x, t)$ of the partial differential equation:

$$u_j^{(m)} = \sin \frac{n\pi x}{L} \left[1 - 2s\left(1 - \cos \frac{n\pi}{N} \right) \right]^{t/\Delta t} \qquad n = 1, 2, \ldots, N - 1$$

$$u(x,\,t) = \sin \frac{n\pi x}{L} e^{-k(n\pi/L)^2 t} \qquad n = 1, 2, \ldots,$$

where $s = k\,\Delta t/(\Delta x)^2$. For the partial differential equation, each wave exponentially decays, $e^{-k(n\pi/L)^2 t}$. For the partial difference equation, the time dependence (corresponding to the spatial part $\sin n\pi x/L$) is

$$Q^m = \left[1 - 2s\left(1 - \cos \frac{n\pi}{N} \right) \right]^{t/\Delta t} \tag{13.3.22}$$

Stability. If $Q > 1$, there is exponential growth in time, while exponential decay occurs if $0 < Q < 1$. The solution is constant in time if $Q = 1$. However, in addition, it is possible for there to be a convergent oscillation in time $(-1 < Q < 0)$, a pure oscillation $(Q = -1)$, and a divergent oscillation $(Q < -1)$. These possibilities are discussed and sketched in Sec. 13.3.5. The value of Q will determine stability. If $|Q| \leq 1$ for all solutions, we say that the numerical scheme is **stable**. Otherwise, the scheme is **unstable**.

We return to analyze $Q^m = Q^{t/\Delta t}$, where $Q = 1 - 2s(1 - \cos n\pi/N)$. Here $Q \leq 1$; the solution cannot be a purely growing exponential in time. However, the solution may be a convergent or divergent oscillation as well as being exponentially decaying. We do not want the numerical scheme to have divergent oscillations in time.* If s is too large, Q may become too negative.

Since $Q \leq 1$, the solution will be "stable" if $Q \geq -1$. To be stable, $1 - 2s(1 - \cos n\pi/N) \geq -1$ for $n = 1, 2, 3, \ldots, N - 1$, or equivalently,

$$s \leq \frac{1}{1 - \cos n\pi/N} \qquad \text{for } n = 1, 2, 3, \ldots, N - 1.$$

---

* Convergent oscillations do not duplicate the behavior of the partial differential equation. However, they at least decay. We tolerate oscillatory decaying terms.

For stability, s must be less than or equal to the smallest, $n = N - 1$,

$$s \leq \frac{1}{1 - \cos (N - 1)\pi/N}.$$

To simplify the criteria, we note that $1 - \cos (N - 1)\pi/N < 2$, and hence we are *guaranteed* that the numerical solution will be stable if $s \leq \frac{1}{2}$:

$$s \leq \frac{1}{2} < \frac{1}{1 - \cos (N - 1)\pi/N}. \tag{13.3.23}$$

In practice, we cannot get much larger than $\frac{1}{2}$ since $\cos (N - 1)\pi/N = -\cos \pi/N$, and hence for π/N large, $1 - \cos (N - 1)\pi/N \approx 2$.

If $s > \frac{1}{2}$, usually $Q < -1$ (but not necessarily) for some n. Then the numerical solution will contain a divergent oscillation. We call this a **numerical instability**. If $s > \frac{1}{2}$, the most rapidly "growing" solution corresponds to a rapid oscillation ($n = N - 1$) in space. **The numerical instability is characterized by divergent oscillation in time ($Q < -1$) of a rapidly oscillatory ($n = N - 1$) solution in space.** Generally, if one observes computer output of this form, one probably has a numerical scheme which is unstable, and hence not reliable. This is what we observed numerically. For $s = \frac{1}{4}$, the solution behaved quite reasonably. However, for $s = 1$ a divergent oscillation was observed in time, rapidly varying in space.

Since $s = k \, \Delta t/(\Delta x)^2$, the restriction $s \leq \frac{1}{2}$ says that

$$\Delta t \leq \frac{\frac{1}{2}(\Delta x)^2}{k}. \tag{13.3.24}$$

This puts a practical constraint on numerical computations. The time steps Δt must not be too large (otherwise the scheme becomes unstable). In fact, since Δx must be small (for accurate computations), (13.3.24) shows that the time step must be exceedingly small. Thus, the forward time, centered spatial difference approximation for the heat equation is somewhat expensive to use.

To minimize calculations, we make Δt as large as possible (maintaining stability). Here $s = \frac{1}{2}$ would be a good value. In this case, the partial difference equation becomes

$$u_j^{(m+1)} = \frac{1}{2}[u_{j+1}^{(m)} + u_{j-1}^{(m)}].$$

The temperature at time Δt later is the average of the temperatures to the left and right.

Convergence. As a further general comparison between the difference and differential equations, we consider the limits of the solution of the partial difference equation as $\Delta x \to 0$ ($N \to \infty$) and $\Delta t \to 0$. We will show that the time dependence of the discretization converges to that of the heat equation if $n/N \ll 1$ (as though we fix n and let $N \to \infty$). If $n/N \ll 1$, then $\cos n\pi/N \approx 1 - \frac{1}{2}(n\pi/N)^2$ from its Taylor series, and hence

$$h_m \approx \left[1 - s\left(\frac{n\pi}{N}\right)^2 \right]^{t/\Delta t} = \left[1 - k \, \Delta t \left(\frac{n\pi}{L}\right)^2 \right]^{t/\Delta t},$$

where $N = L/\Delta x$. Thus, as $\Delta t \to 0$,

$$h_m \longrightarrow e^{-k(n\pi/L)^2 t}$$

since e may be defined as $e = \lim_{z \to 0} (1 + z)^{1/z}$.

However, difficulties may occur in practical computations if n/N is not small. *These are the highly spatially oscillatory solutions of the partial difference equation. For the heat equation these are the only waves that may cause difficulty.*

The relationship between convergence and stability can be generalized. The **Lax equivalency theorem** states: For *consistent* finite difference approximations of time-dependent *linear* partial differential equations which are well posed, the numerical scheme converges if it is stable and it is stable if it converges.

A simplified determination of the stability condition. It is often convenient to analyze the stability of a numerical method quickly. From our analysis (based on the method of separation of variables), we have shown that there are special solutions to the difference equation which oscillate in x:

$$u_j^{(m)} = e^{i\alpha x} Q^{t/\Delta t} \tag{13.3.25}$$

where

$$x = j\,\Delta x \quad \text{and} \quad t = m\,\Delta t.$$

From the boundary conditions α is restricted. Often to simplify the stability analysis, we ignore the boundary conditions and allow α to be any value.* In this case stability follows from (13.3.17) if $s \leq \frac{1}{2}$.

Random walk. The partial difference equation (13.3.15a) may be put in the form

$$u_j^{(m+1)} = su_{j-1}^{(m)} + (1 - 2s)u_j^{(m)} + su_{j+1}^{(m)}. \tag{13.3.26}$$

In the stable region, $s \leq \frac{1}{2}$, this may be interpreted as a probability problem known as a **random walk**. Consider a "drunkard" who in each unit of time Δt stands still or walks randomly to the left or to the right one step Δx. We do not know precisely where that person will be. We let $u_j^{(m)}$ be the probability that a drunkard is located at point j at time $m\,\Delta t$. We suppose that the person is equally likely to move one step Δx to the left or to the right in time Δt, with probability s each. Note that for this interpretation s must be less than or equal to $\frac{1}{2}$. The person cannot move farther than Δx in time Δt; thus, the person stays still with probability $1 - 2s$. The probability that the person will be at $j\,\Delta x$ at the next time $(m + 1)\,\Delta t$ is then given by (13.3.26); it is the sum of the probabilities of the three possible events. For example, the person might have been there at the previous time with probability $u_j^{(m)}$ and did not move with probability $1 - 2s$; the probability of this compound event is $(1 - 2s)u_j^{(m)}$. In addition, the person might have been one step to the left with probability $u_{j-1}^{(m)}$ (or right with probability $u_{j+1}^{(m)}$) and moved one step in the appropriate direction with probability s.

---

* Our more detailed stability analysis showed that the unstable waves occur only for very short wave lengths. For these waves, the boundary is perhaps expected to have the least effect.

The largest time step for stable computations $s = \frac{1}{2}$ corresponds to a random walk problem with zero probability of standing still. If the initial position is known with certainty, then

$$u_j^{(0)} = \begin{cases} 1 & j = \text{initial known location} \\ 0 & j = \text{otherwise.} \end{cases}$$

Thereafter, the person moves to the left or right with probability $\frac{1}{2}$. This yields the binomial probability distribution as illustrated by Pascal's triangle (see Fig. 13.3.7).

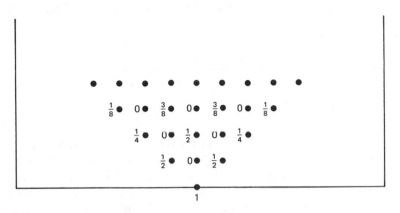

Figure 13.3.7 Pascal's triangle.

13.3.5 Separation of Variables for Partial Difference Equations and Analytic Solutions of Ordinary Difference Equations

The partial difference equation can be analyzed by the same procedure we used for partial differential equations—we separate variables. We begin by assuming that (13.3.15a) has special product solutions of the form,

$$u_j^{(m)} = \phi_j h_m. \tag{13.3.27}$$

By substituting (13.3.16) into (13.3.15a), we obtain

$$\phi_j h_{m+1} = \phi_j h_m + s(\phi_{j+1} h_m - 2\phi_j h_m + \phi_{j-1} h_m).$$

Dividing this by $\phi_j h_m$, separates the variables:

$$\frac{h_{m+1}}{h_m} = 1 + s\left(\frac{\phi_{j+1} + \phi_{j-1}}{\phi_j} - 2\right) = +\lambda,$$

where λ is a separation constant.

The partial difference equation thus yields two ordinary difference equations. The difference equation in discrete time is of first order (meaning involving one difference)

$$h_{m+1} = +\lambda h_m. \tag{13.3.28}$$

The eigenvalue λ (as in partial differential equations) is determined by a boundary value problem, here a second-order difference equation,

$$\phi_{j+1} + \phi_{j-1} = -\left(\frac{-\lambda + 1 - 2s}{s}\right)\phi_j, \tag{13.3.29a}$$

with two boundary conditions from (13.3.15c) and (13.3.15d):

$$\phi_0 = 0 \tag{13.3.29b}$$

$$\phi_N = 0. \tag{13.3.29c}$$

First-order difference equations.   First-order linear homogeneous difference equations with constant coefficients, such as (13.3.28), are easy to analyze. Consider

$$h_{m+1} = \lambda h_m, \tag{13.3.30}$$

where λ is a constant. We simply note that

$$h_1 = \lambda h_0, \, h_2 = \lambda h_1 = \lambda^2 h_0, \text{ etc.}$$

Thus, the solution is

$$h_m = \lambda^m h_0, \tag{13.3.31}$$

where h_0 is an initial condition for the first-order difference equation.

An alternative way to obtain (13.3.31) is to assume that a homogeneous solution exists in the form $h_m = Q^m$. Substitution of this into (13.3.30) yields $Q^{m+1} = \lambda Q^m$ or $Q = \lambda$, rederiving (13.3.31). This latter technique is analogous to the substitution of e^{rt} into constant-coefficient homogeneous differential equations.

The solution (13.3.31) is sketched in Fig. 13.3.8 for various values of λ.

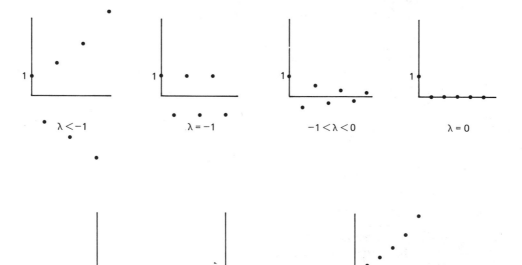

Figure 13.3.8   Solutions of first-order difference equations.

Note that if $\lambda > 1$, then the solution exponentially grows ($\lambda^m = e^{m \log \lambda} = e^{(\log \lambda / \Delta t)t}$ since $m = t/\Delta t$). If $0 < \lambda < 1$, the solution exponentially decays. Furthermore, if $-1 < \lambda < 0$, the solution has an oscillatory (and exponential) decay, known as a **convergent oscillation**. On the other hand, if $\lambda < -1$, the solution has a **divergent oscillation**.

In some situations we might even want to allow λ to be complex. Using the polar form of a complex number, $\lambda = re^{i\theta}$, $r = |\lambda|$ and $\theta = \arg \lambda$ (or angle), we obtain

$$\lambda^m = r^m e^{im\theta} = |\lambda|^m (\cos m\theta + i \sin m\theta). \qquad (13.3.32)$$

For example, the real part is $|\lambda|^m \cos m\theta$. As a function of m, λ^m oscillates (with period $m = 2\pi/\theta = 2\pi/\arg \lambda$). The solution grows in discrete time m if $|\lambda| > 1$ and decays if $|\lambda| < 1$.

We now can summarize (including the complex case). The solution λ^m of $h_{m+1} = \lambda h_m$ remains bounded as m increases (t increases) if $|\lambda| \leq 1$. It grows if $|\lambda| > 1$.

Second-order difference equations. Difference equation (13.3.29a) has constant coefficients [since $(-\lambda + 1 - 2s)/s$ is independent of the step j]. An analytic solution can be obtained easily. For any constant-coefficient difference equation, homogeneous solutions may be obtained by substituting $\phi_j = Q^j$, as we could do for first-order difference equations [see (13.3.31)].

The boundary conditions, $\phi_0 - \phi_N = 0$, suggest that the solution may oscillate. This usually occurs if Q is complex with $|Q| = 1$, in which case an equivalent substitution is

$$\phi_j = (|Q|e^{i\theta})^j = e^{i\theta j} = e^{i\theta(x/\Delta x)} = e^{i\alpha x}, \qquad (13.3.33)$$

since $j = x/\Delta x$, defining $\alpha = \theta/\Delta x = (\arg Q)/\Delta x$. In Exercise 13.3.2 it is shown that (13.3.29) implies that $|Q| = 1$, so that (13.3.33) may be used. Substituting (13.3.33) into (13.3.29a) yields an equation for the wave number α:

$$e^{i\alpha \Delta x} + e^{-i\alpha \Delta x} = \frac{\lambda - 1 + 2s}{s} \quad \text{or} \quad \boxed{2 \cos(\alpha \, \Delta x) = \frac{\lambda - 1 + 2s}{s}.}$$

$$(13.3.34)$$

This yields two values of α (one the negative of the other), and thus instead of $\phi_j = e^{i\alpha x}$ we use a linear combination of $e^{\pm i\alpha x}$, or

$$\phi_j = c_1 \sin \alpha x + c_2 \cos \alpha x. \qquad (13.3.35)$$

The boundary conditions, $\phi_0 = \phi_N = 0$, imply that $c_2 = 0$ and $\alpha = n\pi/L$, where $n = 1, 2, 3, \ldots$. Thus,

$$\boxed{\phi_j = \sin \frac{n\pi x}{L} = \sin \frac{n\pi j \, \Delta x}{L} = \sin \frac{n\pi j}{N}.} \qquad (13.3.36)$$

Further analysis follows that of the preceding subsection.

13.3.6 Matrix Notation

A matrix* notation is often convenient for analyzing the discretization of partial differential equations. For fixed t, $u(x, t)$ is only a function of x. Its discretization $u_j^{(m)}$ is defined at each of the $N + 1$ mesh points (at every time step). We introduce a *vector* $\mathbf{u}$ of dimension $N + 1$, that changes at each time step; it is a function of m, $\mathbf{u}^{(m)}$. The jth component of $\mathbf{u}^{(m)}$ is the value of $u(x, t)$ at the jth mesh point:

$$(\mathbf{u}^{(m)})_j = u_j^{(m)} \tag{13.3.37}$$

The partial difference equation is

$$u_j^{(m+1)} = u_j^{(m)} + s(u_{j+1}^{(m)} - 2u_j^{(m)} + u_{j-1}^{(m)}). \tag{13.3.38}$$

If we apply the boundary conditions, $u_0^{(m)} = u_N^{(m)} = 0$, then

$$u_1^{(m+1)} = u_1^{(m)} + s(u_2^{(m)} - 2u_1^{(m)} + u_0^{(m)}) = (1 - 2s)u_1^{(m)} + su_2^{(m)}.$$

A similar equation is valid for $u_{N-1}^{(m+1)}$. At each time step there are $N - 1$ unknowns. We introduce the $N - 1 \times N - 1$ **tridiagonal matrix A** with all entries zero except for the main diagonal (with entries $1 - 2s$) and neighboring diagonals (with entries s):

$$\mathbf{A} = \begin{pmatrix} 1 - 2s & s & 0 & & 0 \\ s & 1 - 2s & s & & \\ 0 & s & 1 - 2s & & 0 \\ & & & & s \\ 0 & & 0 & s & 1 - 2s \end{pmatrix} \tag{13.3.39}$$

The partial difference equation becomes the following vector equation:

$$\boxed{\mathbf{u}^{(m+1)} = \mathbf{A}\mathbf{u}^{(m)}.} \tag{13.3.40}$$

The vector $\mathbf{u}$ changes in a straightforward way. We start with $\mathbf{u}^{(0)}$ representing the initial condition. By direct computation

$$\mathbf{u}^{(1)} = \mathbf{A}\mathbf{u}^{(0)}$$
$$\mathbf{u}^{(2)} = \mathbf{A}\mathbf{u}^{(1)} = \mathbf{A}^2\mathbf{u}^{(0)},$$

and thus

$$\mathbf{u}^{(m)} = \mathbf{A}^m\mathbf{u}^{(0)}. \tag{13.3.41}$$

---

* This section requires some knowledge of linear algebra.

The matrix $\mathbf{A}$ raised to the mth power describes how the initial condition influences the solution at the mth time step ($t = m\,\Delta t$).

To understand this solution, we introduce the **eigenvalues** μ of the matrix $\mathbf{A}$, the values μ such that there are nontrivial vector solutions ξ:

$$\mathbf{A}\xi = \mu\xi. \tag{13.3.42}$$

The eigenvalues satisfy

$$\det\,[\mathbf{A} - \mu\mathbf{I}] = 0, \tag{13.3.43}$$

where $\mathbf{I}$ is the identity matrix. Nontrivial vectors ξ which satisfy (13.3.42) are called **eigenvectors** corresponding to μ. Since $\mathbf{A}$ is an $(N-1) \times (N-1)$ matrix, $\mathbf{A}$ has $N-1$ eigenvalues. However, some of the eigenvalues may not be distinct, there may be multiple eigenvalues (or **degeneracies**). For a distinct eigenvalue, there is a unique eigenvector (to within a multiplicative constant). In the case of a multiple eigenvalue (of multiplicity k), there may be at most k linearly independent eigenvectors. If for some eigenvalue there are less than k eigenvectors, we say the matrix is **defective**. If $\mathbf{A}$ is real and symmetric [as (13.3.39) is], it is known that any possible multiple eigenvalues are *not* defective. Thus, the matrix $\mathbf{A}$ has $N-1$ eigenvectors (which can be shown to be linearly independent). Furthermore, if $\mathbf{A}$ is real and symmetric, the eigenvalues (and consequently the eigenvectors) are real and the eigenvectors are orthogonal (see Sec. 5.6 Appendix). We let μ_n be the nth eigenvalue and ξ_n the corresponding eigenvector.

We can solve vector equation (13.3.40) (equivalent to the partial difference equation) using the **method of eigenvector expansion**. (This technique is analogous to using an eigenfunction expansion to solve the partial differential equation.) Any vector can be expanded in a series of the eigenvectors:

$$\mathbf{u}^{(m)} = \sum_{n=1}^{N-1} c_n^{(m)}\xi_n. \tag{13.3.44}$$

The vector changes with m (time), and thus the constants $c_n^{(m)}$ depend on m (time):

$$\mathbf{u}^{(m+1)} = \sum_{n=1}^{N-1} c_n^{(m+1)}\xi_n. \tag{13.3.45}$$

However, from (13.3.40),

$$\mathbf{u}^{(m+1)} = \mathbf{A}\mathbf{u}^{(m)} = \sum_{n=1}^{N-1} c_n^{(m)}\mathbf{A}\xi_n = \sum_{n=1}^{N-1} c_n^{(m)}\mu_n\xi_n, \tag{13.3.46}$$

where (13.3.44) and (13.3.42) have been used. By comparing (13.3.45) and (13.3.46), we determine a first-order difference equation with constant coefficients for $c_n^{(m)}$:

$$c_n^{(m+1)} = \mu_n c_n^{(m)}. \tag{13.3.47}$$

This is easily solved,

$$c_n^{(m)} = c_n^{(0)}\,(\mu_n)^m, \tag{13.3.48}$$

and thus

$$\mathbf{u}^{(m)} = \sum_{n-1}^{N-1} c_n^{(0)} (\mu_n)^m \boldsymbol{\xi}_n. \tag{13.3.49}$$

$c_n^{(0)}$ can be determined from the initial condition.

From (13.3.49), the growth of the solution as t increases (m increases) depends on $(\mu_n)^m$, where $m = t/\Delta t$. We recall that since μ_n is real

$$(\mu_n)^m = \begin{cases} \text{exponential growth} & \mu_n > 1 \\ \text{exponential decay} & 0 < \mu_n < 1 \\ \text{convergent oscillation} & -1 < \mu_n < 0 \\ \text{divergent oscillation} & \mu_n < -1. \end{cases}$$

This numerical solution is unstable if any eigenvalue $\mu_n > 1$ or any $\mu_n < -1$.

We need to obtain the $N - 1$ eigenvalues μ of $\mathbf{A}$:

$$\mathbf{A}\boldsymbol{\xi} = \mu\boldsymbol{\xi}. \tag{13.3.50}$$

We let ξ_j be the jth component of $\boldsymbol{\xi}$. Since $\mathbf{A}$ is given by (13.3.39), we can rewrite (13.3.50) as

$$s\xi_{j+1} + (1 - 2s)\xi_j + s\xi_{j-1} = \mu\xi_j \tag{13.3.51a}$$

with

$$\xi_0 = 0 \quad \text{and} \quad \xi_N = 0. \tag{13.3.51b}$$

Equation (13.3.51a) is equivalent to

$$\xi_{j+1} + \xi_{j-1} = \left(\frac{\mu + 2s - 1}{s}\right)\xi_j. \tag{13.3.52}$$

By comparing (13.3.52) with (13.3.29), we observe that the eigenvalues μ of $\mathbf{A}$ are the eigenvalues λ of the second-order difference equation obtained by separation of variables. Thus [see (13.3.17)],

$$\mu = 1 - 2s(1 - \cos(\alpha \Delta x)), \tag{13.3.53}$$

where $\alpha = n\pi/L$ for $n = 1, 2, \ldots, N - 1$. As before, the scheme is usually unstable if $s > \frac{1}{2}$. To summarize this simple case, the eigenvalues can be explicitly determined using Fourier-type analysis.

In more difficult problems it is rare that the eigenvalues of large matrices can be obtained easily. Sometimes the **Gershgorin circle theorem** (see Strang [1980] for an elementary proof) is useful: **Every eigenvalue of A lies in at least one of the circles $c_1, \ldots, c_{N-1}$ in the complex plane where c_i has its center at the ith diagonal entry and its radius equal to the sum of the absolute values of the rest of that row.** If a_{ij} are the entries of $\mathbf{A}$, then all eigenvalues μ lie in at least one of the following circles:

$$|\mu - a_{ii}| \le \sum_{\substack{j=1 \\ (j \ne i)}}^{N-1} |a_{ij}|. \tag{13.3.54}$$

For our matrix $\mathbf{A}$, the diagonal elements are all the same $1 - 2s$ and the rest of the row sums to $2s$ (except the first and last rows, which sum to s). Thus, two circles are $|\mu - (1 - 2s)| < s$ and the other $N - 3$ circles are

$$|\mu - (1 - 2s)| < 2s. \qquad (13.3.55)$$

All eigenvalues lie in the resulting regions [the biggest of which is given by (13.3.55)], as sketched in Fig. 13.3.9. Since the eigenvalues μ are also known to be real, Fig. 13.3.9 shows that

$$1 - 4s \leqslant \mu \leqslant 1.$$

Stability is guaranteed if $-1 \leqslant \mu \leqslant 1$, and thus, the Gershgorin circle theorem implies that the numerical scheme is stable if $s \leqslant \frac{1}{2}$. If $s > \frac{1}{2}$, the Gershgorin circle theorem does *not* imply the scheme is unstable.

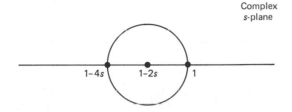

Complex
s-plane

Figure 13.3.9 Gershgorin circles for **A** corresponding to discretization of the heat equation.

13.3.7 Nonhomogeneous Problems

The heat equation with sources may be computed in the same way. Consider

$$\frac{\partial u}{\partial t} = k \frac{\partial^2 u}{\partial x^2} + Q(x, t)$$

$$u(0, t) = A(t)$$

$$u(L, t) = B(t)$$

$$u(x, 0) = f(x).$$

As before, we use a forward difference in time and a centered difference in space. We obtain the following numerical approximation:

$$\frac{u_j^{(m+1)} - u_j^{(m)}}{\Delta t} = \frac{k}{(\Delta x)^2}(u_{j+1}^{(m)} - 2u_j^{(m)} + u_{j-1}^{(m)}) + Q(j\,\Delta x, m\,\Delta t)$$

$$u_0^{(m)} = A(m\,\Delta t)$$

$$u_N^{(m)} = B(m\,\Delta t)$$

$$u_j^{(0)} = f(j\,\Delta x).$$

The solution is easily computed by solving for $u_j^{(m+1)}$. We claim that our stability analysis for homogeneous problems is valid for nonhomogeneous problems. Thus, we compute with $s = k\,\Delta t/(\Delta x)^2 \leqslant \frac{1}{2}$.

13.3.8 Other Numerical Schemes

The numerical scheme for the heat equation, which uses the centered difference in space and forward difference in time, is stable if $s = k\,\Delta t/(\Delta x)^2 \leqslant \frac{1}{2}$. The time step is small [being proportional to $(\Delta x)^2$]. We might desire a less expensive scheme. The truncation error is the sum of terms, one proportional to Δt and the other to $(\Delta x)^2$. If s is fixed (for example $s = \frac{1}{2}$), both errors are $O(\Delta x)^2$ since $\Delta t = s(\Delta x)^2/k$.

Richardson's scheme. For a less expensive scheme one might try a more accurate time difference. Using centered differences in both space and time was first proposed by Richardson (1927):

$$\frac{u_j^{(m+1)} - u_j^{(m-1)}}{\Delta t} = \frac{k}{(\Delta x)^2}(u_{j+1}^{(m)} - 2u_j^{(m)} + u_{j-1}^{(m)}) \tag{13.3.56a}$$

or

$$u_j^{(m+1)} = u_j^{(m-1)} + s(u_{j+1}^{(m)} - 2u_j^{(m)} + u_{j-1}^{(m)}), \tag{13.3.56b}$$

where again $s = k \, \Delta t/(\Delta x)^2$. Here the truncation error is the sum of a $(\Delta t)^2$ and $(\Delta x)^2$ terms. Although in some sense this scheme is more accurate than the previous one, (13.3.56b) should never be used. Exercise 13.3.12(a) shows that this numerical method is always unstable.

Crank–Nicholson scheme. Crank and Nicholson (1947) suggested an alternative way to utilize centered differences. The forward difference in time

$$\frac{\partial u}{\partial t} \approx \frac{u(t + \Delta t) - u(t)}{\Delta t}$$

may be interpreted as the centered difference around $t + \Delta t/2$. The error in approximating $\partial u/\partial t(t + \Delta t/2)$ is $O(\Delta t)^2$. Thus, we discretize the second derivative at $t + \Delta t/2$ with a centered difference scheme. Since this involves functions evaluated at this in-between time, we take the average at t and $t + \Delta t$. This yields the Crank–Nicholson scheme,

$$\frac{u_j^{(m+1)} - u_j^{(m)}}{\Delta t} = \frac{k}{2}\left[\frac{u_{j+1}^{(m)} - 2u_j^{(m)} + u_{j-1}^{(m)}}{(\Delta x)^2} + \frac{u_{j+1}^{(m+1)} - 2u_j^{(m+1)} + u_{j-1}^{(m+1)}}{(\Delta x)^2}\right]. \tag{13.3.57}$$

It is not obvious, but nevertheless true (Exercise 13.3.13), that the truncation error remains the sum of two terms, one $(\Delta x)^2$ and the other $(\Delta t)^2$. The advantage of the Crank–Nicholson method is that the scheme is stable for all $s = k \, \Delta t/(\Delta x)^2$, as is shown in Exercise 13.3.12(b). Δt can be as large as desired. We can choose Δt to be proportional to Δx [rather than $(\Delta x)^2$]. The error is then $O(\Delta x)^2$, an equivalent accuracy as the earlier scheme with much less work (computing). Crank–Nicholson is a practical method. However, the Crank-Nicholson scheme (see Fig. 13.3.10) involves six points (rather than four for the simpler stable method), three of which are at the advanced time. We cannot directly march

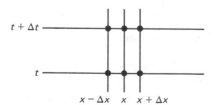

Figure 13.3.10 Implicit Crank–Nicholson scheme.

forward in time with (13.3.57). Instead, to advance one time step, (13.3.57) requires the solution of a linear system of $N - 1$ equations. The scheme (13.3.57) is called **implicit** (while (13.3.12) is called **explicit**). The matrices involved are tridiagonal, and thus the linear system may be solved using Gauss elimination easily (and relatively inexpensively) even if N is large.

13.3.9 Other Types of Boundary Conditions

If $\partial u / \partial x = g(t)$ at $x = 0$ (rather than u being given at $x = 0$), then we must introduce a numerical approximation for the boundary condition. Since the discretization of the partial differential equation has an $O(\Delta x)^2$ truncation error, we may introduce an equal error in the boundary condition by using a centered difference in space:

$$\frac{\partial u}{\partial x} \sim \frac{u(x + \Delta x, t) - u(x - \Delta x, t)}{2\Delta x}.$$

In this case the boundary condition $\partial u / \partial x = g(t)$ at $x = 0$ becomes

$$\frac{u_1^{(m)} - u_{-1}^{(m)}}{2 \, \Delta x} = g(t) = g(m \, \Delta t) = g_m. \tag{13.3.58}$$

We use (13.3.58) to obtain an expression for the temperature at the *fictitious* point ($x_{-1} = -\Delta x$):

$$u_{-1}^{(m)} = u_1^{(m)} - 2 \, \Delta x g_m. \tag{13.3.59}$$

In this way we determine the value at the fictitious point initially, $u_{-1}^{(0)}$. This fictitious point is needed to compute the boundary temperature at later times via the partial difference equation. If we use forward difference in time and centered difference in space, (13.3.15) can now be applied for $j = 0$ to $j = N - 1$. For example, at $x = 0$ ($j = 0$)

$$u_0^{(m+1)} = u_0^{(m)} + s(u_1^{(m)} - 2u_0^{(m)} + u_{-1}^{(m)})$$
$$= u_0^{(m)} + s(u_1^{(m)} - 2u_0^{(m)} + u_1^{(m)} - 2 \, \Delta x g_m),$$

where (13.3.59) has been used. In this way a partial differential equation can be solved numerically with boundary conditions involving the derivative. (The fictitious point is eliminated between the boundary condition and the partial differential equation.)

EXERCISES 13.3

13.3.1. (a) Show that the truncation error for our numerical scheme, (13.3.3), becomes much smaller if $k(\Delta t)/(\Delta x)^2 = \frac{1}{6}$. [*Hint:* u satisfies (13.3.1a).]

(b) If $k \, \Delta t/(\Delta x)^2 = \frac{1}{6}$, determine the order of magnitude of the truncation error.

13.3.2. By letting $\phi_j = Q^j$, show that (13.3.29) is only satisfied if $|Q| = 1$. [*Hint:* First show that

$$Q^2 + \left(\frac{-\lambda + 1 - 2s}{s}\right)Q + 1 = 0.\Big]$$

13.3.3. Define $L(\phi) = \phi_{j+1} + \phi_{j-1} + \gamma\phi_j$.

(a) Show that $uL(v) - vL(u) = w_{j+1} - w_j$, where $w_j = u_{j-1}v_j - v_{j-1}u_j$.

(b) Since summation is analogous to integration, derive the discrete version of Green's formula

$$\sum_{i=1}^{N-1} [uL(v) - vL(u)] = w_N - w_0.$$

(c) Show that the right-hand side of part (b) vanishes if both u and v satisfy the homogeneous boundary conditions (13.3.29b) and (13.3.29c).

(d) Letting $\gamma = (1 - 2s)/s$, the eigenfunctions ϕ satisfy $L(\phi) = (\lambda/s)\phi$. Show that eigenfunctions corresponding to different eigenvalues are orthogonal in the sense that

$$\sum_{i=1}^{N-1} \phi_i\psi_i = 0.$$

*13.3.4. (a) Using Exercise 13.3.3, determine β_n in (13.3.21) from the initial conditions $u_j^{(0)} = f_j$.

(b) Evaluate the normalization constant

$$\sum_{j=1}^{N-1} \sin^2\frac{n\pi j}{N},$$

for each eigenfunction (i.e., fix n). (*Hint:* Use the double-angle formula and a geometric series.)

13.3.5. Show that at each successive mesh point the sign of the solution alternates for the most unstable mode (of our numerical scheme for the heat equation, $s > \frac{1}{2}$).

13.3.6. Evaluate $1/[1 - \cos(N - 1)\pi/N.]$. What conclusions concerning stability do you reach?

(a) $N = 4$ (b) $N = 6$ (c) $N = 8$ *(d) $N = 10$

(e) Asymptotically for large N

13.3.7. Numerically compute solutions to the heat equation with the temperature initially given in Fig. 13.3.4. Use (13.3.15) with $N = 10$. Do for various s (discuss stability):

(a) $s = 0.49$ (b) $s = 0.50$ (c) $s = 0.51$ (d) $s = 0.52$

13.3.8. Under what condition will an initially positive solution $[u(x, 0) > 0]$ remain positive $[u(x, t) > 0]$ for our numerical scheme (13.3.9) for the heat equation?

13.3.9. Consider

$$\frac{d^2u}{dx^2} = f(x)$$

$$u(0) = 0$$

$$u(L) = 0.$$

(a) Using the centered difference approximation for the second-derivative and dividing the length L into three equal mesh lengths (see Sec. 13.3.2), derive a system of linear equations for an approximation to $u(x)$. Use the notation $x_i = i\,\Delta x$, $f_i = f(x_i)$, and $u_i = u(x_i)$. (*Note:* $x_0 = 0$, $x_1 - \frac{1}{3}L$, $x_2 - \frac{2}{3}L$, $x_3 = L$.)

*(b) Write the system as a matrix system $\mathbf{Au} = \mathbf{f}$. What is $\mathbf{A}$?

(c) Solve for u_1 and u_2.

(d) Show that a "Green's function" matrix **G** can be defined:

$$u_i = \sum_j G_{ij} f_j \qquad (\mathbf{u} = \mathbf{Gf}).$$

What is **G**? Show that it is symmetric, $G_{ij} = G_{ji}$.

13.3.10. Suppose that in a random walk, at each Δt the probability of moving to the right Δx is a and the probability of moving to the left Δx is also a. The probability of staying in place is b. ($2a + b = 1$.)

(a) Formulate the difference equation for this problem.

*__(b)__ Derive a partial differential equation governing this process if $\Delta x \to 0$ and $\Delta t \to 0$ such that

$$\lim_{\substack{\Delta x \to 0 \\ \Delta t \to 0}} \frac{(\Delta x)^2}{\Delta t} = \frac{k}{s}.$$

(c) Suppose that there is a wall (or cliff) on the right at $x = L$ with the property that *after* the wall is reached, the probability of moving to the left is a, to the right c, and for staying in place $1 - a - c$. Assume that no one returns from $x > L$. What condition is satisfied at the wall? What is the resulting boundary condition for the partial differential equation? (Let $\Delta x \to 0$ and $\Delta t \to 0$ as before.) Consider the two cases $c = 0$ and $c \neq 0$.

13.3.11. Suppose that, in a two-dimensional random walk, at each Δt it is equally likely to move right Δx or left or up Δy or down (as illustrated in Fig. 13.3.11).

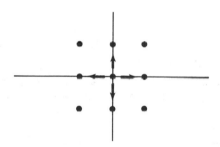

Figure 13.3.11

(a) Formulate the difference equation for this problem.

(b) Derive a partial differential equation governing this process if $\Delta x \to 0$, $\Delta y \to 0$, and $\Delta t \to 0$ such that

$$\lim_{\substack{\Delta x \to 0 \\ \Delta t \to 0}} \frac{(\Delta x)^2}{\Delta t} = \frac{k_1}{s} \quad \text{and} \quad \lim_{\substack{\Delta y \to 0 \\ \Delta t \to 0}} \frac{(\Delta y)^2}{\Delta t} = \frac{k_2}{s}.$$

13.3.12. Use a simplified determination of stability (i.e., substitute $u_j^{(m)} = e^{i\alpha x} Q^{t/\Delta t}$), to investigate:

(a) Richardson's centered difference in space and time scheme for the heat equation, (13.3.56b).

(b) Crank–Nicholson scheme for the heat equation, (13.3.57).

13.3.13. Investigate the truncation error for the Crank–Nicholson method, (13.3.57).

13.3.14. For the following matrices:
1. Compute the eigenvalue.
2. Compute the Gershgorin (row) circles.

3. Compare (1) and (2) according to the theorem.

(a) $\begin{bmatrix} 1 & 0 \\ \frac{1}{2} & 2 \end{bmatrix}$ (b) $\begin{bmatrix} 1 & 0 \\ 3 & 2 \end{bmatrix}$ *(c) $\begin{bmatrix} 1 & 2 & -3 \\ 2 & 4 & -6 \\ 0 & \frac{1}{3} & 2 \end{bmatrix}$

13.3.15. For the examples in Exercise 13.3.14, compute the Gershgorin (column) circles. Show that a corresponding theorem is valid for them.

13.3.16. Using forward differences in time and centered differences in space, analyze carefully the stability of the difference scheme if the boundary condition for the heat equation is

$$\frac{\partial u}{\partial x}(0) = 0 \quad \text{and} \quad \frac{\partial u}{\partial x}(L) = 0.$$

(*Hint:* See Sec. 13.3.9.) Compare your result to the one for the boundary conditions $u(0) = 0$ and $u(L) = 0$.

13.4 TWO-DIMENSIONAL HEAT EQUATION

Similar ideas may be applied to numerically compute solutions of the two-dimensional heat equation

$$\frac{\partial u}{\partial t} = k \left(\frac{\partial^2 u}{\partial x^2} + \frac{\partial^2 u}{\partial y^2} \right).$$

We introduce a two-dimensional mesh (or lattice), where for convenience we assume that $\Delta x = \Delta y$. Using a forward difference in time and the formula for the Laplacian based on a centered difference in both x and y [see (13.2.11)], we obtain

$$\frac{u_{j,l}^{(m+1)} - u_{j,l}^{(m)}}{\Delta t} = \frac{k}{(\Delta x)^2} [u_{j+1,l}^{(m)} + u_{j-1,l}^{(m)} + u_{j,l+1}^{(m)} + u_{j,l-1}^{(m)} - 4u_{j,l}^{(m)}], \qquad (13.4.1)$$

where $u_{j,l}^{(m)} \approx u(j \, \Delta x, l \, \Delta y, m \, \Delta t)$. We march forward in time using (13.4.1).

As before, the numerical scheme may be unstable. We perform a simplified stability analysis, and thus ignore the boundary conditions. We investigate possible growth of spatially periodic waves by substituting

$$u_{j,l}^{(m)} = Q^{t/\Delta t} e^{i(\alpha x + \beta y)} \qquad (13.4.2)$$

into (13.4.1). We immediately obtain

$$Q = 1 + s(e^{i\alpha\Delta x} + e^{-i\alpha\Delta x} + e^{i\beta\Delta y} + e^{-i\beta\Delta y} - 4)$$
$$= 1 + 2s(\cos \alpha\Delta x + \cos \beta\Delta y - 2),$$

where $s = k \, \Delta t/(\Delta x)^2$ and $\Delta x = \Delta y$. To ensure stability, $-1 < Q < 1$, and hence we derive the stability condition

$$s = \frac{k \, \Delta t}{(\Delta x)^2} \leq \frac{1}{4}. \tag{13.4.3}$$

As an elementary example, yet one in which an exact solution is not available, we consider the heat equation on an L-shaped region sketched in Fig. 13.4.1. We assume that the temperature is initially zero. Also, on the boundary $u = 1000$ at $x = 0$, but $u = 0$ on the rest of the boundary. We compute with the largest stable time step, $s = \frac{1}{4}$ $[\Delta t = (\Delta x)^2/4k]$, so that (13.4.1) becomes

$$u_{j,l}^{(m+1)} = \frac{[u_{j+1,l}^{(m)} + u_{j-1,l}^{(m)} + u_{j,l+1}^{(m)} + u_{j,l-1}^{(m)}]}{4}. \tag{13.4.4}$$

In this numerical scheme, the temperature at the next time is the average of the four neighboring mesh (or lattice) points at the present time. We choose $\Delta x = \frac{1}{10}$ ($\Delta t = 1/400k$) and sketch the numerical solution in Figs. 13.4.2–3. We draw contours of approximately equal temperature in order to observe the thermal energy invading the interior of this region.

The partial difference equation is straightforward to apply if the boundary is composed entirely of mesh points. Usually, this is unlikely, in which case some more complicated procedure must be applied for the boundary condition.

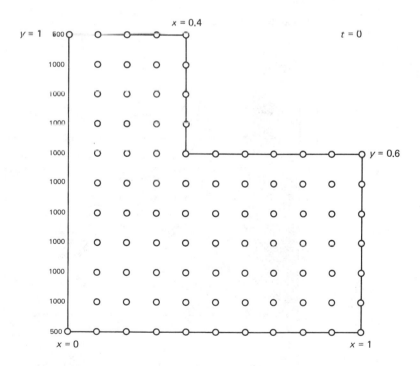

Figure 13.4.1 Initial condition.

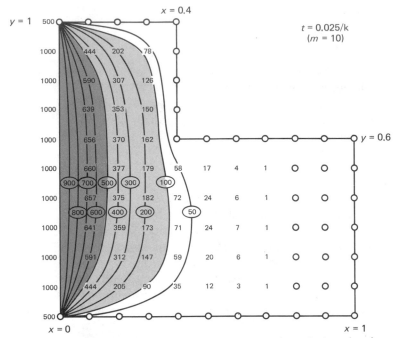

Figure 13.4.2 Numerical computation of temperature in an L-shaped region.

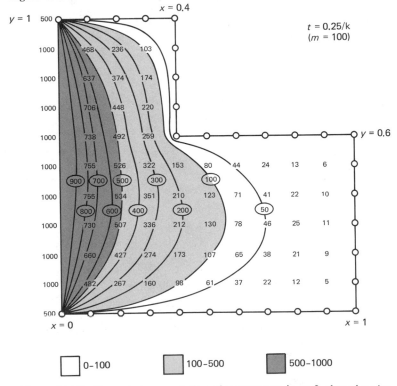

Figure 13.4.3 Numerical computation of temperature in an L-shaped region.

EXERCISES 13.4

*13.4.1. Derive the stability condition (ignoring the boundary condition) for the two-dimensional heat equation if $\Delta x \neq \Delta y$.

13.4.2. Derive the stability condition (including the effect of the boundary condition) for the two-dimensional heat equation with $u(x, y) = 0$ on the four sides of a square if $\Delta x = \Delta y$.

13.4.3. Derive the stability condition (ignoring the boundary condition) for the three-dimensional heat equation if $\Delta x = \Delta y = \Delta z$.

13.4.4. Solve numerically the heat equation on a rectangle $0 < x < 1$, $0 < y < 2$ with the temperature initially zero. Assume that the boundary conditions are zero on three sides, but $u = 1$ on one long side.

13.5 WAVE EQUATION

One may also compute solutions to the one-dimensional wave equation by introducing finite difference approximations. Using centered differences in space and time, the wave equation,

$$\frac{\partial^2 u}{\partial t^2} = c^2 \frac{\partial^2 u}{\partial x^2},$$ (13.5.1)

becomes the following partial difference equation,

$$\frac{u_j^{(m+1)} - 2u_j^{(m)} + u_j^{(m-1)}}{(\Delta t)^2} = c^2 \frac{u_{j+1}^{(m)} - 2u_j^{(m)} + u_{j-1}^{(m)}}{(\Delta x)^2}.$$ (13.5.2)

The truncation error is the sum of a term of $O(\Delta x)^2$ and one of $O(\Delta t)^2$. By solving (13.5.2) for $u_j^{(m+1)}$, the solution may be marched forward in time. Three levels of time are involved in (13.5.2) as indicated in Fig. 13.5.1. $u(x, t)$ is needed "initially" at two values of t (0 and $-\Delta t$) to start the calculation. We use the two initial conditions for the wave equation, $u(x, 0) = f(x)$ and $\partial u/\partial t\,(x, 0) = g(x)$, to compute at $m = 0$ and at $m = -1$. Using a centered difference in time for $\partial u/\partial t$ [so as to maintain an $O(\Delta t)^2$ truncation error] yields

$$u_j^{(0)} = f(x_j) = f(j\,\Delta x)$$ (13.5.3)

$$\frac{u_j^{(1)} - u_j^{(-1)}}{2\,\Delta t} = g(x_j) = g(j\,\Delta x).$$ (13.5.4)

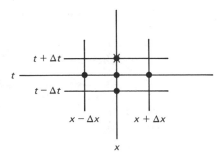

Figure 13.5.1 Marching forward in time for the wave equation.

To begin the calculation we must compute $u_j^{(-1)}$. The initial conditions, (13.5.3) and (13.5.4), are two equations in three unknowns, $u_j^{(-1)}$, $u_j^{(0)}$, $u_j^{(1)}$. The partial difference equation at $t = 0$ provides a third equation:

$$u_j^{(1)} = 2u_j^{(0)} - u_j^{(-1)} + \frac{c^2}{(\Delta x/\Delta t)^2}(u_{j-1}^{(0)} - 2u_j^{(0)} + u_{j+1}^{(0)}). \qquad (13.5.5)$$

$u_j^{(1)}$ may be eliminated from (13.5.4) and (13.5.5) since $u_j^{(0)}$ is known from (13.5.3). In this way we can solve for $u_j^{(-1)}$. Once $u_j^{(-1)}$ and $u_j^{(0)}$ are known, the later values of u may be computed via (13.5.2). Boundary conditions may be analyzed as before.

Our limited experience should already suggest that a stability analysis is important. To determine if any spatially periodic waves grow, we substitute

$$u_j^{(m)} = Q^{t/\Delta t}e^{i\alpha x} \qquad (13.5.6)$$

into (13.5.2), yielding

$$Q - 2 + \frac{1}{Q} = \sigma, \qquad (13.5.7)$$

where

$$\sigma = \frac{c^2}{(\Delta x/\Delta t)^2}(e^{i\alpha\Delta x} - 2 + e^{-i\alpha\Delta x}) = \frac{2c^2}{(\Delta x/\Delta t)^2}[\cos(\alpha \Delta x) - 1]. \qquad (13.5.8)$$

Equation (13.5.7) is a quadratic equation for Q since (13.5.2) involves three time levels and two time differences:

$$Q^2 - (\sigma + 2)Q + 1 = 0 \quad \text{and thus} \quad Q = \frac{\sigma + 2 \pm \sqrt{(\sigma + 2)^2 - 4}}{2}. \qquad (13.5.9a, b)$$

The two roots correspond to two ways in which waves evolve in time. If $-2 < \sigma + 2 < 2$, the roots are complex conjugates of each other. In this case (as discussed earlier)

$$Q^m = (re^{i\theta})^m = r^m e^{im\theta},$$

where $r = |Q|$ and $\theta = \arg Q$. Since

$$|Q|^2 = \frac{(\sigma + 2)^2}{4} + \frac{4 - (\sigma + 2)^2}{4} = 1,$$

the solution oscillates for fixed x as m (time) increases if $-2 < \sigma + 2 < 2$. This is similar to the wave equation itself, which permits time periodic solutions when the spatial part is periodic (e.g., $\sin n\pi x/L \cos n\pi ct/L$). If $\sigma + 2 > 2$ or $\sigma + 2 < -2$, the roots are real with product equal to 1 [see (13.5.9a)]. Thus, one root will be greater than one in absolute value, giving rise to an unstable behavior.

The solution will be stable if $-2 < \sigma + 2 < 2$ or $-4 < \sigma < 0$. From (13.5.8) we conclude that our numerical scheme is stable if

$$\boxed{\frac{c}{\Delta x/\Delta t} \leq 1,} \qquad (13.5.10)$$

known as the **Courant stability condition** (for the wave equation). Here c is the speed of propagation of signals for the wave equation and $\Delta x / \Delta t$ is the speed of propagation of signals for the discretization of the wave equation. Thus, we conclude that for stability, **the numerical scheme must have a greater propagation speed than the wave equation itself.** In this way the numerical scheme will be able to account for the real signals being propagated. The stability condition again limits the size of the time step, in this case

$$\Delta t \le \frac{\Delta x}{c}. \qquad (13.5.11)$$

EXERCISES 13.5

13.5.1. Modify the Courant stability condition for the wave equation to account for the boundary condition $u(0) = 0$ and $u(L) = 0$.

13.5.2. Consider the wave equation subject to the initial conditions

$$u(x, 0) = \begin{cases} 1 & \dfrac{L}{4} < x < \dfrac{3L}{4} \\ 0 & \text{otherwise} \end{cases}$$

$$\frac{\partial u}{\partial t}(x, 0) = 0$$

and the boundary conditions

$$u(0, t) = 0$$
$$u(L, t) = 0.$$

Use nine interior mesh points and compute using centered differences in space and time. Compare to the exact solution.
(a) $\Delta t = \Delta x / 2c$
(b) $\Delta t = \Delta x / c$
(c) $\Delta t = 2\Delta x / c$

13.5.3. For the wave equation, $u(x, t) = f(x - ct)$ is a solution, where f is an arbitrary function. If $c = \Delta x / \Delta t$, show that $u_j^m = f(x_j - ct_m)$ is a solution of (13.5.2) for arbitrary f.

13.5.4. Show that the conclusion of Exercise 13.5.3 is not valid if $c \ne \Delta x / \Delta t$.

13.5.5. Consider the first-order wave equation

$$\frac{\partial u}{\partial t} + c \frac{\partial u}{\partial x} = 0.$$

(a) Determine a partial difference equation by using a forward difference in time and a centered difference in space.
*(b) Analyze the stability of this scheme (without boundary conditions).

*13.5.6. Modify Exercise 13.5.5 for centered difference in space and time.

13.6 LAPLACE'S EQUATION

Introduction.   Laplace's equation

$$\nabla^2 u = 0 \qquad\qquad (13.6.1)$$

is usually formulated in some region such that one condition must be satisfied along the entire boundary. A time variable does not occur, so that a numerical finite difference method must proceed somewhat differently than the heat or wave equations.

Using the standard centered difference discretization, Laplace's equation in two dimensions becomes the following partial difference equation (assuming that $\Delta x = \Delta y$):

$$\frac{u_{j+1,l} + u_{j-1,l} + u_{j,l+1} + u_{j,l-1} - 4u_{j,l}}{(\Delta x)^2} = 0, \qquad (13.6.2)$$

where, hopefully, $u_{j,l} \approx u(j\,\Delta x, l\,\Delta y)$.

The boundary condition may be analyzed in the same way as for the heat and wave equations. In the simplest case u is specified along the boundary (composed of mesh points). The temperatures at the interior mesh points are the unknowns. Equation (13.6.2) is valid at each of these interior points. Some of the terms in (13.6.2) are determined from the boundary conditions, but most terms remain unknown. Equation (13.6.2) can be written as a linear system. Gaussian elimination can be used, but in many practical situations the number of equations and unknowns (equaling the number of interior mesh points) is too large for efficient numerical calculations. This is especially true in three dimensions, where even a coarse $20 \times 20 \times 20$ grid will generate 8000 linear equations in 8000 unknowns.

By rearranging (13.6.2), we derive

$$u_{j,l} = \frac{u_{j+1,l} + u_{j-1,l} + u_{j,l+1} + u_{j,l-1}}{4}. \qquad (13.6.3)$$

The temperature $u_{j,l}$ must be the average of its four neighbors. Thus, the solution of the discretization of Laplace's equation satisfies a mean value property. Also from (13.6.3) we can prove discrete maximum and minimum principles. These properties are analogous to similar results for Laplace's equation itself (see Sec. 2.5.3).

Jacobi iteration.   Instead of solving (13.6.3) exactly, it is more usual to use an approximate iterative scheme. One should not worry very much about the errors in solving (13.6.3) *if they are small* since (13.6.3) already is an approximation of Laplace's equation.

We cannot solve (13.6.3) directly since the four neighboring temperatures are not known. However, the following procedure will yield the solution. We can make an initial guess of the solution, and use the averaging principle (13.6.3) to "update" the solution:

$$u_{j,l}^{(\text{new})} = \frac{1}{4}(u_{j+1,l} + u_{j-1,l} + u_{j,l-1} + u_{j,l+1})^{(\text{old})}.$$

We can continue to do this, a process called **Jacobi iteration**. We introduce the notation $u_{j,l}^{(0)}$ for the initial guess, $u_{j,l}^{(1)}$ for the first iterate (determined from $u_{j,l}^{(0)}$), $u_{j,l}^{(2)}$ for the second iterate (determined from $u_{j,l}^{(1)}$), and so on. Thus, the $(m + 1)$st iterate satisfies

$$u_{j,l}^{(m+1)} = \frac{1}{4}(u_{j+1,l}^{(m)} + u_{j-1,l}^{(m)} + u_{j,l+1}^{(m)} + u_{j,l-1}^{(m)}). \qquad (13.6.4)$$

If the iterates converge, by which we mean, if

$$\lim_{m \to \infty} u_{j,l}^{(m+1)} = v_{j,l},$$

then (13.6.4) shows that $v_{j,l}$ satisfies the discretization of Laplace's equation (13.6.3).

Equation (13.6.4) is well suited for a computer. We cannot allow $m \to \infty$. In practice we stop the iteration process when $u_{j,l}^{(m+1)} - u_{j,l}^{(m)}$ is small (for all j and l). Then $u_{j,l}^{(m+1)}$ will be a reasonably good approximation to the exact solution $v_{j,l}$. (Recall that $v_{j,l}$ itself is only an approximate solution of Laplace's equation.)

The changes that occur at each updating may be emphasized by writing Jacobi iteration as

$$u_{j,l}^{(m+1)} = u_{j,l}^{(m)} + \frac{1}{4}(u_{j+1,l}^{(m)} + u_{j-1,l}^{(m)} + u_{j,l+1}^{(m)} + u_{j,l-1}^{(m)} - 4u_{j,l}^{(m)}). \qquad (13.6.5)$$

In this way Jacobi iteration is seen to be the standard discretization (centered spatial and forward time differences) of the two-dimensional diffusion equation, $\partial u/\partial t = k(\partial^2 u/\partial x^2 + \partial^2 u/\partial y^2)$, with $s = k\,\Delta t/(\Delta x)^2 = \frac{1}{4}$ [see (13.4.4)]. Each iteration corresponds to a time step $\Delta t = (\Delta x)^2/4k$. The earlier example of computing the heat equation on an L-shaped region (see p. 508) is exactly Jacobi iteration. For large m we see the solution approaching values independent of m. The resulting spatial distribution is an accurate approximate solution of the discrete version of Laplace's equation (satisfying the given boundary conditions).

Although Jacobi iteration converges, it will be shown to converge very slowly. To *roughly* analyze the rate of convergence, we investigate the decay of spatial oscillations for an $L \times L$ square ($\Delta x = \Delta y = L/N$). In (13.6.4), m is analogous to time ($t = m\,\Delta t$); (13.6.4) is analogous to the type of partial difference equation analyzed earlier. Thus, we know that there are special solutions

$$u_{j,l}^{(m)} = Q^m e^{i(\alpha x + \beta y)}, \qquad (13.6.6)$$

where $\alpha = n_1\pi/L$, $\beta = n_2\pi/L$, $n_j = 1, 2, \ldots, N - 1$. We assume that in this calculation there are zero boundary conditions along the edge. The solution should converge to zero as $m \to \infty$; we will determine the speed of convergence in order to know how many iterations are expected to be necessary. Substituting (13.6.6) into (13.6.4) yields

$$Q = \frac{1}{4}(e^{i\alpha\Delta x} + e^{-i\alpha\Delta x} + e^{i\beta\Delta y} + e^{-i\beta\Delta y})$$

$$= \frac{1}{2}(\cos \alpha\Delta x + \cos \beta\,\Delta y).$$

Since $-1 < Q < 1$ for all α and β, it follows from (13.6.6) that $\lim_{m\to\infty} u_{j,l}^{(m)} = 0$, as desired. However, the convergence can be very slow. The slowest rate of convergence occurs for Q nearest to 1. This happens for the smallest and largest α and β, $\alpha = \beta = \pi/L$ and $\alpha = \beta = (N-1)\pi/L$, in which case

$$|Q| = \cos\frac{\pi\,\Delta x}{L} = \cos\frac{\pi}{N} \approx 1 - \frac{\pi^2}{2}\frac{1}{N^2}, \tag{13.6.7}$$

since $\Delta x = L/N$ and N is large. In this case $|Q|^m$ is approximately $[1 - \frac{1}{2}(\pi/N)^2]^m$. This represents the error associated with the worst spatial oscillation on a square. If N is large, this error converges slowly to zero. For example, for the error to be reduced by a factor of $\frac{1}{2}$:

$$\left[1 - \frac{1}{2}\left(\frac{\pi}{N}\right)^2\right]^m = \frac{1}{2}.$$

Solving for m using natural logarithms yields

$$m \log\left[1 - \frac{1}{2}\left(\frac{\pi}{N}\right)^2\right] = -\log 2.$$

A simpler formula is obtained since π/N is small. From the Taylor series for x small, $\log(1-x) \approx -x$, it follows that the number of iterations m necessary to reduce the error by $\frac{1}{2}$ is approximately

$$m = \frac{\log 2}{\frac{1}{2}(\pi/N)^2} = N^2\frac{2\log 2}{\pi},$$

using Jacobi iteration. The number of iterations required may be quite large, proportional to N^2 the number of lattice points squared (just to reduce the error in half).

Gauss–Seidel iteration. Jacobi iteration is quite time consuming to compute with. More important, there is a scheme which is easier to implement and which converges faster to the solution of the discretized version of Laplace's equation. It is usual in Jacobi iteration to obtain the updated temperature $u_{j,l}^{(m+1)}$ first in the lower left spatial region. Then we scan completely across a row of mesh points (from left to right) before updating the temperature on the next row above (again from left to right), as indicated in Fig. 13.6.1. For example,

$$u_{3,8}^{(m+1)} = \frac{1}{4}(u_{2,8}^{(m)} + u_{3,7}^{(m)} + u_{3,9}^{(m)} + u_{4,8}^{(m)}).$$

In Jacobi iteration we use the old values $u_{2,8}^{(m)}, u_{3,7}^{(m)}, u_{3,9}^{(m)}, u_{4,8}^{(m)}$ even though new values for two, $u_{3,7}^{(m+1)}$ and $u_{2,8}^{(m+1)}$, already have been calculated. In doing a computer

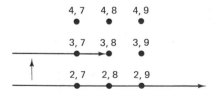

Figure 13.6.1 Gauss–Seidel iteration.

implementation of Jacobi iteration, we cannot destroy immediately the old values (as we have shown some are needed even after new values have been calculated).

The calculation will be easier to program if old values are destroyed as soon as a new one is calculated. We thus propose to use the updated temperatures when they are known. For example,

$$u_{3,8}^{(m+1)} = \frac{1}{4}(u_{2,8}^{(m+1)} + u_{4,8}^{(m)} + u_{3,7}^{(m+1)} + u_{3,9}^{(m)}).$$

In general, we obtain

$$u_{j,l}^{(m+1)} = \frac{1}{4}(u_{j-1,l}^{(m+1)} + u_{j+1,l}^{(m)} + u_{j,l-1}^{(m+1)} + u_{j,l+1}^{(m)}), \qquad (13.6.8)$$

known as **Gauss–Seidel iteration**. If this scheme converges, the solution will satisfy the discretized version of Laplace's equation.

There is no strong reason at this point to believe this scheme converges faster than Jacobi iteration. To investigate the speed at which Gauss–Seidel converges (for a square), we substitute again

$$u_{j,l}^{(m)} = Q^m e^{i(\alpha x + \beta y)}, \qquad (13.6.9)$$

where

$$\alpha = \frac{n_1 \pi}{L}, \qquad \beta = \frac{n_2 \pi}{L}, \qquad n_i = 1, 2, \ldots, N - 1.$$

The result is that

$$Q = \frac{1}{4}[e^{i\alpha\Delta x} + e^{i\beta\Delta y} + Q(e^{-i\alpha\Delta x} + e^{-i\beta\Delta y})]. \qquad (13.6.10)$$

To simplify the algebra, we let

$$z = \frac{e^{i\alpha\Delta x} + e^{i\beta\Delta y}}{4} = \zeta + i\eta, \qquad (13.6.11)$$

and thus obtain $Q = z/(1 - \bar{z})$. Q is complex, $Q = |Q|e^{i\theta}$, $u_{j,l}^{(m)} = |Q|^m e^{i\theta m} e^{i(\alpha x + \beta y)}$. The convergence rate is determined from $|Q|$,

$$|Q|^2 = \frac{z\bar{z}}{(1 - \bar{z})(1 - z)} = \frac{|z|^2}{1 + |z|^2 - 2 \operatorname{Re}(z)} \qquad (13.6.12)$$

$$= \frac{\zeta^2 + \eta^2}{1 + \zeta^2 + \eta^2 - 2\zeta} = \frac{\zeta^2 + \eta^2}{(1 - \zeta)^2 + \eta^2}.$$

Since $|z| < \frac{1}{2}$ and thus $|\zeta| < \frac{1}{2}$, it follows that $|Q| < 1$, yielding the convergence of Gauss–Seidel iteration. However, the rate of convergence is slow if $|Q|$ is near 1. Equation (13.6.12) shows that $|Q|$ is near 1 only if ζ is near $\frac{1}{2}$, which (13.6.11) requires α and β to be as small as possible. For a square, $\alpha = \pi/L$ and $\beta = \pi/L$, and thus

$$\zeta = \frac{1}{2}\cos\frac{\pi \Delta x}{L} \qquad \text{and} \qquad \eta = \frac{1}{2}\sin\frac{\pi \Delta x}{L}.$$

Therefore,

$$|Q|^2 = \frac{1}{5 - 4 \cos \pi \, \Delta x/L} = \frac{1}{5 - 4 \cos \pi/N} \approx \frac{1}{1 + 2(\pi/N)^2} \approx 1 - 2\left(\frac{\pi}{N}\right)^2,$$

since π/N is small. Thus

$$|Q| \approx 1 - \left(\frac{\pi}{N}\right)^2.$$

This $|Q|$ is twice as far from 1 compared to Jacobi iteration [see (13.6.7)]. By doing the earlier analysis, half the number of iterations are required to reduce the error by any fixed fraction. Jacobi iteration should never be used. Gauss–Seidel iteration is a feasible and better alternative.

S-O-R.   Both Jacobi and Gauss–Seidel iterative schemes require the number of iterations to be proportional to N^2, where N is the number of intervals (in one dimension). A remarkably faster scheme is successive overrelaxation or simply S-O-R.

Gauss–Seidel can be rewritten to emphasize the change that occurs at each iteration

$$u_{j,l}^{(m+1)} = u_{j,l}^{(m)} + \tfrac{1}{4}[u_{j-1,l}^{(m)} + u_{j+1,l}^{(m+1)} + u_{j,l-1}^{(m)} + u_{j,l+1}^{(m+1)} - 4u_{j,l}^{(m)}].$$

The bracketed term represents the change after each iteration as $u_{j,l}^{(m)}$ is updated to $u_{j,l}^{(m+1)}$. Historically, it was observed that one might converge faster if larger or smaller changes were introduced. Gauss–Seidel iteration with a **relaxation parameter** ω yields an **S-O-R iteration**:

$$\boxed{u_{j,l}^{(m+1)} = u_{j,l}^{(m)} + \omega[u_{j-1,l}^{(m)} + u_{j+1,l}^{(m+1)} + u_{j,l-1}^{(m)} + u_{j,l+1}^{(m+1)} - 4u_{j,l}^{(m)}].} \qquad (13.6.3)$$

If $\omega = \tfrac{1}{4}$, this reduces to Gauss–Seidel. If S-O-R converges, it clearly converges to the discretized version of Laplace's equation. We will choose ω so that (13.6.13) converges as fast as possible.

We again introduce

$$u_{j,l}^{(m)} = Q^m e^{i(\alpha x + \beta y)}$$

in order to investigate the rate of convergence, where $\alpha = n_1\pi/L$, $\beta = n_2\pi/L$, $n_i = 1, 2, \ldots, N - 1$. We obtain

$$Q = 1 + \omega[e^{i\alpha \Delta x} + e^{i\beta \Delta y} + Q(e^{-i\alpha \Delta x} + e^{-i\beta \Delta y}) - 4].$$

The algebra here is somewhat complicated. We let $z = \omega(e^{i\alpha \Delta x} + e^{i\beta \Delta y}) = \zeta + i\eta$ and obtain $Q = (1 - 4\omega + z)/(1 - \bar{z})$. Q is complex; $|Q|$ determines the convergence rate

$$|Q|^2 = \frac{(1 - 4\omega + \zeta)^2 + \eta^2}{(1 - \zeta)^2 + \eta^2}.$$

Again $|z| < 2\omega$ and thus $|\zeta| < 2\omega$. In Exercise 13.6.1 we show $|Q| < 1$ if $\omega < \tfrac{1}{2}$, guaranteeing the convergence of S-O-R. If $\omega < \tfrac{1}{2}$, $|Q|$ is near 1 only if ζ is near 2ω. This occurs only if α and β are as small as possible; (for a square) $\alpha =$

π/L and $\beta = \pi/L$ and thus $\zeta = 2\omega \cos \pi \, \Delta x/L$ and $\eta = 2\omega \sin \pi \, \Delta x/L$. In this way

$$|Q|^2 = \frac{4\omega^2 + (1 - 4\omega)^2 + (1 - 4\omega)4\omega \cos(\pi/N)}{4\omega^2 + 1 - 4\omega \cos(\pi/N)}.$$

Exercise 13.6.1 shows that $|Q|^2$ is minimized if $\omega = \frac{1}{2} - (\sqrt{2}/2)\sqrt{1 - \cos \pi/N}$. Since π/N is large, we use (for a square) $\omega = \frac{1}{2}(1 - \pi/N)$, in which case

$$|Q| \approx 1 - \frac{\pi}{2N};$$

see Exercise 13.6.2. With the proper choice of ω, $|Q|$, although still near 1, is an order of magnitude further away from 1 than for either Jacobi or Gauss–Seidel iteration. In fact (see Exercise 13.6.2) errors are reduced by $\frac{1}{2}$ in S-O-R with the number of iterations being proportional to N (not N^2).

For nonsquare regions, the best ω for S-O-R is difficult to approximate. However, there exist computer library routines which approximate ω. Thus, often S-O-R is a practical, *relatively* quickly convergent iterative scheme for Laplace's equation.

Other improved schemes have been developed, including the alternating-direction-implicit (ADI) method, which was devised in the mid-1950s by Peaceman, Douglas, and Rachford. More recent techniques exist, and it is suspected that better techniques will be developed in the future.

EXERCISES 13.6

13.6.1. (a) Show that $|Q| < 1$ if $\omega < \frac{1}{2}$ in S-O-R.
    (b) Determine the optimal relaxation parameter ω in S-O-R for a square, by minimizing $|Q|^2$.

13.6.2. (a) If $\omega = \frac{1}{2}(1 - \pi/N)$, show that $|Q| \approx 1 - \pi/2N$ (for large N) in S-O-R.
    (b) Show that with this choice the number of iterations necessary to reduce the error by $\frac{1}{2}$ is proportional to N (not N^2).

13.6.3. Describe a numerical scheme to solve Poisson's equation

$$\nabla^2 u = f(x, y),$$

    (assuming that $\Delta x = \Delta y$) analogous to:
    (a) Jacobi iteration
    (b) Gauss–Seidel iteration
    (c) S-O-R

13.6.4. Describe a numerical scheme (based on Jacobi iteration) to solve Laplace's equation in three dimensions. Estimate the number of iterations necessary to reduce the error in half.

13.6.5. Modify Exercise 13.6.4 for Gauss–Seidel iteration.

13.6.6. Show that Jacobi iteration corresponds to the two-dimensional diffusion equation, by taking the limit as $\Delta x = \Delta y \to 0$ and $\Delta t \to 0$ in some appropriate way.

13.6.7. What partial differential equation does S-O-R correspond to? (*Hint:* Take the limit as $\Delta x = \Delta y \to 0$ and $\Delta t \to 0$ in various ways.) Specialize your result to Gauss–Seidel iteration by letting $\omega = \frac{1}{4}$.

13.6.8. Consider Laplace's equation on a square $0 \le x \le 1$, $0 \le y \le 1$ with $u = 0$ on three sides and $u = 1$ on the fourth.
    (a) Solve using Jacobi iteration (let $\Delta x = \Delta y = \frac{1}{10}$).
    (b) Solve using Gauss–Seidel iteration (let $\Delta x = \Delta y = \frac{1}{10}$).
    (c) Solve using S-O-R iteration [let $\Delta x = \Delta y = \frac{1}{10}$ and $\omega = \frac{1}{2}(1 - \pi/10)$].
    (d) Solve by separation of variables. Evaluate numerically the first 10 or 20 terms.
    (e) Compare as many of the previous parts as you did.

Selected Answers
to Starred Exercises

1.2.3. $\displaystyle\int_0^L cpu\, A\, dx$ **1.2.4. (e)** $u(t) = u_0 \exp\left[-\left(\dfrac{2h}{c\rho r}\right)t\right]$

1.3.2. $K_0(x_{0-})\,\dfrac{\partial u}{\partial x}(x_{0-}, t) = K_0(x_{0+})\,\dfrac{\partial u}{\partial x}(x_{0+}, t)$

1.3.3. $Vc_f\rho_f\,\dfrac{\partial u}{\partial t}(L, t) = -K_0(L)\,\dfrac{\partial u}{\partial x}(L, t)A$, where V is the volume of the bath

1.4.1. (a) $u(x) = \dfrac{Tx}{L}$ **(d)** $u - T + \alpha x$ **(f)** $u(x) = -\dfrac{x^4}{12} + \dfrac{L^3 x}{3} + T$

  (h) $u - T + \alpha(x + 1)$ **1.4.2. (a)** $\dfrac{K_0 L^2}{2}$ **1.4.7. (a)** $\beta = 1 - L$

1.5.2. $\dfrac{\partial u}{\partial t} + \mathbf{v}\cdot\nabla u = k\,\nabla^2 u - u\,\nabla\cdot\mathbf{v}$ (often physically $\nabla\cdot\mathbf{v} = 0$)

1.5.9. (a) $u = \dfrac{T_1 \ln r_2/r + T_2 \ln r/r_1}{\ln r_2/r_1}$ **1.5.11.** $\beta = \dfrac{b}{a}$ **1.5.13.** $u(r) = \dfrac{320}{3}\left(1 - \dfrac{1}{r}\right)$

2.3.1. (a) $\dfrac{1}{r}\dfrac{d}{dr}\left(r\dfrac{d\phi}{dr}\right) = -\lambda\phi,\ \dfrac{dh}{dt} = -\lambda kh$ **(c)** $\dfrac{d^2\phi}{dx^2} = -\lambda\phi,\ \dfrac{d^2 h}{dy^2} = \lambda h$

  (e) $\dfrac{dh}{dt} = \lambda kh,\ \dfrac{d^4\phi}{dx^4} = \lambda\phi$ **(f)** $\dfrac{d^2 h}{dt^2} = -\lambda c^2 h,\ \dfrac{d^2\phi}{dx^2} = -\lambda\phi$

2.3.2. (b) $\lambda = (n\pi)^2,\ n = 1, 2, 3, \ldots$ **(d)** $\lambda = \left[\dfrac{(n - \frac{1}{2})\pi}{L}\right]^2,\ n = 1, 2, 3, \ldots$

  (f) $\lambda = \left(\dfrac{n\pi}{b - a}\right)^2,\ n = 1, 2, 3, \ldots$

2.3.3. (c) $u(x, t) = \displaystyle\sum_{n=1}^{\infty} A_n \sin\dfrac{n\pi x}{L}\, e^{-(n\pi/L)^2 kt}$, where $A_n = \dfrac{2}{L}\displaystyle\int_0^L 2\cos\dfrac{3\pi x}{L}\sin\dfrac{n\pi x}{L}\, dx$

2.3.4. (a) $cpA\displaystyle\sum_{n=1}^{\infty} B_n e^{-k(n\pi/L)^2 t}\left(\dfrac{1 - \cos n\pi}{n\pi/L}\right)$ **(c)** Heat energy equals initial heat energy plus the time integral of the flow *in* of heat energy at the boundaries.

2.3.6. $0 \ (n \neq m), \dfrac{L}{2} \ (n = m \neq 0), L(n = m = 0)$

2.3.8. (a) $u = 0$ (b) $u = e^{-\alpha t} \sum\limits_{n=1}^{\infty} b_n \sin \dfrac{n\pi x}{L} e^{-k(n\pi/L)^2 t}$

2.3.9. (a) If $-\dfrac{\alpha}{k} = \left(\dfrac{n\pi}{L}\right)^2$, $u(x) = A \sin \dfrac{n\pi x}{L}$ (b) If $-\dfrac{\alpha}{k} = \left(\dfrac{\pi}{L}\right)^2$, $u(x, t) \to B \sin \dfrac{\pi x}{L}$ as $t \to \infty$;

If $-\dfrac{\alpha}{k} > \left(\dfrac{\pi}{L}\right)^2$, $u(x, t) \to \infty$ as $t \to \infty$; If $-\dfrac{\alpha}{k} < \left(\dfrac{\pi}{L}\right)^2$, $u(x, t) \to 0$ as $t \to \infty$

2.3.10. (c) $\left[\displaystyle\int_0^L A(x)B(x) \, dx\right]^2 \leq \left(\displaystyle\int_0^L A^2 \, dx\right)\left(\displaystyle\int_0^L B^2 \, dx\right)$

2.4.1. $u = A_0 + \sum\limits_{n=1}^{\infty} A_n \cos \dfrac{n\pi x}{L} e^{-k(n\pi/L)^2 t}$

(a) $A_0 = \dfrac{1}{2}$, $(n \neq 0) \, A_n = \dfrac{-2}{n\pi} \sin \dfrac{n\pi}{2}$ (b) $A_0 = 6, A_3 = 4$, other $A_n = 0$

(c) $A_0 = \dfrac{4}{\pi}$, $(n \neq 0) \, A_n = -\dfrac{4}{L} \displaystyle\int_0^L \sin \dfrac{\pi x}{L} \cos \dfrac{n\pi x}{L} \, dx$ can easily be evaluated

using trigonometric identities or tables of integrals.

2.4.2. $u(x, t) = \sum\limits_{n=1}^{\infty} c_n \cos \dfrac{(n - \frac{1}{2})\pi x}{L} e^{-[(n-1/2)\pi/L]^2 kt}$, where $c_n = \dfrac{2}{L} \displaystyle\int_0^L f(x) \cos \dfrac{(n - \frac{1}{2})\pi x}{L} \, dx$

2.4.3. $\lambda = n^2$, $\phi = \sin nx$ and $\cos nx$, $n = 0, 1, 2, \ldots$

2.5.1. (a) $u(x, y) = A_0 y + \sum\limits_{n=1}^{\infty} a_n \sinh \dfrac{n\pi y}{L} \cos \dfrac{n\pi x}{L}$

(c) $u(x, y) = \sum\limits_{n=1}^{\infty} A_n \cosh \dfrac{n\pi x}{H} \sin \dfrac{n\pi y}{H}$, where $A_n \cosh n\pi L/H = \dfrac{2}{H} \displaystyle\int_0^H g(y) \sin \dfrac{n\pi y}{H} \, dy$

(e) $u(x, y) = \sum\limits_{n=1}^{\infty} A_n h_n(y) \sin \dfrac{n\pi x}{L}$, $A_n h_n(H) = \dfrac{2}{L} \displaystyle\int_0^L f(x) \sin \dfrac{n\pi x}{L} \, dx$,

where $h_n(y) = \cosh \dfrac{n\pi y}{L} + \dfrac{L}{n\pi} \sinh \dfrac{n\pi y}{L}$ **2.5.2.** (a) $\displaystyle\int_0^L f(x) \, dx = 0$

2.5.3. $u(r, \theta) = \sum\limits_{n=0}^{\infty} A_n r^{-n} \cos n\theta + \sum\limits_{n=1}^{\infty} B_n r^{-n} \sin n\theta$

(a) $A_0 = \ln 2$, $A_3 a^{-3} = 4$, other $A_n = 0$, $B_n = 0$ (b) See (2.5.29) with a^n replaced by a^{-n}.

2.5.4. $u(r, \theta) = \dfrac{a^2 - r^2}{2\pi} \displaystyle\int_{-\pi}^{\pi} \dfrac{f(\bar{\theta}) \, d\bar{\theta}}{a^2 + r^2 - 2ar \cos(\theta - \bar{\theta})}$

2.5.5. (a) $u(r, \theta) = \sum\limits_{n=1}^{\infty} A_n r^{2n-1} \cos(2n - 1)\theta$

(c) $u(r, \theta) = \sum\limits_{n=1}^{\infty} A_n r^{2n} \sin 2n\theta$, $A_n = \dfrac{2}{\pi n} \displaystyle\int_0^{\pi/2} f(\theta) \sin 2n\theta \, d\theta$

2.5.6. (a) $u(r, \theta) = \sum\limits_{n=1}^{\infty} B_n r^n \sin n\theta$ **2.5.7.** (b) $u(r, \theta) = \sum\limits_{n=0}^{\infty} A_n r^{3n} \cos 3n\theta$

2.5.8. (a) let $\phi_1(r) = \begin{cases} \ln\left(\dfrac{r}{a}\right) & n = 0 \\[2mm] \left(\dfrac{r}{a}\right)^n - \left(\dfrac{r}{a}\right)^{-n} & \text{so that } \phi_1(a) = 0 \end{cases}$ $\phi_2(r) = \begin{cases} \left(\dfrac{r}{b}\right)^n - \left(\dfrac{r}{b}\right)^{-n} & \text{so that } \phi_2(b) = 0 \\[2mm] \ln\left(\dfrac{r}{b}\right) & n = 0 \end{cases}$

$u(r, \theta) = \sum\limits_{n=0}^{\infty} \cos n\theta [A_n \phi_1(r) + B_n \phi_2(r)] + \sum\limits_{n=1}^{\infty} \sin n\theta [C_n \phi_1(r) + D_n \phi_2(r)]$

$D_n \phi_2(a) = \dfrac{1}{\pi} \displaystyle\int_{-\pi}^{\pi} f(\theta) \sin n\theta \, d\theta, \ldots$

2.5.9. (a) $u(r, \theta) = \sum_{n=1}^{\infty} A_n \left[\left(\dfrac{r}{a} \right)^{2n} - \left(\dfrac{a}{r} \right)^{2n} \right] \sin 2n\theta$, where

$$A_n \left[\left(\dfrac{b}{a} \right)^{2n} - \left(\dfrac{a}{b} \right)^{2n} \right] = \dfrac{4}{\pi} \int_0^{\pi/2} f(\theta) \sin 2n\theta \, d\theta$$

(b) $u(r, \theta) = \sum_{n=1}^{\infty} A_n \sinh\left(\dfrac{n\pi\theta}{\ln b/a} \right) \sin\left(n\pi \dfrac{\ln r/a}{\ln b/a} \right),$

where $A_n \sinh\left(\dfrac{n\pi^2}{2 \ln b/a} \right) = \dfrac{2}{\ln b/a} \int_a^b f(r) \sin\left(n\pi \dfrac{\ln r/a}{\ln b/a} \right) \dfrac{dr}{r}.$

3.2.1. (b)

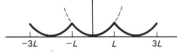

(d)

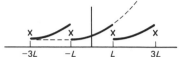

(f)

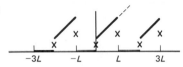

3.2.2. (a)

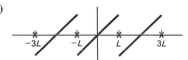

$a_n = 0, \ b_n = \dfrac{2L}{n\pi} (-1)^{n+1}$

(c)

$b_1 = 1$, all others $= 0$

(f)

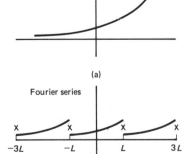

$a_0 = \dfrac{1}{2}$, other $a_n = 0$

$b_n = \dfrac{2}{n\pi}$ (n odd), other $b_n = 0$

3.3.1. (d)

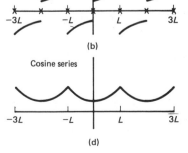

$f(x)$

(a)

Sine series

(b)

Fourier series

(c)

Cosine series

(d)

Selected Answers to Starred Exercises

3.3.2. (d)

(d)

repeat periodically,

$$b_n = \frac{2}{n\pi}\left(1 - \cos\frac{n\pi}{2}\right)$$

3.3.10. $f_e(x) = \frac{1}{2}\begin{cases} x^2 + e^x & x < 0 \\ x^2 + e^{-x} & x > 0 \end{cases}$ $f_0(x) = \frac{1}{2}\begin{cases} x^2 - e^x & x < 0 \\ e^{-x} - x^2 & x > 0 \end{cases}$

3.3.13. $a_n = \frac{2}{L}\int_0^L f(x)\sin\frac{n\pi x}{L}dx$; this is zero for n even since $\sin n\pi x/L$ is odd around $x = L/2$

3.4.1. (a) $\int_a^b u\frac{dv}{dx}dx = uv\Big|_a^b + uv\Big|_{c+}^{c-} - \int_a^b v\frac{du}{dx}dx$

3.4.3. (a) $b_n = \frac{2}{L}(\alpha - \beta)\sin\frac{n\pi x_0}{L} - \frac{n\pi}{L}a_n$ **3.4.9.** $\frac{db_n}{dt} + k\left(\frac{n\pi}{L}\right)^2 b_n = \frac{2}{L}\int_0^L q(x,t)\sin\frac{n\pi x}{L}dx$

3.4.12. $u(x,t) = \sum_{n=0}^{\infty} A_n(t)\cos\frac{n\pi x}{L}$, where $(n \neq 0, 3)\,A_n(t) = A_n(0)e^{-k(n\pi/L)^2 t}$, $A_0(t) = 1 - e^{-t} + A_0(0)$

with $A_0(0) = \frac{1}{L}\int_0^L f(x)\,dx$ and $(n \neq 0)$ $A_n(0) = \frac{2}{L}\int_0^L f(x)\cos\frac{n\pi x}{L}dx$

$$A_3(t) = \left[A_3(0) - \frac{1}{k(3\pi/L)^2 - 2}\right]e^{-k(3\pi/L)^2 t} + \frac{e^{-2t}}{k(3\pi/L)^2 - 2}$$

3.5.1. (c) $\left(\frac{x}{L}\right)^3 = \frac{1}{4} + \sum_{n=1}^{\infty}\left\{6\frac{(-1)^n}{(n\pi)^2} + 12\frac{[1 - (-1)^n]}{(n\pi)^4}\right\}\cos\frac{n\pi x}{L}$ $0 \leq x \leq L$

3.5.4. $b_n = \frac{2n\pi}{L^2 + n^2\pi^2}[1 - (-1)^n\cosh L]$ **3.5.7.** $\frac{\pi^3}{32}$

4.4.1 (a) $\frac{n\pi c}{L}$, $n = 1, 2, \ldots$ **(b)** $\left(m - \frac{1}{2}\right)\pi\frac{c}{H}$, $m = 1, 2, 3, \ldots$

4.4.2. (c) Frequencies are $\sqrt{\dfrac{(n\pi/L)^2 T_0 - \alpha}{\rho}}$

4.4.3. (b) $u(x,t) = e^{-\beta t/2\rho}\sum_{n=1}^{\infty}(a_n\cos\omega_n t + b_n\sin\omega_n t)\sin\frac{n\pi x}{L}$, where $\omega_n = \frac{1}{2}\sqrt{\dfrac{4n^2\pi^2 T_0}{\rho L^2} - \dfrac{\beta^2}{\rho^2}}$

5.3.1. $T_0\dfrac{d^2\phi}{dx^2} + [\lambda\rho_0(x) + \alpha(x)]\phi = 0$

5.3.3. $H(x) = \exp\left[\int^x \alpha(t)\,dt\right] = p(x)$, $\sigma(x) = \beta(x)H(x)$, $q(x) = \gamma(x)H(x)$.

5.3.4. (b) $u(x,t) = \sum_{n=1}^{\infty} a_n e^{-[(n\pi/L)^2 + (v_0/2k)^2]kt}\,e^{(v_0/2k)x}\sin\frac{n\pi x}{L}$, where $a_n = \frac{2}{L}\int_0^L f(x)e^{-(v_0/2k)x}\sin\frac{n\pi x}{L}dx$

5.3.9. (c) $\lambda = \left(\dfrac{n\pi}{\log b}\right)^2$, $n = 1, 2, 3, \ldots$ **5.4.2.** $u(x,t) = \sum_{n=1}^{\infty} a_n\phi_n(x)e^{-\lambda_n t}$ $(\lambda_1 = 0)$

5.4.3. $u(r,t) = \sum_{n=1}^{\infty} a_n\phi_n(r)e^{-\lambda kt}$, where $a_n = \dfrac{\int_0^a f(r)\phi_n(r)r\,dr}{\int_0^a \phi_n^2(r)r\,dr}$

5.4.6. $u(x,t) = \sum_{n=1}^{\infty} A_n\phi_n(x)\cos\sqrt{\lambda_n}\,t$ **5.5.1. (g)** $\alpha\delta - \beta\gamma = 1$

5.5.9. $\lambda = -\int_0^1 \left(\frac{d^2\phi}{dx^2}\right)^2 dx \Big/ \int_0^1 e^x \phi^2 \, dx$ **5.5A.4. (a)** $v(t) = \begin{bmatrix} 2 \\ 1 \end{bmatrix} e^{7t} - \frac{3}{5}\begin{bmatrix} 1 \\ -2 \end{bmatrix} e^{2t}$

5.5A.5. (b) $\lambda = 2 \pm \sqrt{3}$ **5.6.1. (c)** $u_T(x) = 3x - 2x^2, \ \lambda_1 \le 4\frac{1}{8}$

5.7.1. $\frac{1}{2} <$ lowest frequency $< \frac{1}{2}\sqrt{1 + \alpha^2}$

5.8.2. (c) see Fig. 5.8.3a, $(n - 1)\pi < \sqrt{\lambda_n}L < \left(n - \frac{1}{2}\right)\pi$

5.8.3. (b) $(n - 1)\pi < \sqrt{\lambda_n}L < (n - \frac{1}{2})\pi, \ n = 1, 2, 3, \ldots$

5.8.7. (b) $\lambda = \frac{1}{4}$ **(c)** none   **(d)** no   **(e)** yes

5.8.8. (c) $\lambda_n \sim [(n - 1)\pi]^2$ **5.8.10. (a), (b)** $\lambda_1 \approx 4.12$

5.8.13. $\frac{L}{2}\left(1 + \frac{1}{hL}\cos^2\sqrt{\lambda L}\right) = \frac{1}{2}\left(L + \frac{h}{\lambda + h^2}\right)$ **5.9.1. (b)** $\lambda = \left[\left(n + \frac{1}{2}\right)\pi \Big/ \int_0^L (\sigma/p)^{1/2} \, dx_0\right]^2$

5.9.3. (a) $A'' + i\lambda^{1/2}\left(2A'\sigma^{1/2} + \frac{1}{2}\sigma^{-1/2}\sigma'A\right) + qA = 0$

   (e) $A_n = \frac{i}{2}\sigma^{-1/4}\int_0^x \sigma^{-1/4}(A''_{n-1} + qA_{n-1}) \, dx_0$ **5.10.2. (b)** $\frac{\pi^4}{96} = 1 + \frac{1}{3^4} + \frac{1}{5^4} + \frac{1}{7^4} + \cdots$

6.3.1. (a) $u(x, y, t) = \sum_{m=1}^{\infty}\sum_{n=1}^{\infty} a_{nm}e^{-k[(n\pi/L)^2 + (m\pi/H)^2]t}\sin\frac{n\pi x}{L}\sin\frac{m\pi y}{H}$

   (c) $u(x, y, t) = \sum_{m=1}^{\infty}\sum_{n=0}^{\infty} a_{nm}e^{-k[(n\pi/L)^2 + (m\pi/H)^2]t}\cos\frac{n\pi x}{L}\sin\frac{m\pi y}{H}$

6.3.2. (b) $u(x, y, z, t) = \sum_{n=0}^{\infty}\sum_{m=0}^{\infty}\sum_{l=0}^{\infty} A_{nml}\cos\frac{n\pi x}{L}\cos\frac{m\pi y}{H}\cos\frac{l\pi z}{W}e^{-\lambda_{nml}kt}$

   where $\lambda_{nml} = \left(\frac{n\pi}{L}\right)^2 + \left(\frac{m\pi}{H}\right)^2 + \left(\frac{l\pi}{W}\right)^2$

   $u(x, y, z, t) \to a_{000} = \frac{1}{LHW}\int_0^W\int_0^H\int_0^L f(x, y, z) \, dx \, dy \, dz$ as $t \to \infty$

6.3.4. (b) $u(x, y, t) - \sum_{n=0}^{\infty}\sum_{m=0}^{\infty} A_{nm}\cos\frac{n\pi x}{L}\cos\frac{m\pi y}{H}\phi_{nm}(t)$

   where $\phi_{nm}(t) = \begin{cases} t & n = 0, \ m = 0 \\ \sin\omega_{nm}t & \text{otherwise, where } \omega_{nm}^2 = c^2\left[\left(\frac{n\pi}{L}\right)^2 + \left(\frac{m\pi}{H}\right)^2\right] \end{cases}$

   $A_{nm}\phi'_{nm}(0) - \iint f(x, y)\cos\frac{n\pi x}{L}\cos\frac{m\pi y}{H}\, dxdy \Big/ \iint \cos^2\frac{n\pi x}{L}\cos^2\frac{m\pi y}{H}\, dxdy$

6.3.6. (b) $u(x, y, z) = \sum_{n=1}^{\infty}\sum_{m=1}^{\infty} A_{nm}\cosh\sqrt{\lambda_{nm}}\, z \sin\frac{n\pi x}{L}\sin\frac{m\pi y}{W}$, where $\lambda_{nm} = \left(\frac{n\pi}{L}\right)^2 + \left(\frac{m\pi}{W}\right)^2$

6.3.7. (c) (d) $u(x, y, z) = \sum_{n=0}^{\infty}\sum_{m=0}^{\infty} A_{nm}\cosh\beta_{nm}x\cos\frac{n\pi y}{W}\cos\frac{m\pi z}{H}$, where $\beta_{nm}^2 = \left(\frac{n\pi}{W}\right)^2 + \left(\frac{m\pi}{H}\right)^2$

   For **(c)**: Solution exists only if $\int_0^H\int_0^W f(y, z) \, dydz = 0$, in which case A_{00} arbitrary,

   and otherwise $A_{nm}\beta_{nm}\sinh\beta_{nm}L = \iint f\cos\frac{n\pi y}{W}\cos\frac{m\pi z}{H}\, dydz \Big/ \iint \cos^2\frac{n\pi y}{W}\cos^2\frac{m\pi z}{H}\, dydz$;

   For **(d)**: $A_{nm}\cosh\beta_{nm}L = \iint g\cos\frac{n\pi y}{W}\cos\frac{n\pi z}{H}\, dydz \Big/ \iint \cos^2\frac{n\pi y}{W}\cos^2\frac{m\pi z}{H}\, dydz$

6.4.1. (a) $\lambda = \left(\frac{n\pi}{L}\right)^2 + \left(\frac{m\pi}{H}\right)^2, \ n = 0, 1, 2, \ldots$ and $m = 1, 2, 3, \ldots$

6.7.1. $u(r, \theta, t) = \sum_{n=1}^{\infty} c_n J_3(\sqrt{\lambda_{3n}}\, r)\sin 3\theta \sin c\sqrt{\lambda_{3n}}\, t$

6.7.2. (d) $u(r, \theta, t) = \sum\limits_{n=0}^{\infty} \sum\limits_{m=1}^{\infty} (A_{nm} \cos n\theta + B_{nm} \sin n\theta) H_{nm}(r, t)$

where $H_{nm}(r, t) = \begin{cases} t & n = 0, m = 1 \\ \dfrac{J_n(\sqrt{\lambda}r)\sin c \sqrt{\lambda}t}{c\sqrt{\lambda}} & \text{otherwise, with } J'_m(\sqrt{\lambda}\, a) = 0, \quad \lambda > 0; \end{cases}$

$A_{nm} = \displaystyle\iint \beta(r, \theta)\phi_{nm}(r) \cos n\theta \; rdrd\theta \Big/ \iint \phi_{nm}^2 \cos^2 n\theta \; rdrd\theta, \qquad \begin{array}{l} B_{nm} = \text{same as } A_{nm} \\ \text{with } \cos n\theta \to \sin n\theta \end{array}$

where $\phi_{nm}(r) = \begin{cases} 1 & n = 0, m = 1 \\ J_n(\sqrt{\lambda}r) & \text{otherwise} \end{cases}$

6.7.3. (a) $c\sqrt{\lambda}$, where $J_{2m}(\sqrt{\lambda}\, a) = 0$

6.7.5. $c\sqrt{\lambda}$, where $J_{2m}(\lambda^{1/2}a) \, Y_{2m}(\lambda^{1/2}b) = Y_{2m}(\lambda^{1/2}a)J_{2m}(\lambda^{1/2}b)$

6.7.8. $\lim\limits_{t\to\infty} u(r, \theta, t) = \dfrac{1}{\pi a^2} \displaystyle\int_0^a \int_0^{2\pi} f(r, \theta)r \, dr \, d\theta$

6.7.9. (b) $u(r, \theta, t) = \sum\limits_{m=0}^{\infty} \sum\limits_{n=1}^{\infty} A_{mn}\phi_{mn}(r) \cos m\theta \; e^{-\lambda_{mn}kt}$,

where $\phi_{mn}(r) = \begin{cases} 1 & m = 0, n = 1 \; (\lambda = 0) \\ J_m(\sqrt{\lambda_{mn}}\, r) & \text{otherwise } (\lambda > 0) \text{ and where } \dfrac{d}{dr}J_m(\sqrt{\lambda_{mn}}a) = 0 \end{cases}$

6.7.10. $u(r, t) = \sum\limits_{n=1}^{\infty} a_n J_0(\lambda_n^{1/2} r) \, e^{-\lambda nkt}$, where $J_0(\lambda_n^{1/2}\, a) = 0$ **6.7.12. (a)** $y \approx c_1 x^3 + c_2 x^{-2}$

6.7.12. (c) $y \approx c_1\cos(2 \ln x) + c_2\sin(2 \ln x)$ **(e)** $y \approx c_1 \, (x^2 + \cdots) + c_2 \, (x^3 + \cdots)$

6.8.1. (b) $J_m(\lambda^{1/2}) \, Y_m(2\lambda^{1/2}) = J_m(2\lambda^{1/2}) \, Y_m(\lambda^{1/2})$ **(d)** $\dfrac{\pi^2}{2} < \lambda_1 < 2\pi^2$

6.8.2. (d) $u(r, \theta, t) = \sum\limits_{m=1}^{\infty} \sum\limits_{n=1}^{\infty} C_{mn} \sin 2m\theta \, J_{2m}(\sqrt{\lambda_{mn}}\, r)e^{-\lambda mnkt}$, where $J_{2m}(\sqrt{\lambda}\, a) = 0$

6.8.8. $J_{1/2}(z) = \sqrt{\dfrac{2}{\pi z}} \sin z$

6.9.1. (b) $u(r, \theta, z) = \sum\limits_{n=1}^{\infty} A_n \sinh \sqrt{\lambda_{7n}} \, (H - z) \sin 7\theta \, J_7(\sqrt{\lambda_{7n}}\, r)$, where $J_7(\sqrt{\lambda_{7n}}\, a) = 0$

6.9.2. (b) $u(r, \theta, z) = \sum\limits_{n=1}^{\infty} \sum\limits_{m=1}^{\infty} A_{mn}I_m\left[\dfrac{(n - \frac{1}{2})\pi r}{H}\right] \sin m\theta \sin \dfrac{(n - \frac{1}{2})\pi z}{H}$

6.9.3. (b) $u(r, \theta, z, t) = \sum\limits_{\lambda} A_\lambda e^{-\lambda kt} \cos 2m\theta \; \cos \dfrac{n\pi z}{H} \, \phi_\lambda(r)$ $\begin{array}{l} m = 0, 1, 2, \dots \; ; n = 0, 1, 2, \dots \; ; \\ p = 1, 2, 3, \dots \end{array}$

where $\phi_\lambda(r) = \begin{cases} 1 & m = 0 \text{ (for } \cos 0\theta\text{) and } n = 0 \text{ and } p = 1 \\ J_{2m}(\gamma_{mp} \, r/a) & \text{otherwise} \end{cases}$

$\dfrac{d}{dr}J_{2m} \, (\gamma_{mp}) = 0$ and $\lambda = \dfrac{\gamma_{mp}^2}{a^2} + \dfrac{n^2\pi^2}{H^2}$

6.9.4. (a) $u(r, z, t) = \sum\limits_{n=1}^{\infty} \sum\limits_{m=1}^{\infty} A_{nm} J_0(\sqrt{\lambda}r) \sin \dfrac{m\pi z}{H} e^{-[\lambda + (m\pi/H)^2]kt}$,   where $J_0(\sqrt{\lambda}a) = 0$ and

$A_{nm} = \displaystyle\iint f(r, z) \, J_0(\sqrt{\lambda}r) \sin \dfrac{m\pi z}{H} \, r \, drdz \Big/ \iint J_0^2(\sqrt{\lambda}\, r) \sin^2 \dfrac{m\pi z}{H} \, r \, drdz$

7.2.1. (a) $u(x, t) = A + Bx + \sum\limits_{n=1}^{\infty} a_n \sin\left(n - \dfrac{1}{2}\right)\dfrac{\pi x}{L} e^{-[(n - 1/2)\pi/L]^2 kt}$,

where $a_n = \dfrac{2}{L} \displaystyle\int_0^L g(x)\sin\left(n - \dfrac{1}{2}\right)\dfrac{\pi x}{L} \, dx$

7.2.1. (d) $u_E(x) = -\dfrac{x^2}{2} + \left(\dfrac{B - A}{L} + \dfrac{L}{2}\right)x + A$ **7.2.2. (a)** $r(x, t) = A(t)x + \dfrac{[B(t) - A(t)]x^2}{2L}$

7.2.2. (c) $r(x, t) = A(t)x + B(t) - LA(t)$

7.2.6. (a) $u(x, t) = A + (B - A)\dfrac{x}{L} + \displaystyle\sum_{n=1}^{\infty} \sin\dfrac{n\pi x}{L}\left(A_n\cos\dfrac{n\pi ct}{L} + B_n\sin\dfrac{n\pi ct}{L}\right),$

where $A_n = \dfrac{2}{L}\displaystyle\int_0^L \left\{f(x) - \left[A + (B - A)\dfrac{x}{L}\right]\right\}\sin\dfrac{n\pi x}{L}\,dx$ and $B_n = \dfrac{2}{n\pi c}\displaystyle\int_0^L g(x)\sin\dfrac{n\pi x}{L}\,dx$

(d) $u_E(x) = \dfrac{L^2}{c^2\pi^2}\sin\dfrac{\pi x}{L}$ **7.3.1. (c)** $u(x, t) = A(t) + \displaystyle\sum_{n=1}^{\infty} B_n(t)\sin\dfrac{(n - \frac{1}{2})\pi x}{L}$

7.3.1. (f) $u(x, t) = \displaystyle\sum_{n=0}^{\infty} A_n(t)\cos\dfrac{n\pi x}{L}$, where $\dfrac{dA_n}{dt} + k\left(\dfrac{n\pi}{L}\right)^2 A_n = \dfrac{\displaystyle\int_0^L Q(x, t)\cos n\pi x/L\,dx}{\displaystyle\int_0^L \cos^2 n\pi x/L\,dx}$

7.3.3. $u(x, t) = \displaystyle\sum_{n=1}^{\infty} a_n(t)\phi_n(x)$, where $\dfrac{da_n}{dt} = -\lambda_n a_n + \dfrac{\displaystyle\int_0^L \phi_n f\,dx}{\displaystyle\int_0^L \phi_n^2 c\rho\,dx}$

7.3.4. (a) $u_E(x) = A + (B - A)\dfrac{\displaystyle\int_0^x d\bar{x}/K_0(\bar{x})}{\displaystyle\int_0^L d\bar{x}/K_0(\bar{x})}$

7.3.5. $u(r, t) = \displaystyle\sum_{n=1}^{\infty} A_n(t)J_0(\lambda_n^{1/2} r)$, where $J_0(\lambda_n^{1/2}a) = 0$ and where $\dfrac{dA_n}{dt} + k\lambda_n A_n = \dfrac{\displaystyle\int_0^a f(r, t)J_0(\lambda_n^{1/2} r)r\,dr}{\displaystyle\int_0^a J_0^2(\lambda_n^{1/2} r)r\,dr}$

7.3.7. $r(x, t) = \dfrac{x^3}{6L} + \left(\dfrac{t}{L} - \dfrac{L}{6}\right)x$

7.4.1. (b) $u(x, t) = \displaystyle\sum_{n=0}^{\infty} A_n(t)\cos\dfrac{n\pi x}{L}$, where

$A_n(t) = e^{-k(n\pi/L)^2 t}\left(A_n(0) + \displaystyle\int_0^t e^{k(n\pi/L)^2\bar{t}}\left\{q_n(\bar{t}) + \dfrac{I_n k}{L}[(-1)^n B(\bar{t}) - A(\bar{t})]\right\}d\bar{t}\right)$

with $A_n(0) = \dfrac{\displaystyle\int_0^L f(x)\cos\dfrac{n\pi x}{L}\,dx}{\displaystyle\int_0^L \cos^2\dfrac{n\pi x}{L}\,dx}$ and $q_n(t) = \dfrac{\displaystyle\int_0^L Q(x, t)\cos\dfrac{n\pi x}{L}\,dx}{\displaystyle\int_0^L \cos^2\dfrac{n\pi x}{L}\,dx}$ and $I_n = \begin{cases} 1 & n = 0 \\ 2 & n \neq 0 \end{cases}$

7.5.2. (b) $\omega^2 = \left(\dfrac{n\pi c}{L}\right)^2$

7.5.5. (c) $u(r, \theta, t) = \displaystyle\sum_{m=1}^{\infty}\sum_{n=1}^{\infty} A_{nm}(t)J_m(\sqrt{\lambda}r)\sin m\theta$ with $J_m(\sqrt{\lambda}a) = 0$,

where $A_{nm} = \dfrac{1}{c\sqrt{\lambda}}\displaystyle\int_0^t Q_{nm}(\bar{t})\sin c\sqrt{\lambda}(t - \bar{t})\,d\bar{t} + c_{nm}\cos c\sqrt{\lambda}\,t,$

$Q_{nm} = \dfrac{\displaystyle\iint Q(x, y, t)J_m(\sqrt{\lambda}r)\sin m\theta\,r\,drd\theta}{\displaystyle\iint J_m^2(\sqrt{\lambda}r)\sin^2 m\theta\,r\,drd\theta}$, $c_{nm} = \dfrac{\displaystyle\iint f(x, y)J_m(\sqrt{\lambda}r)\sin m\theta\,r\,drd\theta}{\displaystyle\iint J_m^2(\sqrt{\lambda}r)\sin^2 m\theta\,r\,drd\theta}$

7.5.6. (a) $\dfrac{1}{c^2}\dfrac{d^2 a}{dt^2} + \lambda a = \dfrac{\displaystyle\int_0^\pi\int_0^\alpha g\phi r\,dr\,d\theta}{\displaystyle\int_0^\pi\int_0^\alpha \phi^2 r\,dr\,d\theta}$, where $\dfrac{d}{dr}J_m(\lambda^{1/2}\alpha) = 0$

7.6.1. (b) $u(x, y) = \sum_{n=1}^{\infty} \sum_{m=1}^{\infty} A_{nm} \sin \dfrac{n\pi x}{L} \sin \dfrac{m\pi y}{H}$, where $A_{nm} = \dfrac{-Q_{nm} - (2/L)(-1)^n (n\pi/L)^2}{(n\pi/L)^2 + (m\pi/H)^2}$

(d) if $\iint Q \, dxdy = 0$, then $u = \sum_{n=0}^{\infty} \sum_{m=0}^{\infty} A_{nm} \cos \dfrac{n\pi x}{L} \cos \dfrac{m\pi y}{H}$,

where A_{00} is arbitrary and the others are given by

$$-A_{nm} \left[\left(\frac{n\pi}{L} \right)^2 + \left(\frac{m\pi}{H} \right)^2 \right] = \frac{\iint Q \cos \dfrac{n\pi x}{L} \cos \dfrac{m\pi y}{H} \, dxdy}{\iint \cos^2 \dfrac{n\pi x}{L} \cos^2 \dfrac{m\pi y}{H} \, dxdy}$$

7.6.3. (a) $u(r, \theta) = \sum_{m=0}^{\infty} \sum_{n=1}^{\infty} A_{mn} \cos m\theta J_m(\sqrt{\lambda} r) + \sum_{m=1}^{\infty} \sum_{n=1}^{\infty} B_{mn} \sin m\theta J_m(\sqrt{\lambda} r)$,

where $J_m(\sqrt{\lambda} a) = 0$ and $\begin{pmatrix} A_{mn} \\ B_{mn} \end{pmatrix} = \dfrac{-\dfrac{1}{\lambda} \iint Q \begin{pmatrix} \cos m\theta \\ \sin m\theta \end{pmatrix} J_m(\sqrt{\lambda} r) r \, drd\theta}{\iint \begin{pmatrix} \cos^2 m\theta \\ \sin^2 m\theta \end{pmatrix} J_m^2(\sqrt{\lambda} r) r \, drd\theta}$

7.6.6. $u(x, y) = \sum_{n=1}^{\infty} a_n(y) \sin nx$, where $a_n(y) = \frac{1}{3} e^{2y} \delta_{n1} + \alpha_n \sinh ny + \beta_n \cosh ny$ and $\delta_{n1} = \begin{cases} 1 & n = 1 \\ 0 & n \neq 1 \end{cases}$

8.2.1. (d) $G(x, t; x_0, t_0) = \sum_{n=0}^{\infty} \dfrac{1}{I_n} \cos \dfrac{n\pi x}{L} \cos \dfrac{n\pi x_0}{L} e^{-k(n\pi/L)^2(t - t_0)}$, where $I_n = \begin{cases} L & n = 0 \\ L/2 & n \neq 0 \end{cases}$

$$u(x, t) = \int_0^L g(x_0) G(x, t; x_0, 0) \, dx_0 + \int_0^L \int_0^t Q(x_0, t_0) G(x, t; x_0, t_0) \, dt_0 \, dx_0$$

$$+ \int_0^t kB(t_0) G(x, t; L, t_0) \, dt_0 - \int_0^t kA(t) G(x, t; 0, t_0) \, dt_0$$

8.2.3. $G(x, t; x_0, t_0) = \sum_{n=1}^{\infty} \dfrac{2}{L} \sin \dfrac{n\pi x_0}{L} \sin \dfrac{n\pi x}{L} \dfrac{\sin n\pi c(t - t_0)/L}{n\pi c/L}$

$$u(x, t) = \int_0^L \int_0^t Q(x_0, t_0) G(x, t; x_0, t_0) \, dt_0 \, dx_0 + \int_0^L g(x_0) G(x, t; x_0, 0) \, dx_0$$

$$+ \int_0^L f(x_0) \frac{\partial G}{\partial t}(x, t; x_0, 0) \, dx_0$$

8.3.5. (a), (b) $u(x) = \int_0^x (x - x_0) f(x_0) \, dx_0 - x \int_0^L f(x_0) \, dx_0$ **(c)** $G(x, x_0) = \begin{cases} -x & x < x_0 \\ -x_0 & x > x_0 \end{cases}$

8.3.6. (a) See answer to 8.3.5.(c) **(b)**

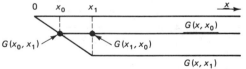

8.3.9. (b) See answer to 8.3.11

8.3.11. (a) $G(x, x_0) = \begin{cases} \dfrac{\sin (x_0 - L) \sin x}{\sin L} & x < x_0 \\ \dfrac{\sin (x - L) \sin x_0}{\sin L} & x > x_0 \end{cases}$ **8.3.13. (b)** $G(x, x_0) = \dfrac{1}{2ik} e^{ik|x - x_0|}$

8.3.14. (d) $u(x) = \int_0^L G(x, x_0) f(x_0) \, dx_0 - \alpha p(0) \dfrac{dG}{dx_0}(x, x_0) \bigg|_{x_0 = 0} - \beta p(L) G(x, L)$

8.3.15. (a) $G(x, x_0) = \begin{cases} \dfrac{1}{k} y_1(x) y_2(x_0) & x < x_0 \\ \dfrac{1}{k} y_1(x_0) y_2(x) & x > x_0 \end{cases}$ where k is a constant

8.3.21. $G(x, x_0) = \begin{cases} 0 & x < x_0 \\ 1 & x > x_0 \end{cases}$ **8.3.25. (b)** $u(x) = \dfrac{1}{6} \int_0^x f(t_0)(x - t_0)^3 \, dt_0$

8.4.2. (a) $0 = \int_0^L \phi_h(x)f(x)\,dx - \alpha p(0)\dfrac{d\phi_h}{dx}\Big|_{x=0} + \beta p(L)\dfrac{d\phi_h}{dx}\Big|_{x=L}$

8.4.3. (b) Infinite number of solutions

8.4.6. (a) $u = 1 + c_1\cos x + c_2\sin x$; no solutions   **(b)** $c_2 = 0$, c_1 arbitrary   **(c)** c_1 and c_2 arbitrary

8.4.8. (a) $u = \frac{1}{2}x\sin x + c_2\sin x$

8.4.10. $G_m(x, x_0) = a\sin x\sin x_0 + \begin{cases} \dfrac{1}{\pi}(x\cos x\sin x_0 + x_0\cos x_0\sin x) - \cos x_0\sin x & x < x_0 \\[2mm] \dfrac{1}{\pi}(x_0\cos x_0\sin x + x\cos x\sin x_0) - \cos x\sin x_0 & x > x_0 \end{cases}$

$u(x) = \int_0^\pi f(x_0)G_m(x, x_0)\,dx_0 - \dfrac{\beta}{\pi}(x\cos x + \sin x) - \alpha\left[\dfrac{1}{\pi}(\sin x + x\cos x) - \cos x\right] + k\sin x,$

where a and k are arbitrary constants

8.4.11. (a), (b) $c = 1$ **(d)** $G_a(x, x_0) = \alpha + \begin{cases} x_0 & x < x_0 \\ x & x > x_0, \end{cases}$ where α is an arbitrary constant

(e) $u(x) = \int_0^L f(x_0)\,G_a(x, x_0)\,dx_0 + k_1$, where k_1 is arbitrary

8.5.3. (c) $G(r, \theta; r_0, \theta_0) = \sum\limits_{m=1}^\infty \sum\limits_{n=1}^\infty \dfrac{\sin m\theta\,\sin m\theta_0\,J_m(\sqrt{\lambda}r)J_m(\sqrt{\lambda}r_0)}{-\lambda \iint J_m^2(\sqrt{\lambda}r)\sin^2 m\theta\,r\,dr\,d\theta}$, where $J_m(\sqrt{\lambda}a) = 0$.

8.5.4. See (8.5.19) with extra integral signs.

8.5.9. (b) $G(\mathbf{x}, \mathbf{x}_0) = \sum\limits_{n=1}^\infty \dfrac{\sin m\theta\,\sin m\theta_0}{m\pi} \begin{cases} \left(\dfrac{r}{a}\right)^m\left[\left(\dfrac{r_0}{a}\right)^m - \left(\dfrac{a}{r_0}\right)^m\right] & r < r_0 \\[3mm] \left(\dfrac{r_0}{a}\right)^m\left[\left(\dfrac{r}{a}\right)^m - \left(\dfrac{a}{r}\right)^m\right] & r > r_0 \end{cases}$

8.5.10. (a) $L = \nabla^2 + k^2$, $\iiint [uL(v) - vL(u)]\,dV = \oiint (u\nabla v - v\nabla u)\cdot \hat{n}\,dS$

(b) $c_2 = 0$, $c_1 = \dfrac{-1}{(4\pi)}$ **(c)** $G = \dfrac{1}{4}[Y_0(kr) - iJ_0(kr)]$

8.5.13. (a) $G(\mathbf{x}, \mathbf{x}_0)$

$= -\dfrac{1}{4\pi}\left[\dfrac{1}{\sqrt{(x - x_0)^2 + (y - y_0)^2 + (z - z_0)^2}} + \dfrac{1}{\sqrt{(x - x_0)^2 + (y + y_0)^2 + (z - z_0)^2}}\right]$

8.5.14. $G(\mathbf{x}, \mathbf{x}_0) = \dfrac{1}{4\pi}\ln\dfrac{[(x - x_0)^2 + (y - y_0)^2][(x + x_0)^2 + (y + y_0)^2]}{[(x - x_0)^2 + (y + y_0)^2][(x + x_0)^2 + (y - y_0)^2]}$

8.5.19. $G(\mathbf{x}, \mathbf{x}_0) = \dfrac{1}{4\pi}\ln\left[a^2\dfrac{r^2 + r_0^2 - 2rr_0\cos(\theta - \theta_0)}{r^2r_0^2 + a^4 - 2rr_0a^2\cos(\theta - \theta_0)}\right]$

$-\dfrac{1}{4\pi}\ln\left[a^2\dfrac{r^2 + r_0^2 - 2rr_0\cos(\theta + \theta_0)}{r^2r_0^2 + a^4 - 2rr_0a^2\cos(\theta + \theta_0)}\right]$

8.5.22. (c) $G(\mathbf{x}, \mathbf{x}_0) = -\dfrac{1}{4\pi}\sum\limits_{n=-\infty}^\infty\left(\dfrac{1}{|\mathbf{x} - \boldsymbol{\alpha}_n|} - \dfrac{1}{|\mathbf{x} - \boldsymbol{\beta}_n|}\right),$

where $\boldsymbol{\alpha}_n = (x_0 + 2Ln, y_0, z_0)$ and $\boldsymbol{\beta}_n = (-x_0 + 2Ln, y_0, z_0)$

9.2.1. $c(\omega) = \begin{cases} \frac{1}{2}[A(-\omega) - iB(-\omega)] & \omega < 0 \\ \frac{1}{2}[A(\omega) + iB(\omega)] & \omega > 0 \end{cases}$ **9.3.1.** $C_m = \dfrac{1}{m\pi\Delta}e^{im\pi/L(x_0 + \Delta/2)}\sin\dfrac{m\pi\Delta}{2L}$

9.4.6., 9.4.7. See table.

9.4.10. (b) $\int_0^L J_0(sr)J_0(s_1r)r\,dr \approx \dfrac{2}{\pi}\dfrac{-\sqrt{s_1/s}\cos sL - \pi/4\,\sin s_1L - \pi/4 + \sqrt{s/s_1}\cos s_1L - \pi/4\,\sin sL - \pi/4}{s^2 - s_1^2}$

(c) $A(s_1) = \int_0^\infty f(r)J_0(s_1r)r\,dr$ **9.4.16.** $\int_0^\infty y^p e^{-ky^n}\,dy = \dfrac{1}{n}k^{-(1 + p)/n}\Gamma\left(\dfrac{1 + p}{n}\right)$

9.5.3. (a) $u(x, t) = \dfrac{1}{\sqrt{4\pi kt}}\int_{-\infty}^\infty f(\bar{x})e^{-(x + ct - \bar{x})^2/4kt}\,d\bar{x}$

9.5.5. (c) $u(x, t) = \dfrac{1}{2\pi} \displaystyle\int_{-\infty}^{\infty} f(\bar{x}) \sqrt{\dfrac{\pi}{kt}}\, e^{-(x-\bar{x})^2/4kt}\, d\bar{x} + \dfrac{1}{2\pi} \displaystyle\int_{0}^{t}\int_{-\infty}^{\infty} Q(\bar{x}, \tau) \sqrt{\dfrac{\pi}{k(t-\tau)}}\, e^{-(x-\bar{x})^2/4k(t-\tau)}\, d\bar{x}\, d\tau$

9.5.6. $A_i(x) = \dfrac{1}{\pi} \displaystyle\int_{0}^{\infty} \cos\left(\dfrac{\omega^3}{3} + \omega x\right) d\omega$ **9.5.7. (c)** $u(x, t) = \dfrac{1}{(3kt)^{1/3}} \displaystyle\int_{-\infty}^{\infty} f(\bar{x}) A_i\left[\dfrac{\bar{x} - x}{(3kt)^{1/3}}\right] d\bar{x}$

9.6.3. $C[e^{-\alpha x^2}] = 2\dfrac{1}{\sqrt{4\pi\,\alpha}}\, e^{-\omega^2/4\alpha}$ **9.6.10.** $C^{-1}(\omega e^{-\omega\alpha}) = \dfrac{(\alpha^2 - x^2)}{(\alpha^2 + x^2)^2}$

9.6.11. $u(x, t) = 1 + \dfrac{1}{\sqrt{4\pi kt}} \displaystyle\int_{0}^{\infty} (f(\bar{x}) - 1)(e^{-(x-\bar{x})^2/4kt} - e^{-(x+\bar{x})^2/4kt})\, d\bar{x}$

9.6.17. (b) $\overline{U}(\omega, t) \approx \dfrac{(2/\pi)\,\omega A e^{i\sigma_0 t}}{\omega^2 + i\sigma_0/k}$

9.7.1. (a) $\bar{u}(\omega, y) = F_2(\omega) \dfrac{\sinh \omega y}{\sinh \omega H} + F_1(\omega) \dfrac{\sinh \omega(H - y)}{\sinh \omega H}$,
where $\bar{u}(\omega, y)$ is the Fourier transform of $u(x, y)$

9.7.2. (b) $\bar{u}(x, \omega) = \dfrac{G_1(\omega)\cosh \omega(L - x)}{\cosh \omega L}$,   where $\bar{u}(x, \omega)$ is the cosine transform of $u(x, y)$

9.7.4. (a) $u(x, y) = \dfrac{1}{2\pi} \displaystyle\int_{0}^{\infty} f(\bar{x}) \ln \dfrac{(x - \bar{x})^2 + y^2}{(x + \bar{x})^2 + y^2}\, d\bar{x}$

9.7.11. (a) $u(x, y, t) = \displaystyle\int_{0}^{\infty}\int_{0}^{\infty} f(x_0, y_0) \dfrac{1}{4\pi kt} \left\{ \exp\left[\dfrac{-(x - x_0)^2 - (y - y_0)^2}{4kt}\right]\right.$

$\quad + \exp\left[\dfrac{-(x + x_0)^2 - (y + y_0)^2}{4kt}\right] - \exp\left[\dfrac{-(x + x_0)^2 - (y - y_0)^2}{4kt}\right]$

$\quad \left. - \exp\left[\dfrac{-(x - x_0)^2 - (y + y_0)^2}{4kt}\right] \right\} dx_0\, dy_0$

9.7.12. (a) $u(x, y, t) = \displaystyle\int_{0}^{\infty} \sum_{n=1}^{\infty} \overline{A}_n(\omega, t) \sin \dfrac{n\pi x}{L} \sin \omega y\, d\omega$, where $\overline{A}_n(\omega, t) = c(\omega) e^{-k[\omega^2 + (n\pi/L)^2]t}$

and $c(\omega) = \dfrac{4}{L\pi} \displaystyle\int_{0}^{\infty}\int_{0}^{L} f(x, y) \sin \dfrac{n\pi x}{L} \sin \omega y\, dx\, dy$

9.7.15. (a) $\overline{U} = F(\omega) e^{-\omega z}$ **(b)** $u(x, y) = \dfrac{z}{2\pi} \displaystyle\int_{-\infty}^{\infty}\int_{-\infty}^{\infty} \dfrac{f(x_0, y_0)\, dx_0\, dy_0}{[(x - x_0)^2 + (y - y_0)^2 + z^2]^{3/2}}$

9.7.16. (b) $u(r, \theta) = \displaystyle\int_{0}^{\infty} A(\omega) \sinh \omega\theta \sin\left(\omega \ln \dfrac{r}{a}\right) d\omega$ **9.7.18.** $u(x, t) = \dfrac{1}{2c} \displaystyle\int_{x-ct}^{x+ct} g(\bar{x})\, d\bar{x}$

10.2.6. (b) t_0

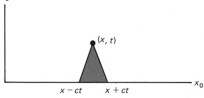

10.2.7. (a) $u(x, t) = \displaystyle\int_{x-ct}^{x+ct} g(x_0) \dfrac{1 - e^{-i\omega(t - |x - x_0|/c)}}{2i\omega c}\, dx_0$

10.2.8. (a) Influence function $= \begin{cases} 0 & \text{if } |\mathbf{x} - \mathbf{x}_0| > ct \\ \dfrac{e^{-i\omega(t - r/c)}}{4\pi c^2 r} & \text{if } |\mathbf{x} - \mathbf{x}_0| < ct, \end{cases}$ where $r = |\mathbf{x} - \mathbf{x}_0|$

10.2.10. (b) $u(x, t) = \begin{cases} 0 & \text{if } x > ct \\ h\left(t - \dfrac{x}{c}\right) & \text{if } x < ct \end{cases}$

10.2.12. (a) $G(\mathbf{x}, t; \mathbf{x}_1, t_1) = \begin{cases} 0 & \text{if } r > c(t - t_1) \\ \dfrac{1}{2\pi c\sqrt{c^2(t - t_1)^2 - r^2}} & \text{if } r < c(t - t_1) \end{cases}$

10.3.2. (a) $H(x) = p\left(u\dfrac{dv}{dx} - v\dfrac{du}{dx}\right) + uv\left(\dfrac{dp}{dx} - r\right)$

10.3.6. (c) If $A = 1$, $u(x, t) = \dfrac{2}{\sqrt{\pi}}\displaystyle\int_{x/\sqrt{4kt}}^{\infty} e^{-\eta^2}\, d\eta$

10.3.7. $G(x, t; x_0, t_0) = \dfrac{1}{\sqrt{4\pi k(t - t_0)}}\left\{\exp\left[\dfrac{-(x - x_0)^2}{4k(t - t_0)}\right] + \exp\left[\dfrac{-(x + x_0)^2}{4k(t - t_0)}\right]\right\}$

11.2.2. $w(x, t) = \cos(x + 3t)$ **11.2.5. (b)** $w(x, t) = t + f(xe^{-t})$

11.2.5. (d) $w(x, t) = e^t f(x - \tfrac{3}{2}t^2)$ **11.2.6.** $x = 2f(x_0)t + x_0$

11.2.8. $u(x, t) = \begin{cases} 1 & x \leq 2t \\ \dfrac{x}{2t} & 2t < x < 4t \\ 2 & x \geq 4t \end{cases}$

11.3.4. (a) $\dfrac{\partial u}{\partial t}(x, 0) = -c\dfrac{dF(x)}{dx}$ **(b)** $\dfrac{\partial u}{\partial x}(0, t) = -\dfrac{1}{c}\dfrac{dF(-ct)}{dt}$

11.4.1. $u(x, t) = \begin{cases} 0 & \text{if } x > ct \\ h\left(t - \dfrac{x}{c}\right) & \text{if } x < ct \end{cases}$ **11.4.2.** $u(x, t) = \begin{cases} \cos x \cos ct & \text{if } x < -ct \\ e^{-(t+x/c)} + \sin x \sin ct & 0 > x > -ct \end{cases}$

11.4.6. $u(x, t) = \begin{cases} 0 & x > ct \\ -c\displaystyle\int_0^{t-x/c} h(\bar{t})\, d\bar{t} & x < ct \end{cases}$

11.4.7. $u(x, t) = \begin{cases} \tfrac{1}{2}[f(x - ct) + f(x + ct)] & \text{if } x > ct \\ \tfrac{1}{2}[f(x + ct) + f(ct - x)] - c\displaystyle\int_0^{t-x/c} h(\bar{t})\, d\bar{t} & \text{if } x < ct \end{cases}$

11.5.1. (b) $u(x, t) = \dfrac{1}{2}[f(x - ct) + f(x + ct)]$

11.6.1. (a) $\rho(x, t) = f(x)$ **(c)** $\rho(x, t) = f(x)e^{-3xt}$ **11.6.2.** $\rho(x, t) = (1 + \sin x)e^{t+x/2}$

11.6.3. (a) $\rho(x, t) = \sin(x - c_0 t)$ **(b)** $\rho(x, t) = \begin{cases} g\left(t - \dfrac{x}{c_0}\right) & x < c_0 t \\ f(x - c_0 t) & x > c_0 t \end{cases}$

11.6.4. (a) $q = u_{max}\rho\left(1 - \dfrac{\rho}{\rho_{max}}\right)$ **(b)** $\rho - \rho_{max}/2$, $u = \dfrac{u_{max}}{2}$, $q = \dfrac{\rho_{max}\, u_{max}}{4}$

11.6.8. (a) $\rho(x, t) = \dfrac{e^{-3(x-ct)}}{3c}(1 - e^{-3ct}) + f(x - ct)$ **(c)** $\rho(x, t) = 5t + f\left(x - \dfrac{1}{2}t^2\right)$

(e) $\rho(x, t) = e^{-t}f\left(x + \dfrac{1}{3}t^3\right)$ **(g)** $\rho(x, t) = \dfrac{1}{2}t^2 + f(xe^{-t})$

11.6.9 (a) $\rho(x, t) = e^{3t}f(x_0)$ where $x = x_0 - \dfrac{1}{6}(e^{6t} - 1)f^2(x_0)$

(c) $\rho(x, t) = e^{-t}f(x_0)$ where $x = x_0 + f(x_0)\displaystyle\int_0^t \tau^2 e^{-\tau}\, d\tau$

11.6.11 $u(x, t) = \begin{cases} e^{3t}f\left(x - t - \dfrac{t^2}{2}\right) & x > t + \dfrac{t^2}{2} \\ e^{3(t-\tau)}g(\tau) & \text{where } x = t + \dfrac{t^2}{2} - \dfrac{3}{2}\tau - \dfrac{\tau^2}{2}, \ x < t + \dfrac{t^2}{2} \end{cases}$

11.6.14. (a) $-Vf' + u_{max}\left(1 - \dfrac{2f}{\rho_{max}}\right)f' = \nu f''$ **(c)** $V = \dfrac{[q]}{[\rho]} = u_{max}\left(1 - \dfrac{\rho_1 + \rho_2}{\rho_{max}}\right)$

11.6.17. (a) $\rho(x, t) = \begin{cases} \dfrac{\rho_{max}}{5} & x < u_{max}t/5 \\ \dfrac{3\rho_{max}}{5} & x > u_{max}t/5 \end{cases}$

11.6.18. (b) $\rho(x, t) = \begin{cases} \rho_1 & x < u_{max}\left(1 - 3\dfrac{\rho_1^2}{\rho_{max}^2}\right)t \\ \dfrac{\rho_{max}}{\sqrt{3}}\sqrt{1 - \dfrac{x}{u_{max}t}} & \text{otherwise} \\ \rho_2 & x > u_{max}\left(1 - \dfrac{3\rho_2^2}{\rho_{max}^2}\right)t \end{cases}$

12.2.4. $\mathcal{L}\left[\displaystyle\int_0^t f(\bar{t})\, d\bar{t}\right] = \dfrac{F(s)}{s}$

12.2.5. (b) $\dfrac{8s}{(s^2 + 16)^2}$ **(d)** $\dfrac{4}{s^2 - 6s + 25}$ **(f)** $e^{-5s}\left(\dfrac{2}{s^3} + \dfrac{10}{s^2} + \dfrac{25}{s}\right) - e^{-8s}\left(\dfrac{2}{s^3} + \dfrac{16}{s^2} + \dfrac{64}{s}\right)$

(h) $\dfrac{24e^{-s}}{s^5}$ **12.2.6. (e)** $\frac{7}{6}e^{-7t} - \frac{1}{6}e^{-t}$

(j) $\frac{2}{9} + \frac{1}{3}\sin 3t - \frac{2}{9}\cos 3t - 5H(t - 4)[\frac{2}{9} + \frac{1}{3}\sin 3(t - 4) - \frac{2}{9}\cos 3(t - 4)]$

12.2.7. (b) $y = 1 + e^{-t}$ **(d)** $y = \begin{cases} \frac{25}{7}e^t - \frac{4}{7}e^{-6t} & 0 < t < 3 \\ -\frac{1}{16}e^{-t} + e^t(\frac{25}{7} + \frac{1}{14}e^{-6}) + e^{-6t}(\frac{1}{35}e^{15} - \frac{4}{7}) & t > 3 \end{cases}$

(f) $y = \frac{1}{3}\sin t - \frac{1}{6}\sin 2t$

12.3.2. $G(t, t_0) = \sin (t - t_0)$ **12.4.3.** $u(x, t) = \sin x \cos ct$ **12.4.4.** $\overline{U}(x, s) = F(s)e^{-\sqrt{s/k}\, x}$

12.5.3. $\overline{U}(x, s) = \dfrac{cB(s)\sinh sx/c}{s \cosh sL/c}$

12.6.4. (a) $\overline{G}(x, s; x_0, t_0) = \dfrac{e^{-st_0}}{\sqrt{sk}}\begin{cases} e^{-\sqrt{s/k}\, x_0}\sinh\sqrt{\dfrac{s}{k}}\, x & x < x_0 \\ e^{-\sqrt{s/k}\, x}\sinh\sqrt{\dfrac{s}{k}}\, x_0 & x > x_0 \end{cases}$

12.7.1. (b) $f(t) = \frac{1}{3}\sin 3t$

12.8.1. $u(x, t) = \displaystyle\sum_{n=1}^{\infty} a_n \sin\dfrac{n\pi x}{L}\cos\dfrac{n\pi ct}{L}$, where $a_n = \dfrac{2}{L}\displaystyle\int_0^L f(x) \sin\dfrac{n\pi x}{L}\, dx$

13.2.6. $\dfrac{\partial^2 u}{\partial x\, \partial y} \approx \dfrac{1}{4(\Delta x)^2}[u(x + \Delta x, y + \Delta y) - u(x - \Delta x, y + \Delta y) - u(x + \Delta x, y - \Delta y) + u(x - \Delta x, y - \Delta y)]$

assuming that $\Delta x = \Delta y$

13.3.4. (a) $\beta_n = \dfrac{\displaystyle\sum_{j=1}^{N-1} f_j \sin n\pi j/N}{\displaystyle\sum_{j=1}^{N-1} \sin^2 n\pi j/N}$ **(b)** $\dfrac{N-1}{2}$ **13.3.6. (d)** Stable if $s < 0.5125$

13.3.9. (b) $A = \dfrac{1}{(\Delta x)^2}\begin{bmatrix} -2 & 1 \\ 1 & -2 \end{bmatrix}$ **13.3.10. (b)** $\dfrac{\partial u}{\partial t} = \dfrac{ka}{s}\dfrac{\partial^2 u}{\partial x^2}$

13.3.14. (c) $\lambda = 0, 3, 4; |\lambda - 1| \leq 5, |\lambda - 4| \leq 8, |\lambda - 2| \leq \frac{1}{3}$

13.4.1. $k\,\Delta t\left[\dfrac{1}{(\Delta x)^2} + \dfrac{1}{(\Delta y)^2}\right] \leq \dfrac{1}{2}$ **13.5.5. (b)** Unstable **13.5.6. (b)** Stable if $\dfrac{c}{\Delta x/\Delta t} \leq 1$

Bibliography

ABRAMOWITZ, M., and STEGUN, I. A., eds. *Handbook of Mathematical Functions*. New York: Dover, 1965.

ANTMAN, S. S., "The equations for large vibrations of strings." *Amer. Math. Monthly 87,* 1980, pp. 359–370.

BENDER, C. M., and ORSZAG, S. A. *Advanced Mathematical Methods for Scientists and Engineers*. New York: McGraw-Hill, 1978.

BERG, P. W., and McGREGOR, J. L. *Elementary Partial Differential Equations*. San Francisco: Holden-Day, 1966.

BOYCE, W. E., and DI PRIMA, R. C. *Elementary Differential Equations and Boundary Value Problems*. New York: Wiley, 1977.

CARRIER, G., and PEARSON, C. *Partial Differential Equations*. New York: Academic Press, 1976.

CARSLAW, H. S., and JAEGER, J. C. *Conduction of Heat in Solids*. 2nd ed. New York: Oxford University Press, 1959.

CHURCHILL, R. V. *Operational Mathematics*. 3rd ed. New York: McGraw-Hill, 1972.

CHURCHILL, R. V., and BROWN, J. W. *Fourier Series and Boundary Value Problems*. 3rd ed. New York: McGraw-Hill, 1978.

COURANT, R., and HILBERT, D. *Methods of Mathematical Physics,* Vols. 1 and 2. New York: Wiley, 1953.

CRC Standard Mathematical Tables, 26th ed. Boca Raton, Fla.: CRC Press, 1981.

DENNEMEYER, R. *Introduction to Partial Differential Equations and Boundary Value Problems*. New York: McGraw-Hill, 1968.

DUFF, G. F. D., and NAYLOR, D. *Differential Equations of Applied Mathematics*. New York: Wiley, 1966.

GARABEDIAN, P. R. *Partial Differential Equations.* New York: Wiley, 1964.

GREENBERG, M. D. *Foundations of Applied Mathematics.* Englewood Cliffs, N.J.: Prentice-Hall, 1978.

HABERMAN, R. *Mathematical Models: Mechanical Vibrations, Population Dynamics, and Traffic Flow.* Englewood Cliffs, N.J.: Prentice-Hall, 1977.

HILDEBRAND, F. B. *Advanced Calculus for Applications.* 2nd ed. Englewood Cliffs, N.J.: Prentice-Hall, 1976.

ISAACSON, E., and KELLER, H. B. *Analysis of Numerical Methods.* New York: Wiley, 1966.

KAPLAN, W. *Advanced Mathematics for Engineers.* Reading, Mass.: Addison-Wesley, 1981.

KEVORKIAN, J., and COLE, J. D. *Perturbation Methods in Applied Mathematics.* New York: Springer-Verlag, 1981.

MATHEWS, J., and WALKER, R. L. *Mathematical Methods of Physics.* New York: W. A. Benjamin, 1970.

MIKHLIN, S. G., ed. *Linear Equations of Mathematical Physics.* New York: Holt, Rinehart and Winston, 1967.

MORSE, P. M., and FESHBACH, H. *Methods of Theoretical Physics,* Parts 1 and 2. New York: McGraw-Hill, 1953.

NAYFEH, A. H. *Perturbation Methods.* New York: Wiley, 1973.

POWERS, D. L. *Boundary Value Problems.* 2nd ed. New York: Academic Press, 1979.

PROTTER, M. H., and WEINBERGER, H. F. *Maximum Principles in Differential Equations.* Englewood Cliffs, N.J.: Prentice-Hall, 1967.

RICHTMYER, R. D., and MORTON, K. W. *Difference Methods for Initial-Value Problems.* 2nd ed. New York: Wiley, 1967.

ROBERTS, G. E., and KAUFMAN, H. *Tables of Laplace Transforms.* New York: W. B. Saunders Co., 1966.

SAGAN, H. *Boundary and Eigenvalue Problems in Mathematical Physics.* New York: Wiley, 1961.

SMITH, G. D. *Numerical Solution of Partial Differential Equations.* New York: Oxford University Press, 1965.

STAKGOLD, I. *Green's Functions and Boundary Value Problems.* New York: Wiley, 1980.

STRANG, G. *Linear Algebra and Its Applications.* 2nd ed. New York: Academic Press, 1980.

STREET, R. L. *The Analysis and Solution of Partial Differential Equations.* Monterey, Calif.: Brooks/Cole, 1973.

WATSON, G. N. *A Treatise on the Theory of Bessel Functions.* Cambridge, England: Cambridge University Press, 1966.

WEINBERGER, H. F. *A First Course in Partial Differential Equations.* Lexington, Mass.: Xerox, 1965.

WHITHAM, G. B. *Linear and Nonlinear Waves.* New York: Wiley, 1974.

YOUNG, E. C. *Partial Differential Equations.* Boston: Allyn and Bacon, 1972.

ZACHMANOGLOU, E. C., and THOE, D. W. *Introduction to Partial Differential Equations with Applications.* Baltimore: Williams & Wilkins, 1976.

ZAUDERER, E. *Partial Differential Equations of Applied Mathematics.* New York: Wiley, 1983.

Index